国外名校名著

Unit Operations of Chemical Engineering
Seventh Edition

化学工程单元操作
（英文改编版）

[美] 沃伦 L. 麦克凯布（Warren L. McCabe）
朱利安 C. 史密斯（Julian C. Smith） 著
彼得·哈里奥特（Peter Harriott）

伍 钦　钟 理
夏 清　熊丹柳　改编

·北京·

图书在版编目(CIP)数据

化学工程单元操作/[美]麦克凯布（McCabe，W.L.），
[美]史密斯（Smith，J.C.），[美]哈里奥特（Harriott，P.）著；
伍钦等改编. —北京：化学工业出版社，2008.7（2024.2重印）
（国外名校名著）
书名原文：Unit Operations of Chemical Engineering
ISBN 978-7-122-03192-1

Ⅰ. 化⋯ Ⅱ. ①麦⋯②史⋯③哈⋯④伍⋯ Ⅲ. 化
工单元操作 Ⅳ. TQ02

中国版本图书馆 CIP 数据核字（2008）第 109413 号

Original language, entitled Unit Operations of Chemical Engineering, seventh edition, ISBN：007-124710-6, by Warren L. McCabe, Julian C. Smith, Peter Harriott, published by The McGraw-Hill Companies, Inc, Copyright ©2005 by The McGraw-Hill Companies, Inc.

All Rights reserved. No part of this publication may be reproduced or distributed by any means, or stored in a database or retrieval system, without the prior written permission of the publisher.

Authorized English language adaptation edition jointly published by McGraw-Hill Education(Asia) Co. and Chemical Industry Press. This edition is authorized for sale in China only. Unauthorized export of this edition is a violation of the Copyright Act. Violation of this Law is subject to Civil and Criminal Penalties.

本书英文改编版由化学工业出版社和美国麦格劳-希尔教育出版（亚洲）公司合作出版。此版本仅限在中华人民共和国境内销售。未经许可出口，视为违反著作权法，将受法律制裁。

未经出版者预先书面许可，不得以任何方式复制或抄袭本书的任何部分。

本书封面贴有 McGraw-Hill 公司仿伪标签，无标签者不得销售。

北京市版权局著作权合同登记号：01-2007-4654

责任编辑：徐雅妮　　　　　　　　　　　装帧设计：郑小红
责任校对：周梦华

出版发行：化学工业出版社（北京市东城区青年湖南街 13 号　邮政编码 100011）
印　　装：涿州市般润文化传播有限公司
787mm×1092mm　1/16　印张 34¼　字数 1031 千字　2024 年 2 月北京第 1 版第 9 次印刷

购书咨询：010-64518888　　　售后服务：010-64518899
网　　址：http：// www. cip. com. cn

凡购买本书，如有缺损质量问题，本社销售中心负责调换。

定　　价：99.00 元　　　　　　　　　　　　　　　　　　　　　　　版权所有　违者必究

前　言

　　McGraw-Hill 公司的《Unit Operations of Chemical Engineering》自 20 世纪 50 年代出版以来至今已修订至第 7 版，是一本在全球范围内很有影响力的教科书。该书内容广泛，结构严谨，逻辑性及实用性强，但该书的内容与知识点之间的衔接与我国化工原理教学大纲有所差别。因此，为了适应我国化工原理双语教学的需要，我们在该书第 7 版的基础上，按照我国化工原理教学大纲的要求进行了改编，将原书的 29 章改编为 12 章。改编后教材的内容顺序基本与中文教材习惯采用的顺序一致，重点介绍了化工单元操作的基本原理、典型设备结构特性及其计算，在内容编排上力求逻辑严谨，同时兼顾工程实用性。

　　由于课时数和篇幅的限制，改编时删除了原书的第 6 章可压缩流体的流动、第 9 章流体的搅拌及混合、第 25 章吸附及固定床分离、第 26 章膜分离过程、第 27 章结晶和第 28 章固体颗粒的性质及处理。改编后各章内容概述如下。

　　第 1 章流体流动，包括流体静力学、流体流动现象、不可压缩流体的流动、管内流动的计算和流体的测量。改编时对原书第 2~8 章的内容进行了调整和删补；为了突出流体流动基本方程中各项的物理概念，方程的导出过程是基于质量守恒和流动系统的第一、第二热力学定律，避免使用复杂的三维微分问题；补充了原书没有的管内流动系统计算的内容，并将颗粒与流体之间相对运动的内容归入第 3 章，流体输送的内容归入第 2 章。

　　第 2 章流体输送，包括液体输送和气体输送，重点介绍管路、泵的结构特点、泵的特性和操作特性，简单介绍气体输送及其输送机械。原书的第 8 章中关于流体输送及流体输送机械的内容较少，改编时增加了离心泵的结构特点、特性曲线、管路特性曲线、离心泵的工作点以及离心泵的串并联等内容。

　　第 3 章非均相体系的流动和分离，内容包括流体与颗粒之间的相对运动、气固分离及分离设备、过滤的基本原理和过滤设备的基本结构、恒速和恒压过滤的计算，是原书第 7 章和第 29 章部分内容的综合，并参考了其它书籍，如《Solid-Liquid Separation》和中文教科书的内容。

　　第 4 章传热，包括热传导、无相变管内外对流传热、相变对流传热、热辐射、热交换过程的计算及换热器。由原书第 10~15 章的内容改编而成，删除了其中的三维传热问题和不稳定传热。

　　第 5 章蒸发，包括蒸发器的分类及结构特点、沸点校正以及单效、多效蒸发器的计算。改编时对温差损失计算、物料衡算和热量衡算等内容都做了必要的补充。

　　第 6 章扩散原理及相间质量传递是在原书第 17 章基础上进行改编，内容包括等分子反向和单向扩散、扩散系数以及相间质量传递原理，删除了质量传递系数的测量等相关内容。

　　第 7 章平衡关系及平衡级，内容包括稀溶液的气液平衡、理想和非理想溶液的气液平衡关系、操作线方程及给定条件下平衡级的计算，在原书第 20 章的基础上增加了气液平衡的内容。

第 8 章气体吸收，包括吸收过程的物料衡算和操作线方程、稀溶液及浓溶液中的吸收操作、最小液气比、传质系数的表示及相互之间的关系以及填料高度的计算。本章是在原书第 18 章的基础上进行改编，原书中有关填料及填料塔设计的内容归入第 12 章，补充了最小液气比和用无溶质基浓度表示的计算法。

第 9 章蒸馏，包括简单蒸馏、精馏原理、有回流比的精馏过程、不稳定精馏以及恒沸精馏和萃取精馏。在原书第 21 章的基础上删去了焓衡算和多组分精馏部分内容，并将筛板塔设计的内容移至第 12 章，补充了微分蒸馏、水蒸气蒸馏、恒沸精馏和萃取精馏以及提馏塔、直接水蒸气加热精馏和多股进料、多侧线抽料等内容。

第 10 章浸提和萃取，包括浸提及萃取设备、液液萃取的原理及计算。改编时只采用原书第 23 章的固液浸提部分，液液萃取部分重新编写。

第 11 章除湿与固体干燥，包括干燥设备、湿空气的性质、湿固体的性质、冷却塔、干燥过程的质量、热量衡算以及干燥原理及干燥时间的计算。改编时采用了原书第 19 章和第 24 章的部分内容。

第 12 章板式塔和填料塔的设计，综合了原书第 18 章和第 21 章有关填料塔和板式塔的内容。

为了计算方便并适应国际单位制的需要，原书的附录全部改为 SI 制并补充了一些计算时常用的数据。本教材尽量采用国际单位制，但是由于图表中单位转换比较困难，因此个别图表仍沿用原书中的英制单位。

原书为教师提供了一些网上资源，包括习题解答、教学 PPT 和与本书相关的基本信息，其网址为 http://www.mhhe.com/mccabe7e。

本书由华南理工大学、天津大学和华东理工大学合作改编完成。第 1 章由天津大学夏清改编；下列各章由华南理工大学教师改编，第 2、3、7、9 章（伍钦）、第 4 章（钟理）、第 5 章（赖万东）、第 8、11 章（杨东杰）、第 12 章（马四朋）、导论和附录（易聪华）；第 6、10 章由华东理工大学熊丹柳改编；部分习题由华南理工大学郑大峰和伍钦提供。全书由伍钦统编。为了满足双语教学的急需，尽管参与改编的教师都有丰富的双语教学经验，但是由于教材内容的调整、删补较多而改编时间较短，故书中内容若存不当之处，恳请各位同行、专家和读者斧正。

本书在改编过程中始终得到化学工业出版社的大力支持，谨此深表谢意！

伍钦
2008 年 4 月于华南理工大学

采用本书作为教材的教师可向 McGraw-Hill 公司北京代表处联系索取登陆网站的密码，传真 010-62790292，电子邮件 webmaster@mcgraw-hill.com.cn。

Contents

Introduction Definitions and Principles ········· 1
  0.1 Unit Operations ·································· 1
  0.2 Unit Systems ······································ 2
      0.2.1 Physical Quantities ···················· 2
      0.2.2 SI Units ···································· 3
      0.2.3 CGS Units ······························· 6
      0.2.4 FPS Engineering Units ·············· 7
      0.2.5 Gas Constant ···························· 8
      0.2.6 Conversion of Units ··················· 9
      0.2.7 Units and Equations ················· 10
  0.3 Dimensional Analysis ························ 12
  PROBLEMS ··· 14

Chapter 1 Fluid Flow ······························ 15
  1.1 Fluid Statics and Its Application ············ 15
      1.1.1 Nature of Fluids ······················· 15
      1.1.2 Hydrostatic Equilibrium ··········· 16
      1.1.3 Applications of Fluid Statics ······ 17
  1.2 Fluid Flow Phenomena ······················ 19
      1.2.1 Newton's Law and Viscosity
             of Fluids ································· 20
      1.2.2 Rheological Properties of Fluids ····· 23
      1.2.3 Types of Fluid Flow and Reynolds
             Number ·································· 25
      1.2.4 Boundary Layers ····················· 29
  1.3 Basic Equations of Fluid Flow ············· 33
      1.3.1 Measures of Flow ···················· 34
      1.3.2 Mass Balance in a Flowing Fluid;
             Continuity ······························· 35
      1.3.3 Overall Energy Balance for Steady-
             state Flow System ···················· 37
      1.3.4 Overall Mechanical Energy Balance
             for Steady-state Flow System ······· 39
      1.3.5 Discussion on the Overall Mechanical
             Energy Balance Equation ············ 40
      1.3.6 Macroscopic Momentum Balances ····· 43
  1.4 Incompressible Flow in Pipes and Channels ··· 47
      1.4.1 Shear Stress and Skin Friction in
             Pipes ······································ 47

      1.4.2 Laminar Flow in Pipes and
             Channels ································· 50
      1.4.3 Turbulent Flow in Pipes and
             Channels ································· 52
      1.4.4 Friction from Changes in Velocity or
             Direction ································ 56
  1.5 Pipe Flow Systems ···························· 62
      1.5.1 Single Pipes ···························· 62
      1.5.2 Multiple Pipe Systems ··············· 64
  1.6 Metering of Fluids ····························· 66
      1.6.1 Insertion Meters ······················ 67
      1.6.2 Full-Bore Meters ····················· 69
  PROBLEMS ··· 76

Chapter 2 Transportation of Fluids ·········· 82
  2.1 Pipe, Fittings, and Valves ···················· 82
      2.1.1 Pipe and Tubing ······················ 82
      2.1.2 Valves ···································· 86
  2.2 Pumps ·· 88
      2.2.1 Developed Head ······················ 89
      2.2.2 Suction Lift and Cavitation of
             Pumps ···································· 91
      2.2.3 Positive-Displacement Pumps ······· 93
      2.2.4 Centrifugal Pumps ··················· 95
      2.2.5 Multistage Centrifugal Pumps ······ 109
      2.2.6 Pump Priming ························ 109
      2.2.7 Pump Selection ······················· 110
  2.3 Fans, Blowers, and Compressors ··········· 111
      2.3.1 Fans ······································ 111
      2.3.2 Blowers and Compressors ·········· 113
      2.3.3 Vacuum pumps ······················· 118
      2.3.4 Comparison of Devices for Moving
             Fluids ···································· 119
  PROBLEMS ··· 120

**Chapter 3 Heterogeneous Flow and
Separation** ··· 124
  3.1 Flow Past Immersed Objects ················ 124
      3.1.1 Drag and Drag Coefficients ········· 124

3.1.2 Flow through Beds of Solids ········128
3.2 Motion of Particles through Fluids ········ 131
3.3 Settling Separation of the Solids from Gases ········ 138
   3.3.1 Gravitational Settling Processes ···138
   3.3.2 Centrifugal Settling Processes ······139
3.4 Filtration Fundamentals ········ 140
   3.4.1 Introduction ········140
   3.4.2 Flow Rate-Pressure Drop Relationships ········145
   3.4.3 Filtration Operations-Basic Equations ········148
   3.4.4 Constant Pressure Filtration ········149
   3.4.5 Constant Rate Filtration ········151
   3.4.6 Constant Rate Followed by Constant Pressure Operation ········151
   3.4.7 Vacuum Filtration—Drum Continuous Filtration ········153
   3.4.8 Washing Filter Cakes ········154
3.5 Introduction to Fluidization ········ 156
PROBLEMS ········ 157

Chapter 4 Heat Transfer and Its Applications ... 161

4.1 Introduction ········ 161
4.2 Heat Transfer by Conduction ········ 163
   4.2.1 Basic Law of Conduction ········163
   4.2.2 Steady-State Conduction ········165
4.3 Principles of Heat Flow in Fluids ········ 169
   4.3.1 Typical Heat-Exchange Equipment ···170
   4.3.2 Energy Balances ········172
   4.3.3 Heat Flux and Heat-Transfer Coefficients ········174
4.4 Heat Transfer to Fluids without Phase Change ········ 184
   4.4.1 Boundary Layers ········184
   4.4.2 Heat Transfer by Forced Convection in Laminar Flow ········187
   4.4.3 Heat Transfer by Forced Convection in Turbulent Flow ········191
   4.4.4 Heat Transfer in Transition Region between Laminar and Turbulent Flow ········196
   4.4.5 Heating and Cooling of Fluid in Forced Convection Outside Tubes ···197
   4.4.6 Natural Convection ········198
4.5 Heat Transfer to Fluids with Phase Change ·· 201
   4.5.1 Heat Transfer from Condensing Vapors ········201
   4.5.2 Heat Transfer to Boiling Liquids ···209
4.6 Radiation Heat Transfer ········ 213
   4.6.1 Emission of Radiation ········214
   4.6.2 Absorption of Radiation by Opaque Solids ········217
   4.6.3 Radiation between Surfaces ········219
4.7 Heat-Exchange Equipment ········ 224
   4.7.1 Shell-and-Tube Heat Exchangers ···225
   4.7.2 Plate-Type Exchangers ········235
   4.7.3 Extended-Surface Equipment ········237
   4.7.4 Condensers ········238
PROBLEMS ········ 240

Chapter 5 Evaporation ········245

5.1 Summarize ········ 245
   5.1.1 Single-and Multiple-Effect Operation ········245
   5.1.2 Vacuum Evaporation ········246
   5.1.3 Liquid Characteristics ········247
5.2 Evaporation Equipment ········ 247
   5.2.1 Circulation Evaporators ········248
   5.2.2 Once-Through Evaporators ········250
   5.2.3 Choice of the Evaporators ········252
5.3 Single-Effect Evaporation ········ 252
   5.3.1 Evaporator Capacity ········252
   5.3.2 Boiling-Point Elevation and Dühring's Rule ········253
   5.3.3 Effect of Liquid Head and Friction on Temperature Drop ········254
5.4 Calculation Methods for Single-Effect Evaporator ········ 255
   5.4.1 Methods of Operation of Evaporators ··255
   5.4.2 Material Balances for Single-Effect Evaporator ········256
   5.4.3 Enthalpy (Heat) Balances for Single-Effect Evaporator ········257
   5.4.4 Heat-Transfer Coefficients ········259
   5.4.5 Single-Effect Calculations ········260
5.5 Multiple-Effect Evaporators ········ 261
   5.5.1 Methods of Feeding ········261

5.5.2 Calculation Methods for
Multiple-Effect Evaporators ·········· 262
5.5.3 Effect of Liquid Head and
Boiling-Point Elevation ············· 266
5.5.4 Optimum Number of Effects ········ 267
5.5.5 Multiple-Effect Calculations ········ 267
5.5.6 Vapor Recompression ················ 270
PROBLEMS ·· 271

Chapter 6　Principles of Diffusion and Mass Transfer Between Phases ················ 274
6.1 Theory of Diffusion ································ 275
6.2 Prediction of Diffusivities ······················· 283
6.3 Mass-Transfer Theories ·························· 289
PROBLEMS ·· 295

Chapter 7　Equilibrium Relations and Equilibrium- Stage Operations ···················· 299
7.1 Equilibrium Relations ····························· 299
7.1.1 Gas-Liquid Equilibrium ··············· 299
7.1.2 Vapor-Liquid Equilibrium Relations ··· 300
7.2 Equilibrium-Stage Operations ················· 305
7.2.1 Equipment for Stage Contacts ······· 305
7.2.2 Principles of Stage Processes ········ 307
PROBLEMS ·· 317

Chapter 8　Gas Absorption ························· 321
8.1 Principles of Absorption ························· 321
8.1.1 Material Balances ······················· 322
8.1.2 Limiting and Optimum Gas-liquid Ratio ··· 324
8.1.3 Rate of Absorption in Packed Towers ······································ 327
8.2 Calculation of Tower Height ···················· 329
8.2.1 Fundamental Calculation Equation of Packing Height ······················· 329
8.2.2 Number of Transfer Units and Height of A Transfer Unit ······················· 330
8.2.3 Alternate Forms of Transfer Coefficients ································ 332
8.2.4 Effect of Pressure ······················· 335
8.2.5 Temperature Variations in Packed Towers ······································ 335
8.3 Multicomponent Absorption ···················· 338
8.4 Desorption or Stripping ··························· 339
8.5 Absorption from Rich Gases ···················· 342
8.6 Absorption in Plate Columns ··················· 344
8.7 Absorption with Chemical Reaction ········· 344
8.8 Cocurrent flow operation ························· 345
PROBLEMS ·· 346

Chapter 9　Distillation ································· 348
9.1 Flash Distillation ···································· 348
9.2 Simple Batch or Differential Distillation ··· 350
9.3 Simple Steam Distillation ························ 353
9.4 Continuous Distillation with Reflux ········· 354
9.4.1 Action on an Ideal Plate ··············· 354
9.4.2 Combination Rectification and Stripping ··································· 355
9.4.3 Material Balances in Plate Columns ··· 357
9.4.4 Number of Ideal Plates; McCabe-Thiele Method ··············· 359
9.4.5 Special Cases for Rectification Using McCabe-Thiele Method ··············· 377
9.4.6 Batch Distillation with Reflux ······· 381
9.5 Azeotropic and Extractive Distillation ······ 382
9.6 Plate Efficiencies ···································· 385
PROBLEMS ·· 388

Chapter 10　Leaching and Extraction ··········· 393
10.1 Leaching ·· 393
10.1.1 Leaching Equipment ··················· 393
10.1.2 Principles of Continuous Countercurrent Leaching ············· 395
10.2 Liquid Extraction ··································· 399
10.2.1 Extraction Equipment ················· 400
10.2.2 Principles of Extraction ··············· 405
PROBLEMS ·· 422

Chapter 11　Drying of Process Materials ······ 426
11.1 Introduction and Methods of Drying ······· 426
11.2 Properties of Moist Air and Humidity Chart ····································· 429
11.2.1 Humidity ·································· 429
11.2.2 Humid Heat of an Air–Water Vapor Mixture ····································· 431
11.2.3 Humid Volume of an Air–Water Vapor Mixture ··························· 432
11.2.4 Total Enthalpy of an Air–Water Vapor Mixture ··························· 432

- 11.2.5 Dew Point of an Air–Water Vapor Mixture ··· 433
- 11.2.6 Adiabatic Saturation Temperatures ··· 433
- 11.2.7 Dry-Bulb Temperature and Wet-Bulb Temperature ··· 434
- 11.2.8 Humidity Chart ··· 437
- 11.2.9 Measurement of Humidity ··· 441
- 11.3 Principles of Drying ··· 442
  - 11.3.1 Temperature patterns in dryers ··· 442
  - 11.3.2 Phase Equilibria ··· 443
  - 11.3.3 Data of Equilibrium Moisture Content for Inorganic and Biological Materials ··· 446
- 11.4 Rate-of-Drying Curves ··· 447
  - 11.4.1 Method Using Experimental Drying Curve ··· 447
  - 11.4.2 Rate of Drying Curves for Constant-Drying Conditions ··· 448
  - 11.4.3 Critical Moisture Content ··· 449
  - 11.4.4 Drying in the Constant-Rate Period ··· 450
  - 11.4.5 Drying in the Falling-Rate Period ··· 450
  - 11.4.6 Moisture Movements in Solids During Drying in the Falling-Rate Period ··· 451
  - 11.4.7 Effect of Shrinkage ··· 452
- 11.5 Calculation Methods for Constant-Rate Drying Period ··· 452
  - 11.5.1 Method Using Experimental Drying Curve ··· 452
  - 11.5.2 Method Using Rate-of-Drying Curve for Constant-Rate Period ··· 453
  - 11.5.3 Method Using Predicted Transfer Coefficients for Constant-Rate Period ··· 453
  - 11.5.4 Effect of Process Variables on Constant-Rate Period ··· 456
- 11.6 Calculation Methods for Falling-rate Drying Period ··· 457
  - 11.6.1 Method Using Numerical Integration ··· 457
  - 11.6.2 Calculation Methods for Special Cases in Falling-Rate Region ··· 458
- 11.7 Equations for Various Types of Dryers ··· 459
  - 11.7.1 Through-Circulation Drying in Packed Beds ··· 459
  - 11.7.2 Tray Drying with Varying Air Conditions ··· 463
  - 11.7.3 Material and Heat Balances for Continuous Dryers ··· 464
  - 11.7.4 Continuous Countercurrent Drying ··· 466
- 11.8 Drying Equipment ··· 468
  - 11.8.1 Dryers for Solids and Pastes ··· 469
  - 11.8.2 Dryers for Solutions and Slurries ··· 474
  - 11.8.3 Selection of Drying Equipment ··· 477
- PROBLEMS ··· 478

Chapter 12 Tray and Packed Tower Design ··· 483
- 12.1 Packed Tower Design ··· 483
  - 12.1.1 Packed Tower and Packings ··· 483
  - 12.1.2 Fluid Mechanics of Packed Tower ··· 486
  - 12.1.3 Calculations of Packed Tower ··· 491
- 12.2 Plate Tower Design ··· 493
  - 12.2.1 Introduction ··· 493
  - 12.2.2 Fluid Mechanics of Plate Columns ··· 495
  - 12.2.3 Plate Efficiencies ··· 504

Appendix ··· 511
1. Unit Conversion ··· 511
2. Dimensionless Groups ··· 512
3. Standard Steel Pipe ··· 513
4. Condenser and Heat-Exchanger Data ··· 514
5. Properties of Liquid Water ··· 518
6. Properties of Saturated Steam and Water ··· 518
7. Viscosities of Gases ··· 520
8. Viscosities of Liquids ··· 522
9. Thermal Conductivities of Metals ··· 524
10. Thermal Conductivities of Various Solids and Insulating Materials ··· 524
11. Thermal Conductivities of Gases and Vapors ··· 526
12. Thermal Conductivities of Liquids Other Than Water ··· 526
13. Specific Heats of Gases ··· 527
14. Specific Heats of Liquids ··· 529
15. Prandtl Numbers for Gases at 1 atm and 100 ℃ ··· 531
16. Prandtl Numbers for Liquids ··· 531

17　Diffusivities and Schmidt Numbers for Gases in Air at 0℃ and 1 atm ·············· 531
18　Collision Integral and Lennard-Jones Force Constants ······················· 532
19　Equilibrium Data for Ethanol-Water System at 101.325 kPa ······················ 533
20　Henry's Law Constants for Gases in Water ································· 534
21　Partial Pressure p of SO_2 in The Gas in Equilibrium with The Mole Fraction x of SO_2 in The Liquid at 20℃ ····················· 535
22　Boiling Point for Inorganic Solution at 1 atm ································ 535

References ·· 536

Introduction

Definitions and Principles

Chemical engineering has to do with industrial processes in which raw materials are changed or separated into useful products. The chemical engineer must develop, design, and engineer both the complete process and the equipment used; choose the proper raw materials; operate the plants efficiently, safely, and economically; and see to it that products meet the requirements set by the customers. Chemical engineering is both an art and a science. Whenever science helps the engineer to solve a problem, science should be used. When, as is usually the case, science does not give a complete answer, it is necessary to use experience and judgment. The professional stature of an engineer depends on skill in utilizing all sources of information to reach practical solutions to processing problems.

The variety of processes and industries that call for the services of chemical engineers is enormous. In the past, the areas of most concern to chemical engineers were ore beneficiation, petroleum refining, and the manufacture of heavy chemicals and organics such as sulfuric acid, methyl alcohol, and polyethylene. Today items such as polymeric lithographic supports for the electronics industry, high-strength composite materials, genetically modified biochemical agents in areas of food processing, and drug manufacture and drug delivery have become increasingly important. The processes described in standard treatises on chemical technology and the process and biochemical industries give a good idea of the field of chemical engineering.

Because of the variety and complexity of modern processes, it is not practicable to cover the entire subject matter of chemical engineering under a single head. The field is divided into convenient, but arbitrary, sectors. This text covers that portion of chemical engineering known as the unit operations.

0.1 Unit Operations

An economical method of organizing much of the subject matter of chemical engineering is based on two facts: (1) Although the number of individual processes is great, each one can be broken down into a series of steps, called operations, each of which in turn appears in process after process; (2) the individual operations have common techniques and are based on the same scientific principles. For example, in most processes solids and fluids must be moved; heat or other forms of energy must be transferred from one substance to another; and tasks such as drying, size reduction, distillation, and evaporation must be performed. The unit operation concept is this: By studying systematically these operations themselves-operations that clearly cross industry and process lines—the treatment of all processes is unified and simplified.

The strictly chemical aspects of processing are studied in a companion area of chemical engineering called reaction kinetics. The unit operations are largely used to conduct the primarily physical steps of preparing the reactants, separating and purifying the products, recycling unconverted reactants, and controlling the energy transfer into or out of the chemical reactor.

The unit operations are as applicable to many physical processes as to chemical ones. For example, the process used to manufacture common salt consists of the following sequence of unit operations: transportation of solids and liquids, transfer of heat, evaporation, crystallization, drying, and screening. No chemical reaction appears in these steps. On the other hand, the cracking of petroleum, with or without the aid of a catalyst, is a typical chemical reaction conducted on an enormous scale. Here the unit operations—transportation of fluids and solids, distillation, and various mechanical separations—are vital, and the cracking reaction could not be utilized without them. The chemical steps themselves are conducted by controlling the flow of material and energy to and from the reaction zone.

Because the unit operations are a branch of engineering, they are based on both science and experience. Theory and practice must combine to yield designs for equipment that can be fabricated, assembled, operated, and maintained. A balanced discussion of each operation requires that theory and equipment be considered together. This book presents such a balanced treatment.

Scientific Foundations of Unit Operations

A number of scientific principles and techniques are basic to the treatment of the unit operations. Some are elementary physical and chemical laws such as the conservation of mass and energy, physical equilibria, kinetics, and certain properties of matter. Their general use is described in the remainder of this chapter. Other special techniques important in chemical engineering are considered at the proper places in the text.

0.2 Unit Systems

The official international system of units is SI (Système International d'Unités). Strong efforts are underway for its universal adoption as the exclusive system for all engineering and science, but older systems, particularly the centimeter-gram-second (cgs) and foot-pound-second (fps) engineering gravitational systems, are still in use and probably will be around for some time. The chemical engineer finds many physiochemical data given in cgs units; that many calculations are most conveniently done in fps units; and that SI units are increasingly encountered in science and engineering. Thus it becomes necessary to be expert in the use of all three systems.

In the following treatment, SI is discussed first, and then the other systems are derived from it. The procedure reverses the historical order, as the SI units evolved from the cgs system. Because of the growing importance of SI, it should logically be given a preference. If, in time, the other systems are phased out, they can be ignored and SI used exclusively.

0.2.1 Physical Quantities

Any physical quantity consists of two parts: a unit, which tells what the quantity is and gives the standard by which it is measured, and a number, which tells how many units are needed to make up the quantity. For example, the statement that the distance between two points is 3 m means all this: A definite length has been measured; to measure it, a standard length, called the meter, has been chosen as a unit; and three 1-m units, laid end to end, are needed to cover the distance. If an integral number of units are either too few or too many to cover a given distance, submultiples, which are fractions of the unit, are defined by dividing the unit into fractions, so that a measurement can be made to any degree of precision in terms of the fractional units. No physical quantity is defined until both the number and the unit are given.

0.2.2 SI Units

The SI system covers the entire field of science and engineering, including electromagnetics and illumination. For the purposes of this book, a subset of the SI units covering chemistry, gravity, mechanics, and thermodynamics is sufficient. The units are derivable from (1) four proportionalities of chemistry and physics; (2) arbitrary standards for mass, length, time, temperature, and the mole; and (3) arbitrary choices for the numerical values of two proportionality constants.

Basic equations

The basic proportionalities, each written as an equation with its own proportionality factor, are

$$F = k_1 \frac{d}{dt}(mu) \tag{0.1}$$

$$F = k_2 \frac{m_a m_b}{r^2} \tag{0.2}$$

$$Q_c = k_3 W_c \tag{0.3}$$

$$T = k_4 \lim_{p \to 0} \frac{pV}{m} \tag{0.4}$$

where F = force
t = time
m = mass
u = velocity
r = distance
W_c = work

Q_c = heat
p = pressure
V = volume
T = thermodynamic absolute temperature
k_1, k_2, k_3, k_4 = proportionality factors

Equation (0.1) is Newton's second law of motion, showing the proportionality between the resultant of all the forces acting on a particle of mass m and the time rate of increase in momentum of the particle in the direction of the resultant force.

Equation (0.2) is Newton's law of gravitation, giving the force of attraction between two particles of masses m_a and m_b a distance r apart.

Equation (0.3) is one statement of the first law of thermodynamics. It affirms the proportionality between the work performed by a closed system during a cycle and the heat absorbed by that system during the same cycle.

Equation (0.4) shows the proportionality between the thermodynamic absolute temperature and the zero-pressure limit of the pressure-volume product of a definite mass of any gas.

Each equation states that if means are available for measuring the values of all variables in that equation and if the numerical value of k is calculated, then the value of k is constant and depends only on the units used for measuring the variables in the equation.

Standards

By international agreement, standards are fixed arbitrarily for the quantities of mass, length, time, temperature, and the mole. These are five of the *base units* of SI. Currently, the standards are as follows.

The standard of mass is the kilogram (kg), defined as the mass of the international kilogram, a platinum cylinder preserved at Sèvres, France.

The standard of length is the meter (m), defined (since 1983) as the length of the path traveled by light in vacuum during a time interval of 1/299,792,458* of a second.

The standard of time is the second (s), defined as 9,192,631.770* frequency cycles of a certain quantum transition in an atom of ^{133}Ce.

The standard of temperature is the Kelvin (K), defined by assigning the value 273.16* K to the temperature of pure water at its triple point, the unique temperature at which liquid water, ice, and steam can exist at equilibrium.

The mole (abbreviated mol) is defined as the amount of a substance comprising as many elementary units as there are atoms in 12* g of ^{12}C. The definition of the mole is equivalent to the statement that the mass of one mole of a pure substance in grams is numerically equal to its molecular weight calculated from the standard table of atomic weights, in which the atomic weight of carbon is given as 12.01115. This number differs from 12* because it applies to the natural isotopic mixture of carbon rather than to pure ^{12}C. In engineering calculations the terms kilogram mole and pound mole are commonly used to designate the mass of a pure substance in kilograms or pounds that is equal to its molecular weight.

The actual number of molecules in one gram mole is given by Avogadro's number, 6.022×10^{23}.

Evaluation of constants

From the basic standards, values of m, m_a, and m_b in Eqs. (0.1) and (0.2) are measured in kilograms, r in meters, and u in meters per second. Constants k_1 and k_2 are not independent but are related by eliminating F from Eqs. (0.1) and (0.2). This gives

$$\frac{k_1}{k_2} = \frac{d(mu)/dt}{m_a m_b / r^2}$$

Either k_1 or k_2 may be fixed arbitrarily. Then the other constant must be found by experiments in which inertial forces calculated by Eq. (0.1) are compared with gravitational forces calculated by Eq. (0.2). In SI, k_1 is fixed at unity and k_2 found experimentally. Equation (0.1) then becomes

$$F = \frac{d}{dt}(mu) \qquad (0.5)$$

The force defined by Eq. (0.5) and also used in Eq. (0.2) is called the *newton* (N).
From Eq. (0.5),

$$1N \equiv 1 \text{ kg} \cdot \text{m/s}^2 \qquad (0.6)$$

Constant k_2 is denoted by G and called the *gravitational constant*. Its recommended value is

$$G = 6.6726 \times 10^{-11} \text{N} \cdot \text{m}^2/\text{kg}^2 \qquad (0.7)$$

Work, energy, and power

In SI, both work and energy are measured in newton-meters, a unit called the *joule* (J), and so

$$1J \equiv 1\text{N} \cdot \text{m} = 1 \text{ kg} \cdot \text{m}^2/\text{s}^2 \tag{0.8}$$

Power is measured in joules per second, a unit called the *watt* (W).

Heat

The constant k_3 in Eq. (0.3) may be fixed arbitrarily. In SI it, like k_1, is set at unity. Equation (0.3) becomes

$$Q_c = W_c \tag{0.9}$$

Heat, like work, is measured in joules.

Temperature

The quantity pV/m in Eq. (0.4) may be measured in $(\text{N}/\text{m}^2)(\text{m}^3/\text{kg})$, or J/kg. With an arbitrarily chosen gas, this quantity can be determined by measuring p and V of m kg of gas while it is immersed in a thermostat. In this experiment, only constancy of temperature, not magnitude, is needed. Values of pV/m at various pressures and at constant temperature can then be extrapolated to zero pressure to obtain the limiting value required in Eq. (0.4) at the temperature of the thermostat. For the special situation in which the thermostat contains water at its triple point, the limiting value is designated by $(pV/m)_0$. For this experiment Eq. (0.4) gives

$$273.16 = k_4 \lim_{p \to 0} (\frac{pV}{m})_0 \tag{0.10}$$

For an experiment at temperature T K, Eq. (0.4) can be used to eliminate k_4 from Eq. (0.10), giving

$$T \equiv 273.16 \frac{\lim_{p \to 0}(\frac{pV}{m})_T}{\lim_{p \to 0}(\frac{pV}{m})_0} \tag{0.11}$$

Equation (0.11) is the definition of the Kelvin temperature scale from the experimental pressure-volume properties of a real gas.

Celsius temperature

In practice, temperatures are expressed on the Celsius scale, in which the zero point is set at the ice point, defined as the equilibrium temperature of ice and air saturated water at a pressure of one atmosphere. Experimentally, the ice point is found to be 0.01K below the triple point of water, and so it is at 273.15 K. The Celsius temperature (℃) is defined by

$$T(\text{℃}) \equiv T(\text{K}) - 273.15 \tag{0.12}$$

On the Celsius scale, the experimentally measured temperature of the steam point, which is the boiling point of water at a pressure of 1 atm, is 100.00℃.

Decimal units

In SI, a single unit is defined for each quantity, but named decimal multiples and submultiples also are recognized. They are listed in Table 0.1. Time may be expressed in the nondecimal units: minutes (min), hours (h), or days (d).

Standard gravity

For certain purposes, the acceleration of free fall in the earth's gravitational field is used. From deductions based on Eq. (0.2), this quantity, denoted by g, is nearly constant. It varies slightly with latitude and height above sea level. For precise calculations, an arbitrary standard g_n has been set, defined by

$$g_n = 9.80665^* \text{ m/s}^2 \qquad (0.13)$$

Table 0.1 SI and cgs prefixes for multiples and submultiples

Factor	Prefix	Abbreviation	Factor	Prefix	Abbreviation
10^{12}	tera	T	10^{-2}	centi	c
10^{9}	giga	G	10^{-3}	milli	m
10^{6}	mega	M	10^{-6}	micro	μ
10^{3}	kilo	k	10^{-9}	nano	n
10^{2}	hecto	h	10^{-12}	pico	p
10^{1}	deka	da	10^{-15}	femto	f
10^{-1}	deci	d	10^{-18}	atto	a

Pressure units

The natural unit of pressure in SI is the newton per square meter. This unit, called the *pascal* (Pa), is inconveniently small, and a multiple, called the *bar*, also is used. It is defined by

$$1 \text{ bar} \equiv 1 \times 10^5 \text{ Pa} = 1 \times 10^5 \text{ N/m}^2 \qquad (0.14)$$

A more common empirical unit for pressure, used with all systems of units, is the *standard atmosphere* (atm), defined by

$$1 \text{ atm} \equiv 1.01325^* \times 10^5 \text{ Pa} = 1.01325 \text{ bars} \qquad (0.15)$$

0.2.3 CGS Units

The older cgs system can be derived from SI by making certain arbitrary decisions.

The standard for mass is the gram (g), defined by

$$1 \text{ g} \equiv 1 \times 10^{-3} \text{ kg} \qquad (0.16)$$

The standard for length is the centimeter (cm), defined by

$$1 \text{ cm} \equiv 1 \times 10^{-2} \text{ m} \qquad (0.17)$$

Standards for time, temperature, and the mole are unchanged.

As in SI, constant k_1 in Eq. (0.1) is fixed at unity. The unit of force is called the *dyne* (dyn), defined by

$$1 \text{ dyn} \equiv 1 \text{ g} \cdot \text{cm/s}^2 \qquad (0.18)$$

The unit for energy and work is the *erg*, defined by

$$1 \text{ erg} \equiv 1 \text{ dyn} \cdot \text{cm} = 1 \times 10^{-7} \text{ J} \qquad (0.19)$$

Constant k_3 in Eq. (0.3) is not unity. A unit for heat, called the *calorie* (cal), is used to convert the unit for heat to ergs. Constant $1/k_3$ is replaced by J, which denotes the quantity called the mechanical equivalent of heat and is measured in joules per calorie. Equation (0.3) becomes

$$W_c = JQ_c \qquad (0.20)$$

Two calories are defined. The *thermochemical calorie* (cal), used in chemistry, chemical engineering thermodynamics, and reaction kinetics, is defined by

$$1 \text{ cal} = 4.1840^* \times 10^7 \text{ ergs} = 4.1840^* \text{ J} \qquad (0.21)$$

The *international steam table calorie* (cal$_{IT}$), used in heat power engineering, is defined by
$$1 \text{ cal}_{IT} \equiv 4.1868^* \times 10^7 \text{ ergs} = 4.1868^* \text{ J} \tag{0.22}$$
The calorie is so defined that the specific heat of water is approximately 1 cal/(g·°C).

The standard acceleration of free fall in cgs units is
$$g_n \equiv 980.665 \text{ cm/s}^2 \tag{0.23}$$

0.2.4 FPS Engineering Units

In some countries a nondecimal gravitational unit system has long been used in commerce and engineering. The system can be derived from SI by making the following decisions.

The standard for mass is the avoirdupois pound (lb), defined by
$$1 \text{ lb} = 0.45359237^* \text{ kg} \tag{0.24}$$

The standard for length is the inch (in.), defined as 2.54^* cm. This is equivalent to defining the foot (ft) as
$$1 \text{ ft} \equiv 2.54 \times 12 \times 10^{-2} \text{ m} = 0.3048^* \text{ m} \tag{0.25}$$

The standard for time remains the second (s).

The thermodynamic temperature scale is called the Rankine scale, in which temperatures are denoted by degrees Rankine and defined by
$$1°R \equiv \frac{1}{1.8}K \tag{0.26}$$

The ice point on the Rankine scale is $273.15 \times 1.8 = 491.67°R$.

The analog of the Celsius scale is the Fahrenheit scale, in which readings are denoted by degrees Fahrenheit. It is derived from the Rankine scale by setting its zero point exactly 32°F below the ice point on the Rankine scale, so that
$$T(°F) \equiv T(°R) - (491.67 - 32) = T(°R) - 459.67 \tag{0.27}$$

The relation between the Celsius and Fahrenheit scales is given by the exact equation
$$T(°F) = 32 + 1.8°C \tag{0.28}$$

From this equation, temperature differences are related by
$$\Delta T(°C) = 1.8 \Delta T(°F) = \Delta T(K) \tag{0.29}$$

The steam point is 212.00°F.

Pound force

The fps system is characterized by a gravitational unit of force, called the *pound force* (lb$_f$). The unit is so defined that a standard gravitational field exerts a force of one pound on a mass of one avoirdupois pound. The standard acceleration of free fall in fps units is, to five significant figures,
$$g_n = \frac{9.80665 \text{m/s}^2}{0.3048 \text{m/ft}} = 32.174 \text{ ft/s}^2 \tag{0.30}$$

The pound force is defined by
$$1 \text{ lb}_f \equiv 32.174 \text{ lb} \cdot \text{ft/s}^2 \tag{0.31}$$

Then Eq. (0.1) gives

Unit Operations of Chemical Engineering

$$F \text{ (lb}_f) \equiv \frac{d(mu)/dt}{32.174} \quad \text{lb} \cdot \text{ft/s}^2 \tag{0.32}$$

Equation (0.1) can also be written with $1/g_c$ in place of k_1:

$$F = \frac{d(mu)/dt}{g_c} \tag{0.33}$$

Comparison of Eqs. (0.32) and (0.33) shows that to preserve both numerical equality and consistency of units in these equations, it is necessary to define g_c, called *Newton's law proportionality factor for the gravitational force unit*, by

$$g_c \equiv 32.174 \text{ lb} \cdot \text{ft}/(\text{s}^2 \cdot \text{lb}_f) \tag{0.34}$$

The unit for work and mechanical energy in the fps system is the *foot-pound force* (ft · lb$_f$). Power is measured by an empirical unit, the *horsepower* (hp), defined by

$$1\text{hp} \equiv 550 \text{ ft} \cdot \text{lb}_f/\text{s} \tag{0.35}$$

The unit for heat is the British thermal unit (Btu), defined by the implicit relation

$$1 \text{ Btu/lb} \cdot {}^\circ\text{F} = 1\text{cal}_{IT}/(\text{g} \cdot {}^\circ\text{C}) \tag{0.36}$$

As in the cgs system, constant k_3 in Eq. (0.3) is replaced by $1/J$, where J is the mechanical equivalent of heat, equal to 778.17 ft · lb$_f$/Btu.

The definition of the Btu requires that the numerical value of specific heat be the same in both systems, and in each case the specific heat of water is approximately 1.0.

0.2.5 Gas Constant

If mass is measured in kilograms or grams, constant k_4 in Eq. (0.4) differs from gas to gas. But when the concept of the mole as a mass unit is used, k_4 can be replaced by the universal gas constant R, which, by Avogadro's law, is the same for all gases. The numerical value of R depends only on the units chosen for energy, temperature, and mass. Then Eq. (0.4) is written

$$\lim_{p \to 0} \frac{pV}{nT} = R \tag{0.37}$$

where n is the number of moles. This equation applies also to mixtures of gases if n is the total number of moles of all the molecular species that make up the volume V.

The accepted experimental value of R is

$$R = 8.31447 \text{ J}/(\text{K} \cdot \text{mol}) = 8.31447 \times 10^7 \text{ ergs}/(\text{K} \cdot \text{mol}) \tag{0.38}$$

Values of R in other units for energy, temperature, and mass are given in Table 0.2.

Table 0.2 Value of the gas constant R

Temperature	Mass	Energy	R
Kelvins	kg mol	J	8,314.47
		cal$_{IT}$	1.9859×10^3
		cal	1.9873×10^3
		m^3-atm	82.056×10^{-3}
	g mol	cm^3-atm	82.056
Degress Rankine	lb mol	Btu	1.9858
		ft·lb$_f$	1,545.3
		hp·h	7.8045×10^{-4}
		kWh	5.8198×10^{-4}

Although the mole is defined as a mass in grams, the concept of the mole is easily extended to other mass units. Thus, the kilogram mole (kg mol) is the usual molecular or atomic weight in kilograms, and the pound mole (lb mol) is that in avoirdupois pounds. When the mass unit is not specified, the gram mole (g mol) is intended. Molecular weight M is a pure number.

Standard molar volume. From Table 0.2, the volume of 1 kg mol of gas at standard conditions (1 atm, 0°C), is $82.056 \times 10^{-3} \times 273 = 22.4$ m^3, or 22.4 (L /g mol). In fps units, the standard volume at 1 atm and 32°F is 359 ft^3/lb mol.

0.2.6 Conversion of Units

Since three unit systems are in common use, it is often necessary to convert the magnitudes of quantities from one system to another. This is accomplished by using conversion factors. Only the defined conversion factors for the base units are required since conversion factors for all other units can be calculated from them. Interconversions between SI and the cgs system are simple. Both use the same standards for time, temperature, and the mole, and only the decimal conversions defined by Eqs. (0.16) and (0.17) are needed. Both SI and the fps system also use the second as the standard for time; the three conversion factors defined for mass, length, and temperature by Eqs. (0.24), (0.25), and (0.26), respectively, are sufficient for all conversions of units between these two systems.

Example 0.1 demonstrates how conversion factors are calculated from the exact numbers used to set up the definitions of units in SI and the fps system. In conversions involving g_c in fps units, the use of the exact numerical ratio 9.80665/0.3048 in place of the fps number 32.1740 is recommended to give maximum precision in the final calculation and to take advantage of possible cancellations of numbers during the calculation.

EXAMPLE 0.1 Using only exact definitions and standards, calculate factors for converting (*a*) newtons to pounds force, (*b*) British thermal units to IT calories, (*c*) atmospheres to pounds force per square inch, and (*d*) horsepower to kilowatts.

Solution
(*a*) From Eqs. (0.6), (0.24), and (0.25),

$$1\,N = 1\,kg \cdot m/s^2 = \frac{1\,lb \cdot ft/s^2}{0.45359237 \times 0.3048}$$

From Eq. (0.30)

$$1\,lb \cdot ft/s^2 = \frac{0.3048}{9.80665}\,lb_f$$

and so

$$1\,N = \frac{0.3048}{9.80665 \times 0.45359237 \times 0.3048}\,lb_f$$

$$= \frac{1}{9.80665 \times 0.45359237}\,lb_f = 0.224809\,lb_f$$

In Appendix 1 it is shown that to convert newtons to pound force, one should multiply by 0.224809. Clearly, to convert from pounds force to newtons, multiply by 1/0.224809 = 4.448221.

(*b*) From Eq. (0.36)

$$1\,Btu = 1\,cal_{IT} \frac{1\,lb}{1\,g} \times \frac{1\,°F}{1\,°C} = 1\,cal_{IT} \frac{1\,lb}{1\,kg} \times \frac{1\,kg}{1\,g} \times \frac{1\,°F}{1\,°C}$$

From Eqs. (0.16), (0.24), and (0.29)

$$1 \text{ Btu} = 1 \text{ cal}_{IT} \frac{0.45359237 \times 1000}{1.8} = 251.996 \text{ cal}_{IT}$$

(c) From Eqs. (0.6), (0.14), and (0.15)

$$1 \text{ atm} = 1.01325 \times 10^5 \text{ kg} \cdot \text{m}/(\text{s}^2 \cdot \text{m}^2)$$

From Eqs. (0.24), (0.25), and (0.34), since 1 ft = 12 in,

$$1 \text{ atm} = 1.01325 \times 10^5 \times \frac{1 \text{ lb/s}^2}{0.45359237} \times \frac{0.3048}{\text{ft}}$$

$$= \frac{1.01325 \times 10^5 \times 0.3048}{32.174 \times 0.45359237 \times 12^2} \text{lb}_f/\text{in}^2$$

$$= 14.6959 \text{ lb}_f/\text{in}^2$$

(d) From Eqs. (0.31) and (0.35)

$$1 \text{ hp} = 550 \text{ ft} \cdot \text{lb}_f/\text{s} = 550 \times 32.174 \text{ ft}^2 \cdot \text{lb/s}^3$$

Using Eqs. (0.24) and (0.25) gives

$$1 \text{ hp} = 550 \times 32.174 \times 0.45359237 \times 0.3048^2 = 745.70 \text{ J/s}$$

Substituting from Eq. (0.8) and dividing by 1,000,

$$1 \text{ hp} = 0.74570 \text{ kW}$$

Although conversion factors may be calculated as needed, it is more efficient to use tables of the common factors. A table for the factors used in this book is given in Appendix 1.

0.2.7 Units and Equations

Although Eqs. (0.1) to (0.4) are sufficient for the description of unit systems, they are but a small fraction of the equations needed in this book. Many such equations contain terms that represent properties of substances, and these are introduced as needed. All new quantities are measured in combinations of units already defined, and all are expressible as functions of the five base units for mass, length, time, temperature, and the mole.

Precision of calculations

In the above discussion, the values of experimental constants are given with the maximum number of significant digits consistent with present estimates of the precision with which they are known, and all digits in the values of defined constants are retained. In practice, such extreme precision is seldom necessary, and defined and experimental constants can be truncated to the number of digits appropriate to the problem at hand, although the advent of the digital computers makes it possible to retain maximum precision at small cost. The engineer should use judgment in setting a suitable level of precision for the particular problem to be solved.

General equations

Except for the appearance of the proportionality factors g_c and J, the equations for all three unit systems are alike. In this text, equations are written for SI units, with a reminder to use g_c and J when working examples in cgs or fps units.

Dimensionless equations and consistent units

Equations derived directly from the basic laws of the physical sciences consist of terms that either have the same units or can be written in the same units by using the definitions of derived quantities to express complex units in terms of the five base ones. Equations meeting this requirement are called *dimensionally homogeneous equations*. When such an equation is divided by any one of its terms, all units in each term cancel and only numerical magnitudes remain. These equations are called *dimensionless equations*.

A dimensionally homogeneous equation can be used as it stands with any set of units provided that the same units for the five base units are used throughout. Units meeting this requirement are called *consistent units*. No conversion factors are needed when consistent units are used.

For example, consider the usual equation for the vertical distance Z traversed by a freely falling body during time t when the initial velocity is u_0:

$$Z = u_0 t + \frac{1}{2} g t^2 \tag{0.39}$$

Examination of Eq. (0.39) shows that the units in each term reduce to that for length. Dividing the equation by Z gives

$$1 = \frac{u_0 t}{Z} + \frac{g t^2}{2Z} \tag{0.40}$$

A check of each term in Eq. (0.40) shows that the units in each term cancel and each term is dimensionless. A combination of variables for which all dimensions cancel in this manner is called a *dimensionless group*. The numerical value of a dimensionless group for given values of the quantities contained in it is independent of the units used, provided they are consistent. Both terms on the right-hand side of Eq. (0.40) are dimensionless groups.

Dimensional equations

Equations derived by empirical methods, in which experimental results are correlated by empirical equations without regard to dimensional consistency, usually are not dimensionally homogeneous and contain terms in several different units. Equations of this type are *dimensional equations*, or dimensionally nonhomogeneous equations. In these equations there is no advantage in using consistent units, and two or more length units, for example, inches and feet, or two or more time units, for example, seconds and minutes, may appear in the same equation. For example, a formula for the rate of heat loss from a horizontal pipe to the atmosphere by conduction and convection is

$$\frac{q}{A} = 0.50 \frac{\Delta T^{1.25}}{(d_0')^{0.25}} \tag{0.41}$$

where q = rate of heat loss, Btu/h
 A = area of pipe surface, ft^2
 ΔT = excess of temperature of pipe wall over that of ambient (surrounding atmosphere), °F
 d_0' = outside diameter of pipe, in

Obviously, the units of q/A are not those of the right-hand side of Eq. (0.41), and the equation is dimensional. Quantities substituted in Eq. (0.41) must be expressed in the units as given, or the equation will give the wrong answer. If other units are to be used, the coefficient must be changed. To express ΔT in degrees Celsius, for example, the numerical coefficient must be changed to $0.50 \times 1.8^{1.25} = 1.042$ since there are 1.8 Fahrenheit degrees in 1 Celsius

12 | Unit Operations of Chemical Engineering

degree of temperature difference.

In this book all equations are dimensionally homogeneous *unless otherwise noted*.

0.3 Dimensional Analysis

Many important engineering problems cannot be solved completely by theoretical or mathematical methods. Problems of this type are especially common in fluid-flow, heat-flow, and diffusion operations. One method of attacking a problem for which no mathematical equation can be derived is that of empirical experimentation. For example, the pressure loss from friction in a long, round, straight, smooth pipe depends on all these variables: the length and diameter of the pipe, the flow rate of the liquid, and the density and viscosity of the liquid. If any one of these variables is changed, the pressure drop also changes. The empirical method of obtaining an equation relating these factors to pressure drop requires that the effect of each separate variable be determined in turn by systematically varying that variable while keeping all others constant. The procedure is laborious, and it is difficult to organize or correlate the results so obtained into a useful relationship for calculations.

There exists a method intermediate between formal mathematical development and a completely empirical study. It is based on the fact that if a theoretical equation does exist among the variables affecting a physical process, that equation must be dimensionally homogeneous. Because of this requirement, it is possible to group many factors into a smaller number of dimensionless groups of variables. The groups themselves rather than the separate factors appear in the final equation.

This method is called *dimensional analysis*, which is an algebraic treatment of the symbols for units considered independently of magnitude. It drastically simplifies the task of fitting experimental data to design equations; it is also useful in checking the consistency of the units in equations, in converting units, and in the scale-up of data obtained in model test units to predict the performance of full-scale equipment.

In making a dimensional analysis, the variables thought to be important are chosen and their dimensions tabulated. If the physical laws that would be involved in a mathematical solution are known, the choice of variables is relatively easy. The fundamental differential equations of fluid flow, for example, combined with the laws of heat conduction and diffusion, suffice to establish the dimensions and dimensionless groups appropriate to a large number of chemical engineering problems. In other situations the choice of variables may be speculative, and testing of the resulting relationships may be needed to establish whether some variables were left out or whether some of those chosen are not needed.

Assuming that the variables are related by a power series, in which the dimension of each term must be the same as that of the primary quantity, an exponential relationship is written in which the exponents relating to any given quantity (for example, length) must be the same on both sides of the equation. The relationship among the exponents is then found algebraically, as shown in Example 0.2.

EXAMPLE 0.2 A steady stream of liquid in turbulent flow is heated by passing it through a long, straight, heated pipe. The temperature of the pipe is assumed to be greater by a constant amount than the average temperature of the liquid. It is desired to find a relationship that can be used to predict the rate of heat transfer from the wall of the liquid.

Solution. The mechanism of this process is discussed in Chap. 4. From the characteristics of the process it may be expected that the rate of heat transfer q/A depends on the quantities listed with their dimensional formulas in Table 0.3. If a theoretical equation for this problem exists, it can be written in the general form

$$\frac{q}{A} = \psi(D, u, \rho, \mu, c_p, k, \Delta T) \tag{0.42}$$

If Eq. (0.42) is a valid relationship, all terms in the function ψ must have the same dimensions as those of the left-hand side of the equation, q/A. Let the phrase the *dimensions of* be shown by the use of square brackets. Then any term in the function must conform to the dimensional formula

$$[\frac{q}{A}] = [D]^a [u]^b [\rho]^c [\mu]^d [c_p]^e [k]^f [\Delta T]^g \tag{0.43}$$

Let an overbar above a symbol denote that it refers to a dimension. Thus $\bar{L}$ refers to the dimension of length. Substituting the dimensions from Table 0.3 gives

$$\bar{H}\bar{L}^{-2}\bar{t}^{-1} = \bar{L}^a \bar{L}^b \bar{t}^{-b} \bar{M}^c \bar{L}^{-3c} \bar{M}^d \bar{L}^{-d} \bar{H}^e \bar{M}^{-e} \bar{T}^{-e} \bar{H}^f \bar{L}^{-f} \bar{t}^{-f} \bar{T}^{-f} \bar{T}^g \tag{0.44}$$

Table 0.3 Quantities and dimensional formulas for Example 0.2

Quantity	Symbol	Dimensions
Heat flow per unit area	q/A	$\bar{H}\,\bar{L}^{-2}\bar{t}^{-1}$
Diameter of pipe (inside)	D	$\bar{L}$
Average velocity of liquid	u	$\bar{L}\,\bar{t}^{-1}$
Density of liquid	ρ	$\bar{M}\,\bar{L}^{-3}$
Viscosity of liquid	μ	$\bar{M}\,\bar{L}^{-1}\bar{t}^{-1}$
Specific heat, at constant pressure, of liquid	c_p	$\bar{H}\,\bar{M}^{-1}\bar{T}^{-1}$
Thermal conductivity of liquid	k	$\bar{H}\,\bar{L}^{-1}\bar{t}^{-1}\bar{T}^{-1}$
Temperature difference between wall and fluid	ΔT	$\bar{T}$

Since Eq. (0.43) is assumed to be dimensionally homogeneous, the exponents of the individual primary units on the left-hand side of the equation must equal those on the right-hand side. This gives the following set of equations:

Exponents of $\bar{H}$: $\qquad\qquad\qquad 1 = e + f \qquad\qquad\qquad$ (0.45a)

Exponents of $\bar{L}$: $\qquad\qquad\qquad -2 = a + b - 3c - d - f \qquad\qquad\qquad$ (0.45b)

Exponents of $\bar{t}$: $\qquad\qquad\qquad -1 = -b - d - f \qquad\qquad\qquad$ (0.45c)

Exponents of $\bar{M}$: $\qquad\qquad\qquad 0 = c + d - e \qquad\qquad\qquad$ (0.45d)

Exponents of $\bar{T}$: $\qquad\qquad\qquad 0 = -e - f + g \qquad\qquad\qquad$ (0.45e)

Here there are seven variables but only five equations. Five of the unknowns may be found in terms of the remaining two. The two letters to be retained must be chosen arbitrarily. The final result is equally valid for all choices, but for this problem it is customary to retain the exponents of the velocity u and the specific heat c_p. The letters b and e will be retained and the remaining five eliminated, as follows.

From Eq. (0.45a):

$$f = 1 - e \tag{0.46a}$$

From Eqs. (0.45e) and (0.46a)

$$g = e + f = e + 1 - e = 1 \tag{0.46b}$$

From Eqs. (0.45c) and (0.46a)
$$d = 1 - b - f = 1 - b - 1 + e = e - b \tag{0.46c}$$

From Eqs. (0.45d) and (0.46c)
$$c = e - d = e - e + b = b \tag{0.46d}$$

From Eqs. (0.45b), (0.46a), (0.46c), and (0.46d)
$$a = -2 - b + 3c + d + f = -2 - b + 3b + e - b + 1 - e = b - 1 \tag{0.46e}$$

By substituting values from Eq. (0.46a) through (0.46e) for the letters a, c, d, and f, Eq. (0.43) becomes

$$[\frac{q}{A}] = [D]^{b-1}[u]^b[\rho]^b[\mu]^{e-b}[c_p]^e[k]^{1-e}[\Delta T] \tag{0.47}$$

Collecting all factors having integral exponents in one group, all factors having exponents b into another group, and those having exponents e into a third gives

$$[\frac{qD}{Ak\Delta T}] = [\frac{Du\rho}{\mu}]^b[\frac{c_p\mu}{k}]^e \tag{0.48}$$

The dimensions of each of the three bracketed groups in Eq. (0.48) are zero, and all groups are dimensionless. Any function whatever of these three groups will be dimensionally homogeneous, and the equation will be a dimensionless one. Let such a function be

$$\frac{qD}{Ak\Delta T} = \Phi(\frac{Du\rho}{\mu}, \frac{c_p\mu}{k}) \tag{0.49}$$

or

$$\frac{q}{A} = \frac{k\Delta T}{D}\Phi(\frac{Du\rho}{\mu}, \frac{c_p\mu}{k}) \tag{0.50}$$

The relationship given in Eqs. (0.49) and (0.50) is the final result of the dimensional analysis. The form of function Φ must be found experimentally, by determining the effects of the groups in the brackets on the value of the group on the left-hand side of Eq. (0.49). The correlations that have been found for this are given in Chap. 4.

Correlating the experimental values of the three groups of variables of Eq. (0.49) is clearly simpler than attempting to correlate the effects of each of the individual factors of Eq. (0.42).

PROBLEMS

0.1 Using defined constants and conversion factors for mass, length, time, and temperature, calculate conversion factors for (*a*) foot-pounds force to kilowatt hours, (*b*) gallons (1 gal = 231 in^3) to liters (10^3 cm^3), (*c*) Btu per pound mole to joules per kilogram mole.

0.2 At what point is the temperature in degrees Fahrenheit equal to the Celsius temperature? Is there any point where the Kelvin temperature is the same as the Rankine temperature?

Chapter 1

Fluid Flow

The behavior of fluids is important to process engineering generally and constitutes one of the foundations for the study of unit operations. An understanding of fluids is essential, not only for accurately treating problems in the movement of fluids through pipes, pumps, and all kinds of process equipment but also for the study of heat flow and the many separation operations that depend on diffusion and mass transfer.

The branch of engineering science that has to do with the behavior of fluids—and fluids are understood to include liquids, gases, and vapors—is called *fluid mechanics*. Fluid mechanics in turn is part of a larger discipline called *continuum mechanics*, which also includes the study of stressed solids.

Fluid mechanics has two branches important to the study of unit operations: *fluid statics*, which treats fluids in the equilibrium state of no shear stress, and *fluid dynamics*, which treats fluids when portions of the fluid are in motion relative to other parts.

Although gases and liquids consist of molecules, it is possible in most cases to treat them as continuous media for the purposes of fluid flow calculations. This treatment as a continuum is valid when the smallest volume of fluid contains a large enough number of molecules so that a statistical average is meaningful and the macroscopic properties of the fluid such as density, pressure, temperature, velocity and so on, vary smoothly or continuously from point to point.

This chapter deals with those areas of fluid mechanics that are important to unit operations. The choice of subject matter is but a sampling of the huge field of fluid mechanics generally. Section 1.1 treats fluid statics and some of its important applications. Section 1.2 discusses the important phenomena appearing in flowing fluids. Section 1.3 deals with the basic quantitative laws and equations of fluid flow. Section 1.4 treats the flow of incompressible fluids through pipes. Section 1.5 deals with various practical pipe flow problems. Section 1.6 deals with measuring and controlling fluids in flow.

1.1 Fluid Statics and Its Application

1.1.1 Nature of Fluids

A fluid is a substance that does not permanently resist distortion. An attempt to change the shape of a mass of fluid results in layers of fluid sliding over one another until a new shape is attained. During the change in shape, shear stresses (Shear is the lateral displacement of one layer of material relative to another layer by an external force. Shear stress is defined as the ratio of this force to the area of the layer. See section 1.2.) exist, the magnitudes of which depend upon the viscosity of the fluid and the rate of sliding; but when a final shape has been reached, all shear stresses will have disappeared. A fluid in equilibrium is free from shear stresses.

Density

At a given temperature and pressure, a fluid possesses a definite density, which in engineering practice is usually measured in kilograms per cubic meter. Although the density of all fluids depends on the temperature and pressure, the variation in density with changes in these variables may be small or large. If the density changes only slightly with moderate changes in temperature and pressure, the fluid is said to be *incompressible*; if the changes in density are significant, the fluid is said to be *compressible*. Liquids are generally considered to be incompressible and gases compressible. The terms are relative, however, and the density of a liquid can change appreciably if pressure and temperature are changed over wide limits. Also, gases subjected to small percentage changes in pressure and temperature act as incompressible fluids, and density changes under such conditions may be neglected without serious error.

Pressure

The pressure in a static fluid is familiar as a surface force exerted by the fluid against a unit area of the walls of its container. Pressure also exists at every point within a volume of fluid. It is a scalar quantity; at any given point its magnitude is the same in all directions.

The pressure at a point within a volume of fluid will be designated as either an *absolute pressure* or a *gage pressure*. Absolute pressure is measured relative to a perfect vacuum (absolute zero pressure), whereas gage pressure is measured relative to the local atmospheric pressure. Thus, a gage pressure of zero corresponds to a pressure that is equal to the local atmospheric pressure. Absolute pressures are always positive, but gage pressures can be either positive or negative depending on whether the pressure is above atmospheric pressure (a positive value) or below atmospheric pressure (a negative value). A negative gage pressure is also referred to as a *suction* or *vacuum pressure*. It is to be noted that pressure differences are independent of the reference, so that no special notation is required in this case.

In addition to the reference used for the pressure measurement, the units used to express the value are obviously of importance. Pressure is a force per unit area. In the SI system the units are N/m^2; this combination is called the pascal and written as Pa (1 N/m^2 = 1 Pa). Pressure can also be expressed as the height of a column of liquid, the units will refer to the height of the column (mm, m, etc.), and in addition, the liquid in the column must be specified (H_2O, Hg, etc.). For example, standard atmospheric pressure can be expressed as 760 mm Hg (abs).

1.1.2 Hydrostatic Equilibrium

In a stationary mass of a single static fluid, the pressure is constant in any cross section parallel to the earth's surface but varies from height to height. Consider the vertical column of fluid shown in Fig. 1.1. Assume the cross-sectional area of the column is S. At a height Z above the base of the column let the pressure be p and the density be ρ. The resultant of all forces on the small volume of fluid of height dZ and cross-sectional area S must be zero. Three vertical forces are acting on this volume: (1) the force from pressure p acting in an upward direction, which is pS; (2) the force from pressure $p+dp$ acting in a downward direction, which is $(p+dp)S$, (3) the force of gravity acting downward, which is $g\rho S dZ$. Then

$$+pS - (p+dp)S - g\rho S dZ = 0 \tag{1.1-1}$$

In this equation, forces acting upward are taken as positive and those acting downward as negative. After simplification and division by S, Eq. (1.1) becomes

$$dp + g\rho dZ = 0 \tag{1.1-2}$$

Equation (1.1-2) cannot be integrated for compressible fluids unless the variation of density with pressure is known throughout the column of fluid. However, it is often satisfactory for engineering calculations to consider ρ to be essentially constant. The density is constant for incompressible fluids and, except for large changes in height, is nearly so for compressible fluids. Integration of Eq. (1.1-2) on the assumption that ρ is constant gives

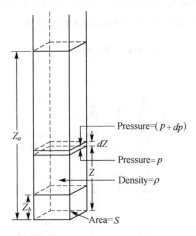

Fig. 1.1 Hydrostatic equilibrium.

$$\frac{p}{\rho} + gZ = \text{const} \tag{1.1-3}$$

or, between the two definite heights Z_a and Z_b, shown in Fig. 1.1,

$$\frac{p_b}{\rho} - \frac{p_a}{\rho} = g(Z_a - Z_b) \tag{1.1-4}$$

Equation (1.1-3) expresses mathematically the condition of hydrostatic equilibrium.

1.1.3 Applications of Fluid Statics

Barometric equation

For an ideal gas, the density and pressure are related by the equation

$$\rho = \frac{pM}{RT} \tag{1.1-5}$$

where M = molecular weight
       T = absolute temperature

Substitution from Eq. (1.1-5) into Eq. (1.1-2) gives

$$\frac{dp}{p} + \frac{gM}{RT}dZ = 0 \tag{1.1-6}$$

Integration of Eq. (1.1-6) between levels a and b, on the assumption that T is constant, gives

$$\ln\frac{p_b}{p_a} = -\frac{gM}{RT}(Z_b - Z_a)$$

or

$$\frac{p_b}{p_a} = \exp\left[-\frac{gM(Z_b - Z_a)}{RT}\right] \tag{1.1-7}$$

Equation (1.1-7) is known as the *barometric equation*.

Methods are available in the literature for estimating the pressure distribution in situations, for example, a deep gas well, in which the gas is not ideal and the temperature is not constant.

Manometers

Simple U tube manometer. The manometer is an important device for measuring pressure differences. Figure 1.2 shows the simplest form of manometer. Assume that the shaded portion of the U tube is filled with liquid A having a density ρ_A and that the arms of the U tube above the liquid are filled with fluid B having a density ρ_B. Fluid B is immiscible with liquid A and less dense than A; it is often a gas such as air or nitrogen.

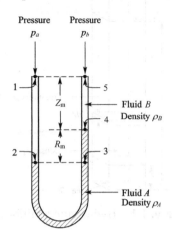

Fig. 1.2 distribution of particle diameter for example

A pressure p_a is exerted in one arm of the U tube and a pressure p_b in the other. As a result of the difference in pressure $p_a - p_b$, the meniscus in one branch of the U tube is higher than in the other, and the vertical distance between the two meniscuses R_m may be used to measure the difference in pressure. To derive a relationship between $p_a - p_b$ and R_m, start at the point 1, where the pressure is p_a, then, as shown by Eq. (1.1-4), the pressure at point 2 is $p_a + g(Z_m + R_m)\rho_B$. By the principles of hydrostatics, this is also the pressure at point 3. The pressure at point 4 is less than that at point 3 by the amount $gR_m\rho_A$, and the pressure at point 5, which is p_b, is still less by the amount $gZ_m\rho_B$. These statements can be summarized by the equation

$$p_a + g[(Z_m + R_m)\rho_B - R_m\rho_A - Z_m\rho_B] = p_b \tag{1.1-8}$$

Simplification of this equation gives

$$p_a - p_b = gR_m(\rho_A - \rho_B) \tag{1.1-9}$$

Note that this relationship is independent of the distance Z_m and of the dimensions of the tube, provided that pressures p_a and p_b are measured in the same horizontal plane. If fluid B is a gas, ρ_B is usually negligible compared to ρ_A and may be omitted from Eq. (1.1-9).

EXAMPLE 1.1 A manometer of the type shown in Fig. 1.2 is used to measure the pressure drop across an orifice (see Fig. 1.33). Liquid A is mercury (density 13590 kg/m^3) and fluid B, flowing through the orifice and filling the manometer leads, is brine (density 1260 kg/m^3). When the pressures at the taps are equal, the level of the mercury in the manometer is 0.9 m below the orifice taps. Under operating conditions, the gauge pressure at the upstream tap is 0.14 bar; the pressure at the downstream tap is 250 mm Hg below atmospheric. What is the reading of the manometer in millimeters?
Solution. Call atmospheric pressure zero; then the numerical data for substitution in Eq. (1.1-9) are

$$p_a = 0.14 \text{bar} = 14000 \text{Pa}$$

From Eq. (1.1-4)

$$p_b = Z_b\rho_A g = -\frac{250}{1000} \times 9.80665 \times 13590 = -33,318 \text{Pa}$$

Substituting in Eq. (1.1-9) gives

$$14000 + 33381 = R_m \times 9.80665 \times (13590 - 12600)$$

$$R_m = 0.391 \text{m}, \quad \text{or } 391 \text{ mm}$$

Two-fluid U tube manometer. In Fig. 1.3 a two-fluid U tube is shown, which is a sensitive device to measure small heads or pressure differences. Let D is the inside diameter of each of the large reservoirs and d is the inside diameter of each of the tubes forming the U. Proceeding and making a pressure balance as for the U tube,

$$p_1 - p_2 = (\rho_A - \rho_C)Rg + \Delta R(\rho_C - \rho_B)g \qquad (1.1\text{-}10)$$

where $\Delta R = R\left(\dfrac{d}{D}\right)^2$ is the height difference between two large reservoirs, R is the reading the manometer, ρ_A is the density of the heavier fluid, and ρ_C is the density of the lighter fluid. Usually (d/D) is made much sufficiently small to be negligible, then

$$p_1 - p_2 = (\rho_A - \rho_C)Rg \qquad (1.1\text{-}11)$$

If ρ_A and ρ_C are close to each other, the reading R is magnified.

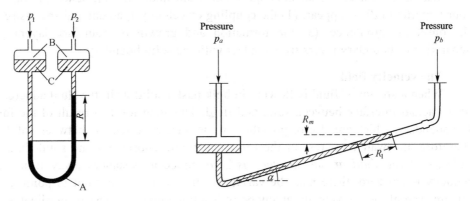

Fig. 1.3 Two-fluid U tube manometer.   **Fig. 1.4** Inclined manometer.

Inclined manometer. For measuring small differences in pressure, the *inclined manometer* shown in Fig. 1.4 may be used. In this type, one leg of the manometer is inclined in such a manner that, for a small magnitude of R_m, the meniscus in the inclined tube must move a considerable distance along the tube. This distance is R_m divided by the sine of α, the angle of inclination. By making α small, the magnitude of R_m is multiplied into a long distance R_1 and a large reading becomes equivalent to a small pressure difference; so

$$p_a - p_b = gR_1(\rho_A - \rho_B)\sin\alpha \qquad (1.1\text{-}12)$$

In this type of pressure gauge, it is necessary to provide an enlargement in the vertical leg so that the movement of the meniscus in the enlargement is negligible within the operating range of the instrument.

1.2 Fluid Flow Phenomena

The behavior of a flowing fluid depends strongly on whether the fluid is under the influence of solid boundaries. In the region where the influence of the wall is small, the shear stress may be negligible and the fluid behavior may approach that of an ideal fluid, one that is incompressible and has zero viscosity. The flow of such an ideal fluid is called *potential flow* and is completely described by the principles of newtonian mechanics and conservation of mass. The mathematical theory of potential flow is highly developed but is outside the scope of this book. Potential flow has two important characteristics: (1) Neither circulations nor eddies can form within the stream, so that potential flow is also called irrotational flow; and (2) friction cannot develop, so that there is no dissipation of mechanical energy into heat.

Potential flow can exist at distances not far from a solid boundary. A fundamental principle of fluid mechanics, originally stated by Prandtl in 1904, is that, except for fluids moving at low velocities or possessing high viscosities, the effect of the solid boundary on the flow is confined to a layer of the fluid immediately adjacent to the solid wall. This layer is called the *boundary layer*, and shear and shear forces are confined to this part of the fluid. Outside the boundary layer, potential flow survives. Most technical flow processes are best studied by considering the fluid stream as two parts, the boundary layer and the remaining fluid. In some situations such as flow in a converging nozzle, the boundary layer may be neglected; and in others, such as flow through pipes, the boundary layer fills the entire channel, and there is no potential flow.

Within the current of an incompressible fluid under the influence of solid boundaries, four important effects appear: (1) the coupling of velocity-gradient and shear-stress fields, (2) the onset of turbulence, (3) the formation and growth of boundary layers, and (4) the separation of boundary layers from contact with the solid boundary.

The velocity field

When a stream of fluid is flowing in bulk past a solid wall, the fluid adheres to the solid at the actual interface between solid and fluid. The adhesion is a result of the force fields at the boundary, which are also responsible for the interfacial tension between solid and fluid. If, therefore, the wall is at rest in the reference frame chosen for the solid-fluid system, *the velocity of the fluid at the interface is zero*. Since at distances away from the solid the velocity is not zero, there must be variations in velocity from point to point in the flowing stream. Therefore, the velocity at any point is a function of the space coordinates of that point, and a velocity field exists in the space occupied by the fluid. The velocity at a given location may also vary with time. When the velocity at each location is constant, the field is invariant with time and the flow is said to be *steady*.

One-dimensional flow. Velocity is a vector, and in general, the velocity at a point has three components, one for each space coordinate. In many simple situations all velocity vectors in the field are parallel or practically so, and only one velocity component, which may be taken as a scalar, is required. This situation, which obviously is much simpler than the general vector field, is called *one-dimensional flow*; an example is steady flow through straight pipe. The following discussion is based on the assumptions of steady one-dimensional flow.

1.2.1 Newton's Law and Viscosity of Fluids

Laminar flow

At low velocities fluids tend to flow without lateral mixing, and adjacent layers slide past one another as playing cards do. There are neither cross-currents nor eddies. This regime is called *laminar flow*. At higher velocities turbulence appears and eddies form, which, as discussed later, lead to lateral mixing. The discussion in section 1.2.1 is limited to laminar flow.

Velocity gradient and rate of shear

Consider the steady one-dimensional laminar flow of an incompressible fluid along a solid plane surface. Figure 1.5(a) shows the velocity profile for such a stream. The abscissa u is the velocity, and the ordinate y is the distance measured perpendicular from the wall and therefore at right angles to the direction of the velocity. At $y = 0, u = 0$, and u increases with distance from the wall but at a decreasing rate. Focus attention on the velocities on two nearby planes, plane A and plane B, a distance Δy apart. Let the velocities along the planes be u_A and u_B, respectively, and assume that $u_B > u_A$. Call $\Delta u = u_B - u_A$. Define the velocity gradient at y_A, du/dy, by

$$\frac{du}{dy} = \lim_{\Delta y \to 0} \frac{\Delta u}{\Delta y} \tag{1.2-1}$$

The velocity gradient is clearly the reciprocal of the slope of the velocity profile of Fig. 1.5(a). The local velocity gradient is also called the shear rate, or time rate of shear. The velocity gradient is usually a function of position in the stream and therefore defines a field, as illustrated in Fig. 1.5(b).

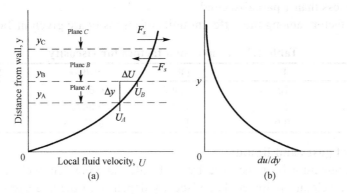

Fig. 1.5 Profiles of velocity and velocity gradient in layer flow:
(a) velocity; (b) velocity gradient or rate of shear.

The shear-stress field

Since an actual fluid resists shear, a shear force must exist wherever there is a time rate of shear. In one-dimensional flow the shear force acts parallel to the plane of the shear. For example, at plane C at distance y_C from the wall, the shear force F_s, shown in Fig. 1.5(a), acts in the direction shown in the figure. This force is exerted by the fluid above plane C on the fluid between plane C and the wall. By Newton's third law, an equal and opposite force $-F_s$ acts on the fluid above plane C from the fluid below plane C. It is convenient to use, not total force F_s, but the force per unit area of the shearing plane, called the *shear stress* and denoted by τ, or

$$\tau = \frac{F_s}{A_s} \tag{1.2-2}$$

where A_s, is the area of the plane. Since τ varies with y, the shear stress also constitutes a field. Shear forces are generated in both laminar and turbulent flow. The shear stress arising from viscous or laminar flow is denoted by τ_v. The effect of turbulence is described later.

Newton's law of viscosity

It has been found experimentally for many fluids that the shear stress arising from viscous or laminar flow is directly proportional to the velocity gradient. Or, as given by Newton's law of viscosity,

$$\tau_v = \mu \frac{du}{dy} \tag{1.2-3}$$

where μ is a proportionality constant called the viscosity of the fluid, in Pa·s or kg/(m·s).

Fluids that follow Newton's law of viscosity, Eq. (1.2-3), are called Newtonian fluids.

Viscosity

In a newtonian fluid, the shear stress is proportional to the shear rate, and the proportionality constant, *viscosity*, is independent of the shear rate.

In SI units τ_v is measured in newtons per square meter and μ in kilograms per meter-second or pascal-second. In the cgs system, viscosity is expressed in grams per centimeter-second, and this unit is called the poise (P). Viscosity data are generally reported in millipascal-seconds or in centipoises (cP = 0.01 P = 1 mPa · s), since most fluids have viscosities much less than 1 pascal-second.

Conversion factors among the different units of viscosity are given in Table 1.1.

Table 1.1 Conversion factors for viscosity

Pa·s	P	cP	lb/ft·s	lb/ft·h
1	10	1,000	0.672	2.420
0.1	1	100	0.0672	242
10^{-3}	0.01	1	6.72×10^{-4}	2.42

Viscosities of gases and liquids

In gases, momentum is transferred by molecules moving relatively large distances to regions where the velocity is lower. The viscosity depends on the average momentum of the molecules, which is proportional to the molecular weight times the average velocity. Since the velocity is proportional to $(T/M)^{1/2}$, the viscosity is proportional to $(MT)^{1/2}$. The viscosity also depends on the mean free path, which decreases as the size of the molecule increases. A simple theory for noninteracting molecules is

$$\mu = 0.00267 \frac{(MT)^{1/2}}{\sigma^2} \tag{1.2-4}$$

where μ = viscosity, cP
  M = molecular weight
  T = absolute temperature, K
  σ = molecular diameter, Å

Note that the predicted effect of molecular weight is small, since the $M^{1/2}$ term in the numerator almost offsets the σ^2 term in the denominator. The average error in using Eq. (1.2-4) to predict viscosity is about 20 percent.

Rigorous theories for gas viscosity include a collision integral Ω_V in the denominator of Eq. (1.2-4) to allow for interactions between colliding molecules. Different approaches to calculating Ω_V are reviewed by Reid. The average error in predicting gas viscosity is 2 to 3 percent. It is more difficult to predict viscosities of gas mixtures, and errors of 6 to 10 percent are typical. Gas viscosities at room temperature are generally between 0.005 and 0.02 cP.

There is no simple correlation with molecular weight. At 20℃ the viscosity is 0.018 cP for air, 0.009 cP for hydrogen, and 0.007 cP for benzene vapor. Values for other substances are given in Appendix 7.

Gas viscosities increase with temperature somewhat more rapidly than predicted by simple kinetic theory. For approximate calculations:

$$\frac{\mu}{\mu_0} = \left(\frac{T}{273}\right)^n \qquad (1.2\text{-}5)$$

where μ = viscosity at absolute temperature T, K
   μ_0 = viscosity at 0℃ (273 K)
   n = constant for a particular gas

Exponent n is approximately 0.65 for air, 0.9 for carbon dioxide, 0.8 for butane, and 1.0 for steam.

The viscosity of a gas is almost independent of pressure in the region where the ideal gas laws apply. The increase in gas density, which leads to an increased flux of molecules across a unit area, is offset by the decrease in the mean free path. At very high pressures the viscosity increases with pressure, especially in the neighborhood of the critical point.

Viscosity of liquids.   The viscosities of liquids are much greater than those of gases at the same temperature, and there is no simple theory to predict liquid viscosity. Molecules in liquids move very short distances between collisions, and most momentum transfer occurs as molecules in a velocity gradient slide past one another. The viscosity generally increases with molecular weight and decreases rapidly with increasing temperature. The main effect of temperature change comes not from the increase in average velocity, as in gases, but from the slight expansion of the liquid, which makes it easier for the molecules to slide past each another. For example, the viscosity of water falls from 1.79 cP at 0℃ to 0.28 cP at 100℃, a 6.4-fold change, though the average molecular velocity increases only by $(373/273)^{0.5}$, or a factor of 1.17.

The viscosity is a strongly nonlinear function of the temperature, but a good approximation for temperatures below the normal boiling point is

$$\ln \mu = A + B/T \qquad (1.2\text{-}6)$$

The absolute viscosities of liquids vary over an enormous range of magnitudes, from about 0.1 cP for liquids near their boiling point to as much as 10^6 P for polymer melts. Most extremely viscous materials are non-newtonian and possess no single viscosity independent of shear rate.

Kinematic viscosity.   The ratio of the absolute viscosity to the density of a fluid μ/ρ is often useful. This property is called the *kinematic viscosity* and designated by v. In SI, the unit for v is square meters per second. In the cgs system, the kinematic viscosity is called the stoke (St), defined as 1 cm^2/s. The fps unit is square feet per second. Conversion factors are

$$1\text{m}^2/\text{s} = 10^4 \text{St} = 10.7639 \text{ft}^2/\text{s}$$

For liquids, kinematic viscosities vary with temperature over a somewhat narrower range than absolute viscosities. For gases, the kinematic viscosity increases more rapidly with temperature than does the absolute viscosity.

1.2.2 Rheological Properties of Fluids

Newtonian and non-newtonian fluids

The relationships between the shear stress and shear rate in a real fluid are part of the science of rheology. Figure 1.6 shows several examples of the rheological behavior of fluids.

The curves are plots of shear stress versus rate of shear and apply at constant temperature and pressure. The simplest behavior is that shown by curve A, which is a straight line passing through the origin. Fluids following this simple linearity are newtonian fluids. In the Newtonian fluid, shear stress is proportional to the shear rate, and can be expressed by Eq. (1.2-3). Gases and most liquids are newtonian. The other curves in Fig. 1.6 represent the rheological behavior of liquids called non-newtonian. Some liquids, for example, sewage sludge, do not flow at all until a threshold shear stress, denoted by τ_0, is attained and then flow linearly, or nearly so, at shear stresses greater than τ_0. Curve B is an example of this relation. Liquids acting this way are called *Bingham plastics*. Line C represents a *pseudoplastic fluid*. The curve passes through the origin, is concave downward at low shear, and becomes nearly linear at high shear. Rubber latex is an example of such a fluid. Curve D represents a *dilatant fluid*. The curve is concave upward at low shear and almost linear at high shear. Quicksand and some sand-filled emulsions show this behavior. Pseudoplastics are said to be *shear rate-thinning* and dilatant fluids *shear rate-thickening*.

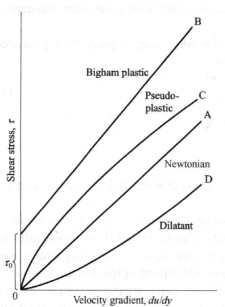

Fig. 1.6 Shear stress versus velocity gradient for newtonian and non-newtonian fluids.

Time-dependent flow

None of the curves in Fig. 1.6 depends on the history of the fluid, and a given sample of material shows the same behavior no matter how long the shearing stress has been applied. Such is not the case for some non-newtonian liquids, whose curves of stress versus rate of shear depend on how long the shear has been active. *Thixotropic* liquids break down under continued shear and on mixing give lower shear stress for a given shear rate; that is, their apparent viscosity decreases with time. *Rheopectic* substances behave in the reverse manner, and the shear stress increases with time, as does the apparent viscosity. The original structures and apparent viscosities are usually recovered on standing.

Table 1.2 Rheological characteristics of fluids

Designation	Effect of increasing shear rate	Time-dependent?	Examples
Pseudoplastic	Thins	No	Polymer solutions, starch suspensions, mayonnaise, paints
Thixotropic	Thins	Yes	Some polymer solutions, shortening, some paints
Newtonian	None	No	Gases, most simple liquids
Dilatant	Thickens	No	Corn flour-sugar solutions, wet beach sand, starch in water
Rheopectic	Thickens	Yes	Bentonite clay suspensions, gypsum suspensions

The rheological characteristics of fluids are summarized in Table 1.2.

Viscoelastic fluids. Viscoelastic fluids show both viscous and elastic properties. They exhibit elastic recovery from deformations that occur during flow, but usually only part of the deformation is recovered upon removal of the stress. Examples of viscoelastic fluids are flour dough, napalm, and certain polymer melts.

Rate of shear versus shear stress for non-newtonian fluids

Bingham plastics, like that represented by curve B in Fig. 1.6 follow a rheological equation of the type

$$\tau_v = \tau_0 + K \frac{du}{dy} \qquad (1.2\text{-}7)$$

where K is a constant. Over some range of shear rates, dilatant and pseudoplastic fluids often follow a power law, also called the *Ostwald-de Waele* equation,

$$\tau_v = K'\left(\frac{du}{dy}\right)^{n'} \qquad (1.2\text{-}8)$$

where K' and n' are constants called *the flow consistency index* and *the flow behavior index*, respectively. Such fluids are known *as power law fluids*. For pseudoplastics (curve C) $n' < 1$, and for dilatant fluids (curve D) $n' > 1$. Clearly $n' = 1$ for newtonian fluids. Values of n' and K' for some pseudoplastic fluids are given in Table 1.3.

Table 1.3 Flow property indexes for pseudoplastic fluids

Fluid	n'	$K' \times 10^{-3}$
1.5% Carboxymethylcellulose in water	0.554	3.13
3.0% Carboxymethylcellulose in water	0.566	9.31
4.0% Paper pulp in water	0.575	20.02
14.3% Clay in water	0.350	0.173
25% Clay in water	0.185	1.59
Applesauce	0.645	0.500
Banana puree	0.458	6.51
Tomato concentrate	0.59	0.2226

1.2.3 Types of Fluid Flow and Reynolds Number

Reynolds experiment

It has long been known that a fluid can flow through a pipe or conduit in two different ways. At low flow rates the pressure drop in the fluid increases directly with the fluid velocity; at high rates it increases much more rapidly, roughly as the square of the velocity. The distinction between the two types of flow was first demonstrated in a classic experiment by Osborne Reynolds, reported in 1883. A horizontal glass tube was immersed in a glass-walled tank filled with water. A controlled flow of water could be drawn through the tube by opening a valve. The entrance to the tube was flared, and provision was made to introduce a fine filament of colored water from the overhead flask into the stream at the tube entrance. Reynolds found that, at low flow rates, the jet of colored water flowed intact along with the mainstream and no cross-mixing occurred. The behavior of the color band showed clearly that the water was flowing in parallel straight lines and that the flow was laminar. When the flow rate was increased, a velocity, called the *critical velocity*, was reached at which the thread of color became wavy and gradually disappeared, as the dye spread uniformly throughout the entire cross section of the stream of water. This behavior of the colored water showed that the water no longer flowed in laminar motion but moved erratically in the form of cross-currents and eddies. This type of motion is *turbulent flow*.

Reynolds number and transition from laminar to turbulent flow

Reynolds studied the conditions under which one type of flow changes to the other and found that the critical velocity, at which laminar flow changes to turbulent flow, depends on four quantities: the diameter of the tube and the viscosity, density, and average linear velocity of the liquid. Furthermore, he found that these four factors can be combined into one group and that the change in the kind of flow occurs at a definite value of the group. The grouping of variables so found was

$$Re = \frac{DV\rho}{\mu} = \frac{DV}{v} \qquad (1.2\text{-}9)$$

where D=diameter of tube
V=average velocity of liquid [Eq. (1.3-3)]
μ=viscosity of liquid
ρ=density of liquid
v=kinematic viscosity of liquid

The dimensionless group of variables denned by Eq. (1.2-9) is called the *Reynolds number* Re. It is one of the named dimensionless groups listed in App. 2. Its magnitude is independent of the units used, provided the units are consistent.

Additional observations have shown that the transition from laminar to turbulent flow actually may occur over a wide range of Reynolds numbers. In a pipe, flow is always laminar at Reynolds numbers below 2100, but laminar flow can persist up to Reynolds numbers well above 24000 by eliminating all disturbances at the inlet. If the laminar flow at such high Reynolds numbers is disturbed, however, say by a fluctuation in velocity, the flow quickly becomes turbulent. Disturbances under these conditions are amplified, whereas at Reynolds numbers below 2100 all disturbances are damped and the flow remains laminar. At some flow rates a disturbance may be neither damped nor amplified; the flow is then said to be neutrally stable. Under ordinary conditions, the flow in a pipe or tube is turbulent at Reynolds numbers above about 4000. Between 2100 and 4000 a *transition region* is found where the flow may be either laminar or turbulent, depending upon conditions at the entrance of the tube and on the distance from the entrance.

Nature of turbulence

Because of its importance in many branches of engineering, turbulent flow has been extensively investigated in recent years, and a large literature has accumulated on this subject. Refined methods of measurement have been used to follow in detail the actual velocity fluctuations of the eddies during turbulent flow, and the results of such measurements have shed much qualitative and quantitative light on the nature of turbulence.

Turbulence may be generated in other ways than by flow through a pipe. In general, it can result either from contact of the flowing stream with solid boundaries or from contact between two layers of fluid moving at different velocities. The first kind of turbulence is called *wall turbulence* and the second kind *free turbulence*. Wall turbulence appears when the fluid flows through closed or open channels or past solid shapes immersed in the stream. Free turbulence appears in the flow of a jet into a mass of stagnant fluid or when a boundary layer separates from a solid wall and flows through the bulk of the fluid. Free turbulence is especially important in mixing.

Turbulent flow consists of a mass of eddies of various sizes coexisting in the flowing stream. Large eddies are continually formed. They break down into smaller eddies, which in turn evolve still smaller ones. Finally, the smallest eddies disappear. At a given time and in a given volume, a wide spectrum of eddy sizes exists. The size of the largest eddy is comparable with the smallest dimension of the turbulent stream; the diameter of the smallest eddies is 10 to 100 μm. Smaller eddies than this are rapidly destroyed by viscous shear. Flow within an eddy is laminar. Since even the smallest eddies contain about 10^{12} molecules, all eddies are of macroscopic size, and turbulent flow is not a molecular phenomenon.

Any given eddy possesses a definite amount of mechanical energy, much like that of a small spinning top. The energy of the largest eddies is supplied by the potential energy of the bulk flow of the fluid. From an energy standpoint, turbulence is a transfer process in which large eddies, formed from the bulk flow, pass their energy of rotation along a continuous series of smaller eddies. This mechanical energy is not appreciably dissipated into heat during the breakup of large eddies into smaller ones, but is passed along almost quantitatively to the smallest eddies. It is finally converted to heat when the smallest eddies are obliterated by viscous action. Energy conversion by viscous action is called *viscous dissipation.*

Deviating velocities in turbulent flow

A typical picture of the variations in the instantaneous velocity at a given point in a turbulent flow field is shown in Fig. 1.7. This velocity is really a single component of the actual velocity vector, all three components of which vary rapidly in magnitude and direction. Also, the instantaneous pressure at the same point fluctuates rapidly and simultaneously with the fluctuations of velocity. Oscillographs showing these fluctuations provide the basic experimental data on which modern theories of turbulence are based.

Although at first sight turbulence seems to be structureless and randomized, studies of oscillographs like that in Fig. 1.7 show that this is not quite so. The randomness and unpredictability of the fluctuations, which are nonetheless constrained between definite limits, exemplify the behavior of certain mathematical "chaotic" nonlinear functions. Quantitative characterization of turbulence, however, is commonly done by statistical analysis of the frequency distributions.

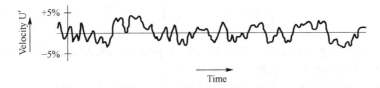

Fig. 1.7 Velocity fluctuations in turbulent flow.
The percentages are based on the constant velocity.

The instantaneous local velocities at a given point can be measured by laser-Doppler anemometers, which are capable of following the rapid oscillations. Local velocities can be analyzed by splitting each component of the total instantaneous velocity into two parts, one a constant part that is the time average, or mean value, of the component in the direction of flow of the stream, and the other, called the *deviating velocity,* the instantaneous fluctuation of the component around the mean. The net velocity is that measured by ordinary flowmeters, such as a pitot tube, which are too sluggish to follow the rapid variations of the fluctuating velocity. The split of a velocity component can be formalized by the following method. Let the three components (in cartesian coordinates) of the instantaneous velocity in directions

28 | Unit Operations of Chemical Engineering

x, y, and z be u_i, v_i, and w_i, respectively. Assume also that the x is oriented in the direction of flow of the stream and that components v_i and w_i are the y and z components, respectively, both perpendicular to the direction of bulk flow. Then the equations defining the deviating velocities are

$$u_i = u + u' \qquad v_i = v' \qquad w_i = w' \qquad (1.2\text{-}10)$$

where u_i, v_i, w_i = instantaneous total velocity components in x, y, and z directions, respectively
u = constant net velocity of stream in x direction
u', v', w' = deviating velocities in x, y, and z directions, respectively

Terms v and w are omitted in Eq. (1.2-10) because there is no net flow in the directions of the y and z axes in one-dimensional flow, and so v and w are zero.

The deviating velocities u', v', and w' all fluctuate about zero as an average. Figure 1.7 is actually a plot of the deviating velocity u', a plot of the instantaneous velocity u_i, however, would be identical in appearance, since the ordinate would everywhere be increased by the constant quantity u.

For pressure,

$$p_i = p + p' \qquad (1.2\text{-}11)$$

where p_i = variable local pressure
p = constant average pressure as measured by ordinary manometers or pressure gauges
p' = fluctuating part of pressure due to eddies

Because of the random nature of the fluctuations, the time averages of the fluctuating components of velocity and pressure vanish when averaged over a time period t_0 of the order of a few seconds. Therefore,

$$\frac{1}{t_0}\int_0^{t_0} u' \, dt = 0 \qquad \frac{1}{t_0}\int_0^{t_0} w' \, dt = 0$$

$$\frac{1}{t_0}\int_0^{t_0} v' \, dt = 0 \qquad \frac{1}{t_0}\int_0^{t_0} p' \, dt = 0 \qquad (1.2\text{-}12)$$

The reason these averages vanish is that for every positive value of a fluctuation there is an equal negative value, and the algebraic sum is zero.

Although the time averages of the fluctuating components themselves are zero, this is not necessarily true of other functions or combinations of these components. For example, the time average of the mean square of any one of these velocity components is not zero. This quantity for component u' is defined by

$$\frac{1}{t_0}\int_0^{t_0} (u')^2 \, dt = \overline{(u')^2} \qquad (1.2\text{-}13)$$

Thus the mean square is not zero, since u' takes on a rapid series of positive and negative values, which, when squared, always give a positive product. Therefore $\overline{(u')^2}$ is inherently positive and vanishes only when turbulence does not exist.

In laminar flow there are no eddies; the deviating velocities and pressure fluctuations do not exist; the total velocity in the direction of flow u_i is constant and equal to u; and v_i and w_i are both zero.

Eddy viscosity

By analogy with Eq. (1.2-3), the relationship between shear stress and velocity gradient in a turbulent stream is used to define an eddy viscosity E_v

$$\tau_t = E_v \frac{du}{dy} \qquad (1.2\text{-}14)$$

Quantity E_v is analogous to μ, the absolute viscosity. Also, in analogy with the kinematic viscosity ν the quantity ε_M, called the *eddy diffusivity of momentum,* is defined as $\varepsilon_M = E_v/\rho$.

The total shear stress in a turbulent fluid is the sum of the viscous stress and the turbulent stress, or

$$\tau = (\mu + E_v)\frac{du}{dy} \qquad (1.2\text{-}15)$$

$$\tau = (\nu + \varepsilon_M)\frac{d(\rho u)}{dy} \qquad (1.2\text{-}16)$$

Although E_v and ε_M are analogous to μ and ν, respectively, in that all these quantities are coefficients relating shear stress and the velocity gradient, there is a basic difference between the two kinds of quantities. The viscosities μ and ν are true properties of the fluid and are the macroscopic result of averaging motions and momenta of myriad molecules. The eddy viscosity E_v and the eddy diffusivity ε_M are not just properties of the fluid but depend on the fluid velocity and the geometry of the system. They are functions of all factors that influence the detailed patterns of turbulence and the deviating velocities, and they are especially sensitive to location in the turbulent field and the local values of the scale and intensity of the turbulence. Viscosities can be measured on isolated samples of fluid and presented in tables or charts of physical properties, as in Apps. 7 and 8. Eddy viscosities and diffusivities are determined (with difficulty, and only by means of special instruments) by experiments on the flow itself.

1.2.4 Boundary Layers

Flow in boundary layers

A boundary layer is defined as that part of a moving fluid in which the fluid motion is influenced by the presence of a solid boundary. As a specific example of boundary layer formation, consider the flow of fluid parallel with a thin plate, as shown in Fig. 1.8. The velocity of the fluid upstream from the leading edge of the plate is uniform across the entire fluid stream. The velocity of the fluid at the interface between the solid and fluid is zero. The velocity increases with distance from the plate, as shown in Fig. 1.8. Each of these curves corresponds to a definite value of x, the distance from the leading edge of the plate. The curves change slope rapidly near the plate; they also show that the local velocity approaches asymptotically the velocity of the bulk of the fluid stream.

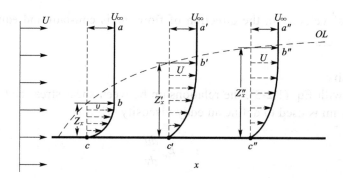

Fig. 1.8 Prandtl boundary layer: x, distance from leading edge; u_∞, velocity of undisturbed stream; Z_x, thickness of boundary layer at distance x, u, local velocity; abc, $a'b'c'$, $a''b''c''$, curves of velocity versus distance from wall at points c, c', c''; OL, outer limit of boundary layer. (The vertical scale is greatly exaggerated.)

In Fig. 1.8 the dashed line OL is so drawn that the velocity changes are confined between this line and the trace of the wall. Because the velocity lines are asymptotic with respect to distance from the plate, it is assumed, in order to locate the dashed line definitely, that the line passes through all points where the velocity is 99 percent of the bulk fluid velocity u_∞. Line OL represents an imaginary surface that separates the fluid stream into two parts: one in which the fluid velocity is constant and the other in which the velocity varies from zero at the wall to a velocity substantially equal to that of the undisturbed fluid. This imaginary surface separates the fluid that is directly affected by the plate from that in which the local velocity is constant and equal to the initial velocity of the approach fluid. The zone, or layer, between the dashed line and the plate constitutes the boundary layer.

The formation and behavior of the boundary layer are important, not only in the flow of fluids but also in the transfer of heat, discussed in Chap. 4, and mass, discussed in Chap. 6.

Laminar and turbulent flow in boundary layers

The fluid velocity at the solid-fluid interface is zero, and the velocities close to the solid surface are, of necessity, small. Flow in this part of the boundary layer very near the surface therefore is essentially laminar. Actually it is laminar most of the time, but occasionally eddies from the main portion of the flow or the outer region of the boundary layer move very close to the wall, temporarily disrupting the velocity profile. These eddies may have little effect on the average velocity profile near the wall, but they can have a large effect on the profiles of temperature or concentration when heat or mass is being transferred to or from the wall. This effect is most pronounced for mass transfer in liquids.

Farther away from the surface the fluid velocities, though less than the velocity of the undisturbed fluid, may be fairly large, and flow in this part of the boundary layer may become turbulent. Between the zone of fully developed turbulence and the region of laminar flow is a transition, or buffer, layer of intermediate character. Thus a turbulent boundary layer is considered to consist of three zones: the viscous sublayer, the buffer layer, and the turbulent zone. The existence of a completely viscous sublayer is questioned by some, since mass-transfer studies suggest that some eddies penetrate all the way through the boundary layer and reach the wall.

Near the leading edge of a flat plate immersed in a fluid of uniform velocity, the boundary layer is thin, and the flow in the boundary layer is entirely laminar. As the layer thickens, however, at distances farther from the leading edge, a point is reached where turbulence appears. The onset of turbulence is characterized by a sudden rapid increase in the thickness of the boundary layer, as shown in Fig. 1.9.

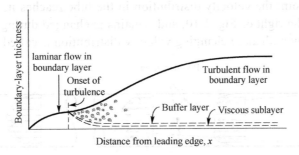

Fig. 1.9 Development of turbulent boundary layer on a flat plate. (The vertical scale is greatly exaggerated.)

When flow in the boundary layer is laminar, the thickness Z_x of the layer increases with $x^{0.5}$, where x is the distance from the leading edge of the plate. For a short time after turbulence appears, Z_x increases with $x^{1.5}$ and then, after turbulence is fully developed, with $x^{0.8}$.

The initial, fully laminar part of the boundary layer may grow to a moderate thickness of perhaps 2 mm with air or water moving at moderate velocities. Once turbulence begins, however, the thickness of the laminar part of the boundary layer diminishes considerably, typically to about 0.2 mm.

Transition from laminar to turbulent flow; Reynolds number. The factors that determine the point at which turbulence appears in a laminar boundary layer are coordinated by the dimensionless Reynolds number denned by the equation

$$Re_x = \frac{x u_\infty \rho}{\mu} \qquad (1.2\text{-}17)$$

where x = distance from leading edge of plate
u_∞ = bulk fluid velocity
ρ = density of fluid
μ = viscosity of fluid

With parallel flow along a plate, turbulent flow first appears at a critical Reynolds number between about 10^5 and 3×10^6. The transition occurs at the lower Reynolds numbers when the plate is rough and the intensity of turbulence in the approaching stream is high, and at the higher values when the plate is smooth and the intensity of turbulence in the approaching stream is low.

Boundary layer formation in straight tubes

Consider a straight, thin-walled tube with fluid entering it at a uniform velocity. As shown in Fig. 1.10, a boundary layer begins to form at the entrance to the tube, and as the fluid moves through the first part of the channel, the layer thickens. During this stage the boundary layer occupies only part of the cross section of the tube, and the total stream consists of a core of fluid flowing in rodlike manner at constant velocity and an annular boundary layer between the wall and the core. In the boundary layer the velocity increases from zero at the wall to the constant velocity existing in the core. As the stream moves farther down the tube, the boundary layer occupies an increasing portion of the cross section. Finally, at a point well downstream from the entrance, the boundary layer reaches the center of the tube, the rodlike core disappears, and the boundary layer occupies the entire cross section of the stream. At this point the velocity distribution in the tube reaches its final form, as shown by the last curve at the right of Fig. 1.10, and remains unchanged during the remaining length of the tube. Such flow with an unchanging velocity distribution is called *fully developed flow*.

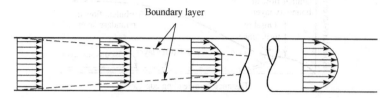

Fig. 1.10 Development of boundary layer flow in pipe.

Transition length for laminar and turbulent flow. The length of the entrance region of the tube necessary for the boundary layer to reach the center of the tube and for fully developed flow to be established is called the *transition length*. Since the velocity varies not only with length of tube but also with radial distance from the center of the tube, flow in the entrance region is two-dimensional.

The approximate length of straight pipe necessary for completion of the final velocity distribution is, for laminar flow,

$$\frac{x_t}{D} = 0.05 Re \qquad (1.2\text{-}18)$$

where x_t = transition length
 D = diameter of pipe

Equation (1.2-18), originally proposed by Nikuradse, was verified experimentally by Rothfus and Prengle. Equation (1.2-18) shows that for a 50-mm- ID pipe and a Reynolds number of 1500, the transition length is 3.75 m. If the fluid entering the pipe is turbulent and the velocity in the tube is above the critical, the transition length is nearly independent of the Reynolds number and is about 40 to 50 pipe diameters, with little difference between the distribution at 25 diameters and that at greater distances from the entrance. For a 50-mm-ID pipe, 2 to 3 m of straight pipe is sufficient when flow is all turbulent. If the fluid entering the tube is in laminar flow and becomes turbulent on entering the tube, a longer transition length, as large as 100 pipe diameters, is needed.

Boundary layer separation and wake formation

In the preceding paragraphs the growth of boundary layers has been discussed. Now consider what happens at the far side of a submerged object, where the fluid leaves the solid surface.

At the trailing edge of a flat plate that is parallel to the direction of flow, the boundary layers on the two sides of the plate have grown to a maximum thickness. For a time after the fluid leaves the plate, the layers and velocity gradients persist. Soon, however, the gradients fade out, the boundary layers intermingle and disappear, and the fluid once more moves with a uniform velocity. This is shown in Fig. 1.11(a).

Suppose, now, the plate is turned at right angles to the direction of flow, as in Fig.1.11(b). A boundary layer forms as before in the fluid flowing over the upstream face. When the fluid reaches the edge of the plate, however, its momentum prevents it from making the sharp turn around the edge, and it separates from the plate and proceeds outward into the bulk of the fluid. Behind the plate is a backwater zone of strongly decelerated fluid, in which large eddies, called *vortices,* are formed. This zone is known as the *wake*. The eddies in the wake are kept in motion by the shear stresses between the wake and the separated current. They consume considerable mechanical energy and may lead to a large pressure loss in the fluid.

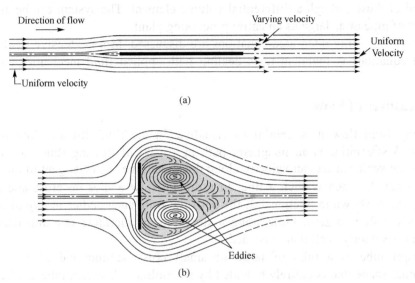

Fig. 1.11 Flow past flat plate: (a) flow parallel with plate; (b) flow perpendicular to plate.

Boundary layer separation occurs whenever the change in velocity of the fluid, in either magnitude or direction, is too large for the fluid to adhere to the solid surface. It is most frequently encountered when there is an abrupt change in the flow channel, such as a sudden expansion or contraction, a sharp bend, or an obstruction around which the fluid must flow. As discussed in Section 1.4, Fig. 1.26, separation may also occur from the velocity decrease in a smoothly diverging channel. Because of the large energy losses resulting from the formation of a wake, it is often desirable to minimize or prevent boundary layer separation. In some cases this can be done by suction, i.e., by drawing part of the fluid into the solid surface at the area of potential separation. Most often, however, separation is minimized by avoiding sharp changes in the cross-sectional area of the flow channel and by streamlining any objects over which the fluid must flow. For some purposes, such as the promotion of heat transfer or mixing in a fluid, boundary layer separation may be desirable.

1.3 Basic Equations of Fluid Flow

In applications of fluid mechanics, the most useful equations are based on the principles of mass balance or continuity; the balances of linear momentum and angular momentum; and the mechanical energy balance. The equations may be written in differential form, showing conditions at a point within a volume element of fluid, or in integrated form applicable to a finite volume or mass of fluid.

Differential equations and shell balances

To be useful in solving engineering problems, the differential equations must be integrated. In some simple cases they can be integrated mathematically, but more often they must be integrated numerically by computer. This would be needed, for example, in as straightforward a situation as the steady flow of air in a large duct containing baffles, or in a highly complex problem such as the transient flow of a high-viscosity non-newtonian polymer melt through a mold.

Useful equations can be derived for some well-defined systems by making a macroscopic shell balance, in which the flows through the boundaries of the entire system are used instead of those through a differential volume element. The system can be as small as a short length of pipe or as large as an entire processing plant.

In this section, only integrated equations are discussed. More extensive treatment of the differential equations is found in texts dealing with applied fluid mechanics and transport processes.

1.3.1 Measures of Flow

In discussing fluid flow it is helpful to visualize, in the fluid stream, fluid paths called streamlines. A streamline is an imaginary path in a mass of flowing fluid so drawn that at every point the vector of the net velocity along the streamline u is tangent to the streamline. There is no net flow across such a line. In turbulent flow eddies do cross and recross the streamline, but as shown in Section 1.2, the net flow from such eddies in any direction other than that of the flow is zero. Flow along a streamline is therefore one-dimensional, and a single term for velocity is all that is needed.

A stream tube is a tube of large or small cross section and of any convenient cross-sectional shape that is entirely bounded by streamlines. A stream tube can be visualized as an imaginary pipe in the mass of flowing fluid through the walls of which there is no net flow. If the tube has a differential cross-sectional area dS, the velocity through the tube can also be denoted by the single term u.

The mass flow through the differential area is

$$d\dot{m} = \rho u dS \tag{1.3-1}$$

To find the total flow through an impermeable conduit of cross-sectional area S, Eq. (1.3-1) is integrated across the entire cross section. In general, the local velocity u varies across the cross section. If the fluid is being heated or cooled, the fluid density also varies, but usually the variation is small and can be neglected. The flow rate through the entire cross section is

$$\dot{m} = \rho \int_S u dS \tag{1.3-2}$$

where ρ is constant across the cross section.

The average velocity V of the entire stream flowing through cross-sectional area S is defined by

$$V \equiv \frac{\dot{m}}{\rho S} = \frac{1}{S}\int_S u dS \tag{1.3-3}$$

Velocity V also equals the total volumetric flow rate of the fluid, divided by the cross-sectional area of the conduit; in fact, it is usually calculated this way. It may be considered to be *the flux of volume*, m^3/m$^2 \cdot$ s. Thus

$$V = \frac{q}{S} \tag{1.3-4}$$

where q is the volumetric flow rate.

Equation (1.3-3) can be written

$$V\rho = \frac{\dot{m}}{S} \equiv G \qquad (1.3\text{-}5)$$

This equation defines the mass velocity G, calculated by dividing the mass flow rate by the cross-sectional area of the channel. In practice, the mass velocity is expressed in kilograms per square meter per second. The advantage of using G is that it is independent of temperature and pressure when the flow is steady (constant $\dot{m}$) and the cross section is unchanged (constant S). This fact is especially useful when compressible fluids are considered, for both V and ρ vary with temperature and pressure. Also certain relationships appear later in this book in which V and ρ are associated as their product, so that the mass velocity represents the net effect of both variables. The mass velocity G can also be described as the mass current density or mass flux, where flux is defined generally as any quantity passing through a unit area in unit time. The average velocity V, as shown by Eq. (1.3-4), can be described as the volume flux of the fluid.

1.3.2 Mass Balance in a Flowing Fluid; Continuity

In any element of fluid (or in any bounded system), the equation for a mass balance is simply

(Rate of mass flow in) − (Rate of mass flow out) = (Rate of mass accumulation)

Average velocities are useful in making a shell balance or flow through a pipe or a piping system. Consider the flow through a conduit of cross-sectional area S_a at the entrance and area S_b at the exit, in which the local fluid velocity varies within the cross section. The average velocity and density at the entrance are V_a and ρ_a; at the exit they are V_b and ρ_b. At steady state the mass flow in equals the mass flow out, and the equation for a mass balance becomes

$$\dot{m} = \rho_a V_a S_a = \rho_b V_b S_b = \rho V S \qquad (1.3\text{-}6)$$

Equation (1.3-6) is also known as the equation of continuity. For the important special case where the flow is through channels of circular cross section

$$\dot{m} = \frac{1}{4}\pi D_a^2 \rho_a V_a = \frac{1}{4}\pi D_b^2 \rho_b V_b$$

from which

$$\frac{\rho_a V_a}{\rho_b V_b} = \left(\frac{D_b}{D_a}\right)^2 \qquad (1.3\text{-}7)$$

where D_a and D_b are the diameters of the channel at the upstream and downstream stations, respectively.

EXAMPLE 1.2 Crude oil, density 887 kg/m³, flows through the piping shown in Fig. 1.12. Pipe A is 50-mm inside diameter, pipe B is 75-mm inside diameter, and each of pipes C is 38-mm inside diameter. An equal quantity of liquid flows through each of the pipes C. The flow through pipe A is 6.65 m³/h. Calculate (a) the mass flow rate in each pipe, (b) the average linear velocity in each pipe, and (c) the mass velocity in each pipe..
Solution. (a) The mass flow rate is the same for pipes A and B and is the product of the density and the volumetric flow rate, or

$$\dot{m} = 6.65 \times 887 = 5898.6 \,\text{kg/h}$$

The mass flow rate through each of pipes C is one-half the total or $5898.6/2 = 2949.3\,\text{kg/h}=0.8192$ kg/s.

(b) Use Eq. (1.3-4). The velocity through pipe A is

$$V_A = \frac{6.65}{3600 \times 0.785 \times 0.050^2} = 0.9413 \text{ m/s}$$

through pipe B is

$$V_B = \frac{6.65}{3600 \times 0.785 \times 0.075^2} = 0.4183 \text{ m/s}$$

and through each of pipes C is

$$V_C = \frac{6.65}{2 \times 3{,}600 \times 0.785 \times 0.038^2} = 0.8148 \text{ m/s}$$

(c) Use Eq. (1.3-5). The mass velocity through pipe A is

$$G_A = \frac{5898.6}{3600 \times 0.785 \times 0.050^2} = 834.9 \text{ kg/(m}^2\cdot\text{s)}$$

through pipe B is

$$G_A = \frac{5898.6}{3600 \times 0.785 \times 0.075^2} = 371.1 \text{ kg/(m}^2\cdot\text{s)}$$

and through each of pipes C is

$$G_A = \frac{5898.6}{3600 \times 2 \times 0.785 \times 0.038^2} = 722.7 \text{ kg/(m}^2\cdot\text{s)}$$

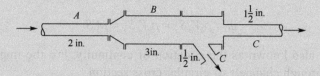

Fig. 1.12 Piping system for Example 1.2.

EXAMPLE 1.3 Air at 20°C and 2-atm absolute pressure enters a finned-tube steam heater through a 50-mm tube at an average velocity of 15 *m/s*. It leaves the heater through a 65-mm tube at 90°C and 1.6 atm absolute. What is the average air velocity at the outlet?
Solution. Let subscript a refer to the heater inlet and b to the outlet. Use Eq. (1.3-7). The quantities needed are

$$D_a = 0.05 \text{ m} \quad D_b = 0.065 \text{ m} \quad\quad p_a = 2 \text{ atm} \quad p_b = 1.6 \text{ atm}$$

$$T_a = 20 + 273.16 = 293.16 \text{ K} \quad\quad T_b = 90 + 273.16 = 363.16 \text{ K}$$

Density is calculated by ideal gas law. The densities at the inlet and outlet are then $\rho_a = Mp_a/(RT_a)$ and $\rho_b = Mp_b/(RT_b)$. Thus

$$\frac{\rho_a}{\rho_b} = \frac{p_a T_b}{p_b T_a}$$

Substituting in Eq. (1.3-7) gives

$$V_b = \frac{V_a \rho_a D_a^2}{\rho_b D_b^2} = \frac{V_a p_a T_b D_a^2}{p_b T_a D_b^2} = \frac{15 \times 2 \times 0.05^2 \times 363.16}{1.6 \times 0.065^2 \times 293.16} = 13.74 \text{ m/s}$$

1.3.3 Overall Energy Balance for Steady-state Flow System

The entity balance for energy is as follows for a control volume.

(Rate of energy input) - (Rate of energy output) = (Rate of energy accumulation) (1.3-8)

The energy E present within a system can be classified in three ways.

(1) Potential energy zg of a unit mass of a fluid is the energy present because of the position of the mass in a gravitational field, where z is the relative height from a reference plane. The unit for zg are J/kg, or m^2/s^2.

(2) Kinetic energy $u^2/2$ of a unit mass of a fluid is the energy present because of the translational or rotational motion of the mass, where u is the velocity in m/s relative to the boundary of the system at a given point. Again in the SI system the units of $u^2/2$ are J/kg, or m^2/s^2.

(3) Internal energy U of a unit mass of a fluid is all of the other energy present, such as rotational and vibrational energy in chemical bonds. Again the units are J/kg, or m^2/s^2.

The total energy of the fluid per unit mass is then

$$E = U + \frac{u^2}{2} + zg \tag{1.3-9}$$

The net energy transported into the system by the flowing stream is therefore

$$-\Delta \left[\left(U + \frac{u^2}{2} + zg \right) \dot{m} \right]$$

where the effect of the minus sign is to make the term read $in - out$. The rate of energy accumulation within the control volume includes this quantity in addition to the heat transfer rate $\dot{Q}$ (heat absorbed from surroundings) and work rate $\dot{W}$ (the work of all kinds obtained from surroundings). The equation (1.3-8) may now be written:

$$\frac{d(mU)}{dt} = -\Delta \left[\left(U + \frac{u^2}{2} + zg \right) \dot{m} \right] + \dot{Q} + \dot{W} \tag{1.3-10}$$

The work $\dot{W}$ may be included work of several forms. First, work is done by fluid as it flows into and out of the control volume. This pressure-volume work per unit mass fluid is pv. Another form of work is the purely mechanical shaft work identified with a rotating shaft crossing the control surface. In addition work may be associated with expansion or contraction of the control volume and there may be stirring work. These forms of work are all included in a rate term represented by $\dot{W}_e$. The equation (1.3-10) becomes:

$$\frac{d(mU)}{dt} = -\Delta \left[\left(U + \frac{u^2}{2} + zg + pv \right) \dot{m} \right] + \dot{Q} + \dot{W}_e \tag{1.3-11}$$

Fluids flowing in pipes exhibit a velocity profile, as shown in Fig. 1.10, which rises from zero at wall to the maximum at the center of the pipe. The kinetic energy of a fluid in a pipe depends on its velocity profile.

Kinetic energy of stream

The term $u^2/2$ in Eq. (1.3-11) is the kinetic energy of a unit mass of fluid all of which is flowing at the same velocity u. When the velocity varies across the stream cross section, the kinetic energy is found in the following manner. Consider an element of cross-sectional area dS. The mass flow rate through this is $\rho u dS$. Each unit mass of fluid flowing through area dS carries kinetic energy in amount $u^2/2$, and the energy flow rate through area dS is therefore

$$d\dot{E}_k = (\rho u dS)\frac{u^2}{2} = \frac{\rho u^3 dS}{2} \tag{1.3-12}$$

where $\dot{E}_k$ represents the time rate of flow of kinetic energy. The total rate of flow of kinetic energy through the entire cross section S is, assuming constant density within the area S,

$$\dot{E}_k = \frac{\rho}{2}\int_S u^3 dS \tag{1.3-13}$$

The total rate of mass flow is given by Eqs. (1.3-2) and (1.3-5), and the kinetic energy per unit mass of flowing fluid, which replaces $u^2/2$ in the equation (1.3-11), is

$$\frac{\dot{E}_k}{\dot{m}} = \frac{\frac{1}{2}\int_S u^3 dS}{\int_S u dS} = \frac{\frac{1}{2}\int_S u^3 dS}{VS} \tag{1.3-14}$$

Kinetic energy correction factor. It is convenient to eliminate the integral of Eq. (1.3-14) by a factor operating on $V^2/2$ to give the correct value of the kinetic energy as calculated from Eq. (1.3-14). This factor, called the kinetic energy correction factor, is denoted by α and is defined by

$$\frac{\alpha V^2}{2} = \frac{\dot{E}_k}{\dot{m}} = \frac{\int_S u^3 dS}{2VS}$$

$$\alpha = \frac{\int_S u^3 dS}{V^3 S} \tag{1.3-15}$$

therefore
$$\dot{E}_k \equiv \alpha\frac{\dot{m}V^2}{2} \tag{1.3-16}$$

If α is known, the average velocity can be used to calculate the kinetic energy from the average velocity by using $\alpha V^2/2$ in place of $u^2/2$. To calculate the value of α from Eq. (1.3-15), the local velocity must be known as a function of location in the cross section, so that the integral in the equation can be evaluated. The same knowledge of velocity distribution is needed to calculate the value of V by Eq. (1.3-3). As shown in next section, α is 2.0 for laminar flow and is about 1.05 for highly turbulent flow. In most cases (except for precise work) the value of α is taken to be 1.0.

Flow processes for which the accumulation term of Eq. (1.3-11) is zero are said to occur at steady state. As discussed with respect to the mass balance, this means the mass of the system within the control volume is constant; it also means that no change occur with time in the properties of the fluid within the control volume nor at its entrances and exits. No expansion of the control volume is possible under these circumstances. The only work of the process is shaft work, and the kinetic energy correction factors are taken as unity. Therefore, the general energy balance, Eq. (1.3-11), becomes:

$$\Delta\left[\left(U+\frac{V^2}{2}+zg+pv\right)\dot{m}\right]=\dot{Q}+\dot{W}_e \tag{1.3-17}$$

A further specialization results when the control volume has but one entrance and one exit. The same mass flowrate $\dot{m}$ then applies to both streams, and Eq. (1.3-17) then reduce to:

$$\Delta\left(U+\frac{V^2}{2}+zg+pv\right)\dot{m}=\dot{Q}+\dot{W}_e \tag{1.3-18}$$

where "Δ" denotes to the change from entrance to exit. Division by $\dot{m}$ gives:

$$\Delta U+\frac{\Delta V^2}{2}+\Delta zg+\Delta pv=\frac{\dot{Q}}{\dot{m}}+\frac{\dot{W}_e}{\dot{m}}=Q+W_e \tag{1.3-19}$$

where Q is the heat absorbed per unit mass of fluid from surroundings, and W_e is the work obtained from the surroundings per unit mass of fluid.
Equation (1.3-19) is the mathematical expression of the energy balance for a steady-state, steady-flow process between one entrance and one exit. All terms represent energy per unit mass of fluid. This is shown in Fig. 1.13.

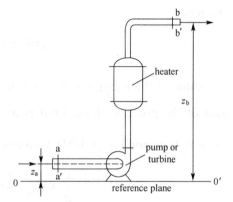

Fig. 1.13 Steady-state flow system for a fluid.

1.3.4 Overall Mechanical Energy Balance for Steady-state Flow System

A more useful type of energy balance for flowing fluids, especially liquids, is a modification of the total energy balance to deal with mechanical energy. Engineers are often concerned with this special type of energy, called mechanical energy, which includes the work term, kinetic energy, potential energy, and flow work. Mechanical energy is a form of energy that is either work or a form that can be directly converted into work. The other terms in the energy-balance equation (1.3-19), heat terms and internal energy, do not permit simple conversion into work because of the second law of thermodynamics and the efficiency of conversion, which depends on the temperatures. Mechanical-energy terms have not such limitation and can be converted almost completely into work. Energy converted to heat or internal energy is lost work or a loss in mechanical energy which is caused by frictional resistance to flow.

Writing the first law of thermodynamics for this case,

$$\Delta U = Q - W' \tag{1.3-20}$$

For the case of steady state flow, when a unit mass of fluid passes from entrance to exit, the work done by the fluid, W', is expressed as

$$W' = \int_{v_a}^{v_b} p\,dv - h_f \tag{1.3-21}$$

where h_f is the sum of all friction loss per unit mass.
Substitution Eq.(1.3-21) into Eq.(1.3-20)

$$\Delta U = Q - \int_{v_a}^{v_b} p\,dv + h_f \tag{1.3-22}$$

then combining the result with Eq. (1.3-19), and using

$$\Delta(pv) = \int_a^b d(pv) = \int_{v_a}^{v_b} p\,dv + \int_{p_a}^{p_b} v\,dp$$

we obtain

$$\Delta V^2/2 + \Delta zg + \int_{p_a}^{p_b} v\,dp = W_e - h_f \tag{1.3-23}$$

Finally, substituting $1/\rho$ for v, the overall mechanical-energy balance equation is obtained:

$$\Delta V^2/2 + \Delta zg + \int_{p_a}^{p_b} \frac{dp}{\rho} = W_e - h_f \tag{1.3-24}$$

The value of the integral in Eq. (1.3-24) depends on the equation of state of the fluid and the path of the process. If the fluid is an incompressible liquid, the integral $\int_{p_a}^{p_b} \frac{dp}{\rho}$ becomes $(p_b - p_a)/\rho$ and Eq. (1.3-24) becomes

$$\frac{p_a}{\rho} + gZ_a + \frac{V_a^2}{2} + W_e = \frac{p_b}{\rho} + gZ_b + \frac{V_b^2}{2} + h_f \tag{1.3-25}$$

1.3.5 Discussion on the Overall Mechanical Energy Balance Equation

Bernoulli equation

In the special case where no mechanical energy is added $W_e = 0$ and for no friction loss $h_f = 0$, then Eq. (1.3-25) become

$$\frac{p_a}{\rho} + gZ_a + \frac{V_a^2}{2} = \frac{p_b}{\rho} + gZ_b + \frac{V_b^2}{2} \tag{1.3-26}$$

Equation (1.3-26) is known as the Bernoulli equation without friction. It is a particular form of a mechanical energy balance.

Each term in Eq. (1.3-26) is a scalar and has the dimensions of energy per unit mass, representing a mechanical energy effect based on a unit mass of the flowing fluid. Terms gZ and $V^2/2$ are the potential and kinetic energy, respectively, of a unit mass of fluid; and p/ρ represents the mechanical work done by forces, external to the stream, on the fluid in pushing it into the tube or the work recovered from the fluid leaving the tube. Equation (1.3-26) shows that in the absence of friction, when the velocity V is reduced, either the height above datum Z or the pressure p or both must increase. When the velocity increases, it does so only at the expense of Z or p. If the height is changed, compensation must be found in a change of either pressure or velocity.

To apply the Bernoulli equation to a specific problem, it is essential to identify the streamline or stream tube and to choose definite upstream and downstream stations. Stations a and b are chosen on the basis of convenience and are usually taken at locations where the most information about pressures, velocities, and heights is available.

> **EXAMPLE 1.4** Brine, specific gravity 60°F /60°F = 1.15, is draining from the bottom of a large open tank through a 50-mm pipe. The drainpipe ends at a point 5 m below the surface of the brine in the tank. Considering a streamline starting at the surface of the brine in the tank and passing through the center of the drain line to the point of discharge, and assuming that friction along the streamline is negligible, calculate the velocity of flow along the streamline at the point of discharge from the pipe.
>
> *Solution.* To apply Eq. (1.3-26), choose station a at the brine surface and station b at the end of the streamline at the point of discharge. Since the pressure at both stations is atmospheric, p_a and p_b are equal, and $p_a/\rho = p_b/\rho$. At the surface of the brine, V_a is negligible, and the term $V_a^2/2$ is dropped. The datum for measurement of heights can be taken through station b, so $Z_b = 0$ and $Z_a = 5m$. Substitution in Eq. (1.3-26) gives
>
> $$5g = \frac{1}{2}V_b^2$$
>
> and the velocity on the streamline at the discharge is
>
> $$V_b = \sqrt{5 \times 2 \times 9.80665} = 9.90 \text{ m/s}$$
>
> Note that this velocity is independent of density and of pipe size.

For the actual fluid, friction manifests itself by the disappearance of mechanical energy, the quantity $\frac{p}{\rho} + \frac{u^2}{2} + gZ$ always decreases in the direction of flow. Also, sometimes provision is made in the equation for the work done on the fluid by a pump.

Fluid Friction

Fluid friction can be defined as any conversion of mechanical energy to heat in a flowing stream. The units of h_f and those of all other terms in Eq. (1.3-25) are energy per unit mass. The term h_f represents all the friction generated per unit mass of fluid (and therefore all the conversion of mechanical energy to heat) that occurs in the fluid between stations a and b. It differs from all other terms in Eq. (1.3-25) in two ways:

(1) The mechanical terms represent conditions at specific locations, namely, the inlet and outlet stations a and b, whereas h_f represents the loss of mechanical energy at all points between stations a and b.

(2) Friction is not interconvertible with the mechanical energy quantities.

The sign of h_f, as defined by Eq. (1.3-25), is always positive. It is zero, of course, in potential flow.

Friction appears in boundary layers because the work done by shear forces in maintaining the velocity gradients in both laminar and turbulent flow is eventually converted to heat by viscous action. Friction generated in unseparated boundary layers is called skin friction. When boundary layers separate and form wakes, additional energy dissipation appears within the wake, and friction of this type is called form friction since it is a function of the position and shape of the solid.

In a given situation both skin friction and form friction may be active in varying degrees.

In the case of Fig. 1.11a, the friction is entirely skin friction; in that of Fig. 1.11b, the friction is largely form friction, because of the large wake, and skin friction is relatively unimportant. The total friction h_f in Eq. (1.3-25) includes both types of friction loss.

Pump work in Bernoulli equation

A pump is used in a flow system to increase the mechanical energy of the flowing fluid, the increase being used to maintain flow, provide kinetic energy, offset friction losses, and sometimes increase the potential energy. Assume that a pump is installed between stations a and b linked by Eq. (1.3-25). Let W_p be the work done by the pump per unit mass of fluid.

Since the Bernoulli equation is a balance of mechanical energy only, account must be taken of friction occurring within the pump. In an actual pump not only are all the sources of fluid friction active, but mechanical friction occurs as well, in bearings and seals or stuffing boxes. The mechanical energy supplied to the pump as negative shaft work must be discounted by these friction losses to give the net mechanical energy actually available to the flowing fluid. Let h_{fp} be the total friction in the pump per unit mass of fluid. Then the net work to the fluid is $W_e = W_p - h_{fp}$. In practice, a pump efficiency denoted by η is used, then, The mechanical energy delivered to the fluid is

$$W_e = \eta W_p \tag{1.3-27}$$

where $\eta < 1$. Equation (1.3-25) corrected for pump work is

$$\frac{p_a}{\rho} + gZ_a + \frac{V_a^2}{2} + \eta W_p = \frac{p_b}{\rho} + gZ_b + \frac{V_b^2}{2} + h_f \tag{1.3-28}$$

Equation (1.3-28) is a final working form of the mechanical energy balance equation for problems on the flow of incompressible fluids.

EXAMPLE 1.5 In the equipment shown in Fig. 1.14, a pump draws a solution of specific gravity 1.84 from a storage tank through a 75-mm inside diameter steel pipe. The efficiency of the pump is 60 percent. The velocity in the suction line is 0.914 m/s. The pump discharges through a 50-mm inside diameter pipe to an overhead tank. The end of the discharge pipe is 15.2 m above the level of the solution in the feed tank. Friction losses in the entire piping system are 29.9 J/kg. What pressure must the pump develop? What is the power delivered to the fluid by the pump?

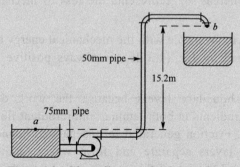

Fig. 1.14 Flow diagram for Example 1.5

Solution. Use Eq. (1.3-28). Take station a at the surface of the liquid in the tank and station *b* at the discharge end of the 50-mm pipe. Take the datum plane for elevations through station *a*. Since the pressure at both stations is atmospheric, $p_a = p_b$. The velocity at station *a* is negligible because of the large diameter of the tank in comparison with that of the pipe. For turbulent flows the kinetic energy factor α can be taken as 1.0 with negligible error. Equation (1.3-28) in SI units becomes

$$W_p \eta = Z_b g + \frac{V_b^2}{2} + h_f$$

The velocity in the 50-mm pipe is

$$V_b = V_a \left(\frac{D_a}{D_b}\right)^2 = 0.914 \left(\frac{0.07}{0.05}\right)^2 = 1.791 \text{m/s}$$

Then

$$0.60 W_p = 15.2 \times 9.81 + \frac{1.791^2}{2} + 29.9$$

and

$$W_p = 301.0 \text{J/kg}$$

The pressure developed by the pump can be found by writing Bernoulli equation over the pump itself. Station *a* is in the suction connection, and station *b* is in the pump discharge. The difference in level between suction and discharge can be neglected, so $Z_a = Z_b$, and Eq. (1.3-28) becomes

$$\frac{p_b - p_a}{\rho} = \frac{V_a^2 - V_b^2}{2} + W_p \eta$$

The pressure developed by the pump is

$$p_b - p_a = 1.84 \times 1000 \left(\frac{0.914^2 - 1.791^2}{2} + 0.6 \times 301.0\right) = 3.301 \times 10^5 \text{Pa} = 330 \text{kPa}$$

The power delivered to the pump is the product of W_p and the mass flow rate. The mass flow rate is

$$\dot{m} = \frac{\pi}{4} D_a^2 V_a \rho = \frac{\pi}{4} \times 0.07^2 \times 0.914 \times 1.84 \times 1000 = 6.469 \text{kg/s}$$

and the power is

$$N = \dot{m} W_p = 6.469 \times 301.0 = 1947 W = 1.95 \text{kW}$$

The power delivered to the fluid is then

$$1.95 \times 0.60 = 1.17 \text{kW}$$

1.3.6 Macroscopic Momentum Balances

An overall momentum balance can be written for the control volume shown in Fig. 1.15, assuming that the flow is steady and unidirectional in the *x* direction. In accordance the basic concept of the momentum balance:

$$\begin{array}{c}\text{Rate of}\\\text{momentum}\\\text{accumulation}\end{array} = \begin{array}{c}\text{Rate of}\\\text{momentum}\\\text{entering}\end{array} - \begin{array}{c}\text{Rate of}\\\text{momentum}\\\text{leaving}\end{array} + \begin{array}{c}\text{Sum of forces}\\\text{acting on the system}\end{array} \quad (1.3\text{-}29)$$

The sum of the forces acting on the fluid in the x direction equals the increase in the momentum flow rate of the fluid, or

$$\sum F = \dot{M}_b - \dot{M}_a \qquad (1.3\text{-}30)$$

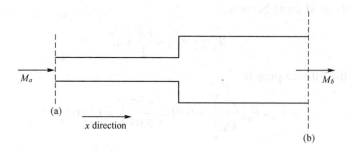

Fig. 1.15   Momentum balance.

Momentum of total stream; momentum correction factor

The momentum flow rate $\dot{M}$ of a fluid stream having a mass flow rate $\dot{m}$ and all moving at a velocity u equals $\dot{m}u$. If u varies from point to point in the cross section of the stream, however, the total momentum flow does not equal the product of the mass flow rate and the average velocity, or $\dot{m}V$; in general it is somewhat greater than this.

The necessary correction factor is best found from the convective momentum flux, that is, the momentum carried by the moving fluid through a unit cross sectional area of the channel in unit time. This is the product of the linear velocity normal to the cross section and the mass velocity (or mass flux). For a differential cross-sectional area dS, then, the momentum flux is

$$\frac{d\dot{M}}{dS} = (\rho u)u = \rho u^2 \qquad (1.3\text{-}31)$$

The momentum flux of the whole stream, for a constant-density fluid, is

$$\frac{\dot{M}}{S} = \frac{\rho \int_S u^2 dS}{S} \qquad (1.3\text{-}32)$$

The momentum correction factor β is defined by the relation

$$\beta = \frac{\dot{M}/S}{\rho V^2} \qquad (1.3\text{-}33)$$

Substituting from Eq. (1.3-32) gives

$$\beta = \frac{1}{S}\int_S \left(\frac{u}{V}\right)^2 dS \qquad (1.3\text{-}34)$$

To find β for any given flow situation, the variation of u with position in the cross section must be known.

Thus Eq. (1.3-30) may be written

$$\sum F = \dot{m}(\beta_b V_b - \beta_a V_a) \qquad (1.3\text{-}35)$$

In using this relation, care must be taken to identify and include in $\sum F$ all force components acting on the fluid in the direction of the velocity component in the equation.

Several such forces may appear: (1) pressure change in the direction of flow; (2) shear stress at the boundary between the fluid stream and the conduit or (if the conduit itself is considered to be part of the system) external forces acting on the solid wall; (3) if the stream is inclined, the appropriate component of the force of gravity. Assuming one-dimensional flow in the x direction, a typical situation is represented by the equation

$$\sum F = p_a S_a - p_b S_b + F_w - F_g \tag{1.3-36}$$

where p_a, p_b = inlet and outlet pressures, respectively
S_a, S_b = inlet and outlet cross sections, respectively
F_w = net force of wall of channel on fluid
F_g = component of force of gravity (written for flow in upward direction)

Layer flow with free surface

In one form of layer flow the liquid layer has a free surface and flows under the force of gravity over an inclined or vertical surface. If such flow is in steady state, with fully developed velocity gradients, the thickness of the layer is constant. Conversely there is so little drag at the free liquid surface that the shear stress there can be ignored. If the flow is laminar and the liquid surface is flat and free from ripples, the fluid motion can be analyzed mathematically.

Consider a layer of a newtonian liquid flowing in steady flow at constant rate and thickness over a flat plate, as shown in Fig. 1.16. The plate is inclined at an angle ϕ with the vertical. The breadth of the layer in the direction perpendicular to the plane of the figure is b, and the thickness of the layer in the direction perpendicular to the plate is δ. Isolate a control volume as shown in Fig. 1.16. The upper surface of the control volume is in contact with the atmosphere, the two ends are planes perpendicular to the plate at a distance L apart, and the lower surface is the plane parallel with the wall at a distance r from the upper surface of the layer.

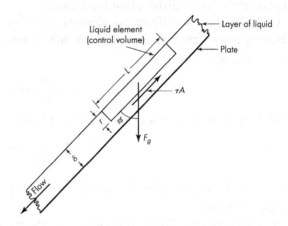

Fig. 1.16 Forces on liquid element in layer flow.

Since the layer is in steady flow with no acceleration, by the momentum principle the sum of all forces on the control volume is zero. The possible forces acting on the control volume in a direction parallel to the flow are the pressure forces on the ends, the shear forces on the upper and lower faces, and the component of the force of gravity in the direction of flow. Since the pressure on the outer surface is atmospheric, the pressures on the control volume at the ends of the volume are equal and oppositely directed. They therefore vanish. Also, by assumption, the shear on the upper surface of the element is neglected. The two forces remaining are therefore the shear force on the lower surface of the control volume and the component of gravity in the direction of flow. Then

$$F_g \cos\phi - \tau A = 0 \tag{1.3-37}$$

where F_g = gravity force

τ = shear stress on lower surface of control volume

A = area of lower surface of control volume

From this equation, noting that $A = bL$ and $F_g = mg = \rho rLbg$,

$$\rho rLbg \cos\phi = \tau Lb$$

or

$$\tau = \rho rg \cos\phi \tag{1.3-38}$$

Since the flow is laminar, $\tau = -\mu du/dr$ and

$$-\mu \frac{du}{dr} = g\rho r \cos\phi \tag{1.3-39}$$

Rearranging and integrating between limits give

$$\int_0^u du = -\frac{g\rho \cos\phi}{\mu} \int_\delta^r r\, dr$$

$$u = \frac{\rho g \cos\phi}{2\mu}(\delta^2 - r^2) \tag{1.3-40}$$

where δ is the total thickness of the liquid layer. Equation (1.3-40) shows that in laminar flow on a plate the velocity distribution is parabolic.

Consider now a differential element of cross-sectional area dS, where $dS = b\,dr$. The differential mass flow rate $d\dot{m}$ through this element equals $\rho u b\, dr$. The total mass flow rate of the fluid then is

$$\dot{m} = \int_0^\delta \rho u b\, dr \tag{1.3-41}$$

Substituting from Eq. (1.3-40) into Eq. (1.3-41) and integrating give

$$\frac{\dot{m}}{b} = \frac{\delta^3 \rho^2 g \cos\phi}{3\mu} = \Gamma \tag{1.3-42}$$

where $\Gamma \equiv \dot{m}/b$ and is called the liquid loading. The units of Γ are kilograms per second per meter of width.

Rearrangement of Eq. (1.3-42) gives, for the thickness of the layer,

$$\delta = \left(\frac{3\mu\Gamma}{\rho^2 g \cos\phi}\right)^{1/3} \tag{1.3-43}$$

The Reynolds number for flow down a flat plate is defined by the equation

$$Re = \frac{4r_H V \rho}{\mu} = 4\delta \frac{\dot{m}}{\delta b \rho} \frac{\rho}{\mu} = \frac{4\Gamma}{\mu} \tag{1.3-44}$$

where r_H = hydraulic radius, defined by Eq. (1.4-11). For flow of a liquid down either the inside or the outside of a pipe, the layer thickness is usually a very small fraction of the pipe diameter and the Reynolds number is the same as for a flat plate, Eq. (1.3-44).

Equation (1.3-43) for the thickness of a falling laminar film was first presented by Nusselt, who used the result to predict heat-transfer coefficients for condensing vapors. Measurements of film thickness on a vertical surface ($\cos\phi = 1$) show that Eq. (1.3-43) is approximately correct for $Re \approx 1000$, but the thickness actually varies with about 0.45 power of the Reynolds number, and the layers are thinner than predicted at low Re and thicker than predicted above $Re = 1000$. The deviations may be due to ripples or waves in the films, which are apparent even at quite low Reynolds numbers.

The transition from laminar to turbulent flow is not as easily detected as in pipe flow since the film is very thin and the ripples make it difficult to observe turbulence in the film. A critical Reynolds number of 2100 has often been used for layer flow, but film thickness measurements indicate a transition at $Re \approx 1200$. Above this point, the thickness increases with about the 0.6 power of the flow rate.

1.4 Incompressible Flow in Pipes and Channels

Chemical engineers in industrial practice are often concerned with the flow of fluids through pipes, tubes, and channels with a noncircular cross section. Usually the pipes are filled with the moving fluid, but some problems involve flow in partially filled pipes, in layers down inclined or vertical surfaces, through beds of solids, or in agitated vessels. This section only deals with the steady flow of incompressible fluids through closed pipes and channels.

1.4.1 Shear Stress and Skin Friction in Pipes

Shear-stress distribution

Consider the steady flow of fluid of constant density in fully developed flow through a horizontal pipe. Visualize a disk-shaped element of fluid, concentric with the axis of the tube, of radius r and length dL, as shown in Fig. 1.17. Assume the element is isolated as a free body. Let the fluid pressure on the upstream and downstream faces of the disk be p and $p + dp$, respectively. Since the fluid possesses a viscosity, a shear force opposing flow will exist on the rim of the element. Apply the momentum equation (1.3-35) between the two faces of the disk. Since the flow is fully developed, $V_b = V_a$, so that $\sum F = 0$. The forces acting on the upstream and downstream faces of the disk are

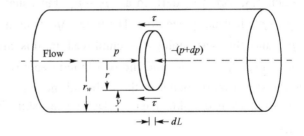

Fig. 1.17   Fluid element in steady flow through pipe.

$$S_a = S_b = \pi r^2 \qquad P_a = P \qquad P_a S_a = \pi r^2 P \qquad P_b S_b = (\pi r^2)(P + dp)$$

where $S_a = S_b = \pi r^2$, is the cross-sectional area of the element of fluid.

The shear force F_s acting on the rim of the element is the product of the shear stress and the cylindrical area, or $(2\pi r dL)\tau$. Therefore

$$\sum F = \pi r^2 p - \pi r^2 (p + dp) - (2\pi r dL)\tau = 0$$

Simplifying this equation and dividing by $\pi r^2 dL$ give

$$\frac{dp}{dL} + \frac{2\tau}{r} = 0 \qquad (1.4\text{-}1)$$

In steady flow, either laminar or turbulent, the pressure at any given cross section of a stream tube is constant, so that dp/dL is independent of r. Equation (1.4-1) can be written for the entire cross section of the tube by taking $\tau = \tau_w$ and $r = r_w$, where τ_w is the shear stress at the wall of the conduit and r_w is the radius of the tube. Equation (1.4-1) then becomes

$$\frac{dp}{dL} + \frac{2\tau_w}{r_w} = 0 \tag{1.4-2}$$

Subtracting Eq. (1.4-1) from Eq. (1.4-2) gives

$$\frac{\tau_w}{r_w} = \frac{\tau}{r} \tag{1.4-3}$$

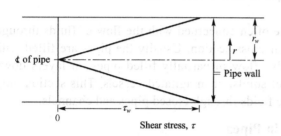

Fig. 1.18 Variation of shear stress in pipe.

Also, when $r = 0, \tau = 0$. The simple linear relation between τ and r in Eq. (1.4-3) is shown graphically in Fig. 1.18. Note that this linear relationship applies in both laminar and turbulent flow, and to both newtonian and non-newtonian fluids.

Relation between skin friction and wall shear

Bernoulli Equation with friction loss can be written over a definite length L of the complete stream

$$\frac{p_a}{\rho} + gZ_a + \frac{V_a^2}{2} = \frac{p_b}{\rho} + gZ_b + \frac{V_b^2}{2} + h_f \tag{1.4-4}$$

In Section 1.3, Δp was defined as $p_b - p_a$, but usually (though not always) $p_a > p_b$, and thus $p_b - p_a$ is usually negative. The term Δp is commonly used for *pressure drop*, that is, $p_a - p_b$, and this terminology is employed in this and subsequent chapters. Here, then, $p_b = p_a - \Delta p$, $Z_b - Z_a = 0$, and the two kinetic energy terms cancel. Also, the only kind of friction is skin friction between the wall and the fluid stream, and the only pressure drop is that resulting from the skin friction. Denote the skin friction by h_{fs}, Then equation (1.4-4) becomes

$$\frac{p_a}{\rho} = \frac{p_b}{\rho} + h_{fs}$$

or

$$\Delta p = \rho h_{fs}$$

Denote pressure drop caused by skin friction h_{fs} by Δp_s:

$$\Delta p_s = \rho h_{fs} \tag{1.4-5}$$

For a definite length L of pipe, dp/dL in Eq.(1.4-2) becomes $-\Delta p_s/L$. Eliminating Δp_s from Eqs. (1.4-2) and (1.4-5) gives the following relation between h_{fs} and τ_w:

$$h_{fs} = \frac{2\tau_w}{\rho r_w} L = \frac{4\tau_w}{\rho D} L \tag{1.4-6}$$

where D is the diameter of the pipe.

The friction factor
Another common parameter, especially useful in the study of turbulent flow, is the *Fanning friction factory* denoted by f and defined as the ratio of the wall shear stress to the product of the density and the velocity head $V^2/2$:

$$f \equiv \frac{\tau_w}{\rho V^2/2} = \frac{2\tau_w}{\rho V^2} \qquad (1.4\text{-}7)$$

Another friction factor is commonly used in the fluid mechanics literature, called the Blasius or Darcy friction factor, and is defined by

$$\lambda = 4f \qquad (1.4\text{-}8)$$

Relations between skin friction parameters
The four common quantities used to measure skin friction in pipes, h_{fs}, Δp_s, τ_w, and f (or λ), are related by the equations

$$h_{fs} = \frac{2\,\tau_w}{\rho\, r_w} L = \frac{\Delta p_s}{\rho} = 4f \frac{L}{D}\frac{V^2}{2} = \lambda \frac{L}{D}\frac{V^2}{2} \qquad (1.4\text{-}9)$$

from which

$$f = \frac{\Delta p_s D}{2L\rho V^2} \qquad (1.4\text{-}10)$$

or

$$\frac{\Delta p_s}{L} = \frac{2f\rho V^2}{D} \qquad (1.4\text{-}10a)$$

Equation (1.4-10a) is the equation usually used to calculate skin friction loss in straight pipe. The subscript s is used in Δp_s and h_{fs} to call attention to the fact that in Eqs. (1.4-9) through (1.4-10a) these quantities, when they are associated with the Fanning friction factor, relate *only to skin friction*. If other terms in the Bernoulli equation are present or if form friction is also active, $p_a - p_b$ differs from Δp_s. If boundary layer separation occurs, h_f is greater than h_{fs}.

Flow in noncircular channels
In evaluating skin friction in channels of noncircular cross section, the diameter in the Reynolds number and in Eq. (1.4-10), the definition of the friction factor, is taken as an *equivalent diameter* D_{eq}, defined as 4 times the hydraulic radius. The hydraulic radius is denoted by r_H and in turn is defined as the ratio of the cross-sectional area of the channel to the wetted perimeter of the channel:

$$r_H \equiv \frac{S}{L_p} \qquad (1.4\text{-}11)$$

where S = cross-sectional area of channel
L_p = perimeter of channel in contact with fluid

Thus, for the special case of a circular tube, the hydraulic radius is

$$r_H = \frac{\pi D^2/4}{\pi D} = \frac{D}{4}$$

The equivalent diameter is $D_{eq} = 4r_H$, or simply D.

An important special case is the annulus between two concentric pipes. Here the hydraulic radius is

$$r_H = \frac{\pi D_o^2/4 - \pi D_i^2/4}{\pi D_o + \pi D_i} = \frac{D_o - D_i}{4}$$

where D_i and D_o are the inside and outside diameters of the annulus, respectively. The equivalent diameter of an annulus is therefore the difference of the diameters. Also the equivalent diameter of a square duct with a width of side b is $D_{eq} = 4(b^2/4b) = b$. For flow between parallel plates, when the distance between them b is much smaller than the width of the plates, the equivalent diameter $D_{eq} = 2b$.

The defining equations for friction factor [Eq. (1.4-9)] and Reynolds number [Eq. (1.2-9)] can be generalized by substituting $2r_H$ for r or D_{eq} for D. The hydraulic radius is especially useful with turbulent flow. It is much less useful with laminar flow, but in many laminar flow situations the fluid flow relationships can be calculated mathematically, as shown in the following section.

1.4.2 Laminar Flow in Pipes and Channels

Equations (1.4-1) through (1.4-10a) apply to both laminar and turbulent flow provided the fluid is incompressible and the flow is steady and fully developed. Because the shear-stress viscosity relationship for laminar flow is simple, derivations from these equations can be made most readily for laminar flow.

Laminar flow of newtonian fluids

The treatment is especially straightforward for a newtonian fluid, for which quantities such as the velocity distribution, the average velocity, and the kinetic energy correction factors are readily calculated.

Velocity distribution. The relation between the local velocity and position in the stream is found as follows. In circular channels, because of symmetry about the axis of the tube, the local velocity u depends only on the radius r. Consider a thin ring of radius r and width dr, forming an element of cross-sectional area dS. Then

$$dS = 2\pi r dr \qquad (1.4\text{-}12)$$

The velocity distribution is found by using the definition of viscosity [Eq. (1.2-3)], written as

$$\mu = -\frac{\tau}{du/dr} \qquad (1.4\text{-}13)$$

The minus sign in the equation accounts for the fact that in a pipe u decreases as r increases. Eliminating τ from Eq. (1.4-3) and (1.4-10a) provides the following ordinary differential equation relating u and r:

$$\frac{du}{dr} = -\frac{\tau_w}{r_w \mu} r \qquad (1.4\text{-}14)$$

Integration of Eq. (1.4-14) with the boundary condition $u = 0$, $r = r_w$ gives

$$\int_0^u du = -\frac{\tau_w}{r_w \mu} \int_{r_w}^r r dr$$

$$u = \frac{\tau_w}{2r_w\mu}(r_w^2 - r^2) \tag{1.4-15}$$

The maximum value of the local velocity is denoted by u_{max} and is located at the center of the pipe. The value of u_{max} is found from Eq. (1.4-15) by substituting 0 for r, giving

$$u_{max} = \frac{\tau_w r_w}{2\mu} \tag{1.4-16}$$

Dividing Eq. (1.4-15) by Eq. (1.4-16) gives the following relationship for the ratio of the local velocity to the maximum velocity:

$$\frac{u}{u_{max}} = 1 - \left(\frac{r}{r_w}\right)^2 \tag{1.4-17}$$

Equation (1.4-17) shows that in laminar flow the velocity distribution with respect to the radius is a parabola with the apex at the centerline of the pipe. The distribution is shown as the dashed line in Fig. 1.19.

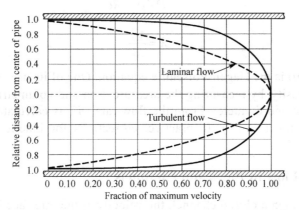

Fig. 1.19 Velocity distribution in pipe, fully developed flow of newtonian fluid, for laminar flow and for turbulent flow at Re = 10000.

Average velocity. Substitution of dS from Eq. (1.4-12), u from Eq. (1.4-15), and πr_w^2 for S into Eq. (1.3-3) gives

$$V = \frac{1}{S}\int_S u\, dS \qquad S = \pi r_w^2 \qquad dS = \pi 2r\, dr$$

$$V = \frac{1}{S}\int_S u\, dS = \frac{\tau_w}{r_w^3 \mu}\int_0^{r_w}(r_w^2 - r^2)r\, dr = \frac{\tau_w r_w}{4\mu} \tag{1.4-18}$$

Comparison of Eqs. (1.4-16) and (1.4-18) shows that

$$\frac{V}{u_{max}} = 0.5 \tag{1.4-19}$$

The average velocity is precisely one-half the maximum velocity.

Kinetic energy correction factor. The kinetic energy factor α is calculated from Eq. (1.3-15), using Eq. (1.4-12) for dS, (1.4-15) for u, and (1.4-18) for V. The final result is $\alpha = 2.0$. The proper term for kinetic energy in the Bernoulli equation [Eq. (1.3-25) or (1.3-28)] for laminar flow in a pipe is therefore V^2.

Hagen-Poiseuille equation

For practical calculations, Eq. (1.4-18) is transformed by eliminating τ_w in favor of Δp_s, by the use of Eq. (1.4-9) and using the pipe diameter in place of the pipe radius. The result is

$$V = \frac{\Delta p_s}{L}\frac{r_w}{2}\frac{r_w}{4\mu} = \frac{\Delta p_s D^2}{32L\mu}$$

Solving for Δp_s, gives

$$\Delta p_s = \frac{32LV\mu}{D^2} \qquad (1.4\text{-}20)$$

and since $\Delta p_s = 4\tau_w L/D$,

$$\tau_w = \frac{8V\mu}{D} \qquad (1.4\text{-}21)$$

Substituting from Eq. (1.4-21) into Eq. (1.4-7) and Eq. (1.4-8) gives

$$f = \frac{16\mu}{DV\rho} = \frac{16}{Re} \qquad (1.4\text{-}22)$$

or

$$\lambda = \frac{64}{Re} \qquad (1.4\text{-}23)$$

Equation (1.4-20) is the Hagen-Poiseuille equation. One of its uses is in the experimental measurement of viscosity, by measuring the pressure drop and volumetric flow rate through a tube of known length and diameter. From the flow rate, V is calculated by Eq. (1.3-3) and μ is calculated by Eq. (1.4-20). In practice, corrections for kinetic energy and entrance effects are necessary.

1.4.3 Turbulent Flow in Pipes and Channels

In turbulent flow through a closed channel, the velocity at the interface between fluid and the solid wall is zero, and (except very infrequently) there are no velocity components normal to the wall. Within a thin volume immediately adjacent to the wall, the velocity gradient is essentially constant and the flow is viscous most of the time. This volume is called the *viscous sublayer*. Formerly it was assumed that this sublayer had a definite thickness and was always free from eddies, but measurements have shown velocity fluctuations in the sublayer caused by occasional eddies from the turbulent fluid moving into this region. Very close to the wall, eddies are infrequent, but there is no region that is completely free of eddies. Within the viscous sublayer only viscous shear is important, and eddy diffusion, if present at all, is minor.

The viscous sublayer occupies only a very small fraction of the total cross section. It has no sharp upper boundary, and its thickness is difficult to define. A transition layer exists immediately adjacent to the viscous sublayer in which both viscous shear and shear due to eddy diffusion exist. The transition layer, which is sometimes called a *buffer layer,* also is relatively thin. The bulk of the cross section of the flowing stream is occupied by entirely turbulent flow called the *turbulent core*. In the turbulent core, viscous shear is negligible in comparison with that from eddy viscosity.

Velocity distribution for turbulent flow

Because of the dependence of important flow parameters on the velocity distribution, considerable study, both theoretical and experimental, has been devoted to determining the velocity distribution in turbulent flow. Although the problem has not been completely solved, useful relationships are available that may be used to calculate the important characteristics of turbulence; and the results of theoretical calculations check with experimental data reasonably well.

A typical velocity distribution for a newtonian fluid moving in turbulent flow in a smooth pipe at a Reynolds number of 10000 is shown in Fig. 1.19. The figure also shows the velocity distribution for laminar flow at the same maximum velocity at the center of the pipe. The curve for turbulent flow is clearly much flatter than that for laminar flow, and the difference between the average velocity and the maximum velocity is considerably less. At still higher Reynolds numbers, the curve for turbulent flow would be even flatter than that in Fig. 1.19.

In turbulent flow, as in laminar flow, the velocity gradient is zero at the centerline. It is known that the eddies in the turbulent core are large but of low intensity, and those in the transition zone are small but intense. Most of the kinetic energy content of the eddies lies in the buffer zone and the outer portion of the turbulent core. At the centerline the turbulence is isotropic. In all other areas of the turbulent flow regime, turbulence is anisotropic; otherwise, there would be no shear.

Relations between maximum velocity and average velocity

The quantities Re and ratio V/u_{max} are useful in relating the average velocity to the maximum velocity in the center of the tube as a function of flow conditions; for example, an important method of measuring fluid flow is the pitot tube (Fig. 1.31), which can be used to measure u_{max}, and this relationship is then used to determine the average velocity from this single observation.

Experimentally measured values of V/u_{max} as a function of the Reynolds number are shown in Fig. 1.20, which covers the range from laminar flow to turbulent flow. For laminar flow the ratio is exactly 0.5, in accordance with Eq. (1.4-19). The ratio changes rapidly from 0.5 to about 0.7, when laminar flow changes to turbulent, and then increases gradually to 0.87 when $Re = 10^6$.

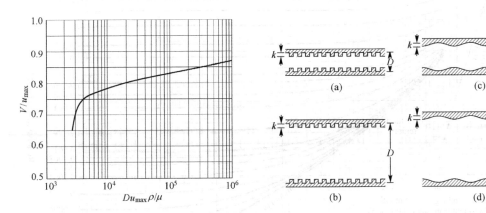

Fig. 1.20 V/u_{max} versus $Re = Du_{max}\rho/\mu$

Fig. 1.21 Types of roughness.

Effect of roughness

The discussion thus far has been restricted to smooth tubes without defining smoothness. It has long been known that in turbulent flow a rough pipe leads to a larger friction factor for a given Reynolds number than a smooth pipe does. If a rough pipe is smoothed, the friction factor is reduced. When further smoothing brings about no further reduction in the friction factor for a given Reynolds number, the tube is said to be *hydraulically smooth*.

Figure 1.21 shows several idealized kinds of roughness. The height of a single unit of roughness is denoted by k and is called the *roughness parameter*. From dimensional analysis, f is a function of both Re and the relative roughness k/D, where D is the diameter of the pipe. For a given kind of roughness, for example, that shown in Fig. 1.21(a) and (b), it can be expected that a different curve of f versus Re would be found for each magnitude of the relative roughness and also that for other types of roughness, such as those shown in Fig. 1.21(c) and (d), a different family of curves of Re versus f would be found for each type of roughness. Experiments on artificially roughened pipe have confirmed these expectations. It has also been found that all clean, new commercial pipes seem to have the same type of roughness and that each material of construction has its own characteristic roughness parameter.

Old, foul, and corroded pipe can be very rough, and the character of the roughness differs from that of clean pipe.

Roughness has no appreciable effect on the friction factor for laminar flow unless k is so large that the measurement of the diameter becomes uncertain.

The friction factor chart

For design purposes, the friction characteristics of round pipe, both smooth and rough, are summarized by the friction factor chart (Fig. 1.22), which is a log-log plot of f versus Re. For laminar flow Eq. (1.4-22), (1.4-23) relates the friction factor to the Reynolds number. A log-log plot of Eq. (1.4-22) is a straight line with a slope of -1. This plot line is shown on Fig. 1.22 for Reynolds numbers less than 2100.

For turbulent flow the lowest line in Fig. 1.22 represents the friction factor for hydraulically smooth tubes and is called von Karman equation which can be described by

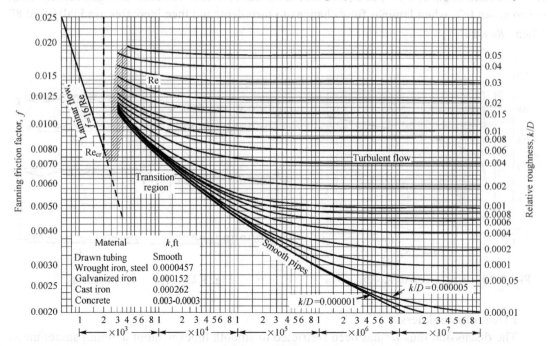

Fig. 1.22 Friction factor plot for circular pipes

$$\frac{1}{\sqrt{f/2}} = 2.5\ln\left(Re\sqrt{\frac{f}{8}}\right) + 1.75 \tag{1.4-24}$$

A much more convenient empirical equation for this line is the relation

$$f = 0.046 Re^{-0.2} \tag{1.4-25}$$

This applies over a range of Reynolds numbers from about 50000 to 1×10^6. Another equation, applicable over a range of Reynolds numbers from 3000 to 3×10^6, is

$$f = 0.0014 + \frac{0.125}{Re^{0.32}} \tag{1.4-26}$$

The other curved lines in the turbulent flow range represent the friction factors for various types of commercial pipe, each of which is characterized by a different value of k. The parameters for several common metals are given in the figure. Clean wrought-iron or steel pipe, for example, has a k value of 4.57×10^{-5} m, regardless of the diameter of the pipe. Drawn copper and brass pipe may be considered hydraulically smooth.

For steel pipe and other rough pipes, the friction factor becomes independent of the Reynolds number for Reynolds numbers greater than 10^6. An empirical equation for this region is

$$f = 0.026(k/D)^{0.24} \tag{1.4-27}$$

For different flow regimes in a given system, the variation of pressure drop with flow rate can be found from Eqs. (1.4-10a), (1.4-22), (1.4-25), and (1.4-27), to give:

For laminar flow ($Re < 2100$) $\Delta p_s / L \propto V$

For turbulent flow ($2500 < Re < 10^6$) $\Delta p_s / L \propto V^{1.8}$

For very turbulent flow ($Re > 10^6$) $\Delta p_s / L \propto V^2$

Figure 1.22 is useful for calculating h_{fs} from a known pipe size and flow rate, but it cannot be used directly to determine the flow rate for a given pressure drop, since Re is not known until V is determined. However, since f changes only slightly with Re for turbulent flow, a trial-and-error solution converges quickly, as shown in Example 1.6.

EXAMPLE 1.6 Water is flowing at 10℃ through a long horizontal plastic pipe, 76.2mm in inside diameter, at a velocity of 2.438 m/s. (*a*) Calculate the pressure drop in Pa per 30.48 m of pipe. (*b*) If the pressure drop must be limited to 13789 Pa per 30.48 m of pipe, what is the maximum allowable velocity of the water?
Solution. (*a*) Use Eq. (1.4-10a),

$$\frac{\Delta p_S}{L} = \frac{2f\rho V^2}{D}$$

The properties of water at 10℃, from App. 5, are $\rho = 1000 \text{kg}/\text{m}^3$ and $\mu = 1.310$ cP. Also $D = 0.0762$m, $V = 2.438$m/s, and $L = 30.48$m.

The Reynolds number is

$$Re = \frac{\rho V D}{\mu} = \frac{1000 \times 2.438 \times 0.0762}{1.310 \times 10^{-3}} = 1.41 \times 10^5$$

From Fig. 1.22, $f = 0.0041$. Substituting in Eq. (1.4-10a) gives

$$\Delta p_S = \frac{2 \times 0.0041 \times 1000 \times 2.438^2 \times 30.48}{0.0762} = 19496 \text{Pa}$$

(b) Since $\Delta p_S = 19496 \text{Pa}$ at 2.438m/s, and $\Delta p_s/L \propto V^{1.8}$ [assuming Eq.(1.4-25) applies], then for $\Delta p_s = 13789 \text{Pa}$

$$\left(\frac{V}{2.438}\right)^{1.8} = \left(\frac{13789}{19496}\right)$$

$$V = 2.011 \text{m/s}$$

This gives a Reynolds number of 1.17×10^5, so Eq. (1.4-25) applies.

Turbulent flow in noncircular channels

As discussed earlier, relationships for turbulent flow in pipes may be applied to noncircular channels by substituting the equivalent diameter D_{eq} (or 4 times the hydraulic radius r_H) for the diameter D in the relevant equations.

1.4.4 Friction from Changes in Velocity or Direction

Whenever the velocity of a fluid is changed, in either direction or magnitude, friction is generated in addition to the skin friction resulting from flow through the straight pipe. Such friction includes form friction resulting from vortices that develop when normal streamlines are disturbed and when boundary layer separation occurs. Often these effects cannot be calculated precisely, and it is necessary to rely on empirical data.

Friction loss from sudden expansion of cross section

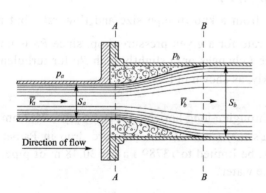

Fig. 1.23 Flow at sudden enlargement of cross section.

If the cross section of the pipe is suddenly enlarged, the fluid stream separates from the wall and issues as a jet into the enlarged section. The jet then expands to fill the entire cross section of the larger conduit. The space between the expanding jet and the conduit wall is filled with fluid in vortex motion characteristic of boundary layer separation, and considerable friction is generated within this space. This effect is shown in Fig. 1.23.

The friction loss h_{fe} from a sudden expansion of cross section is proportional to the velocity head of the fluid in the small conduit and can be written

$$h_{fe} = K_e \frac{V_a^2}{2} \qquad (1.4\text{-}28)$$

where K_e is a proportionality factor called the *expansion loss coefficient* and V_a is the average velocity in the smaller, or upstream, conduit. In this case the calculation of K_e can be made theoretically and a satisfactory result obtained. For turbulent flow in both upstream and downstream pipes, K_e is given by the equation

$$K_e = \left(1 - \frac{S_a}{S_b}\right)^2 \qquad (1.4\text{-}29)$$

where S_a and S_b are the cross-sectional areas of the upstream and downstream conduits, respectively.

Friction loss from sudden contraction of cross section

When the cross section of the conduit is suddenly reduced, the fluid stream cannot follow around the sharp comer and the stream breaks contact with the wall of the conduit. A jet is formed, which flows into the stagnant fluid in the smaller section. The jet first contracts and then expands to fill the smaller cross section, and downstream from the point of contraction the normal velocity distribution eventually is reestablished. The cross section of minimum area at which the jet changes from a contraction to an expansion is called the *vena contracta*. The flow pattern of a sudden contraction is shown in Fig. 1.24. Section *CC* is drawn at the vena contracta. Vortices appear as shown in the figure.

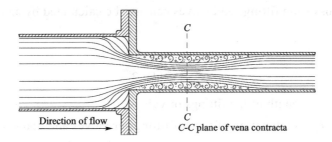

Fig. 1.24 Flow at sudden contraction of cross section.

The friction loss from sudden contraction is proportional to the velocity head in the smaller conduit and can be calculated by the equation

$$h_{fc} = K_c \frac{V_b^2}{2} \qquad (1.4\text{-}30)$$

where the proportionality factor K_c is called the *contraction loss coefficient* and V_b is the average velocity in the smaller, or downstream, section. Experimentally, for laminar flow, $K_c < 0.1$, and the contraction loss h_{fc} is negligible. For turbulent flow, K_c is given by the empirical equation

$$K_c = 0.4\left(1 - \frac{S_b}{S_a}\right) \qquad (1.4\text{-}31)$$

where S_a and S_b are the cross-sectional areas of the upstream and downstream conduits, respectively. When a liquid is discharged through a pipe welded to the wall of a large tank, S_b / S_a is nearly zero and $K_c \cong 0.4$. The value of K_c cannot be accurately predicted from theory, but it corresponds to a vena contracta area of about $0.6 S_b$. The velocity at the vena contracta is then about $V_b / 0.6$ and the kinetic energy is $(1/0.6)^2$ or 2.8 times the kinetic energy of the final stream. However, according to Eq. (1.4-29), the fractional energy loss as the stream expands beyond the vena contracta is only $(1-0.6)^2$, or a loss of $0.16 \times 2.8 \times V_b^2/2$, which corresponds to $K_c = 0.44$.

Effect of fittings and valves

Fittings and valves disturb the normal flow lines and cause friction. In short lines with many fittings, the friction loss from the fittings may be greater than that from the straight pipe. The friction loss h_{ff} from fittings is found from an equation similar to Eqs. (1.4-28) and (1.4-30):

$$h_{ff} = K_f \frac{V_a^2}{2} \tag{1.4-32}$$

where K_f = loss factor for fitting

V_a = average velocity in pipe leading to fitting

Factor K_f is found by experiment and differs for each type of connection. A short list of factors is given in Table 1.4.

The friction loss from fittings and valves can also be calculated by an equation similar to skin friction

$$h_{ff} = 4f \frac{L_e}{D} \frac{V_a^2}{2} \tag{1.4-33}$$

where L_e = equivalent length of the fittings or valves

Equivalent length L_e is a function of type of fittings or valves and diameter of the pipe.

Table 1.4 Loss coefficients for standard threaded pipe fittings

Fitting	K_f	Fitting	K_f
Elbow, standard		Return bend, 180°	1.5
		Gate valve	
45°	0.35	Half open	4.5
90°	0.75	Wide open	0.17
Elbows, welded		Angle valve, wide open	2.0
90° bends, radius 2 × pipe diameter	0.19	Globe valve, wide open	6.0
90° bends, radius 4 × pipe diameter	0.16		
90° bends, radius 6 × pipe diameter	0.21		
Tee			
Straight through	0.4		
Used as elbow	1.0		

Form friction losses in the Bernoulli equation

Form friction losses are incorporated in the h_f term of Eq. (1.3-28). They are combined with the skin friction losses of the straight pipe to give the total friction loss. Consider, for example, the flow of incompressible fluid through the two enlarged headers, the connecting tube, and the open globe valve shown in Fig. 1.25. Let V be the average velocity in the tube,

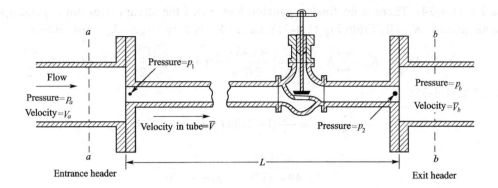

Fig. 1.25 Flow of incompressible fluid through typical assembly.

D the diameter of the tube, and L the length of the tube. The skin friction loss in the straight tube is, by Eq. (1.4-5a), $4f(L/D)(V^2/2)$; the contraction loss at the entrance to the tube is, by Eq. (1.4-30), $K_c(V^2/2)$ the expansion loss at the exit of the tube is, by Eq. (1.4-28), $K_e(V^2/2)$ and the friction loss in the globe valve is, by Eq. (1.4-32), $K_f(V^2/2)$. When skin friction in the entrance and exit headers is neglected, the total friction is

$$h_f = \left(4f\frac{L}{D} + K_c + K_e + K_f\right)\frac{V^2}{2} \tag{1.4-34}$$

To write the Bernoulli equation for this assembly, take station a in the inlet header and station b in the outlet header. Because there is no pump between stations a and b, $W_p = 0$; also the kinetic energy term can be canceled, and Eq. (1.3-28) becomes

$$\frac{p_a - p_b}{\rho} + g(Z_a - Z_b) = \left(4f\frac{L}{D} + k_c + K_e + K_f\right)\frac{V^2}{2} \tag{1.4-35}$$

EXAMPLE 1.7 Crude oil having a specific gravity of 0.93 and a viscosity of 4 cP is draining by gravity from the bottom of a tank. The depth of liquid above the drawoff connection in the tank is 6 m. The line from the drawoff is 0.078m ID steel pipe. Its length is 45 m, and it contains one elbow and two gate valves. The oil discharges into the atmosphere 9 m below the drawoff connection of the tank. What flow rate, in cubic meters per hour, can be expected through the line?
Solution. The quantities needed are

$$\mu = 0.004 \text{ kg/(m·s)} \quad L = 45 \text{ m}$$
$$D = 0.078 \text{ m} \quad \rho = 0.93 \times 998 = 928 \text{ kg/m}^3$$

For fittings, from Table 1.4,

$$\sum K_f = 0.75 + 2 \times 0.17 = 1.09$$

From Eq. (1.3-28), since $p_a = p_b$, and $V_a = 0$, $\eta W_p = 0$

$$\frac{V_b^2}{2} + h_f = g(Z_a - Z_b) = 9.81(6+9) = 147.1 \text{ m}^2/\text{s}^2 \tag{1.4-36}$$

Use Eq. (1.4-34). There is no final expansion loss, since the stream does not expand upon discharge, and $K_e = 0$. From Eq. (1.4-31), since S_a is very large, $K_c = 0.4$. Hence

$$h_f = \left(4f\frac{L}{D} + K_c + \sum K_f\right)\frac{V_b^2}{2} = \left(\frac{4\times 45 f}{0.078} + 0.4 + 1.09\right)\frac{V_b^2}{2} = (2308f + 1.49)\frac{V_b^2}{2}$$

From Eq. (1.4-36),

$$\frac{V_b^2}{2} + h_f = \frac{V_b^2}{2}(1 + 2308f + 1.49) = 147.1$$

$$V_b^2 = \frac{147.1 \times 2}{2.49 + 2308f} = \frac{294.2}{2.49 + 2308f}$$

Use Fig. 1.22 to find f. For this problem

$$Re = \frac{0.078 \times 928 V_b}{0.004} = 18096 V_b$$

$$\frac{k}{D} = \frac{0.0000457}{0.078} = 0.00059$$

Trials give the following:

$V_{b,est}$, m/s	$Re \times 10^{-4}$	f (from Fig. 1.22)	$V_{b,cal}$, m/s
4.00	7.23	0.0056	4.37
4.37	7.91	0.0055	4.40
4.40	7.96	0.0055	4.40

The cross-sectional area of the pipe is 0.00477 m², and the flow rate is $4.40 \times 3600 \times 0.00477 = 75.6$ m³/h.

The velocities of 2 to 4 m/s in Examples 1.6 and 1.7 are typical for liquid flow in medium-size pipes. The optimum velocity or optimum pipe size for a given flow rate depends on pumping costs and capital charges, and an empirical equation for optimum velocity is given in Chap. 2. For gas flows the optimum velocity is much larger than for liquids because of the lower density.

Velocity heads

As shown by Eq. (1.4-34), the friction loss in a complicated flow system can be expressed as a number of *velocity heads,* defined as $V^2/2$. It is a measure of the momentum loss resulting from flow through the pipe or fitting. In a tee used as an elbow, for example, all the momentum in one direction is lost as the stream turns 90°, whereas in a standard 90° elbow only three-fourths of it is lost. In an open globe valve, illustrated in Fig. 1.25, the flow makes two 90° turns and passes through a construction, expanding again on the other side. These changes in velocity and direction lead to a friction loss of 6 velocity heads.

A rapid practical method of estimating friction in straight pipe is as follows. By setting $4f(L/D) = 1.0$, it follows from Eq. (1.4-9) that a length equal to a definite number of pipe diameters generates a friction loss equal to 1 velocity head. Since, in turbulent flow, f varies from about 0.01 to about 0.002, the number of pipe diameters equivalent to a velocity head is from $1/(4 \times 0.01) = 25$ to $1/(4 \times 0.002) = 125$, depending upon the Reynolds number. For ordinary practice, 50 pipe diameters is assumed for this factor. Thus, if the pipe in the system of Fig. 1.25 is standard 2-in. steel (actual ID 2.07in = 0.0526m.) and is 30.48m long, the skin friction is equivalent to $30.48/(0.0526 \times 50) = 12$ velocity heads. In this case, the friction from the single fitting and the expansion and contraction are negligible in comparison with that in the pipe. In other cases, where the pipes are short and the fittings and expansion and contraction losses numerous, the friction loss in the pipes only may be negligible.

Separation from velocity decrease

Boundary layer separation can occur even where there is no sudden change in cross section if the cross section is continuously enlarged. For example, consider the flow of a fluid stream through the trumpet-shaped expander shown in Fig. 1.26. Because of the increase of cross section in the direction of flow, the velocity of the fluid decreases, and by the Bernoulli equation, the pressure must increase. Consider two stream filaments, one, *aa,* very near the wall, and the other, *bb,* a short distance from the wall. The pressure increase over a definite length of conduit is the same for both filaments, because the pressure throughout any single cross section is uniform. The loss in velocity head is, then, the same for both filaments. The initial velocity head of filament *aa* is less than that of filament *bb,* however, because filament *aa* is nearer to the wall. A point is reached, at a definite distance along the conduit, where the velocity of filament *aa* becomes zero but where the velocities of filament *bb* and of all other filaments farther from the wall than *aa* are still positive. This point is point *s* in Fig. 1.26. Beyond point *s* the velocity at the wall changes sign, a backflow of fluid between the wall and filament *aa* occurs, and the boundary layer separates from the wall. In Fig. 1.26, several curves are drawn of velocity *u* versus distance from the wall *y*, and it can be seen how the velocity near the wall becomes zero at point *s* and then reverses in sign. The point *s* is called a *separation point.* Line ,sA is called the *line of zero tangential velocity.*

The vortices formed between the wall and the separated fluid stream, beyond the separation point, cause excessive form friction losses. Separation occurs in both laminar and turbulent flow. In turbulent flow, the separation point is farther along the conduit than in laminar flow. Separation can be prevented if the angle between the wall of the conduit and the axis is made small. The maximum angle that can be tolerated in a conical expander without separation is 7°.

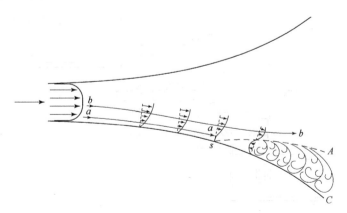

Fig. 1.26 Separation of boundary layer in diverging channel.

Minimizing expansion and contraction losses

A contraction loss can be nearly eliminated by reducing the cross section gradually rather than suddenly. For example, if the reduction in cross section shown in Fig. 1.24 is obtained by a conical reducer or by a trumpet-shaped entrance to the smaller pipe, the contraction coefficient K_c can be reduced to approximately 0.05 for all values of S_b/S_a. Separation and vena contracta formation do not occur unless the decrease in cross section is sudden.

An expansion loss can also be minimized by substituting a conical expander for the flanges shown in Fig. 1.23. The angle between the diverging walls of the cone must be less than 7°, however, or separation may occur. For angles of 35° or more, the loss through a conical expander can become greater than that through a sudden expansion for the same area ratio S_a/S_b because of the excessive form friction from the vortices caused by the separation.

1.5 Pipe Flow Systems

In the previous section, we discussed concepts concerning flow in pipes and ducts. The purpose of this section is to apply these ideas to the solutions of various practical problems. Two classes of pipe system will be considered: those containing a single pipe, and those containing multiple pipes.

Some of the basic components of a typical pipe system are shown in Fig. 1.27. They include the pipes themselves, the various fittings and valves used to connect the individual pipes and control flowrate, and the pumps or turbines that add or remove energy from the fluid. Even the most simple pipe systems are actually quite complex when they are viewed in terms of rigorous analytical considerations. Theoretical analysis method may be only used to treat the simplest pipe flow topics (such as laminar flow in long, straight, constant diameter pipes which is discussed in section 1.4.2). For other complex pipe flow topics, dimensional analysis considerations combined with experimental results will be needed. Such an approach is not unusual in fluid mechanics investigations. When "real-world" effects are important (such as viscous effects in pipe flows), it is often difficult or "impossible" to use only rigorous analysis methods to obtain the desired results. A judicious combination of experimental data with theoretical considerations and dimensional analysis often provides the desired results.

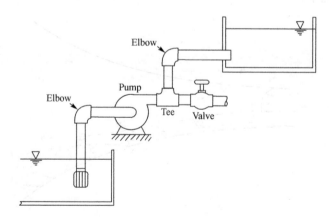

Fig. 1.27 Typical pipe system compoents.

1.5.1 Single Pipes

The single pipe systems are the pipes whose length may be interrupted by various components, or whose diameter may be change in different section.

The three most common types of pipe flow problems are shown in Table 1.5 in terms of the parameters involved. It is assumed the pipe system is defined in terms of the length of pipe sections used and the number of elbows, bends, and valves needed to convey the fluid between the desired locations. In all instances the fluid properties are assumed to be given. The nature of the solution process for the three types of problems can depend strongly on which of the various parameters are independent parameters (the "given") and which is the dependent parameter (the "determine").

In a Type I problem we specify the desired flowrate or average velocity and determine the necessary pressure drop or head loss. Example 1.6 is this kind of problem.

In a Type II problem we specify the applied driving pressure (or, alternatively, the head loss) and determine the flowrate. This type problem requires trial-and-error solution techniques. This is because it is necessary to know the value of the friction factor to carry out the calculations, but the friction factor is a function of the unknown velocity (flowrate) in terms of the Reynolds number. This type problem is shown in Example 1.7.

In a Type III problem we specify the pressure drop and the flowrate and determine the diameter of the pipe needed. This type problem needs trial-and-error solution techniques too. This is, again, because the friction factor is a function of the diameter-through both the Reynolds number and the relative roughness. Thus, neither Re nor k/D are known unless D is known.

Table 1.5   Pipe Flow Types

Variable	Type I	Type II	Type III
a. Fluid			
Density	Given	Given	Given
Viscosity	Given	Given	Given
b. Pipe			
Diameter	Given	Given	Determine
Length	Given	Given	Given
Roughness	Given	Given	Given
c. Flow			
Flowrate or Average Velocity	Given	Determine	Given
d. Pressure			
Pressure Drop or Head Loss	Determine	Given	Given

For the pipe system that contains pipes in series as shown in Fig. 1.28, every fluid particle that passes through the system passes through each of the pipes. Thus, the flowrate (but not the velocity) is the same in each pipe, and the friction loss is the sum of the friction losses in each of the pipes.

The continuity equation can be written as follows:

$$q_1 = q_2 = q_3 = q \tag{1.5-1}$$

and the Bernoulli equation applied from location A to location B of Fig. 1.28 is

$$\frac{p_A}{\rho} + gZ_A + \frac{V_A^2}{2} = \frac{p_B}{\rho} + gZ_B + \frac{V_B^2}{2} + h_f \tag{1.5-2}$$

where h_f is friction loss from point A to point B

$$h_f = h_{f1} + h_{f2} + h_{f3} = \sum_{i=1}^{3} \left(4f_i \frac{L_i}{D_i} + K_{ci} + K_{ei} + K_{fi} \right) \frac{V_i^2}{2} \tag{1.5-3}$$

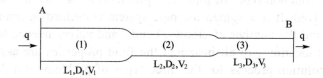

Fig. 1.28   Series pipe systems.

where the subscripts $i(i = 1, 2, 3)$ refer to each of the pipes. In general, the friction factors will be different for each pipe because the Reynolds numbers ($Re_i = \rho V_i D_i / \mu$) and the relative roughnesses (k_i / D_i) will be different. If the flowrate is given, it is a straightforward calculation to determine the head loss or pressure drop (Type I problem). If the pressure drop is given and the flowrate is to be calculated (Type II problem), an iteration scheme is needed. In this situation none of the friction factors f_i, are known, so the calculations may involve more trial-and-error attempts than for corresponding single pipe systems. The same is true for problems in which the pipe diameter (or diameters) is to be determined (Type III problems).

1.5.2   Multiple Pipe Systems

Multiple pipe systems contain more than one pipe. The complex system of tubes in our lungs (beginning with the relatively large-diameter trachea and ending in minute bronchi after numerous branchings) and the maze of pipes in a city's water distribution system are typical of such systems. The governing mechanisms for the flow in multiple pipe systems are the same as for the single pipe systems. However, because of the numerous unknowns involved, additional complexities may arise in solving for the flow in multiple pipe systems. In this section multiple pipe systems containing pipes in parallel, or branching configuration will be discussed.

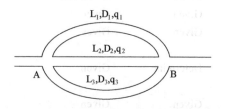

Fig. 1.29   Parallel pipe systems.

Parallel pipe system

One of common multiple pipe system contains pipes in parallel, as is shown in Fig. 1.29. In this system a fluid particle traveling from A to B may take any of the paths available, with the total flowrate equal to the sum of the flowrates in each pipe. The continuity equation for parallel pipes is

$$q = q_1 + q_2 + q_3 \qquad (1.5\text{-}4)$$

However, by writing the Bernoulli equation between points A and B through each parallel pipe, it is found that the friction loss per unit mass experienced by any fluid traveling between points A and B is the same, independent of the path taken. Thus, the Bernoulli equations for parallel pipes are

$$\frac{p_A}{\rho} + gZ_A + \frac{V_A^2}{2} = \frac{p_B}{\rho} + gZ_B + \frac{V_B^2}{2} + h_{fAB} \qquad (1.5\text{-}5)$$

where

$$h_{fAB} = h_{f1} = h_{f2} = h_{f3} \qquad (1.5\text{-}6)$$

Again, the method of solution of these equations depends on what information is given and what is to be calculated.

EXAMPLE 1.8 Find the distribution of flow for three parallel pipes arrangement shown in figure 1.29. The water kinematic viscosity $\nu = 10^{-6} \text{m}^2/\text{s}$. the total discharge is $q = 0.020 \text{m}^3/\text{s}$. The parameters of the parallel pipes are listed in Table 1.6.

Table 1.6 The parameters of the parallel pipes for Example 1.8

Pipe	L/m	D/m	λ	ΣK
1	100	0.05	0.023	10
2	150	0.075	0.025	3
3	200	0.085	0.021	2

Solution. For parallel pipes, the continuity equation can be written as Eq. (1.5-4)

$$q = q_1 + q_2 + q_3$$

And the governing equation of parallel pipes is

$$h_{f1} = h_{f2} = h_{f3} = \left(\lambda \frac{L}{D} + \Sigma K\right) \frac{V^2}{2} \qquad (1.5\text{-}7)$$

Based on $V = q / \left(\frac{\pi}{4} D^2\right)$ and known parameters listed in Table 1.6, Eq. (1.5-7) can be expressed as

$$3.262 q_1^2 = 2.091 q_2^2 = 1.100 q_3^2 \qquad (1.5\text{-}8)$$

Combining Eq.(1.5-4) and Eq. (1.5-8) gives

$$q_1 = 0.00504 \text{m}^3/\text{s}, \qquad q_2 = 0.00629 \text{m}^3/\text{s}, \qquad q_3 = 0.00867 \text{m}^3/\text{s}$$

Branching system

The branching system is shown in Fig. 1.30. Three reservoirs at known elevations are connected together with three pipes of known properties (lengths, diameters, and roughnesses). The problem is to determine the flowrates into or out of the reservoirs. If valve (1) were closed, the fluid would flow from reservoir B to C, the problem became single pipe system, and the flowrate could be easily calculated. Similar calculations could be carried out if valves (2) or (3) were closed with the others open.

For the conditions indicated in Fig. 1.30, if all valves are open, it is clear that fluid flows from reservoir A because the other two reservoir levels are lower. Whether the fluid flows into or out of reservoir B depends on the elevation of reservoirs B and C and the properties (length, diameter, roughness) of the three pipes. In general, it is not necessarily obvious which direction the fluid flows, and the solution process must include the determination of this direction.

The continuity equation at the junction node N of branching pipe system is written as

$$\sum_{i=1}^{3} q_i = 0 \qquad (1.5\text{-}9)$$

and assuming fluid flows from reservoir A to reservoirs B and C, the Bernoulli equation for each branching pipe from the junction node to reservoir could be

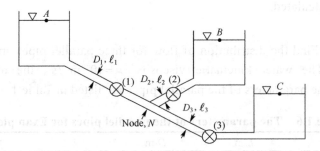

Fig. 1.30 Branching pipe system.

$$\frac{p_A}{\rho} + gZ_A + \frac{V_A^2}{2} = \frac{p_N}{\rho} + gZ_N + \frac{V_N^2}{2} + h_{f1} \tag{1.5-10}$$

$$\frac{p_N}{\rho} + gZ_N + \frac{V_N^2}{2} = \frac{p_B}{\rho} + gZ_B + \frac{V_B^2}{2} + h_{f2} \tag{1.5-11}$$

$$\frac{p_N}{\rho} + gZ_N + \frac{V_N^2}{2} = \frac{p_C}{\rho} + gZ_C + \frac{V_C^2}{2} + h_{f3} \tag{1.5-12}$$

where h_{fi} is the friction loss in branching pipe, subscript $i(i = 1, 2, 3)$ refers to the branching pipe. Subscript N refers to the junction node. Substituting Eq. (1.5-11) into Eq. (1.5-12), we obtained

$$\frac{p_B}{\rho} + gZ_B + \frac{V_B^2}{2} + h_{f2} = \frac{p_C}{\rho} + gZ_C + \frac{V_C^2}{2} + h_{f3} \tag{1.5-13}$$

The steps to solve the above equations are outlined below.

(1) Assume a flowrate q_1 in pipe 1. Established the $\frac{p_N}{\rho} + gZ_N$ at the junction by solving Eqs. (1.5-10).

(2) Compute the remaining flowrate q_i, using the Eq. (1.5-11) and Eq. (1.5-12).

(3) Substitute the q_i, into Eq. (1.5-9) to check for continuity balance. Generally, the flow imbalance at the junction, $\Delta q = q_1 - q_2 - q_3$ will be nonzero.

(4) Adjust the flow q_1 as $q_1 + \Delta q$ and repeat steps 2 and 3 until Δq is within desired limits.

1.6 Metering of Fluids

To control industrial processes, it is essential to know the amount of material entering and leaving the process. Because materials are transported in the form of fluids wherever possible, it is important to measure the rate at which a fluid is flowing through a pipe or other channel. Many different types of meters are used industrially. Selection of a meter is based on the applicability of the instrument to the specific problem, its installed cost and costs of operation, the range of flow rates it can accommodate (its rangeability), and its inherent accuracy. Sometimes a rough indication of the flow rate is all that is needed; at other times a highly accurate measurement, usually of the mass flow rate, is required for such purposes as controlling reactor feeds or transferring custody of the fluid from one owner to another.

A few types of flowmeters measure the mass flow rate directly, but the majority measure the volumetric flow rate or the average fluid velocity, from which the volumetric flow rate can be calculated. To convert the volumetric rate to the mass flow rate requires that the fluid density under the operating conditions be known. Most meters operate on all the fluid in the pipe or channel and are known as full-bore meters. Others, called insertion meters, measure the flow rate, or more commonly the fluid velocity, at one point only. The total flow rate, however, can often be inferred with considerable accuracy from this single-point measurement.

Detailed descriptions of commercial flowmeters, listing their advantages and limitations, are available in the literature.

1.6.1 Insertion Meters

In this type of meter the sensing element, which is small compared to the size of the flow channel, is inserted into the flow stream. A few insertion meters measure the average flow velocity, but the majority measure the local velocity at one point only. The positioning of the sensing element is therefore important if the total flow rate is to be determined. The local measured velocity must bear a constant and known relationship to the average velocity of the fluid.

The point of measurement may be at the centerline of the channel, and the average velocity may be found from the ratio of the average to the maximum velocity. Alternatively, the sensor may be located at the critical point in the channel where the local velocity equals the average velocity. In either case precautions must be taken, usually by providing long calming sections upstream of the meter, to ensure that the velocity profile is fully developed and not distorted.

Pitot tube

The pitot tube is a device used to measure the local velocity along a streamline. The principle of the device is shown in Fig. 1.31. The opening of the impact tube a is parallel to the flow direction. The opening of the static tube b is perpendicular to the direction of flow. The two tubes are connected to the legs of a manometer or equivalent device for measuring small pressure differences. The static tube measures the static pressure p_0 since there is no velocity component perpendicular to its opening. The impact opening includes a stagnation point B at which the streamline AB terminates.

The fluid flows into impact opening at point B, pressure builds up, and then remains stationary at this point, called the stagnation point. The difference in the stagnation pressure p_s at point B and the static pressure p_0 measured by the static tube b represents the pressure rise associated with the deceleration of the fluid. Since the manometer of the pitot tube measures the pressure difference $p_s - p_0$. If the fluid is incompressible, we can write the

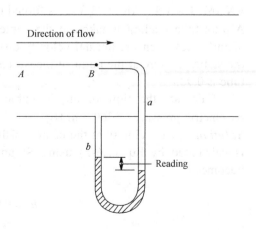

Fig. 1.31 Principle of pitot tube.

Bernoulli equation between point A, where the velocity u_0 is undisturbed before the fluid decelerates, and point B, where the velocity u is zero.

$$\frac{u_0^2}{2} - \frac{u^2}{2} + \frac{p_0 - p_s}{\rho} = 0$$

Setting $u = 0$ and solving for u_0,

$$u_0 = \sqrt{\frac{2(p_s - p_0)}{\rho}} \qquad (1.6\text{-}1)$$

The velocity measured by an ideal pitot tube would conform exactly to Eq. (1.6-1). Well-designed instruments are in error by not more than 1 percent of theory, but when precise measurements are to be made, the pitot tube should be calibrated and an appropriate correction factor applied.

$$u_0 = C\sqrt{\frac{2(p_s - p_0)}{\rho}} \qquad (1.6\text{-}2)$$

where C is a dimensionless correction factor, it generally varies between about 0.98 to 1.0. The value of the pressure drop $p_s - p_0$ in Pa is related to R, the reading on the manometer as follows:

$$p_s - p_0 = R(\rho_A - \rho)g \qquad (1.6\text{-}3)$$

where ρ_A is the density of the fluid in the manometer in kg/m^3, ρ is the density of flowing fluid in kg/m^3, and R is the manometer reading in m.
Substituting Eq. (1.6-3) to Eq. (1.6-2)

$$u_0 = C\sqrt{\frac{2R(\rho_A - \rho)g}{\rho}} \qquad (1.6\text{-}4)$$

The disadvantages of the pitot tube are (1) that most designs do not give the average velocity directly and (2) that its readings for gases are extremely small. When is used for measuring low-pressure gases, some form of multiplying gauge, like at shown in Fig. 1.3, must also be used.

EXAMPLE 1.9 Air at 93.3 °C is forced through a long, circular flue 914 mm in diameter. A pitot tube reading is taken at the center of the flue at a sufficient distance from flow disturbances to ensure normal velocity distribution. The pitot reading is 13.7 mm H$_2$O, and the static pressure at the point of measurement is 387 mm H$_2$O. The coefficient of the pitot tube is 0.98.
  Calculate the flow of air, in cubic meter per minute, measured at 15.6 °C and a barometric pressure of 760 mm Hg.
Solution. The velocity at the center of the flue, which is that measured by the pitot tube, is calculated by Eq. (1.6-2) using SI units and a coefficient of 0.98. Equation (1.6-2) becomes

$$u_0 = 0.98\sqrt{\frac{2(p_s - p_0)}{\rho}} \qquad (1.6\text{-}5)$$

The necessary quantities are as follows. The absolute pressure at the instrument is

$$p = 101330 + 387 \times 9.81 = 105126 \text{Pa}$$

The density of the air in the instrument is

$$\rho = \frac{pM}{RT} = \frac{105126 \times 29 \times 10^{-3}}{8.314(273.15 + 93.3)} = 1.00 \text{kg/m}^3$$

From the manometer reading

$$p_s - p_0 = 13.7 \times 9.81 = 134.4 \text{Pa}$$

By Eq. (1.6-5), the maximum velocity is

$$u_{max} = 0.98\sqrt{\frac{2 \times 134.4}{1.00}} = 16.07 \text{ m/s}$$

To obtain the average velocity from the maximum velocity, Fig. 1.20 is used. The Reynolds number, based on the maximum velocity, is calculated as follows. From App. 7, the viscosity of air at 93.3°C is 0.022cP, and

$$Re_{max} = \frac{1.00 \times 16.07 \times 0.914}{0.022 \times 10^{-3}} = 6.67 \times 10^5$$

The ratio V/u_{max} from Fig. 1.20, is a little greater than 0.86. Using 0.86 as an estimated value gives

$$V = 0.86 \times 16.07 = 13.82 \text{ m/s}$$

The Reynolds number Re is $6.67 \times 10^5 \times 0.86 = 5.74 \times 10^5$, and V/u_{max} is exactly 0.86 as estimated. The volumetric flow rate is

$$q = \frac{\pi}{4} D^2 V \left(\frac{273.15 + 15.6}{273.15 + 93.3} \right) \left(\frac{105126}{101330} \right)$$

$$= \frac{\pi}{4} \times 0.914^2 \times 13.82 \left(\frac{273.15 + 15.6}{273.15 + 93.3} \right) \left(\frac{105126}{101330} \right) = 7.41 \text{m}^3/\text{s} = 444.5 \text{m}^3/\text{min}$$

1.6.2 Full-Bore Meters

The most common types of full-bore meters are venturi and orifice meters and variable-area meters such as rotameters. Other full-bore measuring devices include V-element, magnetic, vortex-shedding, turbine, and positive-displacement meters; ultrasonic meters; and mass flow devices such as Coriolis flowmeters.

Venturi meter

A venturi meter is shown in Fig. 1.32. A short conical inlet section leads to a throat section, then to a long discharge cone. Pressure taps at the start of the inlet section and at the throat are connected to a manometer or differential pressure transmitter.

In the upstream cone the fluid velocity is increased and its pressure decreased. The pressure drop in this cone is used to measure the flow rate. In the discharge cone the velocity is decreased and the original pressure largely recovered. The angle of the discharge cone is made small, between 5° and 15°, to prevent boundary layer separation and to minimize friction. Since there is no separation in a contracting cross section, the upstream cone can be made shorter than the downstream cone. Typically 90 percent of the pressure loss in the upstream cone is recovered.

Although venturi meters can be applied to the measurement of gas flow rates, they are most commonly used with liquids, especially large flows of water where, because of the large pressure recovery, a venturi requires less power than other types of meters.

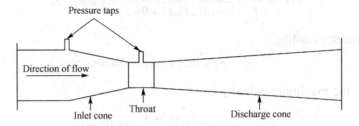

Fig. 1.32 Venturi meter.

The basic equation for a venturi meter is obtained by writing the Bernoulli equation for incompressible fluids across the upstream cone. If V_a and V_b, are the average upstream and downstream velocities, respectively, and ρ is the density of the fluid, Eq. (1.3-28) becomes

$$V_b^2 - V_a^2 = \frac{2(p_a - p_b)}{\rho} \qquad (1.6\text{-}6)$$

The continuity relation (1.3-7) can be written, since the density is constant, as

$$V_a = \left(\frac{D_b}{D_a}\right)^2 V_b = \beta^2 V_b \qquad (1.6\text{-}7)$$

where D_a = diameter of pipe
 D_b, = diameter of throat of meter
 β = diameter ratio D_b/D_a

If V_a is eliminated from Eqs. (1.6-6) and (1.6-7), the result is

$$V_b = \frac{1}{\sqrt{1-\beta^4}} \sqrt{\frac{2(p_a - p_b)}{\rho}} \qquad (1.6\text{-}8)$$

Venturi coefficient. Equation (1.6-8) applies strictly to the frictionless flow of incompressible fluids. To account for the small friction loss between locations a and b, Eq. (1.6-8) is corrected by introducing an empirical factor C_v, and writing

$$V_b = \frac{C_v}{\sqrt{1-\beta^4}} \sqrt{\frac{2(p_a - p_b)}{\rho}} \qquad (1.6\text{-}9)$$

The small effects of the kinetic energy factors α_a, and α_b, are also taken into account in the definition of C_v. The coefficient C_v is determined experimentally. It is called the *venturi coefficient, velocity of approach not included.* The effect of the approach velocity V_a, is accounted for by the term $1/\sqrt{1-\beta^4}$. When D_b is less than $\frac{1}{4}D_a$, the approach velocity and the term β can be neglected, since the resulting error is less than 0.2 percent.

For a well-designed venturi, the constant C_v, is about 0.98 for pipe diameters of 2 to 8 in. and about 0.99 for larger sizes.

Volumetric and mass flow rates. The velocity through the venturi throat V_b, is not the quantity usually desired. The flow rates of practical interest are the volumetric and mass flow rates through the meter. The volumetric flow rate is calculated by substituting V_b, from Eq. (1.6-9) into Eq. (1.3-4) to get

$$q = V_b S_b = \frac{C_v S_b}{\sqrt{1-\beta^4}} \sqrt{\frac{2(p_a - p_b)}{\rho}} \tag{1.6-10}$$

where q = volumetric flow rate

S_b = area of throat

The mass flow rate is obtained by multiplying the volumetric flow rate by the density, or

$$\dot{m} = q\rho = \frac{C_v S_b}{\sqrt{1-\beta^4}} \sqrt{2(p_a - p_b)\rho} \tag{1.6-11}$$

where $\dot{m}$ is the mass flow rate.

Orifice meter

The venturi meter has certain practical disadvantages for ordinary plant practice. It is expensive, it occupies considerable space, and its ratio of throat diameter to pipe diameter cannot be changed. For a given meter and definite manometer system, the maximum measurable flow rate is fixed, so if the flow range is changed, the throat diameter may be too large to give an accurate reading or too small to accommodate the next maximum flow rate. The orifice meter meets these objections to the venturi but at the price of a larger power consumption.

A standard sharp-edged orifice is shown in Fig. 1.33. It consists of an accurately machined and drilled plate mounted between two flanges with the hole concentric with the pipe in which it is mounted. (Off-center or segmental openings are also used on occasion.) The opening in the plate may be beveled on the downstream side. Pressure taps, one above and one below the orifice plate, are installed and are connected to a manometer or differential pressure transmitter. The positions of the taps are arbitrary, and the coefficient of the meter will depend upon the position of the taps. Three of the recognized methods of placing the taps are shown in Table 1.7. Flange taps are the most common. The taps shown in Fig. 1.33 are vena contracta taps.

The principle of the orifice meter is identical with that of the venturi. The reduction of the cross section of the flowing stream in passing through the orifice increases the velocity head at the expense of the pressure head, and the reduction in pressure between the taps is measured by the manometer. Bernoulli's equation provides a basis for correlating the increase in velocity head with the decrease in pressure head.

Table 1.7 Data on orifice taps

Type of tap	Distance of upstream tape from upstream face of orifice	Distance of downstream tap from downstream face
Flange	1 in. (25mm)	1 in. (25mm)
Vena contracta	1 pipe diameter (actual inside)	0.3-0.8 pipe diameter, depending on β
Pipe	$2\frac{1}{2}$ times nominal pipe diameter	8 times nominal pipe diameter

One important complication appears in the orifice meter that is not found in the venturi. Because of the sharpness of the orifice, the fluid stream separates from the downstream side of the orifice plate and forms a free-flowing jet in the downstream fluid. A vena contracta forms, as shown in Fig. 1.33. The jet is not under the control of solid walls, as is the case in the venturi, and the area of the jet varies from that of the opening in the orifice to that of the vena contracta. The area at any given point, for example at the downstream tap, is not easily determinable, and the velocity of the jet at the downstream tap is not easily related to the diameter of the orifice. Orifice coefficients are smaller and more variable than those for the venturi, and the quantitative treatment of the orifice meter is modified accordingly.

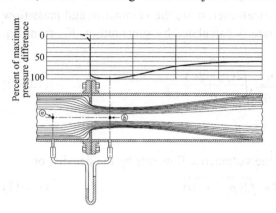

Fig. 1.33 Orifice meter.

Extensive and detailed design standards for orifice meters are available in the literature. They must be followed if the performance of a meter is to be predicted accurately without calibration. For approximate or preliminary design, however, it is satisfactory to use an equation similar to Eq. (1.6-9) as follows:

$$V_o = \frac{C_o}{\sqrt{1-\beta^4}} \sqrt{\frac{2(p_a - p_b)}{\rho}} \qquad (1.6\text{-}12)$$

where V_o = velocity through orifice
β = ratio of orifice diameter to pipe diameter
p_a, p_b = pressures at stations a and b in Fig. 1.33

In Eq. (1.6-12), C_o, is the *orifice coefficient*, velocity of approach not included. It corrects for the contraction of the fluid jet between the orifice and the vena contracta, for friction, and for α_a, and α_b, Coefficient C_o is always determined experimentally. It varies considerably with changes in β and with Reynolds number at the orifice Re_o. This Reynolds number is defined by

$$Re_o = \frac{D_o V_o \rho}{\mu} = \frac{4\dot{m}}{\pi D_o \mu} \qquad (1.6\text{-}13)$$

where C_o is the orifice diameter.

Equation (1.6-12) is useful for design because C_o, is almost constant and independent of β provided Re_0 is greater than about 30000. Under these conditions C_o, maybe taken as 0.61 for both flange taps and vena contracta taps. For process applications, β should be between 0.20 and 0.75. If β is less than 0.25, the term $\sqrt{1-\beta^4}$ differs negligibly from unity. Equations (1.6-10) and (1.6-11) for venturi meters may be used for orifice meters by substituting C_o for C_v, V_o for V_b, and S_o, the cross-sectional area of the orifice, for S_b.

It is especially important that enough straight pipe be provided both upstream and downstream of the orifice to ensure a flow pattern that is normal and undisturbed by fittings, valves, or other equipment. Otherwise the velocity distribution will not be normal, and the orifice coefficient will be affected in an unpredictable manner. Data are available for the minimum length of straight pipe that should be provided upstream and downstream of the orifice to ensure normal velocity distribution. Straightening vanes in the approach line may be used if the required length of pipe is not available upstream of the orifice.

Pressure recovery. Because of the large friction losses from the eddies generated by the reexpanding jet below the vena contracta, the pressure recovery in an orifice meter is poor. The resulting power loss is one disadvantage of the orifice meter. The fraction of the orifice differential that is permanently lost depends on the value of β, and the relationship between the fractional loss and β is shown in Fig. 1.34. For a value of β of 0.5, the lost head is about 73 percent of the orifice differential.

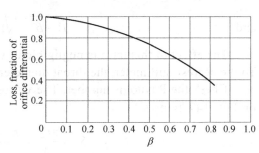

Fig. 1.34 Overall pressure loss in orifice meters.

The pressure difference measured by pipe taps, where the downstream tap is eight pipe diameters below the orifice, is really a measurement of permanent loss rather than of the orifice differential.

EXAMPLE 1.10. An orifice meter with flange taps is to be installed in a 100-mm line to measure the flow of water. The maximum flow rate is expected to be 50 m³/h at 15 ℃. The manometer used to measure the differential pressure is to be filled with mercury, and water is to fill the leads above the surfaces of the mercury. The water temperature will be 15 ℃ throughout, (*a*) If the maximum manometer reading is to be 1.25 m, what diameter, to the nearest millimeter, should be specified for the orifice? (*b*) What will be the power to operate the meter at full load?

Solution

(*a*) Equation (1.6-10) is used to calculate the orifice diameter. The quantities to be substituted are

$$q = \frac{50}{3600} = 0.0139 \text{ m}^3/\text{s}$$

$$\rho = 999 \text{ kg/m}^3 \quad (\text{App.5})$$

$$C_o = 0.61 \quad g = 9.80665 \text{ m/s}^2$$

From Eq. (1.1-9),

$$p_a - p_b = 9.80665 \times 1.25 \times (13.6 - 1.0) \times 999 = 154300 \text{ N/m}^2$$

Substituting these values in Eq. (1.6-10) gives

$$0.0139 = \frac{0.61 S_o}{\sqrt{1-\beta^4}} \sqrt{\frac{2 \times 154,300}{999}}$$

from which

$$\frac{S_o}{\sqrt{1-\beta^4}} = 1.296 \times 10^{-3} = \frac{\pi D_o^2}{4\sqrt{1-\beta^4}}$$

As a first approximation, call $\sqrt{1-\beta^4} = 1.0$. Then

$$D_o = 40.6 \text{ mm} \qquad \beta = \frac{40.6}{100} = 0.406$$

and
$$\sqrt{1-\beta^4} = \sqrt{1-0.406^4} = 0.986$$

The effect of this term is negligible in view of the desired precision of the final result. To the nearest millimeter, the throat diameter should be 41 mm.

Check the Reynolds number. The viscosity of water at 15°C, from App. 5, is 1.147 cP or 0.001147 kg/m·s

$$S_o = \frac{\pi D_o^2}{4} = \frac{\pi \times 0.041^2}{4} = 0.00132 \text{ m}^2$$

$$V_o = \frac{q}{S_o} = \frac{0.0139}{0.00132} = 10.53 \text{ m/s}$$

The Reynolds number, from Eq. (1.6-13), is

$$Re_o = \frac{0.041 \times 10.53 \times 999}{0.001147} = 376000$$

The Reynolds number is large enough to justify the value of 0.61 for C_o.

(b) From Fig. 1.34, for $\beta = 0.406$, the permanent loss in pressure is 81 percent of the differential. Since the maximum volumetric flow rate is 0.0139 m³/s, the power required to operate the orifice meter at full flow is

$$P = 0.81 q(p_a - p_b) = \frac{0.81 \times 0.0139 \times 154300}{1000} = 1.737 \text{ kW}$$

Rotameters — area meters

In the orifice, nozzle, or venturi, the variation of flow rate through a constant area generates a variable pressure drop, which is related to the flow rate. Another class of meters, called area meters, consists of devices in which the pressure drop is constant, or nearly so, and the area through which the fluid flows varies with the flow rate. The area is related, through proper calibration, to the flow rate.

The most important area meter is the rotameter, which is shown in Fig. 1.35. It consists essentially of a gradually tapered glass tube mounted vertically in a frame with the large end up. The fluid flows upward through the tapered tube and suspends freely a float (which actually does not float but is completely submerged in the fluid). The float is the indicating element, and the greater the flow rate, the higher the float rides in the tube. The entire fluid stream must flow through the annular space between the float and the tube wall. The tube is marked in divisions, and the reading of the meter is obtained from the scale reading at the reading edge of the float, which is taken at the largest cross section of the float. A calibration curve must be available to convert the observed scale reading to flow rate. Rotameters can be used for either liquid or gas flow measurement.

The bore of a glass rotameter tube is either an accurately formed, plain conical taper or a taper with three beads, or flutes, parallel with the axis of the tube. The tube shown in Fig. 1.35 is a tapered tube. In the first rotameters, angled notches in the top of the float made it rotate, but the float does not rotate in most current designs. For opaque liquids, for high temperatures or pressures, or for other conditions where glass is impracticable, metal tubes are used. Metal tubes are plain tapered. Since in a metal tube the float is invisible, means must be provided for either indicating or transmitting the meter reading. This is accomplished by attaching a rod, called an extension, to the top or bottom of the float and using the extension as an armature. The extension is enclosed in a fluid-tight tube mounted on one of the fittings. Since the inside of this tube communicates directly with the interior of the rotameter, no box for the extension is needed. The tube is surrounded by external induction coils. The length of the extension exposed to the coils varies with the position of the float. This in turn changes the inductance of the coil, and the variation of the inductance is measured electrically to operate a control valve or to give a reading on a recorder. Also, a magnetic follower, mounted outside the extension tube and adjacent to a vertical scale, can be used as a visual indicator for the top edge of the extension. By such modifications the rotameter has developed from a simple visual indicating instrument using only glass tubes into a versatile recording and controlling device.

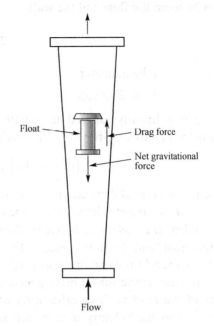

Fig. 1.35 Principle of a rotameter.

Floats may be constructed of metals of various densities from lead to aluminum or from glass or plastic. Stainless-steel floats are common. Float shapes and proportions are also varied for different applications.

Rotameters have a nearly linear relationship between flow and position of the float, compared with a calibration curve for an orifice meter, for which the flow rate is proportional to the square root of the reading. The calibration of a rotameter, unlike that of an orifice meter, is not sensitive to the velocity distribution in the approaching stream, and neither long, straight approaches nor straightening vanes are necessary.

Theory and calibration of rotameters. For a given float of density ρ_f, volume v_f, and a fluid density ρ; For a given flow rate, the equilibrium position of the float in the rotameter is established by a balance of three forces: (1) the gravity of the float: $v_f \rho_f g$; (2) the buoyant force of the fluid on the float: $v_f \rho g$; (3) the drag force on the float: $F_D = A_f C_D \rho \dfrac{V^2}{2}$

where A_f = the projected area of the float.

C_D = drag coefficient when fluid flows through the annular area.

For equilibrium

$$F_D = A_f C_D \rho \frac{V^2}{2} = v_f \rho_f g - v_f \rho g = v_f (\rho_f - \rho) g \qquad (1.6\text{-}14)$$

For a given meter operating on a certain fluid, the right-hand side of equation above is constant and independent of the flow rate. Therefore, when the flow rate increases, the position of the float must change to keep the drag force constant.

If the change in drag coefficient is small, which is usually the case for large rotameters with low or moderate viscosity fluids, and the total flow rate is proportional to the annular area between the float and the wall

$$q = V \frac{\pi}{4}(d_t^2 - d_f^2) \qquad (1.6\text{-}15)$$

where d_t = tube diameter
d_f = float diameter

For a linearly tapered tube with a diameter at the bottom about equal to the float diameter, the area for flow is a quadratic function of the height of the float.

$$(d_t^2 - d_f^2) = (d_f + ah)^2 - d_f^2 = 2d_f ah + a^2 h^2 \qquad (1.6\text{-}16)$$

where h = vertical distance from the entrance
a = constant related to the tube taper

When the clearance between float and tube wall is small, the term $a^2 h^2$ is relatively unimportant and the flow almost a linear function of the height h. Therefore rotameters tend to have a nearly linear relationship between flow rate and position of the float, compared with a calibration curve for an orifice meter, for which the flow rate is proportional to the square root of the reading. The calibration of a rotameter, unlike that of an orifice meter, is not sensitive to the velocity distribution in the approaching stream, and neither long, straight approaches nor straightening vanes are necessary.

For a given float of density ρ_f and a fluid density ρ, the mass flow rate m is given by

$$m = q\rho = \rho K \sqrt{\frac{\rho_f - \rho}{\rho}} \qquad (1.6\text{-}17)$$

Volumetric flow rate q is the reading on the rotameter, and K is a constant which is determined experimentally. If a fluid ρ_B is used instead of ρ, at the same height or reading of q on the rotameter and assuming K does not vary appreciably, the following approximation can be used:

$$\frac{m}{m_B} = \frac{q\rho}{q_B \rho_B} = \sqrt{\frac{(\rho_f - \rho)\rho}{(\rho_f - \rho_B)\rho_B}} \qquad (1.6\text{-}18)$$

For gases where $\rho_f \gg \rho$ and ρ_B

$$q_B = q\sqrt{\frac{\rho}{\rho_B}} \qquad (1.6\text{-}19)$$

PROBLEMS

1.1 What will be the (a) the gauge pressure and (b) the absolute pressure of water at depth 12m below the surface? ρ_{water} = 1000 kg/m^3, and $P_{atmosphere}$ = 101kN/m^2.

1.2 An inclined manometer is required to measure an air pressure of 3mm of water to an accuracy of +/–3%. The inclined arm is 8mm in diameter and the larger arm has a diameter of 24mm. The manometric fluid has density 740 kg/m^3 and the scale may be read to +/– 0.5mm.
What is the angle required to ensure the desired accuracy may be achieved?

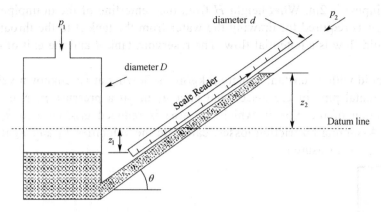

Figure for problem 1.2

1.3 A differential manometer as shown in Fig. is sometimes used to measure small pressure difference. When the reading is zero, the levels in two reservoirs are equal. Assume that fluid B is methane, that liquid C in the reservoirs is kerosene (specific gravity = 0.815), and that liquid A in the U tube is water. The inside diameters of the reservoirs and U tube are 51mm and 6.5mm, respectively. If the reading of the manometer is 145mm, what is the pressure difference over the instrument.

In meters of water, (a) when the change in the level in the reservoirs is neglected, (b) when the change in the levels in the reservoirs is taken into account? What is the percent error in the answer to the part (a)?

1.4 There are two U tube manometers fixed on the fluid bed reactor, as shown in the figure. The readings of two U tube manometers are R_1=400mm, R_2=50mm, respectively. The indicating liquid is mercury. The top of the manometer is filled with the water to prevent from the mercury vapor diffusing into the air, and the height R_3=50mm. Try to calculate the pressure at point A and B.

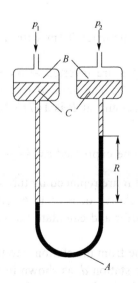
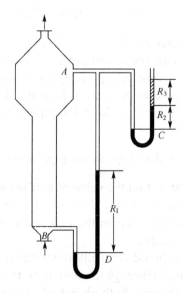

Figure for problem 1.3   Figure for problem 1.4

1.5 Water discharges from the reservoir through the drainpipe, which the throat diameter is d. The ratio of D to d equals 1.25. The vertical distance h between the tank A and axis of the drainpipe is 2m. What height H from the centerline of the drainpipe to the water level in reservoir is required for drawing the water from the tank A to the throat of the pipe? Assume that fluid flow is a potential flow. The reservoir, tank A and the exit of drainpipe are all open to air.

1.6 A liquid with a constant density ρ kg/m^3 is flowing at an unknown velocity V_1 m/s through a horizontal pipe of cross-sectional area A_1 m^2 at a pressure p_1 N/m^2, and then it passes to a section of the pipe in which the area is reduced gradually to A_2 m^2 and the pressure is p_2. Assuming no friction losses, calculate the velocities V_1 and V_2 if the pressure difference $(p_1 - p_2)$ is measured.

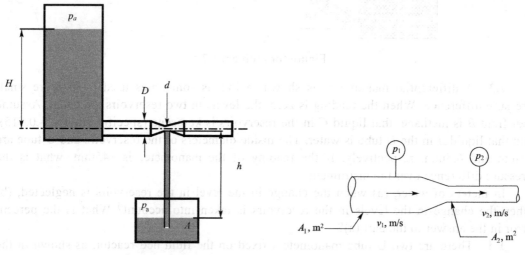

Figure for problem 1.5   Figure for problem 1.6

1.7 A liquid whose coefficient of viscosity is μ flows below the critical velocity for laminar flow in a circular pipe of diameter d and with mean velocity V. Show that the pressure loss in a length of pipe is $\dfrac{32\mu V}{d^2}$.

Oil of viscosity 0.05 kg/ms flows through a pipe of diameter 0.1m with a velocity of 0.6m/s. Calculate the loss of pressure in a length of 120m.

1.8 In a vertical pipe carrying water, pressure gauges are inserted at points A and B where the pipe diameters are 0.15m and 0.075m respectively. The point B is 2.5m below A and when the flow rate down the pipe is 0.02 m^3/s, the pressure at B is 14715 N/m^2 greater than that at A.

Assuming the losses in the pipe between A and B can be expressed as $k\dfrac{V^2}{2}$ where u is the velocity at A, find the value of k. If the gauges at A and B are replaced by tubes filled with water and connected to a U-tube containing mercury of relative density 13.6, give a sketch showing how the levels in the two limbs of the U-tube differ and calculate the value of this difference in metres.

1.9 The liquid vertically flows down through the tube from the station a to the station b, then horizontally through the tube from the station c to the station d, as shown in figure. Two segments of the tube, both ab and cd, have the same length, the diameter and roughness.

Find: (a) the expressions of $\dfrac{\Delta p_{ab}}{\rho g}$, h_{fab}, $\dfrac{\Delta p_{cd}}{\rho g}$ and h_{fcd}, respectively. (b) the relationship between readings R_1 and R_2 in the U tube.

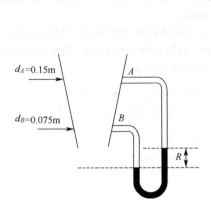
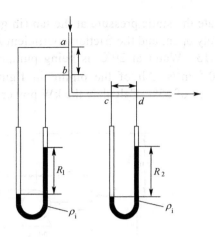

Figure for problem 1.8   Figure for problem 1.9

1.10 Water passes through a pipe of diameter d_i=0.004 m with the average velocity 0.4 m/s, as shown in Figure.

(a) What is the pressure drop $-\Delta p$ when water flows through the pipe length L=2 m, in m H_2O column?

(b) Find the maximum velocity and point r at which it occurs.

(c) Find the point r at which the average velocity equals the local velocity.

(d) if kerosene flows through this pipe, how do the variables above change?

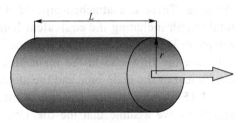

Figure for problem 1.10

(the viscosity and density of Water are 0.001 Pa·s and 1000 kg/m³, respectively; and the viscosity and density of kerosene are 0.003 Pa·s and 800 kg/m³, respectively)

1.11 1250cm³/s of water is to be pumped through a steel pipe, 25mm diameter and 30m long, to tank 12m higher than its reservoir. Calculate approximately the power required. What power motor (in kW) would you provide? (The roughness of a steel pipe will be taken as 0.045mm; the efficiency of the pump is 60%.)

1.12 As shown in the figure, the water level in the reservoir keeps constant. A steel drainpipe (with the inside diameter of 100mm) is connected to the bottom of the reservoir. One arm of the U tube manometer is connected to the drainpipe at the position 15m away from the bottom of the reservoir, and the other is opened to the air, the U tube is filled with mercury and the left-side arm of the U tube above the mercury is filled with water. The distance between the upstream tap and the outlet of the pipeline is 20m.

(a) When the gate valve is closed, R=600mm, h=1500mm; when the gate valve is opened partly, R=400mm, h=1400mm. The friction coefficient λ is 0.025, and the loss coefficient of the entrance is 0.5. Calculate the flow rate of water when the gate valve is opened partly. (in m³/h)

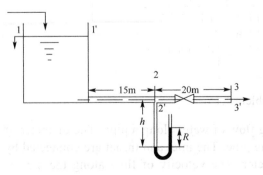

Figure for problem 1.12

(b) When the gate valve is widely open,

calculate the static pressure at the tap (in gauge pressure, N/m²). $l_e/d \approx 15$ when the gate valve is widely open, and the friction coefficient λ is still 0.025.

1.13 Water at 20℃ is being pumped from a tank to an elevated tank at the rate of 5.0×10^{-3} m³/s. All of the piping in figure is 4-in. schedule 40 pipe. The pump has an efficiency of 65%. Calculate the kW power needed for the pump.

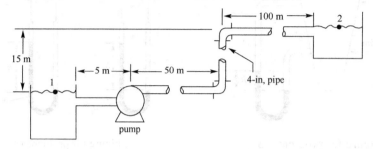

Figure for problem 1.13

1.14 Water at 20℃ passes through a steel pipe with an inside diameter of 300mm and 2m long. There is a attached-pipe ($\Phi 60 \times 3.5$mm) which is parallel with the main pipe. The total length including the equivalent length of all form losses of the attached-pipe is 10m. A rotameter is installed in the branch pipe. When the reading of the rotameter is 2.72m³/h, try to calculate the flow rate in the main pipe and the total flow rate, respectively. The frictional coefficient of the main pipe and the attached-pipe is 0.018 and 0.03, respectively.

1.15 Three reservoirs are connected by three pipes as are shown in figure. For simplicity we assume that the diameter of each pipe is 0.305m, the friction coefficient for each is 0.02, and because of the length-to-diameter ratio, minor losses are negligible. Determine the flowrate into or out of each reservoir.

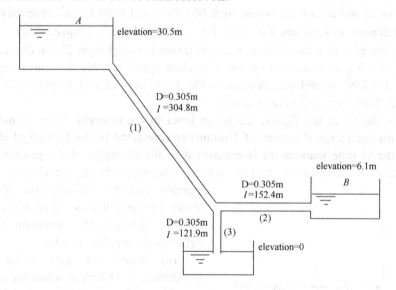

Figure for problem 1.15

1.16 A Venturimeter is used for measuring flow of water along a pipe. The diameter of the Venturi throat is two fifths the diameter of the pipe. The inlet and throat are connected by water filled tubes to a mercury U tube manometer. The velocity of flow along the pipe is found to be $2.5\sqrt{R}$ m/s, where R is the manometer reading in metres of mercury. Determine the loss of head between inlet and throat of the Venturi when *R* is 0.49m. (Relative density of mercury is 13.6.)

1.17 Sulphuric acid of specific gravity 1.3 is flowing through a pipe of 50 mm internal diameter. A thin-lipped orifice, 10mm, is fitted in the pipe and the differential pressure shown by a mercury manometer is 10cm. Assuming that the leads to the manometer are filled with the acid, calculate (a)the weight of acid flowing per second, and (b) the approximate pressure drop caused by the orifice.

The coefficient of the orifice may be taken as 0.61, the specific gravity of mercury as 13.6, and the density of water as 1000 kg/m^3

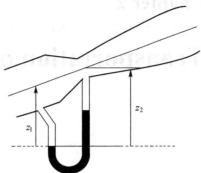

Figure for problem 1.16

1.18 Water flows through an orifice of 25mm diameter situated in a 75mm pipe at the rate of 300cm^3/s. what will be the difference in level on a water manometer connected across the meter? Take the viscosity of water as 1mN s/m^2.

名 词 术 语

英文 / 中文

- absolute pressure / 绝压
- absolute viscosity / 黏度，动力黏度
- bingham plastics / 宾汉塑性流体
- boundary layer / 边界层
- boundary layer separation / 边界层分离
- buffer layer / 过渡层
- compressible fluid / 可压缩流体
- deviating velocity / 脉动速度
- dilatant fluid / 胀塑性流体
- eddy viscosity / 涡流黏度
- equivalent diameter / 当量直径
- fluid dynamics / 流体动力学
- fluid mechanics / 流体力学
- fluid statics / 流体静力学
- form friction / 局部阻力(形体阻力)
- free turbulence / 自由湍流
- fully developed flow / 充分发展了的流动
- gage pressure / 表压
- hydraulic radius / 水力半径
- hydraulically smooth / 水力光滑管
- inclined manometer / 斜管压差计
- incompressible fluid / 不可压缩流体
- kinematic viscosity / 运动黏度
- laminar flow / 层流
- newtonian fluid / 牛顿型流体

英文 / 中文

- non-newtonian fluid / 非牛顿型流体
- orifice meter / 孔板流量计
- pitot tube / 皮托管，测速管
- potential flow / 势流
- pseudoplastic fluid / 假塑性流体
- reference plane / 基准面
- rheology / 流变学
- rheopectic substance / 流凝性流体
- rotameter / 转子流量计
- roughness parameter / 粗糙度
- skin friction / 直管阻力
- thixotropic liquid / 触变性流体
- transition length / 进口段长度
- turbulent flow / 湍流
- turbulent zone / 湍流中心
- two-fluid U tube manometer / 双液 U 形管压差计，微差压差计
- U tube manometer / U 形管压差计
- vacuum pressure / 真空度
- velocity head / 速度头
- venturi meter / 文丘里流量计
- viscoelastic fluid / 黏弹性流体
- viscous sublayer / 层流内层
- wall turbulence / 壁湍流

Chapter 2

Transportation of Fluids

This chapter deals with the transportation of fluids, both liquids and gases. Solids are sometimes handled by similar methods by suspending them in a liquid to form a pumpable slurry or by conveying them in a high-velocity gas stream. It is cheaper to move fluids than solids, and materials are transported as fluids whenever possible. In operations at close to atmospheric pressure, gases are often carried in square or rectangular ducts, but under elevated pressures they are carried in pipe or tubing that is circular in cross section.

2.1 Pipe, Fittings, and Valves

2.1.1 Pipe and Tubing

Fluids are usually transported in pipe or tubing, which is circular in cross section and available in widely varying sizes, wall thicknesses, and materials of construction. There is no clear-cut distinction between the terms *pipe* and *tubing*. Generally speaking, pipe is heavy-walled and relatively large in diameter and comes in moderate lengths of 6 or 12 m; tubing is thin-walled and often comes in coils several hundred feet long. Metallic pipe can be threaded; tubing usually cannot. Pipe walls are usually slightly rough; tubing has very smooth walls. Lengths of pipe are joined by screwed, flanged, or welded fittings; pieces of tubing are connected by compression fittings, flare fittings, or soldered fittings. Finally, tubing is usually extruded or cold-drawn, while metallic pipe is made by welding, casting, or piercing a billet in a piercing mill.

Pipe and tubing are made from many materials, including metals and alloys, wood, ceramics, glass, and various plastics. Polyvinyl chloride, or PVC, pipe is extensively used for wastewater lines. In process plants the most common material is low-carbon steel, fabricated into what is sometimes called black-iron pipe. Wrought-iron and cast-iron pipes are also used for a number of special purposes.

Sizes

Pipe and tubing are specified in terms of their diameter and their wall thickness. With steel pipe the standard nominal diameters, in U.S. practice, range from 1/8 to 30 in. For large pipe, more than 12 in. in diameter, the nominal diameters are the actual outside diameters; for small pipe the nominal diameter does not correspond to any actual dimension. The nominal value is close to the actual inside diameter for 3- to 12-in. pipe, but for very small pipe this is not true. Regardless of wall thickness, the outside diameter of all pipe of a given nominal size is the same, to ensure interchangeability of fittings. Standard dimensions of steel pipe are given in App. 3. Pipe of other materials is also made with the same outside diameters as steel pipe to permit interchanging parts of a piping system. These standard sizes for steel pipe, therefore, are known as IPS (iron pipe size) or NPS (normal pipe size). Thus the designation "2-in. nickel IPS pipe" means nickel pipe having the same outside diameter as standard 2-in. steel pipe.

The wall thickness of pipe is indicated by the *schedule number,* which increases with the thickness. Ten schedule numbers—10, 20, 30, 40, 60, 80, 100, 120, 140, and 160—are in use, but with pipe less than 8 in. in diameter only numbers 40, 80, 120, and 160 are common. In standards for the United Kingdom the nominal size (called the DN number) is expressed in millimeters (for example, 100, 150, up to 600) and the minimum bore is specified for each nominal size.

The size of tubing is indicated by the outside diameter. The normal value is the actual outer diameter, to within very close tolerances. Wall thickness is ordinarily given by the BWG (Birmingham wire gauge) number, which ranges from 24 (very light) to 7 (very heavy). Sizes and wall thicknesses of heat-exchanger tubing are given in App. 4.

Selection of pipe sizes

The pipe size selected for a particular installation depends mainly on the cost of the pipe and fittings and the cost of energy needed for pumping the fluid. The cost of the pipe and the annual capital charges increase with about the 1.5 power of the pipe diameter, while the power cost for turbulent flow varies with the −4.8 power of the diameter. Equations have been presented giving the optimum pipe diameter as a function of flow rate and fluid density, but these can be converted to the optimum velocity, which is nearly independent of the flow rate. For turbulent flow of liquids in steel pipes larger than 1 in. (25 mm) in diameter, the optimum velocity is

$$u_{opt} = \frac{10.745 m^{0.1}}{\rho^{0.36}} \qquad (2.1\text{-}1)$$

where u_{opt} = optimum velocity, m/s
m = mass flow rate, kg/s
ρ = fluid density, kg/m^3

For water and similar fluids u_{opt} is 0.9 to 1.8 m/s; for air or steam at low to moderate pressures u_{opt} is 6 to 24 m/s. For flow in heat exchanger tubes, the optimum design velocity is often higher than that given by Eq. (2.1-1) because of improved heat transfer at high fluid velocities.

When flow is by gravity from overhead tanks or when a viscous liquid is being pumped, low velocities are favored, in the range of 0.06 to 0.24 m/s. The dimensions of standard steel pipe are shown in App. 3.

For large complex piping systems the cost of piping may be a substantial fraction of the total investment, and elaborate computer methods of optimizing pipe sizes are justified.

Joints and fittings

The methods used to join pieces of pipe or tubing depend in part on the properties of the material but primarily on the thickness of the wall. Thick-walled tubular products are usually connected by screwed fittings, by flanges, or by welding. Pieces of thin-walled tubing are joined by soldering or by compression or flare fittings. Pipe made of brittle materials such as glass or carbon or cast iron is joined by flanges or bell-and-spigot joints.

When screwed fittings are used, the ends of the pipe are threaded externally with a threading tool. The thread is tapered, and the few threads farthest from the end of the pipe are imperfect, so that a tight joint is formed when the pipe is screwed into a fitting. Tape of polytetrafluoroethylene is wrapped around the threaded end to ensure a good seal. Threading weakens the pipe wall, and the fittings are generally weaker than the pipe itself; when screwed fittings are used, therefore, a higher schedule number is needed than with other types of joints. Screwed fittings are standardized for pipe sizes up to 12 in. (300 mm), but because of the difficulty of threading and handling large pipe, they are rarely used in the field with pipe larger than 3 in. (75 mm).

Lengths of pipe larger than about 2 in. (50 mm) are usually connected by flanges or by welding. Flanges are matching disks or rings of metal bolted together and compressing a gasket between their faces. The flanges themselves are attached to the pipe by screwing them on or by welding or brazing. A flange with no opening, used to close a pipe, is called a *blind flange* or a *blank flange*. For joining pieces of large steel pipe in process piping, especially for high-pressure service, welding has become the standard method. Welding makes stronger joints than screwed fittings do, and since it does not weaken the pipe wall, lighter pipe can be used for a given pressure. Properly made welded joints are leakproof, whereas other types of joints are not. Environmental protection legislation considers flanged and screwed joints to be sources of emission of volatile materials. Almost the only disadvantage of a welded joint is that it cannot be opened without destroying it.

Allowances for expansion

Almost all pipes are subjected to varying temperatures, and in some high temperature lines the temperature change is very large. Such changes cause the pipe to expand and contract. If the pipe is rigidly fixed to its supports, it may tear loose, bend, or even break. In large lines, therefore, fixed supports are not used; instead the pipe rests loosely on rollers or is hung from above by chains or rods. Provision is also made in all high-temperature lines for taking up expansion, so that the fittings and valves are not put under strain. This is done by bends or loops in the pipe, by packed expansion joints, by bellows or packless joints, and sometimes by flexible metal hose.

Prevention of leakage around moving parts

In many kinds of processing equipment it is necessary to have one part move in relation to another part without excessive leakage of a fluid around the moving member. This is true in packed expansion joints and in valves where the stem must enter the valve body and be free to turn without allowing the fluid in the valve to escape. It is also necessary where the shaft of a pump or compressor enters the casing, where an agitator shaft passes through the wall of a pressure vessel, and in other similar places.

Common devices for minimizing leakage while permitting relative motion are stuffing boxes and mechanical seals. Neither completely stops leakage, but if no leakage whatever of the process fluid can be tolerated, it is possible to modify the device to ensure that only innocuous fluids leak into or escape from the equipment. The motion of the moving part may be reciprocating or rotational or both together; it may be small and occasional, as in a packed expansion joint, or virtually continuous, as in a process pump.

Stuffing boxes. A stuffing box can provide a seal around a rotating shaft and also around a shaft that moves axially. In this it differs from mechanical seals, which are good only with rotating members. The "box" is a chamber cut into the stationary member surrounding the shaft or pipe, as shown in Fig. 2.1(a). Often a boss is provided on the casing or vessel wall to give a deeper chamber. The annular space between the shaft and the wall of the chamber is filled with *packing,* consisting of a rope or rings of inert material containing a lubricant such as graphite. The packing, when compressed tightly around the shaft, keeps the fluid from passing out through the stuffing box and yet permits the shaft to turn or move back and forth. The packing is compressed by a follower ring, or gland, pressed into the box by a flanged cap or packing nut. The shaft must have a smooth surface so that it does not wear away the packing; even so, the pressure of the packing considerably increases the force required to move the shaft. A stuffing box, even under ideal conditions, does not completely stop fluid from leaking out; in fact, when the box is operating properly, there should be small leakage. Otherwise the wear on the packing and the power loss in the unlubricated stuffing box are excessive.

When the fluid is toxic or corrosive, means must be provided to prevent it from escaping from the equipment. This can be done by using a *lantern gland* (Fig. 2.1(b)), which may be looked upon as two stuffing boxes on the same shaft, with two sets of packing separated by a lantern ring. The ring is H-shaped in cross section, with holes drilled through the bar of the H in the direction perpendicular to the axis of the shaft. The wall of the chamber of the stuffing box carries a pipe that takes fluid to or away from the lantern ring. By applying vacuum to this pipe, any dangerous fluid that leaks through one set of packing rings is removed to a safe place before it can get to the second set. Or by forcing a harmless fluid such as water under high pressure into the lantern gland, it is possible to ensure that no dangerous fluid leaks out the exposed end of the stuffing box.

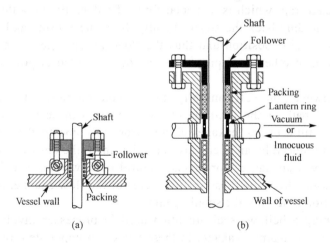

Fig. 2.1 Stuffing boxes: (*a*) simple form; (*b*) with lantern gland.

Mechanical seals. In a rotary, or mechanical, seal the sliding contact is between a ring of graphite and a polished metal face, usually of carbon steel. A typical seal is shown in Fig. 2.2. Fluid in the high-pressure zone is kept from leaking out around the shaft by the stationary graphite ring held by springs against the face of the rotating metal collar. Stationary U-cup packing of rubber or plastic is set in the space between the body of the seal and the chamber holding it around the shaft; this keeps fluid from leaking past the nonrotating part of the seal and yet leaves the graphite ring free to move axially so that it can be pressed tightly against the collar. Rotary seals require less maintenance than stuffing boxes and have come into wide use in equipment handling highly corrosive fluids.

86 | Unit Operations of Chemical Engineering

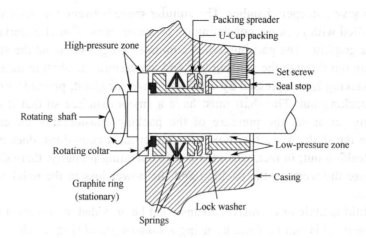

Fig. 2.2 Mechanical seal.

2.1.2 Valves

A typical processing plant contains thousands of valves of many different sizes and shapes. Despite the variety in their design, however, all valves have a common primary purpose: to slow down or stop the flow of a fluid. Some valves work best in on-or-off service, fully open or fully closed. Others are designed to throttle, to reduce the pressure and flow rate of a fluid. Still others permit flow in one direction only or only under certain conditions of temperature and pressure. A steam trap, which is a special form of valve, allows water and inert gas to pass through while holding back the steam. Finally, by using sensors and automatic control systems to adjust the valve position and thus the flow through the valve, the temperature, pressure, liquid level, or other fluid properties can be controlled at points remote from the valve itself.

In all cases, however, the valve initially stops or controls flow. This is done by placing an obstruction in the path of the fluid, an obstruction that can be moved about as desired inside the pipe with little or no leakage of fluid from the pipe to the outside. Where the resistance to flow introduced by an open valve must be small, the obstruction and the opening that can be closed by it are large. For precise control of flow rate, usually obtained at the price of a large pressure drop, the cross sectional area of the flow channel is greatly reduced, and a small obstruction is set into the small opening.

Valves containing a bellows seal are often used in processes involving hazardous or toxic materials to ensure against leakage. In these valves an upper stem raises or lowers the top of an expandable bellows, moving a lower stem that is attached inside the bellows. The lower stem raises or lowers the valve disk. The upper stem may rotate; the lower stem does not. The lower end of the bellows is sealed to the valve body by a gasket or by welding. Bellows valves are available in various alloys in sizes from 1/2 in. (12 mm) to 12 in. (300 mm).

Gate valves and globe valves

The two most common types of valves, gate valves and globe valves, are illustrated in Fig. 2.3. In a gate valve the diameter of the opening through which the fluid passes is nearly the same as that of the pipe, and the direction of flow does not change. As a result, a wide-open gate valve introduces only a small pressure drop. The disk is tapered and fits into a tapered seat; when the valve is opened, the disk rises into the bonnet, completely out of the path of the fluid. Gate valves are not recommended for controlling flow and are usually left fully open or fully closed.

Globe valves (so called because in the earliest designs the valve body was spherical) are widely used for controlling flow. The opening increases almost linearly with stem position, and wear is evenly distributed around the disk. The fluid passes through a restricted opening and changes direction several times, as can be seen by visualizing the flow through the valve illustrated in Fig. 2.3(b). As a result, the pressure drop in this kind of valve is large.

Most automatic control valves are similar to globe valves, but the handwheel is replaced by a spring-diaphragm pneumatic activator or an electric motor, and the valve position depends on a signal from the controller. The simple disk shown in Fig. 2.3(b) may be replaced by a tapered plug or other shape (Fig. 2.3(c)), designed to give certain flow lift characteristics.

Plug cocks and ball valves

For temperatures below 250℃, metallic plug cocks are useful in chemical process lines. As in a laboratory stopcock, a quarter turn of the stem takes the valve from fully open to fully closed; and when it is fully open, the channel through the plug may be as large as the inside of the pipe itself, and the pressure drop is minimal. In a ball valve the sealing element is spherical, and the problems of alignment and "freezing" of the element are less than with a plug cock.

In both plug cocks and ball valves, the area of contact between the moving element and seat is large, and both can therefore be used in throttling service. Ball valves find occasional applications in flow control.

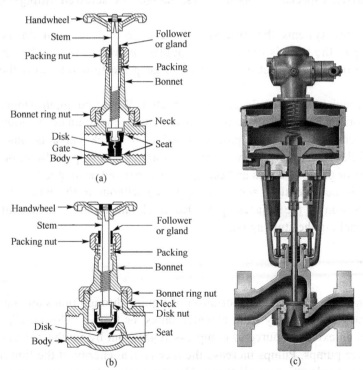

Fig. 2.3 Common valves: (*a*) gate valve; (*b*) globe valve; (*c*) control valve with pneumatic valve activator.

Check valves

A check valve permits flow in one direction only. It is opened by the pressure of the fluid in the desired direction; when the flow stops or tends to reverse, the valve automatically closes by gravity or by a spring pressing against the disk. Common types of check valves are shown in Fig. 2.4. The movable disk is shaded.

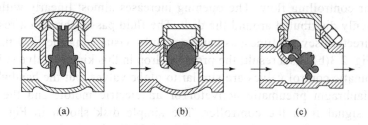

Fig. 2.4 Check valves: (*a*) lift check; (*b*) ball check; (*c*) swing check.

Recommended practice

In designing and installing a piping system, many details must be given careful attention, for the successful operation of the entire plant may turn upon a seemingly insignificant feature of the piping arrangement. Some general principles are important enough to warrant mention. In installing pipe, for example, the lines should be parallel and contain, as far as possible, right-angle bends. In systems where the process lines are likely to become clogged, provision should be made for opening the lines to permit cleaning them out. Unions or flanged connections should be generously included, and tees or crosses with their extra openings closed with plugs should be substituted for elbows in critical locations. With hazardous materials, especially volatile ones, flanged or screwed fittings should be used sparingly.

In gravity flow systems the pipe should be oversized and contain as few bends as possible. Fouling of the lines is particularly troublesome where flow is by gravity, since the pressure head on the fluid cannot be increased to keep the flow rate up if the pipe becomes restricted.

Leakage through valves should also be expected. Where complete stoppages of flow is essential, therefore, where leakage past a valve would contaminate a valuable product or endanger the operators of the equipment, a valve or check valve is inadequate. In this situation a blind flange set between two ordinary flanges will stop all flow; or the line can be broken at a union or pair of flanges and the open ends capped or plugged.

Valves should be accessible and well supported without strain, with suitable allowance for thermal expansion of the adjacent pipe. Room should be allowed for fully opening the valve and for repacking the stuffing box.

2.2 Pumps

This section deals with the transportation of liquids through pipes and channels. Liquids are sometimes moved by gravity from elevated tanks, or from a "blowcase"(a storage vessel pressurized from an external source of compressed gas), but by far the most common devices for the purpose are pumps. Pumps increase the mechanical energy of the liquid, increasing its velocity, pressure, or elevation or all three. The two major classes are positive-displacement pumps and centrifugal pumps. Positive-displacement units apply pressure directly to the liquid by a reciprocating piston, or by rotating members which form chambers alternately filled by and emptied of the liquid. Centrifugal pumps generate high rotational velocities, then, convert the resulting kinetic energy of the liquid to pressure energy. In pumps the density of liquid does not change appreciably and may be considered constant.

2.2.1 Developed Head

A typical pump application is shown diagrammatically in Fig. 2.5. The pump is installed in a pipeline to provide the energy needed to draw liquid from a reservoir and discharge a constant volumetric flow rate at the exit of the pipeline, Z above the level of the liquid. At the pump itself, the liquid enters the suction connection at station a and leaves the discharge connection at station b. A mechanical energy balance equation can be written between stations a and b. Equation (1.3-25) serves for this.

Since the only friction is that occurring in the pump itself and is accounted for by the mechanical efficiency η, $h_f = 0$. Then Eq. (1.3-25) can be written

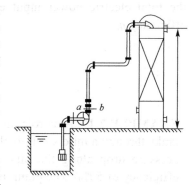

Fig. 2.5 pump flow system

$$W = \left(\frac{p_b}{\rho} + gZ_b + \frac{u_b^2}{2}\right) - \left(\frac{p_a}{\rho} + gZ_a + \frac{u_a^2}{2}\right) \qquad (2.2\text{-}1)$$

The Equation (2.2-1) can be divided by g, gives

$$H = \left(\frac{p_b}{\rho g} + Z_b + \frac{u_b^2}{2g}\right) - \left(\frac{p_a}{\rho g} + Z_a + \frac{u_a^2}{2g}\right) \qquad (2.2\text{-}2)$$

The quantity H is called total head, which each term has the dimension of length. Often the head developed by a pump is expressed in meters of fluid.

Power requirement

Using the total mechanical-energy-balance Equation(2.2-1) on a pump and piping system, the actual or theoretical mechanical energy W J/kg, also called as hydraulic power, added to the fluid by the pump can be calculated. The power delivered to the fluid is calculated from the mass flow rate m and the head developed W by pump. It is denoted by N_e and defined by

$$N_e = mW \qquad (2.2\text{-}3)$$

where m is the mass flow rate in kg/s. If η is the fractional pump efficiency and the power supplied to the pump drive from an external source is denoted by N. It is calculated from N_e.

$$N = \frac{N_e}{\eta} \qquad (2.2\text{-}4)$$

The mechanical energy W in J/kg added to the fluid is often expressed as the developed head H of the pump in m of fluid being pumped, as Equation(2.2-2), The power delivered to the fluid is also calculated from Equation (2.2-5)

$$N_e = mgH \qquad (2.2\text{-}5)$$

To calculate the power of a fan where the pressure difference is on the order of a few hundred *mm* of water, a linear average density of the gas between the inlet and outlet of the fan is used to calculate W.

Electric Motor Efficiency

Since most pumps are driven by electric motors, the efficiency of the electric motor must be taken into account to determine the total electric power to the motor. Typical efficiencies η_e of electric motors are as follows: 77% for 0.5 kW motors, 82% for 2 kW, 85% for 5 kW, 88% for 20 kW, 90% for 50 kW, 91% for 100 kW, and 93% for 500 kW and larger. Hence, the total electric power input equals the shaft power divided by the electric motor drive efficiency η_e:

$$total\ electric\ power = \frac{N}{\eta_e} \qquad (2.2\text{-}6)$$

EXAMPLE 2.1 A petroleum fraction is pumped 2 km from a distillation plant to storage tanks through a mild steel pipeline, 150 mm in diameter, at the rate of 0.04m³/s. what is the pressure drop along the pipe and the power supplied to the pumping unit if it has an efficiency of 50%? The pump impeller is eroded and the pressure at its delivery falls to one half. By how much is the flow rate reduced? (Assuming that the friction coefficient λ maintains unchanged)
Specific gravity of the liquid =0.705, viscosity of the liquid=0.5mPa·s, roughness of pipe surface ε=0.004mm.
Solution. Cross-sectional area of pipe

$$S=(\pi/4)\times 0.15^2=0.0177m^2$$

and velocity in the pipe

$$u=0.04/0.0177=2.26m/s$$

$$Re = \frac{du\rho}{\mu} = \frac{0.015\times 2.26\times 705}{0.5\times 10^{-3}} = 4.78\times 10^5$$

$$\varepsilon=0.004mm,\quad \varepsilon/d=0.004/150=0.000027$$

From figure λ=0.0134(f=0.0034), the energy needed is to overcome the friction loss

$$W = h_f = \lambda \frac{l}{d}\times\frac{u^2}{2} = 0.0134\times\frac{2000}{0.15}\times\frac{2.26^2}{2} = 456.3 J/kg$$

The power required if the pump has an efficiency of 50% is

$$N = \frac{Wm}{\eta} = \frac{456.3\times 0.04\times 705}{0.5} = 25734W = 25.7kW$$

If, due to impeller erosion, the delivery pressure is halved, the new flowrate can be found from

$$\frac{h_f}{2} = W' = h'_f = \lambda\frac{l}{d}\times\frac{u^2}{2}$$

Assuming the friction coefficient is unchanged

$$u = \sqrt{\frac{dh_f}{\lambda l}} = \sqrt{\frac{0.15\times 456.3}{0.0134\times 2000}} = 1.6m/s$$

The volumetric flowrate is now

$$Q_v=1.6\times 0.0177=0.028m^3/s$$

2.2.2 Suction Lift and Cavitation of Pumps

The power calculated by Eq.(2.2-6) depends on the difference in pressure between discharge and suction and is independent of the pressure level. From energy considerations it is immaterial whether the suction pressure is below atmospheric pressure or well above it, as long as the fluid remains liquid. However, if the suction pressure is only slightly greater than the vapor pressure, some liquid may flash to vapor inside the pump, a process called cavitation, which greatly reduces the pump capacity and causes severe erosion. If the suction pressure is actually less than the vapor pressure, there will be vaporization in the suction line, and no liquid can be drawn into the pump.

The satisfactory operation of a pump requires that vaporization of the liquid being pumped does not occur at any condition of operation. This is so desired because when a liquid vaporizes its volume increases very much. For example, 1 m^3 of water at room temperature becomes 1700 m^3 of vapor at the same temperature. This makes it clear that if we are to pump a fluid effectively, it must be kept always in the liquid form.

To avoid cavitation, the pressure at the pump inlet must exceed the vapor pressure and exceed it by a certain value called the *net positive suction head* (NPSH). NPSH is one of the most widely used and least understood terms associated with pumps. Pump manufacturers measure these values experimentally and include them with the pumps furnished. Understanding the significance of NPSH is very much essential during installation as well as operation of the pumps.

The required value of NPSH is about 2 to 3 m for small centrifugal pumps(<0.4m^3/min); but it increases with pump capacity, impeller speed, and discharge pressure, and values up to 15 m are recommended for very large pumps. For water below 100℃ at 1750 rpm and centrifugal pumps, typical values of NPSH are as follows: For pressures 3500 kPa or below: up to 0.75 m^3/min, 1.5 m; 2 m^3/min, 2.1 m; 4 m^3/min, 3 m; 8 m^3/min, 5.5m.

Suction lift

Cavitation must be avoided to assure long machine life. For a given pump, cavitation can be avoided by accurate suction lift. From an energy balance, shown in Fig. 2.6, the least height of the suction station 2 of pump above the liquid level in the reservoir may be found as

$$H_g = \frac{p_1 - p_2}{\rho g} - \frac{u_2^2}{2g} - H_f \tag{2.2-7}$$

where H_g is height between the open surface of liquid and pump centerline, called as the suction lift in m, p_1, p_2 are the pressure at the station 1 and 2, respectively, in Pa, H_f friction loss in suction line to pump in m, and $\frac{u_2^2}{2g}$ is velocity head in m, ρ is density of liquid in kg/m^3.

To avoid cavitation, the pressure in station 2 must exceed the vapor pressure by a certain value NPSH. NPSH is defined as the difference between the absolute stagnation pressure in the flow at the pump suction and liquid vapor pressure and given by

$$\text{NPSH} = \frac{p_2}{\rho g} + \frac{u_2^2}{2g} - \frac{p_v}{\rho g} \tag{2.2-8}$$

where p_v is vapor pressure of fluid at the given temperature in N/m^2. Substitution of Eq.(2.2-8) into Eq.(2.2-7) for p_2 gives

$$H_g = \frac{p_1 - p_v}{\rho g} - \text{NPSH} - H_f \tag{2.2-9}$$

For a pump taking suction from a reservoir, the pressure on the surface of liquid is p_a, like that shown in Figure 2.6, so, the suction lift is calculated by

$$H_g = \frac{p_a - p_v}{\rho g} - \text{NPSH} - H_f \tag{2.2-10}$$

The available NPSH is customarily calculated as

$$\text{NPSH} = \frac{p_a - p_v}{\rho g} - H_f - H_g \tag{2.2-11}$$

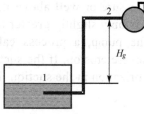

Fig. 2.6 Diagram for NPSH available in a pumping system.

The manufacturer usually tests the pump with water at different capacities, created by throttling the suction side. When the first signs of vaporization induced cavitation occur, the suction pressure is noted. This pressure is converted into the head.

For the special situation where the liquid is practically nonvolatile (p_v=0), the friction negligible (h_f=0), and pressure at station 1 atmospheric, the maximum possible suction lift can be obtained by subtracted the required NPSH from the barometric head.

EXAMPLE 2.2 Benzene at 37.8℃ is pumped through the system of Fig. 2.7 at the rate of 9.09 m³/h. The reservoir is at atmospheric pressure. The gauge pressure at the end of the discharge line is 3.45 kN/m². The discharge is 3.05 m, and the pump suction is 1.22m above the level in the reservoir. The discharge line is Φ48×3.5mm. The friction in the suction line is known to be 3.45 kN/m², and that in the discharge line is 37.9 kN/m². The mechanical efficiency of the pump is 0.60. The density of benzene is 865 kg/m³, and its vapor pressure at 37.8℃ is 26.2 kN/m². Calculate (a) the developed head of the pump and (b) the total power input. (c) If the pump manufacturer specifies a required NPSHR of 3.05 m, will the pump be suitable for this service?

Solution

(a) The pump work W is found by using Eq. (1.3-25). The upstream station a' is at the level of the liquid in the reservoir, and the downstream station b' is at the end of the discharge line, as shown in Fig. 2.7. When the level in the tank is chosen as the datum of heights and it is noted that $u_{a'}$=0, Eq. (1.3-25) gives

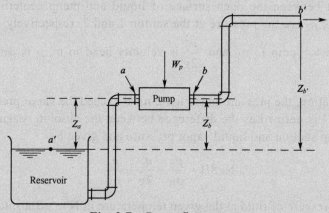

Fig. 2.7 Pump flow system.

$$W = \left(\frac{p_{b'}}{\rho} + gZ_{b'} + \frac{u_{b'}^2}{2} + h_f\right) - \frac{p_{a'}}{\rho}$$

The exit velocity $u_{b'}$ is found by using data from the diameter of pipe,

$$u_{b'} = \frac{4 \times 9.09}{3.14 \times 0.041^2 \times 3600} = 1.91 \text{m/s}$$

Eq. (1.3-25) gives

$$W = \left(\frac{p_{b'}}{\rho} + gZ_{b'} + \frac{u_{b'}^2}{2} + h_f\right) - \frac{p_{a'}}{\rho}$$

$$= \frac{345 \times 10^3}{865} + 9.81 \times 3.05 + \frac{1.91^2}{2} + \frac{(37.9+3.45) \times 10^3}{865}$$

$$= 478.4 \text{J/kg}$$

(b) The mass flow rate is

$$m = \frac{9.09 \times 865}{3600} = 2.18 \text{kg/s}$$

The power input is, from Eq.(2.2-4),

$$N = \frac{Wm}{\eta} = \frac{478.4 \times 2.18}{0.6} = 1738 \text{J/s} = 1.738 \text{kW}$$

(c) Use Eq.(2.2-10), $p_{a'}/\rho = 26200/865 = 30.29 \text{J/kg}$. The vapor pressure corresponds to a head of

$$\frac{p_{a'}}{\rho g} = \frac{30.29}{9.81} = 3.87 \text{m}$$

The friction in the suction line is

$$\frac{\Delta p}{\rho g} = \frac{3450}{865 \times 9.81} = 0.41 \text{m}$$

The value of the available NPSH from Eq.(2.2-10) is

$$\text{NPSH} = \frac{101300}{865 \times 9.81} - 3.87 - 0.41 - 1.22 = 6.44 \text{m}$$

The available NPSH is considerably larger than the minimum required value of 3.05m, so the pump should be suitable for the proposed service.

2.2.3 Positive-Displacement Pumps

Whereas the total dynamics head developed by a centrifugal or axial-pump is uniquely determined for any given flow by the speed at which it rotates, positive-displacement pump and those which approach positive displacement will ideally produce maximum head attainable determined by the power available in the drive and the strength of the pump parts.

In the first major class of pumps, a definite volume of liquid is trapped in a chamber, which is alternately filled from the inlet and emptied at a higher pressure through the discharge. There are two subclasses of positive-displacement pumps. In reciprocating pumps, the chamber is a stationary cylinder that contains a piston or plunger; in rotary pumps, the chamber moves from inlet to discharge and back to the inlet.

Reciprocating pumps

Piston pumps, plunger pumps, and diaphragm pumps are examples of reciprocating pumps. In a piston pump, liquid is drawn through an inlet check valve into the cylinder by the withdrawal of a piston and then is forced out through a discharge check valve on the return stroke. Most piston pumps are double-acting with liquid admitted alternately on each side of the piston so that one part of the cylinder is being filled while the other is being emptied, shown in Fig.2.8. Often two or more cylinders are used in parallel with common suction and discharge headers, and the configuration of the pistons is adjusted to minimize fluctuations in the discharge rate. The piston may be motor-driven through reducing gears, or a steam cylinder may be used to drive the piston rod directly. The maximum discharge pressure for commercial piston pumps is about 50 atm.

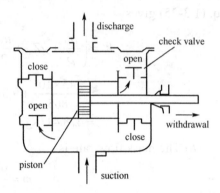

Fig. 2.8 Double-acting reciprocating pump.

For higher pressures, plunger pumps are used. A heavy-walled cylinder of small diameter contains a close-fitting reciprocating plunger, which is merely an extension of the piston rod. At the limit of its stroke the plunger fills nearly all the space in the cylinder. Plunger pumps are single-acting and usually are motor-driven. They can discharge against a pressure of 1500 atm or more.

In a diaphragm pump, the reciprocating member is a flexible diaphragm of metal, plastic, or rubber. This eliminates the need for packing or seals exposed to the liquid being pumped, a great advantage when handling toxic or corrosive liquids. Diaphragm pumps handle small to moderate amounts of liquid, up to about $0.4 m^3/min$, and can develop pressures in excess of 100 atm.

The mechanical efficiency of reciprocating pumps varies from 40 to 50 percent for small pumps to 70 to 90 percent for large ones. It is nearly independent of speed within normal operating limits and decreases slightly with an increase in discharge pressure because of added friction and leakage.

Volumetric efficiency. The ratio of the volume of fluid discharged to the volume swept by the piston or plunger is called the *volumetric efficiency*. In positive displacement pumps the volumetric efficiency is nearly constant with increasing discharge pressure, although it drops a little because of leakage. Because of the constancy of volume flow, plunger and diaphragm pumps are widely used as "metering pumps," injecting liquid into a process system at controlled but adjustable volumetric rates.

Rotary pumps

A wide variety of rotary positive-displacement pumps are available. They bear such names as gear pumps, lobe pumps, screw pumps, cam pumps, and vane pumps. Two examples of gear pumps are shown in Fig. 2.9. Unlike reciprocating pumps, rotary pumps contain no check valves. Close tolerances between the moving and stationary parts minimize leakage from the discharge space back to the suction space; they also limit the operating speed. Rotary pumps operate best on clean, moderately viscous fluids, such as light lubricating oil. Discharge pressures up to 200 atm or more can be attained.

In the spur-gear pump (Fig. 2.9(a)) intermeshing gears rotate with close clearance inside the casing. Liquid entering the suction line at the bottom of the casing is caught in the spaces between the teeth and the casing and is carried around to the top of the casing and forced out the discharge. Liquid cannot short-circuit back to the suction because of the close meshing of the gears in the center of the pump.

In the internal-gear pump (Fig. 2.9(b)) a spur gear, or pinion, meshes with a ring gear with internal teeth. Both gears are inside the casing. The ring gear is coaxial with the inside of the casing, but the pinion, which is externally driven, is mounted eccentrically with respect to the center of the casing. A stationary metal crescent fills the space between the two gears. Liquid is carried from inlet to discharge by both gears, in the spaces between the gear teeth and the crescent.

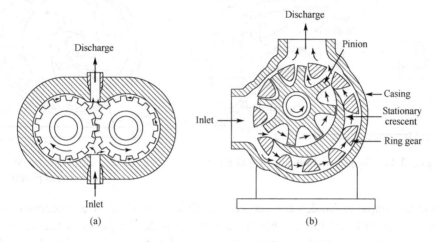

Fig. 2.9 Gear pumps: (*a*) spur-gear pump; (*b*) internal-gear pump.

Peristaltic pumps

In the production of biochemicals, small leak proof peristaltic pumps are frequently used. Such a pump consists of a length of flexible tubing squeezed by a succession of moving rollers, trapping the liquid and causing it to move along the tubing, shown as in Fig. 2.10. The discharge rate is almost constant, unlike that of a diaphragm pump. Peristaltic pumps can be used only for small flow rates but are often the best choice when liquids must be moved without the possibility of leakage or exposure to the air.

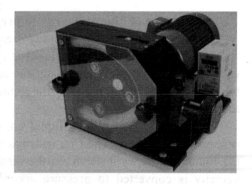

Fig. 2.10 Peristaltic pump.

2.2.4 Centrifugal Pumps

In the second major class of pumps, the mechanical energy of the liquid is increased by centrifugal action. A simple, but very common, example of a centrifugal pump is shown in Fig. 2.11. The liquid enters through a suction connection concentric with the axis of a high-speed rotary element called the *impeller,* which carries radial vanes integrally cast in it. Liquid flows outward in the spaces between the vanes and leaves the impeller at a considerably greater velocity with respect to the ground than at the entrance to the impeller.

In a properly functioning pump, the space between the vanes is completely filled with liquid flowing without cavitation. The liquid leaving the outer periphery of the impeller is collected in a spiral casing called the *volute* and leaves the pump through a tangential discharge connection. In the volute the velocity head of the liquid from the impeller is converted to pressure head. The power is applied to the fluid by the impeller and is transmitted to the impeller by the torque of the driveshaft, which usually is driven by a direct-connected motor at constant speed, commonly at 1750 or 3450 r/min.

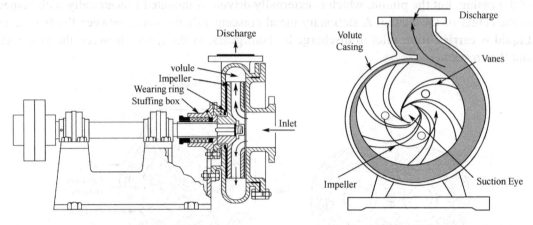

Fig. 2.11 Single-suction centrifugal pump.   **Fig. 2.12** Liquid flow path inside a centrifugal pump

Because the impeller blades are curved, the fluid is pushed in a tangential and radial direction by the centrifugal force. This force acting inside the pump is the same one that keeps water inside a bucket that is rotating at the end of a string. Figure 2.12 below depicts a side cross-section of a centrifugal pump indicating the movement of the liquid.

2.2.4.1 Structure of Centrifugal Pump

Conversion of kinetic energy to pressure energy
The key idea is that the energy created by the centrifugal force is *kinetic energy*. The amount of energy given to the liquid is proportional to the square of *velocity* at the edge or vane tip of the impeller. The faster the impeller revolves or the bigger the impeller is, then the higher will be the velocity of the liquid at the vane tip and the greater the energy imparted to the liquid.

This kinetic energy of a liquid coming out of an impeller is harnessed by creating a *resistance* to the flow. The first resistance is created by the pump volute (casing) that catches the liquid and slows it down. In the discharge nozzle, the liquid further decelerates and its velocity is converted to pressure according to Bernoulli's principle. Therefore, the head (pressure in terms of height of liquid) developed is approximately equal to the velocity energy at the periphery of the impeller expressed by the following well-known formula:

$$H = \frac{u^2}{2g} \qquad (2.2\text{-}12)$$

where H total head developed in meter, u velocity at periphery of impeller in m/s.

The energy changes occur by virtue of two main parts of the pump, the impeller and the volute or diffuser. The impeller is the rotating part that converts driver energy into the kinetic energy. A rotating component comprised of an impeller and a shaft. The volute or diffuser is the stationary part that converts the kinetic energy into pressure energy.

A stationary component comprised of a casing, casing cover, and bearings.

The general components, both stationary and rotary, are depicted in Fig. 2.13. The main components are discussed in brief below.

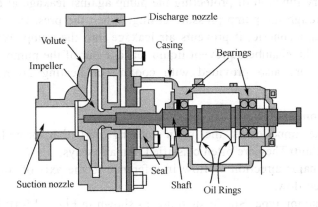

Fig. 2.13 General components of Centrifugal Pump

Stationary components

Casings are generally of two types: volute and circular. The impellers are fitted inside the casings.

Volute casings build a higher head; *circular casings* are used for low head and high capacity.

A *volute* is a curved funnel increasing in area to the discharge port as shown in Figure 2.12. As the area of the cross-section increases, the volute reduces the speed of the liquid and increases the pressure of the liquid. One of the *main purposes of a volute casing* is to help balance the hydraulic pressure on the shaft of the pump. However, this occurs best at the manufacturer's recommended capacity. Running volute-style pumps at a lower capacity than the manufacturer recommends can put lateral stress on the shaft of the pump, increasing wear-and-tear on the seals and bearings, and on the shaft itself. Double-volute casings are used when the radial thrusts become significant at reduced capacities.

Circular casing have stationary diffusion vanes surrounding the impeller periphery that convert velocity energy to pressure energy. Conventionally, the diffusers are applied to multi-stage pumps.

Suction and discharge nozzle

The suction and discharge nozzles are part of the casings itself. They commonly have the following configurations.

End suction/Top discharge. The suction nozzle is located at the end of and concentric to, the shaft while the discharge nozzle is located at the top of the case perpendicular to the shaft. This pump is always of an overhung type and typically has lower NPSH because the liquid feeds directly into the impeller eye.

Top suction /Top discharge nozzle. The suction and discharge nozzles are located at the top of the case perpendicular to the shaft. This pump can either be an overhung type or between-bearing type but is always a radially split case pump.

Seal chamber and stuffing box

Seal chamber and Stuffing box both refer to a chamber, either integral with or separate from the pump case housing that forms the region between the shaft and casing where sealing media are installed. When the sealing is achieved by means of a mechanical seal, the chamber is commonly referred to as a Seal Chamber. When the sealing is achieved by means of packing, the chamber is referred to as a Stuffing Box. Both the seal chamber and the stuffing box have the primary function of protecting the pump against leakage at the point where the shaft passes out through the pump pressure casing. When the pressure at the bottom of the chamber is below atmospheric, it prevents air leakage into the pump. When the pressure is above atmospheric, the chambers prevent liquid leakage out of the pump. The seal chambers and stuffing boxes are also provided with cooling or heating arrangement for proper temperature control.

rotating components

Impeller. The impeller is the main rotating part that provides the centrifugal acceleration to the fluid. They are often classified in many ways.

(1) Based on major direction of flow in reference to the axis of rotation: Radial flow, axial flow, and mixed flow.

(2) Based on suction type: Single-suction, as shown in Fig. 2.14(a): Liquid inlet on one side; Double-suction: Liquid inlet to the impeller symmetrically from both sides, as shown in Fig.2.14(b).

A common type uses a double-suction impeller, which accepts liquid from both sides, as shown in Fig. 2.14(b).

(3) Based on mechanical construction: Open, no shrouds or wall to enclose the vanes, Semi-open or vortex type, and Shrouds or sidewall enclosing the vanes.

An open impeller is characterized by impeller blades that are supported almost entirely by the impeller hub, as shown in Fig.2.15. This is the simplest impeller style and it is primarily applied to clean, non-abrasive, low horsepower applications. An open impeller is lighter in weight than its shrouded counterpart. Less impeller weight reduces shaft deflection and enables the use of a smaller diameter shaft, at a lower cost, than an equivalent shrouded impeller.

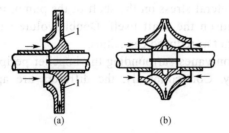

Fig. 2.14 Suction types of impeller:
(*a*) Single-suction; (*b*) Double-suction.

Fig.2.15 Open impeller.

An *open impeller* typically operates at a higher efficiency than a shrouded impeller of the same specific speed. The largest contributor to efficiency loss in an enclosed radial impeller is disc friction caused by the front and back impeller shrouds turning in close proximity to the stationary casing walls. Removing the shrouds eliminates the disc friction.

One drawback of the open impeller is that it is more susceptible to abrasive wear than a shrouded impeller. High velocity fluid on the impeller blades in close proximity to the casing walls establishes rotating vortices that accelerate wear when abrasives are present.

A tight clearance between the impeller and the front and back casing walls is necessary to maximize efficiency. As the impeller wears, these clearances open and efficiency drops rapidly. The tight operating clearances required on both sides of an open impeller for efficient operation precludes adjustment of the impeller axial position to compensate for wear.

A *semi-open impeller* is a compromise between an open and an enclosed impeller. It incorporates a single shroud, usually located on the back of the impeller, as shown in Fig. 2.16. A semi-open impeller has a solids passing capability similar to that found in an open impeller. With only a single shroud a semi-open impeller is easy to manufacture and completely accessible for applying surface hardening treatments. For moderately abrasive slurries, especially if plugging is a concern, a semi-open impeller is a good choice.

A semi-open impeller operates more efficiently than an enclosed impeller because of lower disc friction and tighter axial clearances. It has an advantage over an open impeller in that it can be adjusted axially to compensate for casing wear.

Fig.2.16 Semi-open impeller.

High axial thrust is the primary drawback of a semi-open impeller design. Axial thrust balance is manageable through design for both open and enclosed impellers. On a semi-open impeller, the entire backside surface of the shroud is subject to the full impeller discharge pressure. The front side of the shroud is at suction pressure at the eye of the impeller and increases along the impeller radius due to centrifugal action. The differential between the pressure profiles along the two sides of the shroud creates the axial thrust imbalance. This can be managed somewhat through the use of pump-out vanes on the back side of the shroud, but the vanes will start to lose effectiveness if the impeller is moved forward in the casing to compensate for wear. Some manufacturers have integrated an adjustable wear-plate into the casing design so that clearance adjustments can be made. Combined with hard materials or surface hardening treatments, this option provides a good design in lightly to moderately abrasive applications.

An obvious question is why use a semi-open impeller in a solids application if an open impeller with an adjustable wear plate could be used instead? It might seem logical that an open impeller of hard metal construction, used in conjunction with an adjustable wear liner, would combine good solids handling characteristics, with low thrust imbalance, light weight, and adjustability for wear. Unfortunately, true open impellers lack the structural support to prevent blade collapse or deformation under the demands of most industrial applications. A semi-open impeller is well suited for handling solids in applications where the blades might encounter high impact loads from rocks and the like, or in higher power applications. In both situations the shroud provides additional structural support and reinforcement to protect against blade collapse or deformation.

One improvement that has been made to the semi-open impeller is the use of a partial shroud, as shown in Fig.2.17. Most of the pressure developed by the impeller, and most of the shroud surface area, is in the outer diameter region of the impeller. Elimination of the shroud in this area reduces the axial thrust in a semi-open impeller without compromise to the structure support provided by the full back shroud.

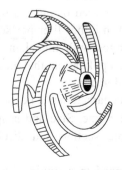

Fig. 2.17 Semi-open impeller with partial shroud.

Open and semi-open impellers are less likely to clog, but need manual adjustment to the volute or back-plate to get the proper impeller setting and prevent internal re-circulation. Vortex pump impellers are great for solids and "stringy" materials but they are up to 50% less efficient than conventional designs.

Shrouds or sidewall enclosing the vanes. Closed impellers require wear rings and these wear rings present another maintenance problem.

An **enclosed impeller** incorporates a full front and back shroud, as shown in Fig. 2.18. Fluid flows through the internal impeller passages without hydraulic interaction with the stationary casing walls. In a well designed enclosed impeller, the relative velocity between the impeller and the fluid at any given radius is quite small. This results in less wear than other impeller styles.

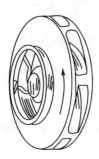

Fig.2.18 Enclosed impeller.

A portion of the fluid exiting an enclosed impeller leaks back to the pump suction by traveling through the gap between the front impeller shroud and the casing. An enclosed impeller typically has wear rings or radial pump-out vanes to control this leakage. A centrifugal pump with an enclosed impeller is usually not dependent on tight axial clearances to manage leakage. Therefore an enclosed impeller pump can tolerate moderate wear with little adverse effect on overall performance and efficiency.

Open, enclosed, and semi-open impellers are available in pumps and should be chosen based upon the characteristics of the application.

The number of impellers determines the number of stages of the pump. A single stage pump has one impeller only and is best for low head service. A two-stage pump has two impellers in series for medium head service. A multi-stage pump has three or more impellers in series for high head service.

Shaft. The basic purpose of a centrifugal pump shaft is to transmit the torques encountered when starting and during operation while supporting the impeller and other rotating parts. It must do this job with a deflection less than the minimum clearance between the rotating and stationary parts.

Definition of important terms

The key performance parameters of centrifugal pumps are capacity, head, BHP (Brake horse power), BEP (Best efficiency point) and specific speed. The pump curves provide the operating window within which these parameters can be varied for satisfactory pump operation. The following parameters or terms are discussed in detail in this section.

Capacity. Capacity means the flow rate with which liquid is moved or pushed by the pump to the desired point in the process. It is commonly measured in either liter per second (l/s) or cubic meters per hour (m^3/h). The capacity usually changes with the changes in operation of the process. For example, a boiler feed pump is an application that needs a constant pressure with varying capacities to meet a changing steam demand.

The capacity depends on a number of factors like: Process liquid characteristics i.e. density, viscosity; Size of the pump and its inlet and outlet sections; Impeller size, Size and shape of cavities between the vanes; Impeller rotational speed RPM; Pump suction and discharge temperature and pressure conditions

For a pump with a particular impeller running at a certain speed in a liquid, the only items on the list above that can change the amount flowing through the pump are the pressures at the pump inlet and outlet. The effect on the flow through a pump by changing the outlet pressures is graphed on a pump curve.

Head. The pressure at any point in a liquid can be thought of as being caused by a vertical column of the liquid due to its weight. The height of this column is called the static head and is expressed in terms of meter of liquid.

The same *head* term is used to measure the kinetic energy created by the pump. In other words, head is a measurement of the height of a liquid column that the pump could create from the kinetic energy imparted to the liquid. Imagine a pipe shooting a jet of water straight up into the air, the height the water goes up would be the head.

The head is not equivalent to pressure. Head is a term that has units of a length and pressure has units of force per unit area. The main reason for using head instead of pressure to measure a centrifugal pump's energy is that the pressure from a pump will change if the specific gravity (weight) of the liquid changes, but the head will not change. Since any given centrifugal pump can move a lot of different fluids, with different specific gravities, it is simpler to discuss the pump's head and forget about the pressure. A given pump with a given impeller diameter and speed will raise Q liquid to a certain height regardless of the weight of the liquid.

So a centrifugal pump's performance on any Newtonian fluid, whether it's heavy (sulfuric acid) or light (gasoline) is described by using the term 'head'. The pump performance curves are mostly described in terms of head.

2.2.4.2 Centrifugal Pump Theory

Except in very small pumps the impeller vanes are not truly radial but are curved backward, opposite to the direction of rotation, as shown in Figs. 2.19. The blade tips are at an angle β with the tangent to the circular rim of the impeller. Angle β is almost always less than 90°, if it is greater than 90°, with forward curving blades, flow in the piping system may become unstable.

Figure 2.19 shows diagrammatically how the liquid flows through a centrifugal pump. The liquid enters axially at the suction connection, station *a*. In the rotating eye of the impeller, the liquid spreads out radially and enters the channels between the vanes at station 1. It flows through the impeller, leaves the periphery of the impeller at station 2, is collected in the volute, and leaves the pump discharge at station *b*.

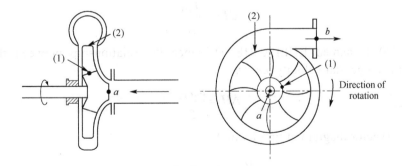

Fig. 2.19 Centrifugal pump showing Bernoulli stations.

Ideal pump

Two idealizations are now accepted. It is assumed, first, all liquid flowing across the periphery of the impeller is moving at the same velocity, so the numerical value(but not the vector direction) is v_2 at all points; second, it is assumed that the angle between the vector v_2 and the tangent is the actual vane angle β_2. This assumption in turn is equivalent to an assumption that there are an infinite number of vanes, of zero thickness, at an infinitesimal distance apart. This ideal state is referred to as *perfect guidance*.

As shown in the vector diagram for a single vane, Fig.2.20, liquid leaves the impeller at a absolute velocity V_2 at an angle α with the tangent to the impeller rim, absolute velocity as sum of velocity v_2 relative to blade and rotor velocity u_2. Velocity V_2 has a tangential component V_{t2} and a radial component V_{r2}. Component V_{t2} is assumed equal to the radial velocity of the tip of the blades; V_{r2} equals the volumetric flow rate Q divided by the peripheral area of the impeller A.

Theoretical head

The angular momentum of an object moving about a center of rotation is the vector product of the position vector and the tangential momentum vector of the object (its mass m times its tangential component of velocity V_t). Figure2.20 shows the rotating for the situation involving two-dimensional flow. The fluid at the eye of impeller is moving about tip of the impeller at a velocity V. The angular momentum of a mass m of fluid is therefore rmV_t. The power input to the impeller, and therefore the power required by the pump, can be calculated from the angular-momentum equation for steady flow.

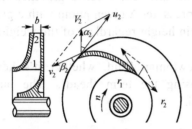

Fig. 2.20 Velocity vectors at discharge of vanes in a centrifugal pump.

$$T = m(r_2 V_{t2} - r_1 V_{t1}) \tag{2.2-13}$$

Since work done per second $N=T\omega$ (ω angular velocity), the power equation for an ideal pump is (where $\alpha=90°, V_{t1}=0$)

$$N_e = T\omega = mr_2 V_{t2} \omega \tag{2.2-14}$$

Work done per second by the impeller on the fluid equates the rate of energy transfer

$$N_e = mg\Delta H \tag{2.2-15}$$

Therefore, combining Eqs(2.2-14)and(2.2-15) gives

$$\Delta H = \frac{r_2 V_{t2} \omega}{g} \tag{2.2-16}$$

Equation(2.2-16) is named EULER HEAD. Since the relationship between the angular velocity and the tangential velocity $\omega r_2 = u_2$

$$\Delta H = \frac{u_2 V_{t2}}{g} \tag{2.2-17}$$

From Fig.2.20 (vector diagram at tip of vane)

$$V_{t2} = u_2 - \frac{V_{r2}}{\tan \beta_2} \tag{2.2-18}$$

and

$$\Delta H = \frac{u_2 (u_2 - V_{r2}/\tan \beta_2)}{g} \tag{2.2-19}$$

ΔH is the ideal theoretical head developed. In practice, the total head across the pump is less than this due to energy dissipation in eddies and in friction.

The relation between the ideal theoretical developed head and the capacity of the pump

For the peripheral area of the impeller A, the volume flow rate Q is

$$Q = V_{r2}A \tag{2.2-20}$$

Where A is the cross section area of the impeller periphery, therefore the relation between head and volumetric flow. Substituting from Eq.(2.2-20) into Eq.(2.2-19) gives

$$\Delta H = \frac{u_2(u_2 - Q/A\tan\beta_2)}{g} \tag{2.2-21}$$

Since u_2, A, and β_2 are constant for a given rotating speed and a given size of the pump, Eq.(2.2-21) shows that the relation between head and volumetric flow rate is linear, then

$$\Delta H_\infty = k - BQ \tag{2.2-22}$$

The equation is for a frictionless or ideal pump, where $\eta=0$. The plot of theoretical head ΔH_∞ versus capacity Q is shown in Figure 2.21. The developed head of an actual pump is considerably less than that calculated from the ideal pump relation. Lower curve in Figure 2.21 indicates the relation between the actual head and capacity of centrifugal pump.

Pump performance also depends on β_2, as shown in Fig.2.22. If outlet blade faces direction of rotation it is called forward facing. If it faces away it is called backward facing. If radial it is called radial. $\beta_2 > 90°$ are forward facing, $\beta_2 = 90°$ are radial and $\beta_2 < 90°$ are backward facing. $\beta_2 > 90°$ are forward facing and head increases with Q, $\beta_2 = 90°$ are radial and head is independent of Q and $\beta_2 < 90°$ are backward facing and head decreases with increasing Q.

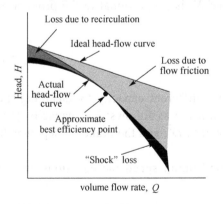

Fig. 2.21 Thoerectical and actual characteristic curve.

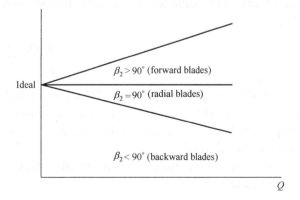

Fig. 2.22 Effects of Angle β on pump performance.

Difference between Theoretical and Actual Head Developed

The Euler head represents a theoretical head. The **Actual Head, H** is the actual head developed by the pump, and can be measured by a manometer connected to the inlet and outlet pipes of a pump. The actual developed head is considerably less and drops precipitously to zero as the rate increases to a certain value in any given pump. The difference between the theoretical and actual curves results primarily from circulatory flow. Other contributing factors to the head loss are fluid friction in the passages and channels of the pump, and shock losses from the sudden change in direction of the liquid leaving the impeller and joining the stream of liquid traveling circumferentially around the casing.

Many complicating factors determine the actual efficiency and performance characteristics of a pump. Hence, the actual experimental performance of the pump is usually employed. The performance is usually expressed by the pump manufacturer in terms of curves called *characteristic curves*, which are usually for water. The head H in m produced will be the same for any liquid of the same viscosity. The pressure produced, $p = H\rho g$, will be in proportion to the density. Viscosities of less than 0.05 Pa·s (50 cp) have little effect on the head produced. The brake kW varies directly as the density.

2.2.4.3 The Affinity Laws

The affinity laws are mathematical expressions that define changes in pump capacity, head, and BHP when a change is made to pump speed, impeller diameter, or both.

When the rotational speed n of a pump is increased, tip speed u_2 rises proportionally; in an ideal pump velocity V_2, V_{t2}, and V_{r2} also increase directly with n. Hence, as rough approximations, the following relationships, called *affinity laws*, can be used for a given pump. The *capacity* Q_1 in m³/s is directly proportional to the rpm n_1. According to the equation(2.2-20):

$$\frac{Q_1}{Q_2} = \frac{n_1}{n_2} \tag{2.2-23}$$

Head H changes in direct proportion to the square of impeller Q ratio, or the square of speed n ratio:

$$\frac{H_1}{H_2} = \left(\frac{Q_1}{Q_2}\right)^2 = \left(\frac{n_1}{n_2}\right)^2 \tag{2.2-24}$$

According to the equation(2.2-24), the *power consumed* N_1 is proportional to the product of $H_1 Q_1$, or

$$\frac{N_1}{N_2} = \frac{H_1 Q_1}{H_2 Q_2} = \left(\frac{n_1}{n_2}\right)^3 \tag{2.2-25}$$

A given pump can be modified when needed for a different capacity by changing the impeller size. Then the *affinity laws* for a constant-rpm n are as follows: The capacity Q is proportional to the diameter D, the head H is proportional to D^2, and the brake horsepower W is proportional to D^3.

If changes are made to both impeller diameter and pump speed the equations can be combined to:

$$\frac{Q_1}{Q_2} = \frac{n_1 D_1}{n_2 D_2} \tag{2.2-26}$$

$$\frac{H_1}{H_2} = \left(\frac{n_1 D_1}{n_2 D_2}\right)^2 \tag{2.2-27}$$

$$\frac{N_1}{N_2} = \left(\frac{n_1 D_1}{n_2 D_2}\right)^3 \tag{2.2-28}$$

This equation is used to hand-calculate the impeller trim diameter from a given pump performance curve at a bigger diameter.

The Affinity Laws are valid only under conditions of constant efficiency.

2.2.4.4 Characteristic Curves; Head-Capacity Relation

Before you select a pump model, examine its performance curve, which is indicated by its head-flow rate or operating curve. The curve shows the pump's capacity plotted against total developed head. It also shows efficiency, required power input, and suction head requirements (net positive suction head requirement) over a range of flow rates. The plots of actual head, total power consumption, and efficiency vs. volumetric flow rate are called the characteristic curves of a pump, shown in Figure2.23. For publication purpose, it is more convenient to draw several curves in a single graph. This presentation method shows a number of Head-Capacity curves for one speed and several impeller diameters or one impeller diameter and several different speeds for the same pump.

The capacity and pressure needs of any system can be defined with the help of a graph called a *system curve*. Similarly the capacity *vs.* pressure variation graph for a particular pump defines its characteristic *pump performance curve*.

The pump suppliers try to match the system curve supplied by the user with a pump curve that satisfies these needs as closely as possible. A pumping system operates where the pump curve and the system resistance curve intersect. The intersection of the two curves defines the operating point of both pump and process. However, it is impossible for one operating point to meet all desired operating conditions. For example, when the discharge valve is throttled, the system resistance curve shift left and so does the operating point.

Developing a Pump Performance Curve

A pump's performance is shown in its characteristics performance *curve* where its capacity i.e. flow rate is plotted against its developed head. The pump performance curve also shows its efficiency (BEP), required input power (in BHP), NPSH, speed (in RPM), and other information such as pump size and type, impeller size, etc. This curve is plotted for a constant speed (rpm) and a given impeller diameter (or series of diameters). It is generated by tests performed by the pump manufacturer. Pump curves are based on a specific gravity of 1.0. Other specific gravities must be considered by the user.

Fig.2.24 shows a typical characteristic curve of a centrifugal pump. At any fixed speed the pump will operate along this curve and no other points. It is not possible to reduce or increase the capacity at a given head unless the discharge is throttled. The characteristic curve of a pump can be change by changing both the revolution speed and diameter of pump.

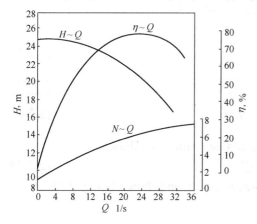

Fig. 2.23 Characteristic curve of centrifugal pump.

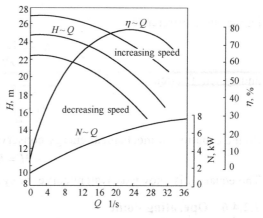

Fig. 2.24 Effects of revolution on the characteristic curve of centrifugal pump.

- On pumps with variable-speed drivers such as steam turbines or transducers, it is possible to change the characteristic curve. The characteristic curve moves upward with increasing the speed of the centrifugal pump, by contraries, the curve moves downward as decreasing the speed, shown in Fig. 2.24.
- Characteristic curve can also be changed as the impeller diameter of centrifugal pump is changed.

2.2.4.5 Developing a System Curve

The *system resistance or system head curve* is the change in flow with respect to head of the system. It must be developed by the user based upon the conditions of service. These include physical layout, process conditions, and fluid characteristics. It represents the relationship between flow and hydraulic losses in a system in a graphic form and, since friction losses vary as a square of the flow rate, the system curve is parabolic in shape. Hydraulic losses in piping systems are composed of pipe friction losses, valves, elbows and other fittings, entrance and exit losses, and losses from changes in pipe size by enlargement or reduction in diameter.

In addition to the pump design, the operational performance of a pump depends upon factors such as:
- the load characteristics,
- upstream and downstream
- pipe friction, and valve performance

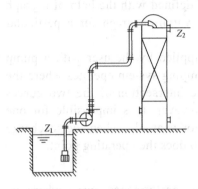

Fig.2.25 Pump flow system.

A Bernoulli equation can be written between station 1 and 2 in Fig. 2.25

$$H = \left(\frac{p_2}{\rho g} + z_2 + \frac{u_2^2}{2g}\right) - \left(\frac{p_1}{\rho g} + z_1 + \frac{u_1^2}{2g}\right) + H_f \qquad (2.2\text{-}29)$$

where the system friction H_f

$$H_f = \left(\lambda \frac{l}{d} + \lambda \frac{l_e}{d}\right)\frac{u^2}{2g} \qquad (2.2\text{-}30)$$

Rearrange the Eq.(2.2-29)

$$H = \left(\frac{p_2 - p_1}{\rho g} + z_2 - z_1\right) + \lambda\left(\frac{l + l_e}{d}\right)\frac{u^2}{2g} \qquad (2.2\text{-}31)$$

let the static difference

$$K = \frac{p_2 - p_1}{\rho g} + z_2 - z_1 \qquad (2.2\text{-}32)$$

and system friction

$$\lambda\left(\frac{l + l_e}{d}\right)\frac{u^2}{2g} = BQ^2 \qquad (2.2\text{-}33)$$

So the equation of mechanical energy conservation can be written as follow

$$H = K + BQ^2 \qquad (2.2\text{-}34)$$

The equation is known as system curve, or system-head curve, and shown in Fig. 2.26.

2.2.4.6 Operating Point

The pump suppliers try to match the system curve supplied by the user with a pump curve that satisfies these needs as closely as possible. A pumping system operates where the pump curve and the system resistance curve intersect. The intersection of the two curves defines the operating point of both pump and process. However, it is impossible for one operating point to meet all desired operating conditions.

A system-head curve can be superimposed on the head-capacity curve of a centrifugal pump. The point of intersection between the characteristic curve of centrifugal pump and system head curve of pump flow system is called operating point. The point A in the Fig. 2.26 is the operating point for a given pump at the specific speed.

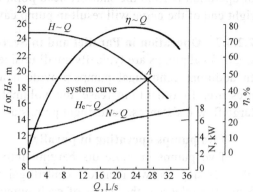

Fig. 2.26 System head curve

Operating point change

The point at which a pump operates in a given piping system depends on the flow rate and head loss of that system. For a given system, volumetric flow rate is compared to system head loss on a system characteristic curve. By graphing a system characteristic curve and the pump characteristic curve on the same coordinate system, the point at which the pump must operate is identified. For example, in Fig. 2.26, the operating point for the centrifugal pump in the original system is designated by the intersection of the pump curve and the system curve. Several methods can be used to change operating point:

• Change the system-head curve
• Change the centrifugal pump characteristic curve

It is convenient to change the system-head curve by a valve in discharge. The operating point can be changed as adjusting the opening of valve (Fig.2.27). For example, when the discharge valve is throttled, the system resistance curve shift left and so does the operating point. The operating point can also be changed as the impeller diameter or revolution speed of centrifugal pump is changed (Fig.2.28).

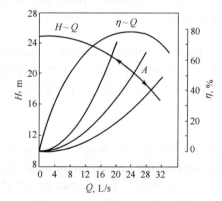

Fig. 2.27 Change of operating point by changing system curve.

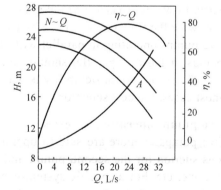

Fig. 2.28 Change of operating point by changing characteristic curve.

Normal Operating Range

A typical performance curve is a plot of Total Head vs. Flow rate for a specific impeller diameter. The plot starts at zero flow. The head at this point corresponds to the shut-off head point of the pump. The curve then decreases to a point where the flow is in maximum and the head minimum. This point is sometimes called the run-out point. The pump curve is relatively flat and the head decreases gradually as the flow increases. This pattern is common for radial flow pumps. Beyond the run-out point, the pump cannot operate. The pump's range of operation is from the shut-off head point to the run-out point. Trying to run a pump off the right end of the curve will result in pump cavitation and eventually destroy the pump.

2.2.4.7 Operation in Parallel and in Series of Centrifugal Pump

Centrifugal pumps are typically small in size and can usually be built for a relatively low cost. In addition, centrifugal pumps provide a high volumetric flow rate with a relatively low pressure. In order to increase the volumetric flow rate in a system or to compensate for large flow resistances, centrifugal pumps are often used in parallel or in series.

Two pumps operating in parallel

Two pumps or more discharging into a common line are said to operate in parallel. Fig.2.29 depicts two identical centrifugal pumps operating at the same speed in parallel. Since the inlet and the outlet of each pump shown in Fig. 2.29 are at identical points in the system, each pump must produce the same pump head. The total flow rate in the system, however, is the sum of the individual flow rates for each pump. For a specific system-head curve, two identical pumps can be selected to discharge, double the capacity of each at the same head. The performance curve for two identical pumps operating in parallel is in Fig. 2.29.

When the system characteristic curve is considered with the curve for pumps in parallel, the operating point at the intersection of the two curves represents a higher volumetric flow rate than for a single pump and a greater system head loss. As shown in Fig. 2.29, a greater system head loss occurs with the increased fluid velocity resulting from the increased volumetric flow rate. Because of the greater system head, the volumetric flow rate is actually less than twice the flow rate achieved by using a single pump.

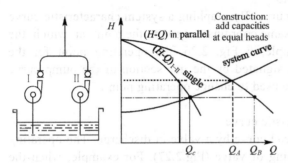

Fig. 2.29 Two pumps operating in parallel.

Two pumps operating in parallel will be more effective with a flatter system head curve. Two pumps or more in parallel should be selected in case one by itself can not perform as requirement for, or certain flexibility is necessary. Otherwise one pump by itself with a spare is, in most cases, the best selection.

Two pumps operating in series

Two pumps or more are said to operate in series when the discharge of the first pump serves as suction for the second pump, and the discharge of the second pump as suction for the third one, etc. For a specific system-head curve, two identical pumps can be selected to double the head of the first pump at the same capacity. The performance curve for two identical pumps operating in series is in Fig. 2.30.

Centrifugal pumps are used in series to overcome a larger system head loss than one pump can compensate for individually. As illustrated in Fig. 2.30, two identical centrifugal pumps operating at the same speed with the same volumetric flow rate contribute the same pump head. Since the inlet to the second pump is the outlet of the first pump, the head produced by both pumps is the sum of the individual heads. The volumetric flow rate from the inlet of the first pump to the outlet of the second remains the same.

As shown in Fig. 2.30, using two pumps in series does not actually double the resistance to flow in the system. The two pumps provide adequate pump head for the new system and also maintain a slightly higher volumetric flow rate. The operation of two pumps in series a steeper system curve will be more effective than with a flatter system curve

2.2.5 Multistage Centrifugal Pumps

The maximum head that is practicable to generate in a single impeller is limited by the peripheral speed reasonably attainable. A so-called high-energy centrifugal pump can develop a head of more than 200 m in a single

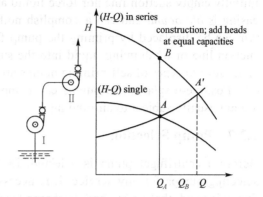

Fig. 2.30 Two pumps operating in series.

stage; but generally when a head greater than about 60 m is needed, two or more impellers can be mounted in series on a single shaft and a multistage pump so obtained. The discharge from the first stage provides suction for the second, the discharge from the second provides suction for the third, and so forth, as shown in Fig.2.31. The developed heads of all stages add to give a total head several times that of a single stage.

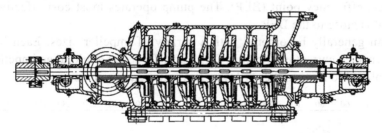

Fig. 2.31 Seven-stage diffuser-type multistage centrifugal pump.

These pumps are used for services requiring heads (pressures) higher than can be generated by a single impeller. All impellers are in series, the liquid passing from one impeller to the next and finally to the pump discharge. The total head then is the summation of the heads of the individual impellers. Deep-well pumps, high-pressure water-supply pumps, boiler-feed pumps, fire pumps and charge pumps for refinery processes are examples of multistage pumps required for various services.

Multistage pumps may be of the volute type with singe- or double-suction impellers, or of the diffuser type. They may have horizontally split casings or for extremely high pressure, 20 to 40 MPa, vertically split barrel-type exterior casings with inner casings containing diffusers, interstage passages, etc.

2.2.6 Pump Priming

If the impeller speed, the radius of the impeller, and the velocity of the fluid leaving the impeller are constant, the developed head is the same for fluids of all densities and is the same for liquids and gases.

The increase in pressure, however, is the product of the developed head and the fluid density. If a pump develops, say, a head of 30.5m and is full of water, the increase in pressure is 2.95 atm.

If the pump is full of air at ordinary density, the pressure increase is about 0.0035atm. A centrifugal pump trying to operate on air, then, can neither draw liquid upward from an initially empty suction line nor force liquid along a full discharge line. A pump with air in its casing is *air bound* and can accomplish nothing until the air has been replaced by a liquid. Air can be displaced by priming the pump from an auxiliary priming tank connected to the suction line or by drawing liquid into the suction line by an independent source of vacuum. Also, several types of self-priming pumps are available.

Positive-displacement pumps can compress a gas to a required discharge pressure and are not usually subject to air binding.

2.2.7 Pump Selection

Before a centrifugal pump is selected, its application must be clearly understood. When selecting pumps for any service, it is necessary to know the liquid to be handled, the total dynamic head, the suction and discharge heads and, in most cases, the temperature, viscosity, vapor pressure, and specific gravity. In chemical industry, the task of pumps election is frequently further complicated by the presence of solids in the liquid and liquid corrosion characteristics requiring special materials of construction.

The performance curve of centrifugal pump shows the pump's capacity plotted against total developed head. It also shows efficiency, required power input, and suction head requirements (net positive suction head requirement) over a range of flow rates. It also shows the pump's best efficiency point (BEP). The pump operates most cost effectively when the operating point is close to the BEP.

Pumps can generally be ordered with a variety of impeller sizes. Each impeller has a separate performance curve (as shown in Fig. 2.32). To select a midrange impeller that can be trimmed or replaced to meet higher or lower flow rate requirements.

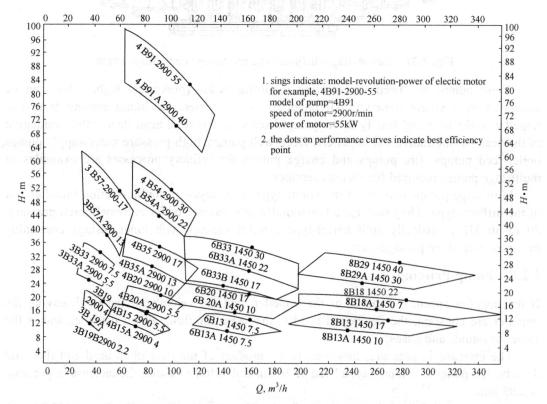

Fig. 2.32 Characteristic curves of centrifugal pumps.

To minimize pumping system energy consumption, select a pump so the system curve intersects the pump curve within 20% of its BEP. Select a pump with high efficiency contours over your range of expected operating points. A few points of efficiency improvement can save significant energy over the life of the pump.

> **Example 2.3** A process requires 56.8m³/m at a total operating head of 45.72m. Assume the centrifugal pump will be powered by a 522.388kW motor, operate for 8000 hours annually, and transport fluid with a specific gravity of 1.0. One candidate pump has an efficiency (η_1) of 81% at the operating point; a second is expected to operate at 78% efficiency (η_2). What are the energy savings given selection of the first pump?
>
> $$\text{Reduced Power Requirements} = HQ\rho g \times \left(\frac{1}{\eta_1} - \frac{1}{\eta_2}\right)$$
>
> **Solution.** Assuming an efficiency of 96% for the pump drive motor, the annual energy savings are:
> Reduced Power Requirements =
>
> $$HQ\rho g \times \left(\frac{1}{\eta_1} - \frac{1}{\eta_2}\right) = \frac{45.72 \times 56.8 \times 1000 \times 9.81}{60 \times 1000}\left(\frac{1}{0.81} - \frac{1}{0.78}\right) = 20.16\text{kW}$$
>
> Energy Savings = 20.16×8000 hours/year / 0.96 = 168000 kW·h/year
>
> These savings are valued at $8400 per year at an energy price of 5 cents per kW·h. Assuming a 15-year pump life, total energy savings are $126000. With an assumed cost differential between the two pumps of $5000, the simple payback for purchasing the first pump will be approximately 7 months.

2.3 Fans, Blowers, and Compressors

These are machines that move and compress gases. Fans discharge large volumes of gas (usually air) into open spaces or large ducts. They are low-speed machines that generate very low pressure, on the order of 0.04 atm. Blowers are high-speed rotary devices (using either positive displacement or centrifugal force) that develop a maximum pressure of about 2 atm. Compressors, which are also positive displacement or centrifugal machine, discharge at pressures from 2 atm to several thousand atmospheres. Note that while *pump* generally refers to a device for moving a liquid, the terms *air pump* and *vacuum pump* designate machines for compressing a gas.

In fans the density of the fluid does not change appreciably and may be assumed constant. In blowers and compressors, however, the density change is too great to justify this assumption, and in discussing these devices compressible flow theory is required.

2.3.1 Fans

Large fans are usually centrifugal, operating on exactly the same principle as centrifugal pumps. Their impeller blades, however, may be curved forward; this would lead to instability in a pump, but not in a fan. Typical fan impellers are shown in Fig. 2.33; they are mounted inside light sheet-metal casings. Clearances are large and discharge heads low, from 130 to 1500 mm H2O. Sometimes, as in ventilating fans, nearly all the added energy is converted to velocity energy and almost none to pressure head. In any case, the gain in velocity absorbs an appreciable fraction of the added energy and must be included in estimating the efficiency and power. The total efficiency, where the power output is credited with both pressure and velocity heads, is about 70 percent.

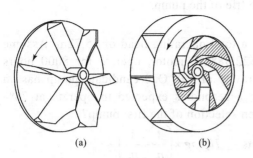

Fig. 2.33 Impellers for centrifugal fans.

Since the change in density in a fan is small, the incompressible flow equations used in the discussion of centrifugal pumps are adequate. One difference between pumps and gas equipment recognizes the effect of pressure and temperature on the density of the gas entering the machine. Gas equipment is ordinarily rated in terms of *standard cubic mete*. A volume in standard cubic meter is that measured at a specified temperature and pressure regardless of the actual temperature and pressure of the gas to the machine. Various standards are used in different industries, but a common one is based on a pressure of 760mmHg and a temperature of 0℃. This corresponds to a molal volume of 22.4m³/kmol. Another is based on 762mmHg and 15.6°C, with a molal volume of 23.6m³/kmol.

EXAMPLE 2.4 It is desired to use 28.32 m³/min of air (metered at a pressure of 101.3 kPa and 294.1 K) in a process. This amount of air, which is at rest, enters the fan suction at a pressure of 741.7 mmHg and a temperature of 366.3 K and is discharged at a pressure of 769.6 mmHg and a velocity of 45.7 m/s. A centrifugal fan having a fan efficiency of 60% is to be used. Calculate the brake-kW power needed.

Solution. Incompressible flow can be assumed, since the pressure drop is only (27.9/741.7)100, or 3.8% of the upstream pressure. The average density of the flowing gas can be used in the mechanical-energy-balance equation.

The density at the suction, point 1, is

$$\rho_1 = \frac{28.97}{22.414} \times \frac{273.2}{366.3} \times \frac{741.7}{760} = 0.940 \, \text{kg/m}^3$$

(The molecular weight of 28.97 for air, the volume of 22.414 m³/kg mol at 101.3 kPa, and 273.2 K) The density at the discharge, point 2, is

$$\rho_2 = 0.940 \frac{769.6}{741.7} = 0.975 \, \text{kg/m}^3$$

The average density of the gas is

$$\rho_{av} = \frac{\rho_1 + \rho_2}{2} = \frac{0.940 + 0.975}{2} = 0.958 \, \text{kg/m}^3$$

The mass flow rate of the gas is

$$m = \frac{28.32}{60} \times \frac{273.2}{294.1 \times 22.414} \times 28.97 = 0.5667 \, \text{kg/s}$$

The developed pressure head is

$$\frac{p_2 - p_1}{\rho_{av}} = \frac{(769.6 - 741.7)/760}{0.958} \times 101.33 \times 10^3 = 3883 \, \text{J/kg}$$

The developed velocity head for $u_1 = 0$ is

$$u_2^2 = \frac{45.7^2}{2} = 1044 \text{ J/kg}$$

Writing the mechanical-energy-balance equation,

$$z_1 g + \frac{u_1^2}{2} + \frac{p_1}{\rho} + W = z_2 g + \frac{u_2^2}{2} + \frac{p_2}{\rho} + h_f$$

Setting $z_1 = 0$, $z_2 = 0$, $u_1 = 0$, and $h_f = 0$, and solving for W,

$$W = \frac{u_2^2}{2} + \frac{p_2 - p_1}{\rho_{av}} = 1044 + 3883 = 4927 \text{ J/kg}$$

Substituting into Eq.(2.2-4),

$$N = \frac{Wm}{\eta} = \frac{4927 \times 0.5667}{0.6} = 4654 \text{W} = 4.654 \text{kW}$$

2.3.2 Blowers and Compressors

When the pressure on a compressible fluid is increased adiabatically, the temperature of the fluid also increases. The temperature rise has a number of disadvantages. Because the specific volume of the fluid increases with temperature, the work required for compressing a kilogram of fluid is larger than if the compression were isothermal. Excessive temperatures lead to problems with lubricants, stuffing boxes, and materials of construction. The fluid may be one that cannot tolerate high temperatures without decomposing.

For the isentropic (adiabatic and frictionless) pressure change of an ideal gas, the temperature relation is, using Eq.(2.3-1),

$$\frac{T_2}{T_1} = \left(\frac{p_2}{p_1}\right)^{1-\frac{1}{\gamma}} \tag{2.3-1}$$

where T_1, T_2 = inlet and outlet absolute temperatures, respectively
p_1, p_2 = corresponding inlet and outlet pressures
γ = ratio of specific heats c_p/c_v

For a given gas, the temperature ratio increases with an increase in the compression ratio p_2/p_1. This ratio is a basic parameter in the engineering of blowers and compressors. In blowers with a compression ratio below about 3 or 4, the adiabatic temperature rise is not large, and no special provision is made to reduce it. In compressors, however, where the compression ratio may be as high as 10 or more, the isentropic temperature becomes excessive. Also, since actual compressors are not frictionless, the heat from friction is absorbed by the gas, and temperatures well above the isentropic temperature are attained. Compressors, therefore, are cooled by jackets through which cold water or refrigerant is circulated. In small cooled compressors, the exit gas temperature may approach that at the inlet, and isothermal compression is achieved. In very small ones, air cooling by external fins cast integrally with the cylinder is sufficient. In larger units, where cooling capacity is limited, a path different from isothermal or adiabatic compression, called polytropic compression, is followed.

Positive-displacement blower

A positive-displacement blower is shown in Fig. 2.34(a). These machines operate as gear pumps do except that, because of the special design of the "teeth," the clearance is only a few thousandths of an inch. The relative position of the impellers is maintained precisely by heavy external gears. A single-stage blower can discharge gas at 0.4 to 1 atm gauge, a two-stage blower at 2 atm. The blower shown in Fig. 2.34(a) has two lobes. Three-lobe machines are also common.

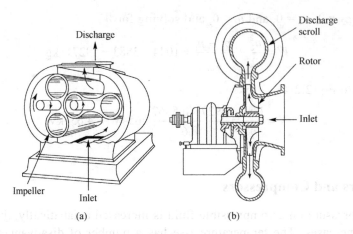

Fig. 2.34 Typical blowers: (*a*) positive-displacement two-lobe blower; (*b*) single-suction centrifugal blower.

Centrifugal blowers

A single-stage centrifugal blower is shown in Fig. 2.34(*b*). In appearance it resembles a centrifugal pump, except that the casing is narrower and the diameters of the casing and discharge scroll are relatively larger than in a centrifugal pump. The operating speed is high—3600 r/min or more. High speeds and large impeller diameters are required because very high heads, in meters or feet of low-density fluid, are needed to generate modest pressure ratios. Thus the velocities appearing in a diagram like Fig. 2.20 are, for a centrifugal blower, approximately 10 times those in a centrifugal pump.

Positive-displacement compressors

Rotary positive-displacement compressors can be used for discharge pressures up to about 6 atm. These devices include sliding-vane, screw-type, and liquid piston compressors. (See Ref. 6.) For high to very high discharge pressures and modest flow rates, reciprocating compressors are the most common type. An example of a single-stage compressor is shown in Fig. 2.35. These machines operate mechanically in the same way as reciprocating pumps, with the differences that leak prevention is more difficult and the temperature rise is important. The cylinder walls and cylinder heads are cored for cooling jackets using water or refrigerant. Reciprocating compressors are usually motor-driven and are nearly always double acting.

When the required compression ratio is greater than can be achieved in one cylinder, multistage compressors are used. Between each stage are coolers, which are tubular heat exchangers cooled by water or refrigerant. Intercoolers have sufficient heat-transfer capacity to bring the interstage gas streams to the initial suction temperature. Often an after cooler is used to cool the high-pressure gas from the final stage.

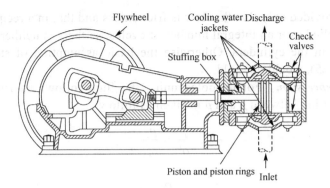

Fig. 2.35 Reciprocating compressor.

Centrifugal compressors

Centrifugal compressors are multistage units containing a series of impellers on a single shaft rotating at high speeds in a massive casing. Internal channels lead from the discharge of one impeller to the inlet of the next. These machines compress enormous volumes of air or process gas—up to 340000 m³/h at the inlet—to an outlet pressure of 20 atm. Smaller-capacity machines discharge at pressures up to several hundred atmospheres. Interstage cooling is needed on the high-pressure units. Figure 2.36 shows a typical centrifugal compressor. Axial-flow machines handle even larger volumes of gas, up to 1×10^6 m³/h, but at lower discharge pressures of 2 to 12 atm. In these units the rotor vanes propel the gas axially from one set of vanes directly to the next. Interstage cooling is normally not required.

Fig. 2.36 Interior of centrifugal compressor. (*MAN Turbomachinery Inc. USA.*)

Equations for blowers and compressors

Because of the change in density during compressible flow, the integral form of the Bernoulli equation is inadequate. Equation (1.3-25), however, can be written differentially and used to relate the shaft work to the differential change in pressure head. In blowers and compressors the mechanical, kinetic, and potential energies do not change appreciably, and the velocity and static head terms can be dropped. Also, on the assumption that the compressor is frictionless, $\eta=1.0$ and $h_f=0$. With these simplifications, Eq. (1.3-25) becomes

$$dW = \frac{dp}{\rho}$$

Integration between the suction pressure p_1 and the discharge pressure p_2 gives the work of compression of an ideal frictionless gas

$$W = \int_{p_1}^{p_2} \frac{dp}{\rho} \tag{2.3-2}$$

To use Eq.(2.3-2), the integral must be evaluated, which requires information on the path followed by the fluid in the machine from suction to discharge. The procedure is the same whether the compressor is a reciprocating unit, a rotary positive-displacement unit, or a centrifugal unit, provided only that the flow is frictionless and that in a reciprocating machine the equation is applied over an integral number of cycles, so there is neither accumulation nor depletion of fluid in the cylinders. Otherwise the basic assumption of steady flow, which underlies Eq. (1.3-25), would not hold.

Adiabatic compression. For uncooled units, the fluid follows an isentropic path. For ideal gases, the relation between p and ρ is given as follow

$$\frac{p}{\rho^\gamma} = \frac{p_1}{\rho_1^\gamma}$$

or
$$\rho = \frac{\rho_1}{p_1^{1/\gamma}} p^{1/\gamma} \tag{2.3-3}$$

Substituting ρ from Eq.(2.3-3) into Eq.(2.3-2) and integrating give

$$W = \frac{p_1^{1/\gamma}}{\rho_1} \int_{p_1}^{p_2} \frac{dp}{p^{1/\gamma}} = \frac{p_1^{1/\gamma}}{(1-1/\gamma)\rho_1}(p_2^{1-1/\gamma} - p_1^{1-1/\gamma})$$

By multiplying the coefficient by $p_1^{1-1/\gamma}$ and dividing the terms in the parentheses by the same quantity, this equation becomes

$$W = \frac{p_1 \gamma}{(\gamma-1)\rho_1}\left[\left(\frac{p_2}{p_1}\right)^{1-1/\gamma} - 1\right] = \frac{\gamma}{\gamma-1} \times \frac{RT_1}{M}\left[\left(\frac{p_2}{p_1}\right)^{1-1/\gamma} - 1\right] \tag{2.3-4}$$

Equation(2.3-4) shows the importance of the compression ratio p_2/p_1.

Isothermal compression. When cooling during compression is complete, the temperature is constant and the process is isothermal. The relation between p and ρ then is simply

$$\frac{p}{\rho} = \frac{p_1}{\rho_1} \tag{2.3-5}$$

Eliminating ρ from Eqs.(2.3-2) and (2.3-5) and integrating give

$$W = \frac{p_1}{\rho_1}\int_{p_1}^{p_2}\frac{dp}{p} = \frac{p_1}{\rho_1}\ln\frac{p_2}{p_1} = \frac{2.3026 RT_1}{M}\ln\frac{p_2}{p_1} \tag{2.3-6}$$

For a given compression ratio and suction condition, the work requirement in isothermal compression is less than that for adiabatic compression. This is one reason why cooling is useful in compressors.

A close relation exists between the adiabatic and isothermal cases. By comparing the integrands in the equation above, it is clear that if $\gamma =1$, the equations for adiabatic and for isothermal compression are identical.

Polytropic compression. In large compressors the path of the fluid is neither isothermal nor adiabatic. The process may still be assumed to be frictionless, however. It is customary to assume that the relation between pressure and density is given by

$$\frac{p}{\rho^n} = \frac{p_1}{\rho_1^n} \tag{2.3-7}$$

where n is a constant. Use of this equation in place of Eq.(2.3-3) obviously yields Eq.(2.3-4) with the replacement of γ by n.

The value of n is found empirically by measuring the density and pressure at two points on the path of the process, for example, at suction and discharge. The value of n is calculated by the equation

$$n = \frac{\ln(p_2/p_1)}{\ln(\rho_2/\rho_1)} \qquad (2.3\text{-}8)$$

This equation is derived by substituting p_2 for p and ρ_2 for ρ in Eq.(2.3-7) and taking logarithms.

Compressor efficiency. The ratio of the theoretical work (or fluid power) to the actual work (or total power input) is, as usual, the efficiency and is denoted by η. The maximum efficiency of reciprocating compressors is about 80 to 85 percent; it can be up to 90 percent for centrifugal compressors.

Power equation. To calculate the brake power when the efficiency is η,

$$N = \frac{Wm}{\eta} \qquad (2.3\text{-}9)$$

where m = kg gas/s and W = J/kg.

The power required by an adiabatic compressor is readily calculated from Eq.(2.3-4). The *dimensional* formula is

$$N = \frac{m\gamma}{(\gamma-1)\eta}\frac{RT_1}{M}\left[\left(\frac{p_2}{p_1}\right)^{1-1/\gamma} - 1\right] \qquad (2.3\text{-}10)$$

where N = power, W
  m = mass of gas compressed, kg gas/s
  T_1 = inlet temperature, K
For isothermal compression

$$N = \frac{2.3026 RT_1 m}{M\eta}\ln\frac{p_2}{p_1} \qquad (2.3\text{-}11)$$

EXAMPLE 2.5 A three-stage reciprocating compressor is to compress 306 std m³/h of methane from 0.95 to 61.3 atm abs. The inlet temperature is 26.7°C. For the expected temperature range the average properties of methane are

C_p = 38.9 J/g mol °C γ = 1.31

(a) What is the brake horsepower if the mechanical efficiency is 80 percent? (b) What is the discharge temperature from the first stage? (c) If the temperature of the cooling water is to rise 11.1°C, how much water is needed in the intercoolers and after cooler for the compressed gas to leave each cooler at 26.7°C? Assume that jacket cooling is sufficient to absorb frictional heat.

Solution
(a) For a multistage compressor it can be shown that the total power is a minimum if each stage does the same amount of work. By Eq.(2.3-4) this is equivalent to the use of the same compression ratio in each stage. For a three-stage machine, therefore, the compression ratio of one stage should be the cube root of the overall

$$\frac{p_b}{p_a} = \left(\frac{61.3}{0.95}\right)^{\frac{1}{3}} = 4 \qquad m = \frac{306}{3600 \times 22.414} \times 16 = 0.0607 \text{kg/s}$$

The power required for each stage is, by Eq.(2.3-10),

$$N = \frac{m\gamma}{(\gamma-1)\eta} \frac{RT_1}{M}\left[\left(\frac{p_2}{p_1}\right)^{1-\frac{1}{\gamma}} - 1\right] =$$

$$= \frac{0.0607 \times 1.31}{(1.31-1) \times 0.8} \frac{8314.3(273.2+26.7)}{16}\left[4^{1-1/1.31} - 1\right]$$

$$= 49967.8\left(4^{0.2366} - 1\right) = 49967.8 \times 0.388 = 19387.5 \text{J/s}$$

The total power for all stages is 3 × 19387.5= 58162.5J/s=58.2 kW.
(b) From Eq.(2.3-1), the temperature at the exit of each stage is

$$\frac{T_2}{T_1} = \left(\frac{p_2}{p_1}\right)^{1-\frac{1}{\gamma}} = 4^{1-1/1.31} = 4^{0.2366} = 1.388$$

$$T_2 = 1.388 T_1 = 1.388 \times (273.2+26.7) = 416.3 \text{K}$$

(c) The heat load in each cooler is

$$mc_p(T_2 - T_1) = 0.0607 \times 38.9(416.3 - 299.9)/16 = 17.18 \text{kJ/s}$$

The total heat load is 3×17.18=51.53kJ/s. The cooling water requirement is

$$m_w c_p \Delta t = 51.53$$

$$m_w = \frac{51.53}{c_p \Delta t} = \frac{51.53}{4.197 \times 11.1} = 1.11 \text{kg/s}$$

2.3.3 Vacuum pumps

A compressor that takes suction at a pressure below atmospheric and discharges against atmospheric pressure is called a *vacuum pump*. Any type of blower or compressor—reciprocating, rotary, or centrifugal—can be adapted to vacuum practice by modifying the design to accept very low-density gas at the suction and attain the large compression ratios necessary. As the absolute pressure at the suction decreases, the volumetric efficiency drops and approaches zero at the lowest absolute pressure attainable by the pump. Usually the mechanical efficiency is also lower than that for compressors. The required displacement increases rapidly as the suction pressure falls, so a large machine is needed to move much gas. The compression ratio used in vacuum pumps is higher than that in compressors, ranging up to 100 or more, with a correspondingly high adiabatic discharge temperature. Actually, however, the compression is nearly isothermal because of the low mass flow rate and the effective heat transfer from the relatively large area of exposed metal.

Jet ejectors
An important kind of vacuum pump that does not use moving parts is the jet ejector, shown in Fig. 2.37, in which the fluid to be moved is entrained in a high-velocity stream of a second fluid. The motive fluid and the fluid to be moved may be the same, such as when

compressed air is used to move air, but usually they are not. Industrially greatest use is made of steam-jet ejectors, which are valuable for drawing a fairly high vacuum. As shown in Fig. 2.37, steam at about 7 atm is admitted to a converging-diverging nozzle, from which it issues at supersonic velocity into a diffuser cone. The air or other gas to be moved is mixed with the steam in the first part of the diffuser, lowering the velocity to acoustic velocity or below; in the diverging section of the diffuser the kinetic energy of the mixed gases is converted to pressure energy, so that the mixture can be discharged directly to the atmosphere. Often it is sent to a water-cooled condenser, particularly if more than one stage is used. Otherwise each stage would have to handle all the steam admitted to the preceding stages. As many as five stages are used in industrial processing.

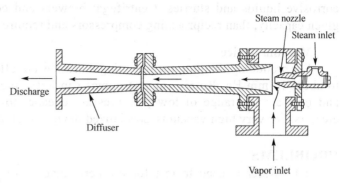

Fig. 2.37 Steam-jet ejector.

Jet ejectors require very little attention and maintenance and are especially valuable with corrosive gases that would damage mechanical vacuum pumps. For difficult problems the nozzles and diffusers can be made of corrosion-resistant metal, graphite, or other inert material. Ejectors, particularly when multistage, use large quantities of steam and water. They are rarely used to produce absolute pressures below 1 mm Hg. Steam jets are no longer as popular as they once were, because, of the great increase in the cost of steam. In many instances where corrosion is not a serious consideration, they have been replaced by mechanical vacuum pumps, which use much less energy for the same service.

2.3.4 Comparison of Devices for Moving Fluids

In all types of fluid-moving equipment, the flow capacity, power requirements, and mechanical efficiency are all highly important. Reliability and ease of maintenance are also highly desirable, often essential. In small units, simplicity and trouble-free operation are usually more important than high mechanical efficiency with its saving of a few kilowatts of power.

Positive-displacement machines

Positive-displacement machines, in general, handle smaller quantities of fluids at higher discharge pressures than do centrifugal machines. Positive-displacement pumps are not subject to air binding and are usually self-priming. In both positive-displacement pumps and blowers, the discharge rate is nearly independent of the discharge pressure, so that these machines are extensively used for controlling and metering flow. Reciprocating devices require considerable maintenance but can produce the highest pressures. They deliver a pulsating stream. Rotary pumps work best on fairly viscous lubricating fluids, discharging a steady stream at moderate to high pressures. They cannot be used with slurries. Rotary blowers usually discharge gas at a maximum pressure of 2 atm from a single stage. The discharge line of a positive-displacement pump cannot be closed without stalling or breaking the pump, so that a bypass line with a pressure relief valve is required.

Centrifugal machines

Centrifugal pumps, blowers, and compressors all deliver fluid at a uniform pressure without shocks or pulsations. They run at higher speeds than positive-displacement machines and are connected to the motor drive directly instead of through a gearbox. The discharge line can be completely closed without damage. Centrifugal pumps can handle a wide variety of corrosive liquids and slurries. Centrifugal blowers and compressors are much smaller, for given capacity, than reciprocating compressors and require less maintenance.

Vacuum devices

For producing vacuum, reciprocating machines are effective for absolute pressures down to about 10 mm Hg. Rotary vacuum pumps can lower the absolute pressure to 0.01 mm Hg and over a wide range of low pressures are cheaper to operate than multistage steam-jet ejectors. For very high vacuums, specialized devices such as diffusion pumps are needed.

PROBLEMS

2.1 Water is used to test for the performances of pump. The gauge pressure at the discharge connection is 152 kPa and the reading of vacuum gauge at the suction connection of the pump is 24.7 kPa as the flow rate is 26m³/h. The shaft power is 2.45kW while the centrifugal pump operates at the speed of 2900r/min. If the vertical distance between the suction connection and discharge connection is 0.4m, the diameters of both the suction and discharge line are the same. Calculate the mechanical efficiency of pump and list the performance of the pump under this operating condition.

2.2 Water is transported by a pump from reactor, which has 200 mm Hg vacuum, to the tank, in which the gauge pressure is 0.5 kgf/cm², as shown in Fig. The total equivalent length of pipe is 200 m including all local frictional loss. The pipeline is $\phi 57\times 3.5$ mm , the orifice coefficient of C_o and orifice diameter d_o are 0.62 and 25 mm, respectively. Frictional coefficient λ is 0.025. Calculate: Developed head H of pump, in m (the reading R of U pressure gauge in orifice meter is 168 mm Hg).

2.3 . A centrifugal pump is to be used to extract water from a condenser in which the vacuum is 640 mm of mercury, as shown in figure. At the rated discharge, the net positive suction head must be at least 3m above the cavitation vapor pressure of 710mm mercury vacuum. If losses in the suction pipe accounted for a head of 1.5m. What must be the least height of the liquid level in the condenser above the pump inlet?

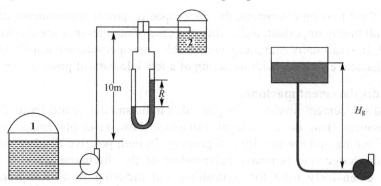

Figure for problem 2.2　　　　　Figure for problem 2.3

2.4 Sulphuric acid is pumped at 3 kg/s through a 60m length of smooth 25 mm pipe. Calculate the drop in pressure. If the pressure drop falls by one half, what will the new flowrate be ?
- Density of acid　　1840kg/m³
- Viscosity of acid　　25×10^{-3}Pa • s

2.5 An oil pump (type of 65Y-60B) is to be used to pump oil (density of 760kg/m³) through a pipe of 53mm diameter at the rate of 15m³/h from the oil tank (opening to the air) to an equipment in which the gauge pressure is 15mH₂O. The level of oil in the tank is kept constant. The center line of exit is 5m above the level of oil in the oil tank. Under the operating condition, the vapor pressure of the oil is 80kPa. If the friction losses in the suction line and discharge line are 1m and 4m, respectively. Try to calculate whether the pump can work in order if the oil pump is installed at 1.2m below the level of oil in the tank? Suppose the local atmosphere pressure is 101.33kPa.

The performance of 65Y-60B oil pump is listed as follows

Flow rate, m³/h	19.8	Efficiency, %	55
Total head, m	38	NPSH, m	2.6
Shaft power, kW	3.75		

2.6 The fluid is pumped through the horizontal pipe from section A to B with the $\varphi 38 \times 2.5$mm diameter and length of 30 meters, shown as figure. The orifice meter of 16.4mm diameter is used to measure the flow rate. Orifice coefficient $C_o = 0.63$, the permanent loss in pressure is $3.5 \times 10^4 \text{N/m}^2$, the friction coefficient $\lambda = 0.024$. Find:

(a) What is the pressure drop along the pipe AB?

(b) What is the ratio of power obliterated in pipe AB to total power supplied to the fluid when the shaft work is 500W, 60% efficiency? (The density of fluid is 870kg/m³)

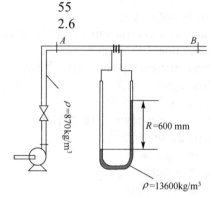

Figure for problem 2.6

2.7 The flow of water through a 50mm pipe is measured by means of an orifice meter with a hole of 40mm diameter. The pressure difference recorded is 150mm on a mercury-under-water manometer and the orifice coefficient of the meter is 0.6. What is the pressure drop as the water flows through over a 30m length of the pipe? What type of pump would you use in light of service requiement? See Figure for problem 2.7. And what type of pump would you use if the flowrate is increased by 40%(assuming that friction coefficient $\lambda = 0.024$ and remains constant, density of mercury = 13600kg/m³)

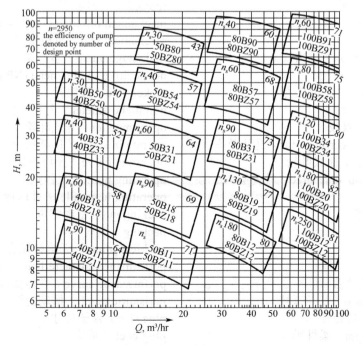

Figure for problem 2.7

名 词 术 语

英文 / 中文

actual head / 实际压头
adiabatic compression / 绝热压缩
affinity law / 相似定律
agitator / 搅拌器
air bound / 气缚
automatic control valve / 自动控制阀
axial flow / 轴向流
best efficiency point / 最佳效率点
black-iron / 黑铁
blowcase / 加压箱
brake horse power / 制动功率
cam pump / 凸轮泵
capacity / 流率
carbon steel / 碳钢
cast-iron / 铸铁
cavitation / 气蚀
centrifugal blower / 离心鼓风机
centrifugal compressor / 离心压缩机
centrifugal pump / 离心泵
characteristic curve / 特性曲线
check valve / 止回阀
circular casing / 环壳
circulatory flow / 环流
cold-drawn / 冷拔
developed head / 扬程
diaphragm pump / 隔膜泵
diffuser / 播散叶
discharge connection / 排出管
discharge nozzle / 排出口
discharge pressure / 排出压力
double-acting / 双动
double-suction / 双吸
efficiency of the electric motor / 马达效率
enclosed impeller / 闭式叶轮
end suction / Top discharge / 轴向吸入 / 顶部排出
expansion / 膨胀
fan / 通风机
fittings / 管件
flange / 法兰
flanged cap / 法兰盖
flanged connection / 法兰联接
flare fitting / 扩口管件
flexible diaphragm / 柔性膜
flexible metal hose / 挠性金属管

英文 / 中文

follower ring / 从动环
fouling / 污垢
fully close / 全关
fully open / 全开
gate valve / 闸式阀
gear pump / 齿轮泵
globe valve / 球心阀
graphite / 石墨
handwheel / 手轮
head / 压头
head-capacity relation / 压头流量关系
heat-exchanger / 热交换器
heavy-walled / 厚壁
ideal pump / 理想泵
impeller / 叶轮
internal-gear pump / 内啮合齿轮泵
isothermal compression / 等温压缩
jet ejector / 喷射器
kinetic energy / 动能
lantern gland / 液封环
lantern ring / 压盖
leakage / 泄漏
lobe pump / 凸轮旋转泵
low-carbon steel / 低碳钢
lubricant / 滑润剂
mechanical efficiency / 机械效率
mechanical seal / 机械密封
multistage pump / 多级泵
net positive suction head (NPSH) / 气蚀余量
nickel pipe / 镀镍管
nominal diameter / 公称直径
open impeller / 开式叶轮
operate in series / 串联操作
operating curve / 操作曲线
operating point / 工作点
operation in parallel / 并联操作
packing nut / 密封螺母
packing / 填料
packless joints / 无填料接合
performance characteristics / 性能特性
performance curve / 性能曲线
peristaltic pump / 蠕动的泵
pipe size / 管子尺寸
plug cock / 旋塞
plunger pump / 柱塞泵

英文 / 中文	英文 / 中文
polytropic compression / 多变压缩	spur-gear pump / 正齿轮泵
Polyvinyl chloride, or PVC / 聚氯乙烯	stationary diffusion vane / 静态导叶
positive-displacement blower / 正位移鼓风机	steam trap / 疏水器
	stoppage / 中断
positive-displacement pump / 正位移泵	stuffing box / 填料箱
pressure head / 压头	suction and discharge nozzle / 吸入和排出口
pump priming / 泵的启动	suction connection / 吸入管
pumpable / 可用泵输送的	suction lift / 安装高度
radial flow / 径向流	suction line / 吸入管
reciprocating compressor / 往复式压缩机	system head curve / 管路特性曲线
reciprocating piston / 往复活塞	theoretical head / 理论压头
required input power / 输入功率(制动功率)	top suction，top discharge nozzle / 顶部吸入 / 顶部排出
rotary positive-displacement compressor / 旋转式正位移压缩机	transportation of fluid / 流体输送
rotary pump / 旋转泵	vacuum pump / 真空泵
rotating member / 回转空腔	vane pump / 叶片泵
screw pump / 螺杆泵	vapor pressure / 饱和蒸汽压
screwed fitting / 螺纹接头	variable-speed driver / 变速器
sealed chamber / 密封室	volumetric efficiency / 体积效率
semi-open impeller / 半开式叶轮	volumetric flow rate / 体积流量
single-acting / 单动	volute / 蜗壳
single-suction / 单吸	wastewater line / 废水管路
soldered fitting / 焊接管件	wrought-iron / 熟铁

Chapter 3

Heterogeneous Flow and Separation

3.1 Flow Past Immersed Objects

In many problems, the effect of the fluid on the solid is of interest. The fluid may be at rest and the solid moving through it; the solid may be at rest and the fluid flowing past it; or both may be moving. The situation in which the solid is immersed in, and surrounded by, fluid is the subject of this chapter. It is generally immaterial which phase, solid or fluid, is assumed to be at rest, and it is the relative velocity between the two that is important. An exception to this is seen in some situations when the fluid stream has been previously influenced by solid walls and is in turbulent flow. The scale and intensity of turbulence then may be important parameters in the process.

3.1.1 Drag and Drag Coefficients

The force in the direction of flow exerted by the fluid on the solid is called *drag*. By Newton's third law of motion, an equal and opposite net force is exerted by the object on the fluid. When the wall of the object is parallel with the direction of flow, as in the case of the thin flat plate shown in Fig. 1.11(a), the only drag force is the wall shear τ_w. More generally, however, the wall of an immersed object makes an angle with the direction of flow. Then the component of the wall shear in the direction of flow contributes to drag. Another contribution comes from the fluid pressure which acts in a direction normal to the wall; drag comes from the pressure component in the direction of flow. The total drag on an element of area is the sum of the two components. An extreme example is the drag of a flat plate perpendicular to the flow, as shown in Fig. 1.11(b), where the drag is due entirely to the pressure component.

Fig.3.1 shows the pressure and shear forces acting on an element of area dA inclined at an angle of $90° - \alpha$ to the direction of flow. The drag from wall shear is $\tau_w \sin \alpha \, dA$, and that from pressure is $p \cos \alpha \, dA$. The total drag on the object is the sum of the integrals of these quantities, each evaluated over the entire surface of the body in contact with the fluid. The total integrated drag from wall shear is called *wall drag,* and the total integrated drag from pressure is called *form drag*.

In potential flow, $\tau_w = 0$, and there is no wall drag. Also, the pressure drag in the direction of flow is balanced by an equal force in the opposite direction, and the integral of the form drag is zero. There is no net drag in potential flow.

The phenomena causing both wall and form drag in actual fluids are complicated, and in general the drag cannot be predicted. For spheres and other regular shapes at low fluid velocities, the flow patterns and drag forces can be estimated from published correlations or by numerical calculations using the general momentum balance equation. For irregular shapes and high velocities they are most easily determined by experiment.

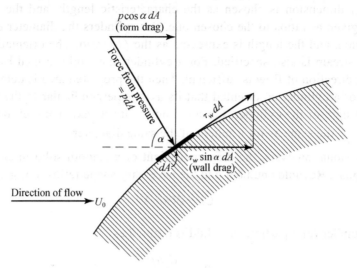

Fig. 3.1 Wall drag and form drag on immersed object.

Drag coefficients

In treating fluid flow through pipes and channels, a friction factor, defined as the ratio of the shear stress to the product of the velocity head and density, was shown to be useful. An analogous factor, called the *drag coefficient,* is used for immersed solids. Consider a smooth sphere immersed in a flowing fluid and at a distance from the solid boundary of the stream sufficient for the approaching stream to be at a uniform velocity. Define the projected area of the solid object as the area obtained by projecting the object on a plane perpendicular to the direction of flow, as shown in Fig. 3.2. Denote the projected area by A_P. For a sphere, the projected area is that of a great circle, or $(\pi/4)\,d_p^2$, where d_p is the diameter. If F_D is the total drag, the average drag per unit projected area is F_D/A_P. Just as the friction factor f is defined as the ratio of τ_w to the product of the density of the fluid and the velocity head, so the drag coefficient C_D is defined as the ratio of F_D/A_P to this same product, or

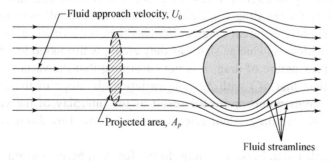

Fig. 3.2 Flow past immersed sphere.

$$C_D \equiv \frac{F_D/A_p}{\rho u^2/2} \qquad (3.1\text{-}1)$$

where u is the velocity of the approaching stream (by assumption u is constant over the projected area).

For particles having shapes other than spherical, it is necessary to specify the size and geometric form of the object and its orientation with respect to the direction of flow of the fluid. One major dimension is chosen as the characteristic length, and the other important dimensions are given as ratios to the chosen one. For cylinders the diameter d is taken as the defining dimension, and the length is expressed as the L/d ratio. The orientation between the particle and the stream is also specified. For a cylinder, the angle formed by the axis of the cylinder and the direction of flow is sufficient. Then the projected area is determinate and can be calculated. For a cylinder so oriented that its axis is perpendicular to the flow, A_p is Ld_p, where L is the length of the cylinder. For a cylinder with its axis parallel to the direction of flow, A_p is $(\pi/4)d_p^2$, the same as for a sphere of the same diameter.

From dimensional analysis, the drag coefficient of a smooth solid in an incompressible fluid depends upon a Reynolds number and the necessary shape ratios. For given shape

$$C_D = \phi(Re_p)$$

The Reynolds number for a particle in a fluid is defined as

$$Re_p = \frac{d_p u \rho}{\mu} \tag{3.1-2}$$

where d_p = characteristic length

A different C_D-versus-Re_p relation exists for each shape and orientation. The relation must in general be determined experimentally, although a well-substantiated theoretical equation exists for smooth spheres at low Reynolds numbers.

Drag coefficients for compressible fluids increase with an increase in the Mach number when the latter becomes greater than about 0.6. Coefficients in supersonic flow are generally greater than in subsonic flow.

Drag coefficients of typical shapes

In Fig. 3.3, curves of C_D versus Re_p are shown for spheres, long cylinders, and disks. The curves are based on drag measurements for objects held in a fixed position in a flowing stream. The axis of the cylinder and the face of the disk are perpendicular to the direction of flow, and these curves apply only when this orientation is maintained. When a disk or cylinder falls by gravity through a quiescent fluid, it will twist and turn as it falls, giving quite different values for the drag coefficient. Spheres in free fall may rotate and follow a spiral path, giving slightly different drag coefficients than for a fixed sphere. The behavior of drops and bubbles, which can change shape as they move, is discussed in a later section.

From the complex nature of drag, it is not surprising that the variation of C_D with Re_p is more complicated than that of f with Re. The variations in slope of the curves of C_D versus Re_p at different Reynolds numbers are the result of the interplay of the various factors that control form drag and wall drag. Their effects can be followed by discussing the case of the sphere.

For low Reynolds numbers, the drag force for a sphere conforms to a theoretical equation called *Stokes' law,* which may be written

$$F_D = 3\pi\mu d_p u \tag{3.1-3}$$

The equation for the terminal velocity of a sphere at low Reynolds numbers, Eq. (3.2-16), is also called Stokes' law.

From Eq. (3.1-3), the drag coefficient predicted by Stokes' law, using Eq. (3.1-1), is

$$C_D = \frac{24}{d_p u \rho / \mu} = \frac{24}{Re_p} \qquad (3.1\text{-}4)$$

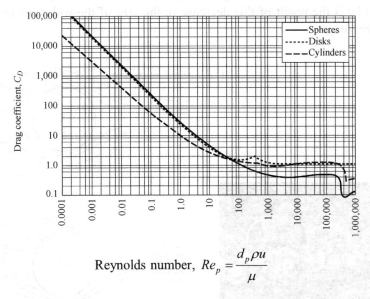

Fig. 3.3 Drag coefficients for spheres, disks, and cylinders.

In theory, Stokes' law is valid only when Re_p is considerably less than unity. Practically, as shown by the left-hand portion of the graph of Fig. 3.3, Eqs. (3.1-3) and (3.1-4) may be used with small error for all Reynolds numbers less than 1. At the low velocities at which the law is valid, the sphere moves through the fluid by deforming it. The wall shear is the result of viscous forces only, and inertial forces are negligible. The motion of the sphere affects the fluid at considerable distances from the object, and if there is a solid wall within 20 or 30 diameters of the sphere, Stokes' law must be corrected for the wall effect. The type of flow treated in this law is called *creeping flow*. The law is especially valuable for calculating the resistance of small particles, such as dust or fogs, moving through gases or liquids of low viscosity or for the motion of larger particles through highly viscous liquids.

As the Reynolds number increases beyond $Re_p = 1$, the flow pattern behind the sphere becomes different from that in front of the sphere, and the drag coefficient becomes greater than the Stokes' law limit of $24/Re_p$. At a Reynolds number of about 20, a zone of recirculating flow develops near the rear stagnation point. The recirculation zone or wake increases in size with increasing Reynolds number, and at $Re_p = 100$ the wake covers nearly one-half of the sphere. The large eddies or vortices in the wake dissipate considerable mechanical energy and cause the pressure to be much less than the upstream pressure. This makes the form drag quite large relative to the drag caused by wall shear.

At moderate Reynolds numbers of 200 to 300, oscillations develop in the wake and vortices disengage from the wake in a regular fashion, forming in the downstream fluid a series of moving vortices or a "vortex street," illustrated in Fig. 3.4. A similar vortex street is formed when a fluid flows across wires or long cylinders and is responsible for the hum of telephone wires in the wind. The frequency of vibration can be used to measure flow rate.

For $Re_p = 10^3$ to 10^5 the drag coefficient is nearly constant at 0.40 to 0.45, changing only slightly as the point of boundary layer separation slowly shifts toward the nose of the sphere. Fig.3.5(a) shows the flow pattern for $Re_p \cong 10^5$, where the boundary layer on the front part of the sphere is still laminar and the angle of separation is 85°. When the front boundary layer becomes turbulent at $Re_p \cong 300{,}000$, the separation point moves toward the rear of the sphere and the wake shrinks, as shown in Fig. 3.5(b). The remarkable drop in the drag coefficient from 0.45 to 0.10 is the result of this decrease in the size of the wake and the corresponding decrease in form drag.

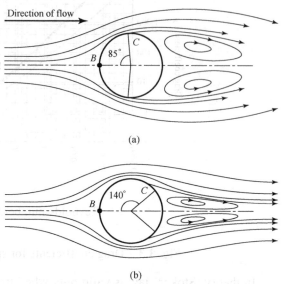

Fig. 3.4 Vortex street downstream of a vertical cylinder (at left edge of figure).

Fig. 3.5 Flow past single sphere, showing separation and wake formation: (*a*) laminar flow in boundary layer; (*b*) turbulent flow in boundary layer; *B*, stagnation point; *C*, separationpoint.

3.1.2 Flow through Beds of Solids

In many technical processes, liquids or gases flow through beds of solid particles. Important examples are filtration and the two-phase countercurrent flow of liquid and gas through packed towers. In filtration, the bed of solids consists of small particles that are removed from the liquid by a filter cloth or fine screen. In other equipment, such as ion-exchange or catalytic reactors, a single fluid (liquid or gas) flows through a bed of granular solids. Filtration is discussed in section 3.4 and packed towers in chapter 8. The present treatment is restricted to the flow of a single fluid phase through a column of stationary solid particles.

The resistance to the flow of a fluid through the voids in a bed of solids is the resultant of the total drag of all the particles in the bed. Depending on the Reynolds number, $dpu\rho/\mu$, laminar flow, turbulent flow, form drag, separation, and wake formation occur. As in the drag of a single solid particle, there is no sharp transition between laminar and turbulent flow like that occurring in flow through pipes and channels of constant cross section.

The most common methods of calculating the pressure drop through a bed of solids are based on estimates of total drag on the solid boundaries of the tortuous channels through the bed. The actual channels are irregular in shape, have a variable cross section and orientation, and are highly interconnected. However, to calculate an equivalent channel diameter, it is assumed that the bed has a set of uniform circular channels whose total surface area and void volume match those of the bed. The total surface area is the surface area per particle times the number of particles, but it is more convenient to base the calculation on the volume fraction particles in the bed and the surface-volume ratio for the particles. This ratio is $6/d_p$ for a sphere, since $s_p = \pi d_p^2$ and $v_p = \frac{1}{6}\pi d_p^3$. For other shapes or irregular particles, the equation for surface-volume ratio includes a sphericity Φs, defined as the surface-volume ratio for a sphere of diameter d_p divided by the surface-volume ratio for the particle whose nominal size is d_p. Thus

$$\Phi_s = \frac{6/d_p}{s_p/v_p}$$

or

$$\frac{s_p}{v_p} = \frac{6}{\Phi_s d_p} \tag{3.1-5}$$

Values of the sphericity of several shapes are given in Table 3.1. Although cubes and short cylinders have a sphericity of 1.0, they have more surface area than spheres of the same volume. A short cylinder with the same volume as a sphere of diameter d has a nominal size of $0.874d$ and a surface area $3.60d^2$, compared to the sphere with area $3.14\ d^2$.

The volume fraction particles in the bed is $1 - \varepsilon$, where ε is the porosity or void fraction. If the particles are porous, the pores are generally too small to permit any significant flow through them, so ε is taken to be the external void fraction of the bed and not the total porosity.

Table 3.1 Sphericity of miscellaneous materials

Material	Sphericity	Material	Sphericity
Spheres, cubes, short cylinders ($L = d_p$)	1.0	Ottawa sand	0.95
Raschig rings ($L = d_p$)		Rounded sand	0.83
		Coal dust	0.73
$L = d_o$, $d_i = 0.5\ d_o$	0.58	Flint sand	0.65
$L = d_o$, $d_i = 0.75 d_o$	0.33	Crushed glass	0.65
Berl saddles	0.3	Mica flakes	0.28

To determine the equivalent channel diameter d_e, the surface area for n parallel channels of length L is set equal to the surface-volume ratio times the particle volume $S_0L(1 - \varepsilon)$, where S_0 is the cross-sectional area of the bed:

$$n\pi d_e L = SL(1-\varepsilon)\frac{6}{\Phi_s d_p} \tag{3.1-6}$$

The void volume in the bed is the same as the total volume of the n channels:

$$SL\varepsilon = \frac{1}{4}n\pi d_e^2 L \tag{3.1-7}$$

Combining Eqs. (3.1-6) and (3.1-7) gives an equation for d_e:

$$d_e = \frac{2}{3}\Phi_s d_p \frac{\varepsilon}{1-\varepsilon} \tag{3.1-8}$$

For the typical void fraction of 0.4, $d_e = 0.44\Phi s d_p$, or the equivalent diameter is roughly one-half the particle size.

The pressure drop depends on the average velocity in the channels u', which is proportional to the superficial or empty-tower velocity u and inversely proportional to the porosity:

$$u' = \frac{u}{\varepsilon} \qquad (3.1\text{-}9)$$

With the average velocity and channel size now expressed in terms of the measurable parameters u, d_p, and ε, the channel model can be used to predict the form of the correlation for pressure drop. For flow at very low Reynolds numbers, the pressure drop should vary with the first power of the velocity and inversely with the square of the channel size, in accordance with the Hagen-Poiseuille equation for laminar flow in straight tubes, Hagen-Poiseuille equation. The equations for u' and d_e are used in Hagen-Poiseuille equation, and a correction factor λ_1 is added to account for the fact that the channels are actually tortuous and not straight and parallel:

$$\frac{\Delta p}{L} = \frac{32 u' \mu}{d^2} = \frac{32 \lambda_1 u \mu}{\frac{4}{9}\varepsilon \Phi_s^2 d_p^2} \times \frac{(1-\varepsilon)^2}{\varepsilon^2} \qquad (3.1\text{-}10)$$

or

$$\frac{\Delta p}{L} = \frac{72 \lambda_1 u \mu}{\Phi_s^2 d_p^2} \times \frac{(1-\varepsilon)^2}{\varepsilon^3} \qquad (3.1\text{-}11)$$

Several studies have shown that the form of Eq. (3.1-11) is correct, and experiments give an empirical constant of 150 for $72\lambda_1$:

$$\frac{\Delta p}{L} = \frac{150 \mu u}{\Phi_s^2 d_p^2} \times \frac{(1-\varepsilon)^2}{\varepsilon^3} \qquad (3.1\text{-}12)$$

Equation (3.1-12) is called the *Kozeny-Carman* equation and is applicable for flow through beds at particle Reynolds numbers up to about 1.0. There is no sharp transition to turbulent flow at this Reynolds number, but the frequent changes in shape and direction of the channels in the bed lead to significant kinetic energy losses at higher Reynolds numbers. The constant 150 corresponds to $\lambda_1 = 2.1$, which is a reasonable value for the tortuosity factor. For a given system, Eq. (3.1-12) indicates that the flow is proportional to the pressure drop and inversely proportional to the fluid viscosity. This statement is also known as *Darcy's law*, which is often used to describe flow of liquids through porous media.

As the flow rate through a packed bed increases, the slope of the p-versus-u plot gradually increases, and at very high Reynolds numbers, p varies with the 1.9 or 2.0 power of the superficial velocity. An empirical correlation for pressure drop in packed beds at high Reynolds number ($Re_p > 1000$) is the *Burke-Plummer* equation

$$\frac{\Delta p}{L} = \frac{1.75 \rho u^2}{\Phi_s d_p} \times \frac{1-\varepsilon}{\varepsilon^3} \qquad (3.1\text{-}13)$$

An equation covering the entire range of flow rates can be obtained by assuming that the viscous losses and the kinetic energy losses are additive. The result is called the *Ergun* equation:

$$\frac{\Delta p}{L} = \frac{150 \mu u}{\Phi_s^2 d_p^2} \times \frac{(1-\varepsilon)^2}{\varepsilon^3} + \frac{1.75 \rho u^2}{\Phi_s d_p} \times \frac{(1-\varepsilon)}{\varepsilon^3} \qquad (3.1\text{-}14)$$

Ergun showed that Eq. (3.1-14) fitted data for spheres, cylinders, and crushed solids over a wide range of flow rates. He also varied the packing density for some materials to verify the $(1-\varepsilon)^2/\varepsilon^3$ term for the viscous loss part of the equation and the $(1-\varepsilon)/\varepsilon^3$ term for the kinetic energy part. Note that a small change in ε has a very large effect on p, which makes it difficult to predict p accurately and to reproduce experimental values after a bed is repacked.

3.2 Motion of Particles through Fluids

Many processing steps, especially mechanical separations, involve the movement of solid particles or liquid drops through a fluid. The fluid may be gas or liquid, and it may be flowing or at rest. Examples are the elimination of dust and fumes from air or flue gas, the removal of solids from liquid wastes, and the recovery of acid mists from the waste gas of an acid plant.

Mechanics of particle motion

The movement of a particle through a fluid requires an external force acting on the particle. This force may come from a density difference between the particle and the fluid, or it may be the result of electric or magnetic fields. In this section only gravitational or centrifugal forces, which arise from density differences, are considered.

Three forces act on a particle moving through a fluid: (1) the external force, gravitational or centrifugal; (2) the buoyant force, which acts parallel with the external force but in the opposite direction; and (3) the drag force, which appears whenever there is relative motion between the particle and the fluid. The drag force acts to oppose the motion and acts parallel with the direction of movement but in the opposite direction.

In the general case, the direction of movement of the particle relative to the fluid may not be parallel with the direction of the external and buoyant forces, and the drag force then makes an angle with the other two. In this situation, which is called *two-dimensional motion*, the drag must be resolved into components, and this complicates the treatment of particle mechanics. Equations are available for two dimensional motion, but only the one-dimensional case, where the lines of action of all forces acting on the particle are collinear, is considered in this book.

Equations for one-dimensional motion of particle through fluid

Consider a particle of mass m moving through a fluid under the action of an external force F_e. Let the velocity of the particle relative to the fluid be u. Let the buoyant force on the particle be F_b, and let the drag be F_D. Then the resultant force on the particle is $F_e - F_b - F_D$, the acceleration of the particle is du/dt, and by Eq. (0.33), since m is constant,

$$m\frac{du}{dt} = F_e - F_b - F_D \tag{3.2-1}$$

The external force can be expressed as a product of the mass and the acceleration a_e of the particle from this force, and

$$F_e = ma_e \tag{3.2-2}$$

The buoyant force is, by Archimedes's principle, the product of the mass of the fluid displaced by the particle and the acceleration from the external force. The volume of the particle is m/ρ_p, where ρ_p is the density of the particle, and the particle displaces this same volume of fluid. The mass of fluid displaced is $(m/\rho_p)\rho$, where ρ is the density of the fluid. The buoyant force is then

$$F_b = \frac{m\rho a_e}{\rho_p} \tag{3.2-3}$$

The drag force, is, from Eq. (3.1-1),

$$F_D = \frac{C_D u^2 \rho A_p}{2} \tag{3.2-4}$$

where C_D = dimensionless drag coefficient
A_p = projected area of particle measured in plane perpendicular to direction of motion of particle.

Substituting the forces from Eqs. (3.2-2) to (3.2-4) into Eq. (3.2-1) gives

$$\frac{du}{dt} = a_e - \frac{\rho a_e}{\rho_p} - \frac{C_D u^2 \rho A_p}{2m} = a_e \frac{\rho_p - \rho}{\rho_p} - \frac{C_D u^2 \rho A_p}{2m} \tag{3.2-5}$$

Motion from gravitational force. If the external force is gravity, a_e is g, the acceleration of gravity, and Eq. (3.2-5) becomes

$$\frac{du}{dt} = g\frac{\rho_p - \rho}{\rho_p} - \frac{C_D u^2 \rho A_p}{2m} \tag{3.2-6}$$

Motion in a centrifugal field. A centrifugal force appears whenever the direction of movement of a particle is changed. The acceleration from a centrifugal force from circular motion is

$$a_e = r\omega^2 \tag{3.2-7}$$

where r=radius of path of particle
ω=angular velocity, rad/s
Substituting into Eq. (3.2-5) gives

$$\frac{du}{dt} = r\omega^2 \frac{\rho_p - \rho}{\rho_p} - \frac{C_D u^2 \rho A_p}{2m} \tag{3.2-8}$$

In this equation, u is the velocity of the particle relative to the fluid and is directed outwardly along a radius.

Terminal velocity

In gravitational settling, g is constant. Also, the drag always increases with velocity. Equation (3.2-6) shows that the acceleration decreases with time and approaches zero. The particle quickly reaches a constant velocity, which is the maximum attainable under the circumstances and which is called the *terminal velocity*. The equation for the terminal velocity u_t is found, for gravitational settling, by taking $du/dt = 0$. Then from Eq. (3.2-6),

$$u_t = \sqrt{\frac{2g(\rho_p - \rho)m}{A_p \rho_p C_D \rho}} \tag{3.2-9}$$

In motion from a centrifugal force, the velocity depends on the radius, and the acceleration is not constant if the particle is in motion with respect to the fluid. In many practical uses of centrifugal force, however, du/dt is small in comparison with the other two terms in Eq.(3.2-8); and if du/dt is neglected, a terminal velocity at any given radius can be defined by the equation

$$u_t = \omega\sqrt{\frac{2r(\rho_p - \rho)m}{A_p \rho_p C_D \rho}} \tag{3.2-10}$$

Drag coefficient

The quantitative use of Eqs. (3.2-5) to (3.2-10) requires that numerical values be available for the drag coefficient C_D. Figure 3.3, which shows the drag coefficient as a function of Reynolds number, indicates such a relationship. Part of the curve of C_D versus Re_p for spheres is reproduced in Fig. 3.6, as well as a curve for crushed limestone particles. These curves, however, apply only under restricted conditions. The particle must be solid; it must be far from other particles and from the vessel walls, so that the flow pattern around the particle is not distorted; and it must be moving at its terminal velocity with respect to the fluid. The drag coefficients for accelerating particles are appreciably greater than those shown in Fig. 3.6, so a particle dropped in a still fluid takes longer to reach terminal velocity than would be predicted using the steady-state values of C_D. Particles injected into a fast-flowing stream also accelerate more slowly than expected, and the drag coefficients in this case are therefore less than the normal values. However, for most processes involving small particles or drops, the time for acceleration to the terminal velocity is still quite small and is often ignored in analysis of the process.

Variations in particle shape can be accounted for by obtaining separate curves of C_D versus Re_p for each shape, as shown in Fig. 3.3 for cylinders and disks. As pointed out earlier, however, the curves for cylinders and disks in Fig. 3.3 apply only to a specified orientation of the particle. In the free motion of nonspherical particles through a fluid, the orientation is constantly changing. This change consumes energy, increasing the effective drag on the particle, and C_D is greater than for the motion of the fluid past a fixed particle. For example, the drag coefficients of crushed limestone, as shown in Fig. 3.6, are more than twice as large as those for spheres of the same nominal diameter. As a result, the terminal velocity, especially with disks and other platelike particles, is less than would be predicted from curves for a fixed particle orientation.

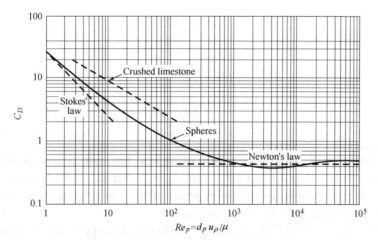

Fig. 3.6 Drag coefficients for spheres and irregular particles

In the following treatment the particles are assumed to be spherical, for once the drag coefficients for free-particle motion are known, the same principles apply to any shape.

When the particle is at sufficient distance from the boundaries of the container and from other particles, so that its fall is not affected by them, the process is called *free settling*. If the motion of the particle is impeded by other particles, which will happen when the particles are near one another even though they may not actually be colliding, the process is called *hindered settling*. The drag coefficient in hindered settling is greater than that in free settling.

If the particles are very small, brownian movement appears. This is a random motion imparted to the particle by collisions between the particle and the molecules of the surrounding fluid. This effect becomes appreciable at a particle size of about 2 to 3 μm and predominates over the force of gravity with a particle size of 0.1μm or less. The random movement of the particle tends to suppress the effect of the force of gravity, so settling does not occur. Application of centrifugal force reduces the relative effect of brownian movement

Motion of spherical particles

If the particles are spheres of diameter d_p,

$$m = \frac{1}{6}\pi d_p^3 \rho_p \quad (3.2\text{-}11)$$

and

$$A_p = \frac{1}{4}\pi d_p^2 \quad (3.2\text{-}12)$$

Substitution of m and A_p from Eqs. (3.2-11) and (3.2-12) into Eq. (3.2-9) gives the equation for gravity settling of spheres

$$u_t = \sqrt{\frac{4g(\rho_p - \rho)d_p}{3C_D\rho}} \quad (3.2\text{-}13)$$

In the general case, the terminal velocity can be found by trial and error after guessing Re_p to get an initial estimate of C_D. For the limiting cases of very low or very high Reynolds numbers, equations can be used to get ut directly.

At low Reynolds numbers, the drag coefficient varies inversely with Re_p, and the equations for C_D, F_D, and u_t are

$$C_D = \frac{24}{Re_p} \quad (3.2\text{-}14)$$

$$F_D = 3\pi\mu u_t d_p \quad (3.2\text{-}15)$$

$$u_t = \frac{d_p^2(\rho_p - \rho)g}{18\mu} \quad (3.2\text{-}16)$$

Equation (3.2-16), like Eq. (3.1-3), is a form of Stokes' law, which applies when the particle Reynolds number is less than 1.0. At $Re_p = 1.0$, $C_D = 26.5$ instead of 24, as predicted from Eq. (3.2-14), and since the terminal velocity depends of the square root of the drag coefficient, Stokes' law is about 5 percent in error at this point.

For $1{,}000 < Re_p < 200{,}000$, the drag coefficient is approximately constant, and the equations are

$$C_D = 0.44 \quad (3.2\text{-}17)$$

$$F_D = 0.055\pi d_p^2 u_t^2 \rho \quad (3.2\text{-}18)$$

$$u_t = 1.75\sqrt{\frac{gd_p(\rho_p - \rho)}{\rho}} \qquad (3.2\text{-}19)$$

Equation (3.2-19) is *Newton's law* and applies only for fairly large particles falling in gases or low-viscosity fluids.

As shown by Eqs. (3.2-16) and (3.2-19), the terminal velocity *ut* varies with d_p^2 in the Stokes' law range, whereas in the Newton's law range it varies with $d_p^{0.5}$.

Equation (3.2-16) can be modified to predict the settling velocity of a small sphere in a centrifugal field by substituting $r\omega^2$ for g.

$$u_t = \frac{d_p^2(\rho_p - \rho)r\omega^2}{18\mu} \qquad (3.2\text{-}20)$$

EXAMPLE 3.1 Mixture of galena and limestone in the proportion of 1 to 4 by weight is subjected to elutriation by an upward current of water flowing at 5mm/s. Assuming that the size distribution for each material is the same, and as shown by the following figure. Estimate the percentage of galena in the material carried away and in the material left behind. Take the viscosity of water as 1mPa and use Stokes' equation (the densities of galena and limestone are 7500 and 2700kg/m³, respectively).

Solution. It is necessary to determine the size of particle which has a settling velocity equal to that of the upward velocity of fluid, 5mm/s taking largest particle, d=100μm

$$Re = \frac{ud\rho}{\mu} = \frac{5\times 10^{-3} \times 100 \times 10^{-6} \times 1000}{1\times 10^{-3}} = 0.5$$

Thus the bulk of particles the flow will be within region of laminar flow and the settling velocity is given by Stokes' equation, Eq. (3.2-16)

$$u_t = \frac{d_p^2(\rho_p - \rho)g}{18\mu}$$

For a particle of galena settling at 5mm/s

$$5\times 10^{-3} = \frac{d_p^2(7500-1000)\times 9.81}{18\times 10^{-3}} = 3.54\times 10^6 d_p^2$$

$$d_p = 3.76\times 10^{-5} \text{m} = 37.6\mu\text{m}$$

For a particle of limestone settling at 5mm/s

$$5\times 10^{-3} = \frac{d_p^2(2700-1000)\times 9.81}{18\times 10^{-3}} = 9.27\times 10^6 d_p^2$$

$$d_p = 7.35\times 10^{-5} \text{m} = 73.5\mu\text{m}$$

Thus the particles of galena less than 37.6μm and particles of limestone less than 73.5 μm will be removed in the water stream.

The data are given by Fig 3.7 shows that 40% of the galena and 75% of the limestone will be removed in this way. In 5 kg feed, there is 1kg galena and 4kg limestone.

Therefore galena removed=0.4kg, leaving 0.6kg limestone removed=4×0.75=

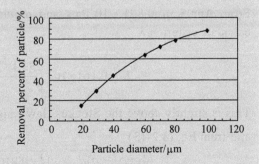

Fig. 3.7 Removal of particle for Example 3.1.

3kg, leaving 1kg.

Hence, in the material removed, Percentage galena=0.4/3.4=11.8%; in the material remaining, Percentage galena=0.6/1.6=37.5%.

Criterion for settling regime

To identify the range in which the motion of the particle lies, the velocity term is eliminated from the Reynolds number by substituting u_t from Eq. (3.2-16) to give, for the Stokes' law range,

$$Re_p = \frac{d_p u_t \rho}{\mu} = \frac{d_p^3 (\rho_p - \rho) \rho g}{18 \mu^2} \tag{3.2-21}$$

If Stokes' law is to apply, Re_p must be less than 1.0. To provide a convenient criterion K, let

$$K = d_p \left[\frac{\rho(\rho_p - \rho)g}{\mu^2} \right]^{1/3} \tag{3.2-22}$$

Then, from Eq. (3.2-21), $Re_p = \frac{1}{18} K^3$. Setting Re_p equal to 1.0 and solving give $K = 18^{1/3} = 2.6$.

If the size of the particle is known, K can be calculated from Eq. (3.2-22). If K so calculated is less than 2.6, Stokes' law applies. Substitution for u_t from Eq. (3.2-19) shows that for Newton's law range $Re_p = 1.75 K^{1.5}$. Setting this equal to 1000 and solving gives $K = 68.9$. Thus if K is greater than 68.9 but less than 2360, Newton's law applies. When K is greater than 2360, the drag coefficient may change abruptly with small changes in fluid velocity. Under these conditions, as well as in the range between Stokes' law and Newton's law (2.6 < K < 68.9), the terminal velocity is calculated from Eq. (3.2-13) using a value of C_D found by trial from Fig. 3.6.

EXAMPLE 3.2(a) Estimate the terminal velocity for 80-mesh to 100-mesh particles of limestone (ρ_p=2800kg/m³) velocity falling in water at 30℃. (b) How much higher would the velocity be in a centrifugal separator there the acceleration is 50g?
Solution.
(a) d_p for 100-mesh =0.147mm
d_p for 80-mesh =0.175mm
Average diameter $\bar{d}_p = 0.161$mm

From App.5, μ=0.801×10³Pa.s and ρ=995.7 kg/m³, to find which settling law applies calculate criterion K.

$$K = 0.161 \times 10^{-3} \left[\frac{9.81 \times 995.7 (2800 - 995.7)}{(0.801 \times 10^{-3})} \right]^{\frac{1}{3}} = 4.86$$

This is slightly above the Stockes' law range. Assume Re_p=5; then from Fig.3.6, $C_D \approx 14$, and from Eq. (3.2-13)

$$u_t = \sqrt{\frac{4g(\rho_p - \rho)d_p}{3C_D \rho}} = \sqrt{\frac{4 \times 9.81(2800 - 995.7)0.161 \times 10^{-3}}{3 \times 14 \times 995.7}} = 0.0165 \text{m/s}$$

Check:

$$Re_p = \frac{0.161 \times 10^{-3} \times 0.0165 \times 995.7}{0.801 \times 10^{-3}} = 3.30$$

Since C_D at Re_p=3.30 is greater than 14, the revised u_t and Re_p will be less than the above values, so guess a lower value of Re_p, guess

$$Re_p = 2.5 \qquad C_D \approx 20$$

$$u_t = 0.0165\sqrt{\frac{14}{20}} = 0.0138 \text{m/s}$$

$$Re_p = 3.30\left(\frac{0.0138}{0.0165}\right) = 2.76$$

This is close enough to the value of 2.5, and

$$u_t \approx 0.014 \text{ m/s}$$

(b) using a_e=50g in replace of g in Eq (3.2-22), since only the acceleration changes, K= $4.86 \times 50^{1/3}$ =17.90. This is still in the intermediate settling range. Estimate Re_p=40; from Fig. 3.6, C_D =4.1 and

$$u_t = \sqrt{\frac{4 \times 50 g(\rho_p - \rho)d_p}{3C_D \rho}} = \sqrt{\frac{4 \times 50 \times 9.81 \times (2800 - 995.7) \times 0.161 \times 10^{-3}}{3 \times 4.1 \times 995.7}} = 0.216 \text{m/s}$$

Check:

$$Re_p = \frac{0.161 \times 10^{-3} \times 0.216 \times 995.7}{0.801 \times 10^{-3}} = 43, \text{close to } 40$$

$$u_t \approx 0.22 \text{m/s}$$

Hindered settling

In hindered settling, the velocity gradients around each particle are affected by the presence of nearby particles, so the normal drag correlations do not apply. Also, the particles in settling displace liquid, which flows upward and makes the particle velocity relative to the fluid greater than the absolute settling velocity. For a uniform suspension, the settling velocity u_s can be estimated from the terminal velocity for an isolated particle using the empirical equation of Maude and Whitmore

$$u_s = u_t \varepsilon^n \qquad (3.2-23)$$

Exponent n changes from about 4.6 in the Stokes' law range to about 2.5 in Newton's law region. For very small particles, the calculated ratio u_s/u_t is 0.62 for $\varepsilon = 0.9$ and 0.095 for $\varepsilon = 0.6$. With large particles the corresponding ratios are $u_s/u_t = 0.77$ and 0.28; the hindered settling effect is not as pronounced because the boundary layer thickness is a smaller fraction of the particle size. In any case, Eq. (3.2-23) should be used with caution, since the settling velocity also depends on the particle shape and size distribution. Experimental data are needed for accurate design of a settling chamber.

If particles of a given size are falling through a suspension of much finer solids, the terminal velocity of the larger particles should be calculated using the density and viscosity of the fine suspension. Equation (3.2-23) may then be used to estimate the settling velocity with ε taken as the volume fraction of the fine suspension, not the total void fraction.

Suspensions of very fine sand in water are used in separating coal from heavy minerals, and the density of the suspension is adjusted to a value slightly greater than that of coal to make the coal particles rise to the surface, while the mineral particles sink to the bottom.

3.3 Settling Separation of the Solids from Gases

Separations are extremely important, much processing equipment is devoted to separating one phase or one material from another.

Mechanical separations are applicable to heterogeneous mixtures not to homogeneous solutions. The techniques are based on physical differences between the particles such as size, shape, or density. They are applicable to separating solids from gases, liquid drops from gases, and solids from liquids.

Many mechanical separations are based on the sedimentation of solid particles or liquid drops through a fluid, impelled by the force of gravity or by centrifugal force. The fluid may be a gas or liquid; it may be flowing or at rest. In some situations the subjective of the process is to remove particles from a stream in order to eliminate contaminants from the fluid or to recover the particles, as in the elimination of dust and fumes from air or flue gas or the removal of solids from liquid wastes.

3.3.1 Gravitational Settling Processes

If a particle starts at rest with respect to the fluid in which it is immersed and is then moved through the fluid by a gravitational force, its motion can be divided into two stages. The first stage is a short period of acceleration, during which the velocity increases from zero to the terminal velocity, the second stage is the period during which the particle is at its terminal velocity.

Since the period of initial acceleration is short, usually on the order of tenths of a second or less, initial acceleration affects are short-range. Terminal velocities, on the other hand, can be maintained as long as the particle is under treatment in the equipment.

Particles heavier than the suspending fluid may be remove from a gas or liquid in a large settling box or settling tank, in which the fluid velocity is low and particles have ample time to settle out.

Dust particles may be removed from gases by a variety of methods. For coarse solid particles, larger than about 325-mesh (43μm diameter), which may be removed by a gravity settling.

Settling chamber design

The principal quantities to be specified in designing a settling chamber are the cross-sectional area and the depth. The area is usually based on data from batch settling tests, even though such tests do not simulate very well the action in a continuous settling chamber. A gravity settling chamber like is shown schematically in Fig. 3.8.

In the absence of air currents, particles settle toward the floor at a speed equal to their terminal velocities. To prevent the air stream from lifting the particles from the floor and re-entraining them, the air velocity should not be greater than about 3m/s.

In the following treatment the particles are assumed to be spherical for once the drag coefficient are known, the

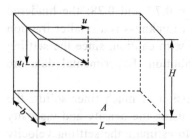

Fig. 3.8 Settling chamber.

size of a settling chamber can be calculated. If the particle is at sufficient distance from the boundaries of the container and from other particles, so that its fall is not affected by them, the settling process is free settling. When the settling of a particle is followed Stokes' law in a settling chamber, Eq. (3.2-16) can be applied to the design of gravitational settling chamber:

$$u_t = \frac{g(\rho_p - \rho)d_p^2}{18\mu}$$

When the height of the settling chamber is denoted by H, then, the particle will take time t_v to approach to the floor

$$t_v = \frac{H}{u_t} \tag{3.3-1}$$

The particle flows through the settling chamber with the fluid at the velocity u, assuming that the length of the settling chamber is denoted by L, then, the time that the particle moves horizontally from the entrance to the boundary is

$$t_h = \frac{L}{u} \tag{3.3-2}$$

The particle removed must take time $t_h \geqslant t_v$, from Eq. (3.3-1) and Eq. (3.3-2)

$$\frac{L}{u} \geqslant \frac{H}{u_t} \tag{3.3-3}$$

Critical situation

$$\frac{L}{u} = \frac{H}{u_t} \tag{3.3-4}$$

If the capacity of the settling chamber is denoted by Q in m³/s, and the width of the settling container b, the relationship among the capacity of the settling chamber, the velocity of fluid flow, and the dimension of settling chamber is

$$Q = Hbu \tag{3.3-5}$$

Solving Eq. (3.3-4) and Eq. (3.3-5) for the dimension of the settling chamber for a given capacity and given heterogeneous mixtures system.

$$Q = bLu_t \tag{3.3-6}$$

3.3.2 Centrifugal Settling Processes

The gravity settling chamber is probably the simplest and earliest type of dust-collection equipment. Practically, however, its industrial utility is limited to removing particles larger than 325 mesh (43μm diameter). A given particle in a given fluid settles under gravitational force at a fixed maximum rate. To increase the settling rate the force of gravity acting on the particle may be replaced by a much stronger centrifugal force.

The ratio of the terminal velocity from the centrifugal force to the terminal velocity of from the gravitational force is derived from Eq. (3.2-16) and Eq. (3.2-20):

$$K_{cg} = \frac{r\omega^2}{g} = \frac{u_T^2}{gr} \tag{3.3-7}$$

where r is the radius, u_T is the tangential velocity of the particle. For a cylone 0.3m in diameter with a tangential velocity of 15m/s near the wall, the ratio K_{cg} called the separation factor is 225/ (0.3×9.81)=76. A large-diameter cyclone has a much lower separation factor at the same velocity, and velocities above 15 to 20m/s are usually impractical because of the high pressure drop and increased abrasive wear. Small-diameter cyclones may have separation factors as high as 2500.

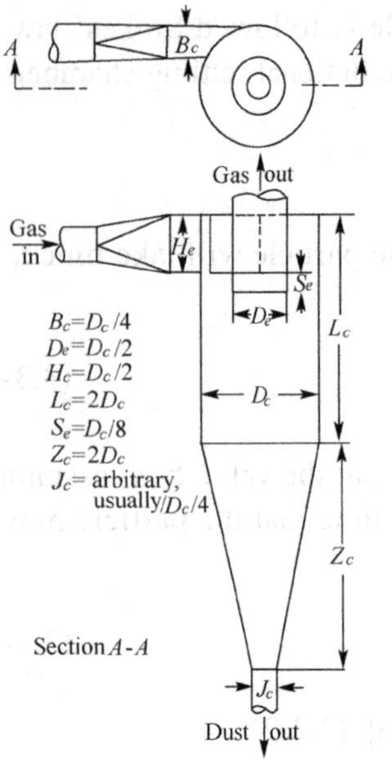

Most centrifugal separators for removing particles from gas streams contain no moving parts. They are typified by the cyclone separator shown in Fig. 3.9.

The incoming dust-laden air receives a rotating motion on entrance to the cylinder and is accelerated radially, but the force on a particle is not constant because of the change in r and because the tangential velocity in the votex varies with r and with distance below the inlet. The vortex so formed develops centrifugal force, which acts to throw the particles radially toward the wall.

The centrifugal force in a cyclone is from 5 times gravity in large, low-velocity units to 2500 times gravity in small, high-pressure units.

In a cyclone, dust particles quickly reach the terminal velocities. The radial acceleration in a cyclone depends on the radius of the path being followed by the air, and given by the following equation.

$B_c = D_c/4$
$D_e = D_c/2$
$H_c = D_c/2$
$L_c = 2D_c$
$S_e = D_c/8$
$Z_c = 2D_c$
J_c = arbitrary, usually $D_c/4$

Fig. 3.9 Cyclone.

$$a_e = \omega^2 r = \left(\frac{u_i}{r}\right)^2 r = \frac{u_i^2}{r} \tag{3.3-8}$$

So terminal velocity is

$$u_t = \frac{(\rho_p - \rho)d_p^2}{18\mu} \times \frac{u_i^2}{r} \tag{3.3-9}$$

Critical diameter. The critical diameter of particle removed by the cyclone can be derived in terms of both the terminal velocity and tangential velocity u_i. From equation (3.3-9), the settling time is written as follow

$$\theta_t = \frac{B}{u_t} = \frac{18\mu rB}{(\rho_p - \rho)d_p^2 u_i^2} \tag{3.3-10}$$

Residence time depends on the tangential velocity u_i and the numbers of spirals N

$$\theta_r = \frac{2\pi rN}{u_i} \tag{3.3-11}$$

Combining two equations, the critical diameter of particle removed by the cyclone is

$$d_c = \sqrt{\frac{9\mu B}{\pi N(\rho_p - \rho)u_i}} \tag{3.3-12}$$

The equation contains effects of cyclone size, velocity, viscosity, and density of solid.

3.4 Filtration Fundamentals

3.4.1 Introduction

Filtration may be defined as the separation of solids from liquids by passing a suspension through a permeable medium which retains the particles. A filtration system can be shown schematically in Fig. 3.10.

In order to obtain fluid flow through the filter medium, a pressure drop Δp has to be applied across the medium; it is immaterial from the fundamental point of view how this pressure drop is achieved but there are four types of driving force:

Gravity, Vacuum, Pressure, Centrifugal

Before introducing the basic filtration relationships it is worthwhile examining the actual process of particle removal.

There are basically two types of filtration used in practice: the so-called surface filters are used for cake filtration in which the solids are deposited in the form of a

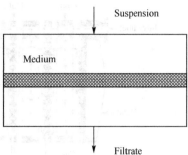

Fig.3.10 Schematic diagram of a filtration system.

cake on the up-stream side of a relatively thin filter medium, while depth filter are used for deep bed filtration in which particle deposition takes place inside the medium and cake deposition on the surface is undesirable.

Surface filter

In a surface filter, the filter medium has a relatively low initial pressure drop and, as seen in Fig. 3.11, particles of the same size as or larger than the openings wedge into the openings and create smaller passages which remove even smaller particles from the fluid. A filter cake is thus formed, which in turn functions as a medium for the filtration of subsequent input suspension. In order to prevent blinding of the medium, filter aids are used.

Surface filters are usually used for suspensions with higher concentrations of solids, say above 1% by volume, because of the blinding of the medium that occurs in the filtration of dilute suspensions. This can be, however, sometimes be avoided by an artificial increase of the input concentration, in particular by adding a filter aid; as filter aids are very porous their presence in the cake improves permeability and often makes cake filtration of dilute and generally difficult slurries possible.

Note that the model described above and shown in Fig. 3.11. is that of conventional batch cake filtration where both the particles and the liquid approach the medium at an angle of 90° and no attempt is made to disturb the cake or prevent its formation.

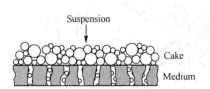

Fig. 3.11 Mechanism of cake filtration.

Fig. 3.12 Frames and plates.

Discontinuous pressure filters

Pressure filters can apply a large pressure differential across the septum to give economically rapid filtration with viscous liquid or fine solids. The most common types of

142 | Unit Operations of Chemical Engineering

pressure filters are presses filter and shell-and –leaf filters.

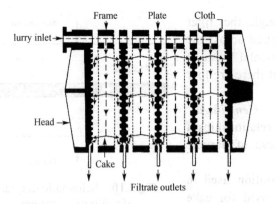

Fig. 3.13 Diagrams of plate-and-frame filter presses.

Filter press. The structure of plate-and-frame press filter is shown in Fig.3.12. A filter press contains a set of frame designed to provide a series of chambers or compartments in which solids may collect, shown in Fig. 3.13. The frames are covered with a filter medium such as canvas. Slurry is admitted to each compartment under pressure; liquor passes through the canvas and out a discharge pipe, leaving a wet cake of solids behind.

After assembly of the press, slurry is admitted to one end of the assembly of plates and frames from a pump or blow case under a pressure of 3 to 10 atm. Auxiliary channels carry slurry from the main inlet channel into each frame. Here the solids are deposited on the cloth-covered faces of the plates. Liquor passes through the cloth, down grooves or corrugations in the plate faces, and out of the press. Filtration is continued until liquor no longer flows out the discharge or the filtration pressure suddenly rises. These occur when the frames are full of solid and no more slurry can enter.

The plates of a filter press may be square or circular, vertical or horizontal. Most commonly the compartments for solids are formed by recesses in the faces of molded polypropylene plates. In other designs, they are formed as in the *plate-and-frame press* shown in Fig. 3.14, in which square plates 150 mm to 2 m on a side alternate with open frames. The plates are 6 to 50 mm thick, the frames 6 to 200 mm) thick. Plates and frames sit vertically in a metal rack, with cloth covering the face of each plate, and are squeezed tightly

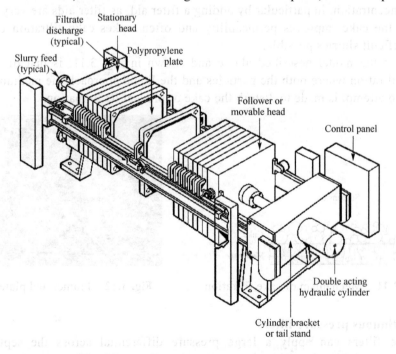

Fig. 3.14 Filter press equipped for automatic operation.

together by a screw or a hydraulic ram. Slurry enters at one end of the assembly of plates and frames. It passes through a channel running lengthwise through one corner of the assembly. Auxiliary channels carry slurry from the main inlet channel into each frame. Here the solids are deposited on the cloth-covered faces of the plates. Liquor passes through the cloth, down grooves or corrugations in the plate faces, and out of the press.

Shell-and-leaf filters. The filter press is useful for many purposes but is not economical for handling large quantities of sludge or for efficient washing with a small amount of wash water. The wash water often channels in the cake and large volumes of wash water may be needed. For filtering under higher pressures than are possible in a plate-and-frame press, to economize on labor, or where better washing of the cake is needed,

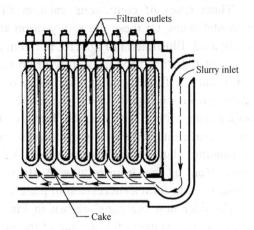

Fig 3.15 Leaf filter.

a shell-and-leaf filter may be used. The leaf filter shown in Fig.3.15 was developed for larger volumes of slurry and more efficient washing. Each leaf is a hollow wire framework covered by a sack of filter cloth.

A number of these leaves are hung in parallel in a closed tank. The slurry enters the tank and is forced under pressure through the filter cloth, where the cake deposits on the outside of the leaf. The filtrate flows inside the hollow framework and out a header. The wash liquid follows the same path as the slurry. Hence, the washing is more efficient than the through-washing in plate-and-frame filter presses. To remove the cake, the shell is opened. Sometimes air is blown in the reverse direction into the leaves to help in dislodging the cake. If the solids are not wanted, water jets can be used to simply wash away the cakes without opening the filter.

Leaf filters also suffer from the disadvantages of batch operation. They can be automated for the filtering, washing, and cleaning cycle. However, they are still cyclical and are used for batch processes and relatively modest throughput processes.

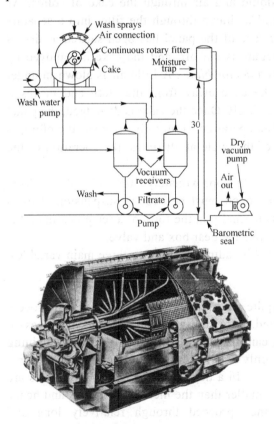

Fig. 3.16 Rotary drum filtration system.

Continuous pressure filters

Batch filters often require considerable operating labor, which in large-scale processing may be prohibitively expensive. The continuous vacuum filters described below were developed to reduce the labor required for filtration.

Three types of continuous vacuum filters are: rotary drum, rotary disc, and rotary horizontal drum. Of these the rotary drum and disc types, in that order, are by far the most widely used. Illustrated in Fig.3.16, horizontal drum with a slotted face turns at 0.1 to 2 r/min in an agitated slurry trough. Rotary drum types of filter range in size from 0.05 m^2 to well over $100m^2$. A filter medium, such as canvas, covers the face of the drum, which is partly submerged in the liquid. Under the slotted cylindrical face of the main drum is a second, smaller drum with a solid surface. Between the two drums are radial partitions dividing the annular space into separate compartments, each connected by an internal pipe to one hole in the rotating plate of the rotary valve. Vacuum and air are alternately applied to each compartment as the drum rotates. A strip of filter cloth covers the exposed face of each compartment to form a succession of panels.

Consider now the panel shown in Fig. 3.16. It is just about to enter the slurry in the trough. As it dips under the surface of the liquid, vacuum is applied through the rotary valve. A layer of solids builds up on the face of the panel as liquid is drawn through the cloth into the compartment, through the internal pipe, through the valve, and into a collecting tank. As the panel leaves the slurry and enters the washing and drying zone, vacuum is applied to the panel from a separate system, sucking wash liquid and air through the cake of solids. As shown in the flow sheet of Fig.3.16, wash liquid is drawn through the filter into a separate collecting tank. After the cake of solids on the face of the panel has been sucked as dry as possible, the panel leaves the drying zone, vacuum is cut off, and the cake is removed by scraping it off with a horizontal knife known as a *doctor blade*. A little air is blown in under the cake to belly out the cloth. This cracks the cake away from the cloth and makes it unnecessary for the knife to scrape the drum face itself. Once the cake is dislodged, the panel reenters the slurry and the cycle is repeated. The operation of any given panel, therefore, is cyclic; but since some panels are in each part of the cycle at all times, the operation of the filter as a whole is continuous.

The simplest type of drum filter operates with approximately 30%-40% of the drum submerged in the tank. The slurry in the filter trough must be kept in suspension. This is achieved by installation of agitator gear which is normally in the form of a reciprocating rake, either rotating about the drum shaft or pivoted above the gear box and valve.

The performance of drum filters is controlled by adjustments of in three main variables: drum speed, vacuum and submergence.

Deep bed filtration

The principles of deep bed filtration are quite well-known. It is a clarification process using a deep bed of granular media, usually sand. As a sewage tertiary treatment process it can frequently produce filtrates containing only 5mg/l or less of suspend matter.

In a depth filter (Fig.3.17), particles are smaller than the medium openings and hence they proceed through relatively long and tortuous pores where they are collected by a number of mechanisms (gravity, diffusion and inertia) and attached to the medium by molecular and electrostatic forces.

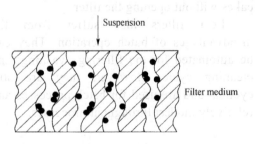

Fig 3.17 Mechanism of deep bed filtration.

The initial pressure drop across the depth filter is generally small than that across a surface filter of comparable efficiency but the build-up of pressure drop as particles are collected is more gradual for a depth filter. Depth filters are commonly used for clarification, i.e. for the separation of fine particles from very dilute suspension, say less than 0.1% by volume.

Rates of filtration are normally in the range of 5 to 15 m/h (100 to 300 gall/ft²h), with a tend toward the higher rates. The thickness of the layers may vary, but overall depths are usually about 0.6 to 1m, except for up-flow filters which are considerably deeper.

Open gravity filters, shown as in Fig. 3.18, usually operate to a head loss limit of 2.5m water gauge, but pressure filters may use up to 2 or 3 times this value. Backwashing is normally employed at a rate sufficient to fluidize the media and to ensure sufficient separation for multiple layer filters. Washing rate of up to 0.6m/min can be achieve this, although this depends on the media, and water temperature. Normally, backwash water would represent about 1% of filtrate production; 3% would be high and quantities over 5% would be excessive.

Of the two types of filtration, cake filtration has the wider application and the following discussion will, on the while, be concerned with cake filtration and surface filter. The main characteristics of filters are the flow rate-pressure drop relationship and other performance characteristics.

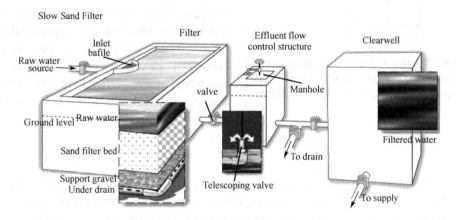

Fig.3.18 Open gravity filter.

3.4.2 Flow Rate-Pressure Drop Relationships

Principles of cake filtration

Filtration is a special example of flow through porous media, which was discussed for cases in which the resistances to flow are constant. In filtration the flow resistances increase with time as the filter medium becomes clogged or a filter cake builds up, and the equations given in section 3.1 must be modified to allow for this. The chief quantities of interest are the flow rate through the filter and the pressure drop across the unit. As time passes during filtration, either the flow rate diminishes or the pressure drop rises. In what is called *constant-pressure filtration* the pressure drop is held constant and the flow rate allowed to fall with time; less commonly, the pressure drop is progressively increased to give what is called *constant-rate filtration*.

In cake filtration the liquid passes through two resistances in series: that of the cake and that of the filter medium. The filter medium resistance, which is the only resistance in clarifying filters, is normally important only during the early stages of cake filtration. The cake resistance is zero at the start and increases with time as filtration proceeds. If the cake is

Pressure drop through filter cake

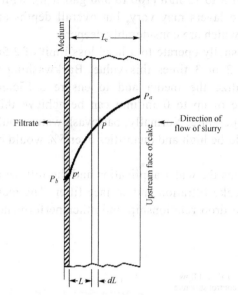

Fig. 3.19 Section through filter medium and cake, showing pressure gradients; p, fluid pressure; L, distance from filter medium.

Fig.3.19 shows diagrammatically a section through a filter cake and filter medium at a definite time t from the start of the flow of filtrate. At this time the thickness of the cake, measured from the filter medium, is L_c. The filter area, measured perpendicularly to the direction of flow, is A. Consider the thin layer of cake of thickness dL lying in the cake at a distance L from the medium. Let the pressure at this point be p. This layer consists of a thin bed of solid particles through which the filtrate is flowing. In a filter bed the velocity is sufficiently low to ensure laminar flow. Accordingly, as a starting point for treating the pressure drop through the cake, So, the relationship between the superficial velocity (or empty-tower velocity) u and the pressure drop can be described by Kozeny-Carman equation. If the superficial velocity of the filtrate is designated as u, Eq. (3.1-12) becomes

$$\Delta p = \frac{150\mu u L}{d_p^2 \Phi^2} \times \frac{(1-\varepsilon)^2}{\varepsilon^3} \tag{3.4-1}$$

Fig.3.19 shows a nonlinear pressure gradient in the cake, which is typical because of the lower cake porosity near the filter medium. Often the pressure drop is expressed as a function of the surface to volume ratio instead of the particle size. Substitution of $6(v_p/s_p)$ for $\Phi_s d_p$ [Eq. (3.1-5)] gives

$$\Delta p = \frac{4.17\mu u L(1-\varepsilon)^2 (s_p/v_p)^2}{\varepsilon^3} \tag{3.4-2}$$

where Δp=pressure gradient at thickness L
μ=viscosity of filtrate
u=linear velocity of filtrate, based on filter area
s_p=surface of single particle
v_p=volume of single particle
ε=porosity of cake

The linear velocity u is given by the equation

$$u = \frac{dV}{Adt} \tag{3.4-3}$$

where V is the volume of filtrate collected from the start of the filtration to time t. Since the filtrate must pass through the entire cake, V/A is the same for all layers and u is independent of L. Substituting u from Eq. (3.4-3) into Eq. (3.4-2) gives

$$u = \frac{dV}{Adt} = \frac{\varepsilon^3}{4.17(s_p/v_p)^2(1-\varepsilon)^2} \times \frac{\Delta p}{\mu L} \qquad (3.4\text{-}4)$$

The filtration rate is

$$\frac{dV}{dt} = \frac{\varepsilon^3}{4.17(s_p/v_p)^2(1-\varepsilon)^2} \times \frac{A\Delta p}{\mu L} \qquad (3.4\text{-}5)$$

Let

$$k = \frac{\varepsilon^3}{4.17(s_p/v_p)^2(1-\varepsilon)^2} \qquad (3.4\text{-}6)$$

where k is a constant referred to as the permeability of the bed and substitutes Eq. (3.4-6) into Eq(3.4-5)

$$\frac{dV}{dt} = k\frac{A\Delta p}{\mu L} \qquad (3.4\text{-}7)$$

This equation relating the flow rate of a filtrate of viscosity μ through a bed of thickness L and face area A to the driving pressure Δp is called Darcy's basic filtration equation. Eq. (3.4-7) is often written in the form

$$\frac{dV}{dt} = \frac{A\Delta p}{\mu R} \qquad (3.4\text{-}8)$$

where R is called the resistance (and is equal to L/k, the medium thickness divided by the permeability of the bed).

Filter medium resistance

In the cake filtration, two resistances are presented in series, one of which, the cake resistance R increases and other, the medium resistance R_m may be assumed constant with time. Eq. (3.4-8) becomes:

$$\frac{dV}{dt} = \frac{A\Delta p}{\mu(R+R_m)} \qquad (3.4\text{-}9)$$

In practice, however, the assumption made above that the medium resistance is constant is rarely true because some penetration and blocking of the medium inevitably occurs when particles impinge on the medium. As the resistance of the cake may be assumed to be directly proportional to the amount of cake deposited (only true for incompressible cakes) it follows that for a given filtration area A.

$$R = rL \qquad (3.4\text{-}10)$$

Where L is the width of cake deposited in m and r is the specific cake resistance. Similarly for the filtration medium of width L_m

$$R_m = rL_m \qquad (3.4\text{-}11)$$

Substitution of Eq. (3.4-10) for R and Eq. (3.4-11) for R_m in Eq. (3.4-9) gives

$$\frac{dV}{dt} = \frac{A\Delta p}{\mu r(L + L_m)} \qquad (3.4\text{-}12)$$

Eq.(3.4-12) relates the flow rate of filtrate to the pressure drop, the volume of cake deposited L and other parameters, some of which can, in certain circumstances, be assumed to be constant. These parameters are discussed briefly.

Pressure drop. The pressure drop Δp may be constant or variable with time depending on the characteristics of the pump used or on the driving force applied. If it varies with time the function $\Delta p = f(t)$ is usually known.

Face area of the filter medium. The face area of the medium A is usually constant, but with a few exceptions such as in the case of equipment with an appreciable cake build-up on a tubular medium or a rotary drum.

Liquid viscosity. The liquid viscosity μ is constant provided that the temperature remains constant during the filtration cycle and that the liquid is Newtonian.

Specific cake resistance. The specific cake resistance r should be constant for incompressible cakes but it may change with time as a result of possible flow consolidation of the cake and also, in the case of variable rate filtration, because of variable approach velocity.

Most cakes, however, are compressible and their specific resistance changes with the pressure drop across the cake. An experimental empirical relationship can sometimes be used over a limited pressure drop range.

$$r = r_0 \Delta p^n \qquad (3.4\text{-}13)$$

Where r_0 is the resistance at unit applied pressure drop and n is a compressibility index(equal to zero for incompressible cakes) obtained from experiments.

Volume of the cake deposited per unit volume of filtrate obtained. The volume of the cake deposited per unit volume of filtrate is defined by c. It can be related to the cumulative volume of filtrate V and the width of cake L in time t by

$$L = \frac{cV}{A} \qquad (3.4\text{-}14)$$

Where c is constant, which depends on the concentration of solids in the slurry and the porosity of cake.

Medium resistance. The medium resistance R_m should normally be constant but it may be vary with time as a result of some penetration of solids into the medium and sometimes it may also change with applied pressure because of the compression of fibres in the medium.

3.4.3 Filtration Operations-Basic Equations

The general filtration equation, Eq.(3.4-12), after substitution of L from Eq.(3.4-14), becomes

$$\frac{dV}{dt} = \frac{A^2 \Delta p}{\mu r c (V + V_m)} \qquad (3.4\text{-}15)$$

where V_m is the medium resistance, which is equal to $\dfrac{L_m A}{c}$. Eq.(3.4-15)can be rewritten

$$(V + V_m)dV = \frac{A^2 \Delta p}{rc\mu} dt \qquad (3.4\text{-}16)$$

3.4.4 Constant Pressure Filtration

For incompressible cakes r is independent of the pressure drop and of position in the cake. If Δp is constant, the Eq. (3.4-16) can be integrated.

$$V^2 + 2VV_m = \frac{2\Delta p A^2}{\mu r c} t \qquad (3.4\text{-}17)$$

Using Eq. (3.4-17) either t or V can be calculated from the value of the other variable provided all the constants are known. It is useful for the mathematical simplicity of the final equations to define a constant K.

$$K = \frac{2\Delta p}{\mu r c} \qquad (3.4\text{-}18)$$

Substitution of K from Eq. (3.4-18) in Eq.(3.4-17) gives the equation of constant pressure filtration:

$$V^2 + 2VV_m = KA^2 t \qquad (3.4\text{-}19)$$

The equation of the constant pressure filtration (3.4-19) can also be in the other form

$$q^2 + 2qq_m = Kt \qquad (3.4\text{-}19a)$$

where the volume of filtrate per unit filtration area $q = \dfrac{V}{A}$ in m.

For experimental determination of filtration constant K and specific resistance r, Eq. (3.4-19) is often put in the form

$$\frac{\Delta t}{\Delta V} = \frac{2}{KA^2} V + \frac{2V_m}{KA^2} \qquad (3.4\text{-}20)$$

Which gives a straight line if $\Delta t/\Delta V$ is plotted against V, is shown in Fig.3.20. The Eq. (3.4-20) and the plot in Fig.3.20 are used to predict the filtration constants. The actual procedure for the evaluation of results is best explained in an example.

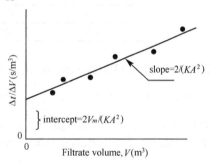

Fig. 3.20 Determination of constants in a constant-pressure filtration run.

EXAMPLE3.3 Evaluate the filtration and specific cake resistance r from pilot scale tests. Filtration tests were carried out with a plate and frame filter press under the following *conditions.*
solids: ρ_p=2710 kg/m^3;
liquid: water at 20℃, μ=0.001N·s/m^2;
filter: plate and frame press, 1 frame, dimensions 430×430×30mm;
suspension: c=1.04, the frame was full of cake at V_c=0.56m^3. The value of 0.3m^3 corresponding to 3686s was chosen as a starting point for constant pressure operation, i.e.
$$V_s=0.3\text{m}^3; \qquad t_s=3686\text{s}$$
The data obtained during the filtration experiment are shown in Table 3.2; the initial stages of filtration were controlled manually before constant pressure filtration at 150kPa was carried out. Determine the filtration constant K and the specific cake resistance r.

150 | Unit Operations of Chemical Engineering

Table 3.2 Data from filtration experiment, Example 3.3.

$10^{-5}\Delta p$, N/m²	t, s	V, m³	$\Delta t/\Delta V$, s/m³
0.4	447	0.04	12458
0.5	851	0.07	12326
0.7	1262	0.10	12120
0.8	1516	0.13	12765
1.1	1886	0.16	12857
1.3	2167	0.19	13809
1.3	2552	0.22	14175
1.3	2909	0.25	15540
1.5	3381	0.28	15250
1.5	3686	0.30	
1.5	4043	0.32	17850
1.5	4398	0.34	17800
1.5	4793	0.36	18450
1.5	5190	0.38	18800
1.5	5652	0.40	19660
1.5	6117	0.42	20258
1.5	6610	0.44	20886
1.5	7100	0.46	21337
1.5	7608	0.48	21789
1.5	8136	0.50	22250
1.5	8680	0.52	22700
1.5	9256	0.54	23208

Solution. Eq. (3.4-20) can be used to evaluate the constant. $\Delta t/\Delta V$ is plotted against V, as in Fig. 3.21 and the slope $\dfrac{2}{KA^2}$ and the intercept $\dfrac{2V_m}{KA^2}$ on the vertical axis of the best straight line drawn through the part of the graph that corresponds to constant pressure operation, i.e. for $V \geqslant V_s$ (i.e. 0.3m³) are measured

$$\frac{2}{KA^2} = 24511, \quad \frac{2V_m}{KA^2} = 9739$$

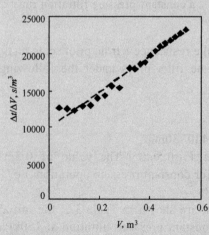

Fig. 3.21 $\Delta t/\Delta V$ is plotted against V.

From which K can be calculated, using $A = 0.43 \times 0.43 \times 2 = 0.37\text{m}^3$

$$K = \frac{2}{24511 \times 0.37^2} = 6.0 \times 10^{-4}$$

From Eq.(3.4-18)

$$r = \frac{2\Delta p}{\mu K c} = \frac{2 \times 1.5 \times 10^5}{0.001 \times 6.0 \times 10^{-4} \times 0.104} = 4.8 \times 10^{12}$$

The medium resistance V_m

$$V_m = 9739 \frac{KA^2}{2}$$

$$= \frac{9739 \times 6.0 \times 10^{-4} \times 0.37^2}{2} = 0.4$$

3.4.5 Constant Rate Filtration

If the flow rate $Q=dV/dt$ is keep constant and the pressure Δp varied, Eq.(3.4-15) becomes

$$Q = \frac{dV}{dt} = \frac{A^2 \Delta p}{\mu rc(V+V_m)} = constant$$

Where V is simply

$$V = Qt \qquad (3.4\text{-}21)$$

Thus

$$\Delta p = \mu rc \frac{Q^2}{A^2}t + \mu rc \frac{Q}{A^2}V_m \qquad (3.4\text{-}22)$$

A plot of Δp against t as in Fig.3.22, will, from Eq.(3.4-22), be a straight line.

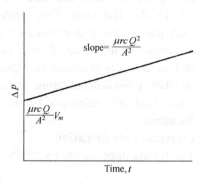

Fig. 3.22 Plot of Δp versus t for constant rate filtration.

3.4.6 Constant Rate Followed by Constant Pressure Operation

In many cases, for example a frame and plate or a leaf press operated with the suspension supplied from a centrifugal pump, the early stages of filtration are conducted at a nearly constant rate. As the cake becomes thicker and offers more resistance to the flow, the pressure developed by the pump becomes a limiting factor and the filtration proceeds at a nearly constant pressure. For such a combined operation, the plot Δp versus time is as shown in Fig.3.23. The equations are

$$\Delta p = \mu rc \frac{Q^2}{A^2}t + \mu rc \frac{Q}{A^2}V_m \qquad \text{for } t < t_s$$

and $\qquad\qquad\qquad \Delta p = constant \qquad\qquad\qquad \text{for } t \geq t_s$

The plot of $\frac{\Delta t}{\Delta V} = f(V)$ will be as shown in Fig.3.24. The equations are

$$V = Qt \qquad (3.4\text{-}21)$$

and

$$V^2 + 2VV_m = KA^2 t \qquad (3.4\text{-}19)$$

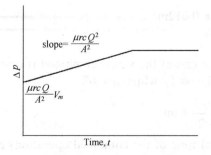

Fig. 3.23 Plot of Δp versus t for constant rate followed be constant pressure filtration.

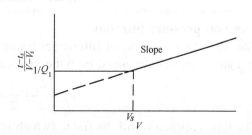

Fig. 3.24 Plot of t/V versus V for constant rate followed by constant pressure filtration.

EXAMPLE 3.4 A filtration experiment is carried out on a single cloth of area 0.02m² to which a slurry is fed at a constant rate, yielding 4×10^{-5} m³/s of filtrate. Readings are taken during the test show that after 100s the pressure is 4×10^4 Pa and after 500s the pressure is 1.2×10^5 Pa. The same filter cloth material is to be used in a plate and frame filter press, each frame having dimensions 0.5m×0.5m×0.08m, to filter the same slurry. The flow rate of slurry per unit area of cloth during the initial constant rate period is to be the same as that used in the preliminary experiment and the constant rate period is to be followed by constant pressure operation once the pressure reaches 8×10^4 Pa. if the volume of cake formed per unit volume of filtrate, c is 0.02, calculate the time required to fill the frame.

Solution.
Constant rate filtration
The filtrate approach velocity for the constant rate period is

$$\frac{Q}{A} = \frac{4\times10^{-5}}{0.02} = 2\times10^{-3}\,\text{m/s}$$

The filtrate approach velocity is the same for both the experimental and the filter press. As two experimental values of Δp and t are known, the constant can be determined by substituting the values of Δp and t in Eq. (3.4-22)

$$4\times10^4 = \mu rc \times 4\times10^{-6} \times 100 + \mu rcV_m \times 2\times10^{-3} \times \frac{1}{0.02}$$

$$12\times10^4 = \mu rc \times 4\times10^{-6} \times 500 + \mu rcV_m \times 2\times10^{-3} \times \frac{1}{0.02}$$

Giving $\mu rc = 5\times10^7$
and $\mu rcV_m = 1\times10^7$, $V_m = 0.004$
so that $\Delta p = 200t + 2\times10^4$ (this is applicable to both the experimental and the filter press), from which t_s can be determined because from the experimental data $\Delta p_s = 8\times10^4$, therefore

$$8\times10^4 = 200t_s + 2\times10^4$$

$$t_s = 300\,\text{s}$$

The cumulative volume V_s passed through one frame in 300s can be calculate from Eq.(3.4-21)

$$V_s = Qt_s = 2\times10^{-3}At_s = 2\times10^{-3} \times 0.5 \times 300 = 0.3\,\text{m}^3$$

And the resulting thickness L_s of cake is

$$L_s = c\frac{V_s}{A} = 0.02 \times \frac{0.3}{0.5} = 0.012\,\text{m}$$

Constant pressure filtration
The final (total) volume of filtrate per frame V_f at the end of the whole combined filtration operation can be determined from the final cake thickness L_f, which is 0.08

$$V_f = \frac{L_f}{c} \times \frac{A}{2} = \frac{0.08 \times 0.5}{0.02 \times 2} = 1\,\text{m}^3$$

The time required to fill the frame (which is the total time of the combined operation) can be determined from Eq.(3.4-17). V_s and V_f are obtained at the starting t_s and the end t_f of constant pressure operation, respectively, institution of these variables in Eq.(3.4-17) gives

$$(V_f^2 - V_s^2) + 2(V_f - V_s)V_m = \frac{2\Delta p A^2}{\mu r c}(t_f - t_s)$$

$$(1^2 - 0.3^2) + 2(1 - 0.3) \times 0.2 = \frac{2 \times 8 \times 10^4 \times 0.5^2}{5 \times 10^7}(t_f - 300)$$

$$t_f = 300 + 1144.5 = 1444.5 s$$

3.4.7 Vacuum Filtration—Drum Continuous Filtration

In a continuous filter, say, of the rotary-drum type, the feed, filtrate, and cake move at steady constant rates. For any particular element of the filter surface, however, conditions are not steady but transient. Follow, for example, an element of the filter cloth from the moment it enters the pond of slurry until it is scraped clean once more. It is evident that the process consists of several steps in series—cake formation, washing, drying, and discharging—and that each step involves progressive and continual change in conditions. The pressure drop across the filter during cake formation is, however, held constant. Thus the foregoing equations for discontinuous constant-pressure filtration may, with some modification, be applied to continuous filters.

If t is the actual filtering time (i.e., the time any filter element is immersed in the slurry), then from Eq. (3.4-19)

$$(V + V_m)^2 = KA^2 t + V_m^2 \quad (3.4\text{-}23)$$

where V is the volume of filtrate collected during time t. Solving Eq. (3.4-23) for V, as a quadratic equation, gives

$$V = \sqrt{KA^2 t + V_m^2} - V_m \quad (3.4\text{-}24)$$

where V is the volume of filtrate collected during time t that it is the actual filtration time. If the fraction of the drum submerged is φ, and drum speed is n in r/s

$$t = \frac{\varphi}{n} \quad (3.4\text{-}25)$$

Substituting t from Eq. (3.4-25) in Eq. (3.4-24) for the volume of filtrate

$$V = \sqrt{KA^2 \frac{\varphi}{n} + V_m^2} - V_m \quad (3.4\text{-}26)$$

The capacity of drum filter Q in m³/s can be derived by Eq.(3.4-26)

$$Q = nV = \sqrt{KA^2 n\varphi + n^2 V_m^2} - nV_m \quad (3.4\text{-}27)$$

In continuous filtration, the resistance of the filter medium is generally negligible compared with the cake resistance. So in Eq. (3.4-27), $V_m = 0$.

$$Q = A\sqrt{K\varphi n} \quad (3.4\text{-}28)$$

In general, the filtration rate increases as the drum speed increases, because the cake formed on the drum face is thinner than at low drum speeds. At speeds above a certain critical value, however, the filtration rate no longer increases with speed but remains constant, and the cake

3.4.8 Washing Filter Cakes

At the end of the filtration period, the filter cake is like a packed bed with the spaces between the particles filled with solution. The cake may be washed in place with water, or sometimes with solvent, to remove solutes that would otherwise remain as impurities on the solid product after drying. If the solid is a waste product, washing may still be necessary to satisfy regulations for disposal, or to recover valuable solutes for reuse. The rate of flow of the wash liquid and the volume of liquid needed to reduce the solute content of the cake to the desired degree are important quantities for the design and operation of a filter. Although the following principles apply to the problem, the best operating conditions cannot be chosen without some experimental data.

In most filters the wash liquid follows the same path as that of the filtrate. But in a filter press the wash passes through the entire thickness of the cake. The last filtrate passes through only one – half the final cake.

For a shell-and-leaf filter the flow rate of the wash liquid, in principle, equals to the final filtering rate because the flow path through which the washing liquid flows is similar to that of filtering during filtration, provided the pressure drop remains unchanged. So the flow rate of washing can be obtained from Eq. (3.4-19). The Eq.(3.4-19) can be written by taking derivative

$$\frac{dV}{dt} = \frac{KA^2}{2(V+V_m)_e} \tag{3.4-29}$$

Thus, the relationship between the washing rate $\left(\frac{dV}{dt}\right)_W$ and the final rate of filtration $\left(\frac{dV}{dt}\right)_E$ is

$$\left(\frac{dV}{dt}\right)_W = \left(\frac{dV}{dt}\right)_E = \frac{KA^2}{2(V+V_m)} \tag{3.4-30}$$

For a filter press the flow rate of washing

$$\left(\frac{dV}{dt}\right)_W = \frac{K(A')^2}{2(V+V_m)} \tag{3.4-31}$$

where the cross-section of the frame $A'=A/2$, then, substitution of $A'=A/2$ in Eq. (3.4-31) gives the washing rate for a plate and frame filter press

$$\left(\frac{dV}{dt}\right)_W = \frac{K(A')^2}{2(V+V_m)} = \frac{KA^2}{8(V+V_m)} \tag{3.4-32}$$

Hence the relationship between the flow rate of washing and that of last filtration in a filter press is

$$\left(\frac{dV}{dt}\right)_W = \frac{1}{4}\left(\frac{dV}{dt}\right)_E \tag{3.4-33}$$

EXAMPLE 3.5 A slurry is filtered in a plate and frame press containing 12 frames, each 0.3m square and 25mm thick. During the first 200s, the filtration pressure is slowly raised to the final value of 400kN/m²(gauge pressure), and during this period the rate of filtration is maintained constant, and volume of filtrate collected was $0.01937 m^3$. After the initial period, filtration is carried out at constant pressure and the cakes are completely formed after a further 900s. The cake is then washed at $274 kN/m^2$ for 600s. What is the volume of filtrate collected per cycle and how much wash water is used? A sample of the slurry had previously been tested, using a vacuum leaf filter of $0.05m^2$ filtering surface and pressure difference $71kN/m^2$. The volume of filtrate collected in the first 300s was $250cm^3$ and, after a further 300s, an additional $150cm^3$ was collected. Assume the cake to be incompressible and the cloth resistance to be the same in the leaf as in the filter press.

Solution. In the leaf filter, filtration is at constant pressure from start, thus

$$V^2 + 2VV_m = KA^2 t$$

For the leaf filter, when $\quad t=300s, \quad V=250cm^3$

and when $\quad t=600s, \quad V=400cm^3$

$$A=0.05m^2 \quad \text{and} \quad \Delta p = 71 kPa$$

Thus

$$(0.025)^2 + 2 \times 0.025 V_m = K \times 0.05^2 \times 300$$

and

$$(0.04)^2 \times + 2 \times 0.04 V_m = K \times 0.05^2 \times 600$$

hence $\quad V_m = 0.0175 \quad \text{and} \quad K = 0.002$

For the filter press

$$A = 2.16 m^2 \quad \Delta p' = 400 kPa$$

In the filter press, a volume V_1 of filtrate is obtained under constant rate conditions in time t_1, and filtration is then carried out at constant pressure

$$(V^2 - V_1^2) + 2(V - V_1)V_m = K'A^2 (t - t_1)$$

The total volume of filtrate collected is therefore given by

$$(V^2 - 0.01937^2) + 2(V - 0.01937) \times 0.0175 = K' \times 2.16^2 \times 900$$

$$K' = K \times \Delta p'/\Delta p = 0.002 \times 400/71 = 0.0112$$

The total volume of filtrate is

$$V = 6.84 m^3$$

Relation between the final rate of filtration and the washing rate is given by Eq. (3.4-33)

$$\left(\frac{dV}{dt}\right)_W = \frac{1}{4}\left(\frac{dV}{dt}\right)_E$$

$$K'' = K \times \Delta p''/\Delta p = 0.002 \times 274/71 = 0.0077$$

Therefore the washing rate is

$$\left(\frac{dV}{dt}\right)_W = \frac{1}{4}\left(\frac{dV}{dt}\right)_E = \frac{K''A^2}{8(V+V_m)} = \frac{0.0077 \times 2.16^2}{8 \times (6.84 + 0.0175)} = 0.000656$$

Washing water used

$$V_W = t_W \left(\frac{dV}{dt}\right)_W = 600 \times 0.000656 = 0.39 m^3$$

3.5 Introduction to Fluidization

When a liquid or a gas is passed at very low velocity up through a bed of solid particles, the particles do not move. If the fluid velocity is steadily increased, the pressure drop and the drag on individual particles increase, and eventually the particles start to move and become suspended in the fluid. The terms *fluidization* and *fluidized bed* are used to describe the condition of fully suspended particles, since the suspension behaves as a dense fluid. If the bed is tilted, the top surface remains horizontal and large objects will either float or sink in the bed depending on their density relative to the suspension. The fluidized solids can be drained from the bed through pipes and valves just as a liquid can, and this fluidity is one of the main advantages of the use of fluidization for handling solids.

Conditions for fluidization

Consider a vertical tube partly filled with a fine granular material such as catalytic cracking catalyst, as shown schematically in Fig. 3.25. The tube is open at the top and has a porous plate at the bottom to support the bed of catalyst and to distribute the flow uniformly over the entire cross section. Air is admitted below the distributor plate at a low flow rate and passes upward through the bed without causing any particle motion. If the particles are quite small, flow in the channels between the particles will be laminar and the pressure drop across the bed will be proportional to the superficial velocity u [Eq. (3.1-12)]. As the velocity is gradually increased, the pressure drop increases, but the particles do not move and the bed height remains the same. At a certain velocity, the pressure drop across the bed counterbalances the force of gravity on the particles or the weight of the bed, and any further increase in velocity causes the particles to move. This is point A on the graph. Sometimes the bed expands slightly with the grains still in contact, since just a slight increase in ε can offset an increase of several percent in u and keep p constant. With a further increase in velocity, the particles become separated enough to move about in the bed, and true fluidization begins (point B).

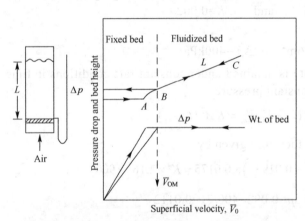

Fig. 3.25 Pressure drop and bed height versus superficial velocity for a bed of solids.

Once the bed is fluidized, the pressure drop across the bed stays constant, but the bed height continues to increase with increasing flow. The bed can be operated at quite high velocities with very little or no loss of solids, since the superficial velocity needed to support a bed of particles is much less than the terminal velocity for individual particles, as will be shown later.

If the flow rate to the fluidized bed is gradually reduced, the pressure drop remains constant, and the bed height decreases, following the line BC which was observed for increasing velocities. However, the final bed height may be greater than the initial value for the fixed bed, since solids dumped in a tube tend to pack more tightly than solids slowly settling from a fluidized state. The pressure drop at low velocities is then less than that in the original fixed bed. On starting up again, the pressure drop offsets the weight of the bed at point B, and this point, rather than point A, should be considered to give the minimum fluidization velocity u_M. To measure u_M, the bed should be fluidized vigorously and allowed to settle with the gas turned off, and the flow rate increased gradually until the bed starts to expand. More reproducible values of u_M can sometimes be obtained from the intersection of the graphs of pressure drop in the fixed bed and the fluidized bed.

Minimum fluidization velocity

An equation for the minimum fluidization velocity can be obtained by setting the pressure drop across the bed equal to the weight of the bed per unit area of cross section, allowing for the buoyant force of the displaced fluid:

$$\Delta p = g(1-\varepsilon_M)(\rho_p - \rho)L \tag{3.5-1}$$

At incipient fluidization, ε is the minimum porosity ε_M (If the particles themselves are porous, ε is the external void fraction of the bed.) Thus

$$\frac{\Delta p}{L} = g(1-\varepsilon_M)(\rho_p - \rho) \tag{3.5-2}$$

Using Eq. (3.5-2) and the Ergun equation [Eq. (3.1-14)] for $\Delta p/L$ at the point of incipient fluidization gives a quadratic equation for the minimum fluidization velocity u_M

$$\frac{150\mu u_M}{D_p^2 \Phi_s^2} \times \frac{(1-\varepsilon_M)}{\varepsilon_M^3} + \frac{1.75\rho u_M^2}{D_p \Phi_s} \times \frac{1}{\varepsilon_M^3} = g(\rho_p - \rho) \tag{3.5-3}$$

For very small particles, only the laminar flow term of the Ergun equation is significant. With $Re_p < 1$, the equation for minimum fluidization velocity becomes

$$u_M \approx \frac{g(\rho_p - \rho)}{150\mu} \times \frac{\varepsilon_M^3}{1-\varepsilon_M} \Phi_s^2 D_p^2 \tag{3.5-4}$$

The equations derived for minimum fluidization velocity apply to liquids as well as to gases, but beyond u_M the appearance of beds fluidized with liquids or gases is often quite different. When fluidizing sand with water, the particles move farther apart and their motion becomes more vigorous as the velocity is increased, but the average bed density at a given velocity is the same in all sections of the bed. This is called *particulate fluidization* and is characterized by a large but uniform expansion of the bed at high velocities.

PROBLEMS

3.1 "Falling ball method" is used to measure the viscosity of liquid. Liquid is placed in the glass container. It takes 7.32s for the steel ball with a diameter of 6.35mm to travel downward 200mm in the liquid. As known, the density of steel is 7900 kg/m³, and the density of the liquid is 1300 kg/m³, try to work out the viscosity of liquid.

3.2 A spherical quartzose particle with a density of 2650 kg/m³ settles freely in the 20℃ air, try to calculate the maximum diameter obeying Stocks' law and the minimum diameter obeying Newton's law.

3.3 It is desired to remove dust particles 50 microns in diameter from 226.5m³/min of air, using a settling chamber for the purpose. The temperature and pressure are 21℃ and 1 atm. The particle density is 2403kg/m³. What minimum dimensions of the chamber are consistent with these conditions? (The maximum permissible velocity of the air is 3m/s.)

3.4 A standard cyclone is to be used to separate the dust of density of 2300 kg/m³ from the gas. The flow rate of gas is 1000m³/h, the viscosity of the gas is 3.6×10^{-5} N·s/m², and the density is 0.674kg/m³. If the diameter of cyclone is 400 mm, attempt to estimate the critical diameter.

3.5 A standard cyclone separator, 0.5m in diameter and inlet width $B=D/4$, is used to remove the particles from dust gas. If the gas enters at 15m/s and the numbers of spirals N in the cyclone takes as 5, at what critical diameter of particle will occur?
(viscosity of air=0.018mPa and density=1.3 kg/m³; density of particle=2700 kg/m³.)

3.6 The particles in waste air of 2700m³/h are separated by a settling chamber with dimensions: length 2m, and breadth 1.5m. The density and viscosity of air is 0.946kg/m³, 0.0219mPa·s, respectively, and density of particle is 2400kg/m³. Find: (a) minimum diameter of particle removed by the settling chamber; (b) critical diameter of particle can be separated if a cyclone with diameter of 0.4m is used to replace the settling chamber (The standard cyclone: $A=D/2$, $B=D/4$, and assuming that the number of spirals $N=4$)

3.7 A filter press of 0.1m² filtering area is used for filtering a sample of the slurry. The filtration is carried out at constant pressure with a vacuum 500mmHg. The volume of filtrate collected in the first 5min was one liter and, after a further 5min, an additional 0.6 liter was collected. How much filtrate will be obtained when the filtration has been carried out for 15 min on assuming the cake to be incompressible?

3.8 The following data are obtained for a filter press of 0.0093m² filtering area in the test.

Pressure difference, kg$_f$/cm²	Filtering time, s	Filtrate, m³
1.05	50	2.27×10^{-3}
	660	9.10×10^{-3}
3.50	17.1	2.27×10^{-3}
	233	9.10×10^{-3}

Calculate:
(1) filtration constant K, V_m at the pressure difference of 1.05;
(2) if the frame of the filter is filled with the cake at 660s, what is the final rate of filtration $\left(\dfrac{dV}{dt}\right)_E$;
(3) and what is the compressible constant of cake n?

3.9 A slurry if filtered by a filter press of 0.1m² filtering area at constant pressure, the equation for a constant pressure filtration is as follows

$$(q+10)^2 = 250(t+0.4)$$

where q=filtrate volume per unit filtering area, in l/m², t= filtering time, in min.
Calculate:
(a) how much filtrate will be gotten after 249.6min?
(b) if the pressure difference is double and both the resistances of the filtration medium and cake are constant, how much filtrate will be obtained after 249.6min?

3.10 Filtration is carried out in a plate and frame filter press, with 20 frames 0.3m square and 50mm thick. At a constant pressure difference of 248.7kN/m², one-quarter of the total filtrate per cycle is obtained for the first 300s. Filtration is continued at a constant pressure for a further 1800s, after which the frames are full. The total volume of filtrate per cycle is 0.7m³ and dismantling and refitting of the press takes 500s.

It is decided to use a rotary drum filter, 1.5m long and 2.2m in diameter, in place of the filter press. Assuming that the resistance of the cloth is the same and that the filter cake is incompressible, calculate the speed of rotation of the drum which will result in the same overall rate of filtration as was obtained with the filter press. The filtration in the rotary filter is carried out at a constant pressure difference of 70kN/m^2 and with 25% of the drum submerged.

3.11 A press filter of 0.093m^2 filtering surface was used to separate the suspension containing calcium carbonate, operating in constant pressure. The volume of filtrate collected was 2.27×10^{-3} m^3 during the 50s, volume of filtrate collected was 3.35×10^{-3} m^3 during the 100 s. How much will the filtrate be obtained as filtering for 200s?

3.12 A press filter of 0.093m^2 filtering surface was used to separate the suspension containing 0.2kg solid(specific gravity 3.0) per kilogram of water, operating in constant pressure. The volume of filtrate collected was 2.27×10^{-3} m^3 during the 50s. What is the rate of filtration and how much volume of cake can be obtained as filtering for 100s? (on assumption that the cake has a voidage of 0.5 and the resistance of filtering medium is negligible)

3.13 A slurry with incompressible cake is filtered by a 1m^2 filter press at constant pressure. If the filtering constant K is 10 m^2/min, operating pressure difference Δp is 2atm and the septum resistance can be ignored. Find

(a) the filtrate volume, in m^3 when the filtering time is 10 min;

(b) if the septum resistance must be considered and V_m is 1m^3, what is the filtrate volume V, in m^3 when the filtering time is 10 min ?

(c) what is the filtration rate dV/dt when the filtering time is 10min for case (1)?

3.14 Slurry is filtered by filter press operating at constant pressure, volume of filtrate collected is 4m^3 during the 20 minute, then dismantling and refitting of the filter press takes 30 minute, find: (a) the capacity of filter press in m^3/h; (b) what capacity could be obtained with doubling in operating pressure during the same time, on assumption that the cake is imcompressible and the resistance of filter medium is negligible; (c) how much could the volume of filtrate be collected during one cycle if filter press is replaced by rotary drum filter having the revolution of 1r/min, assuming that the operating pressure different is the same as (a).

名 词 术 语

英文 / 中文

air current / 空气对流
Archimedes's principle / 阿基米德原理
backwashing / 反洗
boundary layer separation / 边界层分离
boundary layer thickness / 边界层厚度
brownian movement / 布朗运动
cake filtration / 滤饼过滤
cake resistance / 滤饼阻力
canvas / 帆布
catalytic reactor / 催化反应器
constant-pressure filtration / 恒压过滤
constant-rate filtration / 恒速过滤
contaminant / 污染物

英文 / 中文

continuous pressure filter / 连续压滤机
correction factor / 校正因子
corrugation / 波纹
creeping flow / 爬流
criterion / 判断标准
critical diameter / 临界直径
cross-sectional area / 横截面积
cyclone separator / 旋风分离器
deep bed filtration / 深床过滤
density difference / 密度差
depth filter / 深床过滤机
drag coefficient / 阻力系数
drag force / 曳力

英文 / 中文	英文 / 中文
eddy or vortices / 旋涡	plate-and-frame press filter / 板框压滤机
equivalent channel diameter / 孔道的当量直径	pressure drop / 压降
filter area / 过滤面积	pressure filter or presses filter / 压滤机
filter cake / 滤饼	projected area / 投影面积
filter cloth / 过滤机滤布	residence time / 停留时间
filter medium / 过滤介质	rotary drum / 转鼓
filter medium resistance / 过滤介质阻力	sedimentation / 沉降
filtrate / 滤液	separation factor / 分离因子
filtration / 过滤	settling chamber / 沉降室
filtration constant / 过滤常数	settling tank / 沉降槽
filtration equation / 过滤方程	settling time / 沉降时间
fluidization / 流态化	shell-and-leaf filter / 叶滤机
fluidized bed / 流化床	slurry / 滤浆
form drag / 形体阻力	smooth sphere / 光滑球体
free settling / 自由沉降	specific cake resistance / 滤饼比阻
granular media / 颗粒介质	sphericity / 球形度
gravitational settling / 重力沉降	stagnation point / 驻点
heterogeneous mixture / 非均相混合物	superficial velocity / 表观速度
hindered settling / 干扰沉降	surface filter / 表面过滤机
horizontal knife / 水平刮刀	surface-volume ratio / 比表面
incompressible cake / 不可压缩滤饼	suspensions / 悬浮物
ion-exchange / 离子交换	tangential velocity / 切向速度
low reynolds number / 低雷诺数	terminal velocity / 沉降速度
mechanical separation / 机械分离	tortuosity factor / 曲率因子
minimum fluidization velocity / 最小流化速度	tortuous channel / 弯弯曲曲的流道
molecular and electrostatic forces / 分子力和静电力	tortuous pore / 弯曲孔道
motion from gravitational force / 重力场作用下的运动	total void fraction / 总空隙率
motion in a centrifugal field / 离心力场作用下的运动	transition / 过渡
nominal size / 公称尺寸	two-dimensional motion / 二维运动
nonspherical particle / 非球形颗粒	two-phase countercurrent flow / 两相逆流流动
open gravity filter / 敞口式重力过滤器	up-flow filter / 上流式过滤器
packed tower / 填料塔	up-stream side / 上游一侧
particulate fluidization / 颗粒流化	void volume / 空隙
permeable medium / 可渗透介质	volume fraction / 体积分数
	wake formation / 尾流的形成
	wall drag / 壁面阻力
	washing / 洗涤

Chapter 4

Heat Transfer and Its Applications

4.1 Introduction

Practically all the operations that are carried out by the chemical engineer involve the production or absorption of energy in the form of heat. The laws governing the transfer of heat and the types of apparatus that have for their main object the control of heat flow are therefore of great importance. This section of the book deals with heat transfer and its applications in process engineering.

Nature of heat flow.   When two objects at different temperatures are brought into contact, heat flows from the object at the higher temperature to that at the lower temperature. The net flow is always in the direction of the temperature decrease. The mechanisms by which the heat may flow are three: conduction, convection, and radiation.

Conduction.   If a temperature gradient exists in a continuous substance, heat can flow unaccompanied by any observable motion of matter. Heat flow of this kind is called conduction, and according to Fourier's law, the heat flux is proportional to the temperature gradient and opposite to it in sign. For one-dimensional heat flow, Fourier's law is

$$\frac{dq}{dA} = -k\frac{dT}{dx} \qquad (4.1\text{-}1)$$

where q =rate of heat flow in direction normal to surface
  A =surface area
  T =temperature
  x =distance normal to surface
  k =proportionality constant or thermal conductivity

In metals, thermal conduction results from the motion of free electrons, and there is close correspondence between thermal conductivity and electrical conductivity. In solids that are poor conductors of electricity and in most liquids, thermal conduction results from momentum transfer between adjacent vibrating molecules or atoms. In gases, conduction occurs by the random motion of molecules, so that heat is "diffused" from hotter regions to colder ones. The most common example of pure conduction is heat flow in opaque solids such as the brick wall of a furnace or the metal wall of a heat exchanger tube. Conduction of heat in liquids or gases is often influenced by flow of the fluids, and both conductive and convective processes are lumped together under the term convection or convective heat transfer.

Convection.   Convection can refer to the flow of heat associated with the movement of a fluid, such as when hot air from a furnace enters a room, or to the transfer of heat from a hot surface to a flowing fluid. The second meaning is more important for unit operations, as it includes heat transfer from metal walls, solid particles, and liquid surfaces. The convective flux is usually proportional to the difference between the surface temperature and the temperature of the fluid, as stated in Newton's law of cooling

$$\frac{q}{A} = h(T_s - T_f) \qquad (4.1\text{-}2)$$

where T_s=surface temperature
T_f=bulk temperature of fluid, far from surface
h =heat transfer coefficient

Note that the linear dependence on the temperature driving force T_s-T_f is the same as that for pure conduction in a solid of constant thermal conductivity, as can be shown by integrating Eq. (4.1-1). Unlike thermal conductivity, the heat-transfer coefficient is not an intrinsic property of the fluid, but depends on the flow patterns determined by fluid mechanics as well as on the thermal properties of the fluid. If $T_f-T_s > 0$, heat will be transferred from the fluid to the surface.

Natural and forced convection. When currents in a fluid result from buoyancy forces created by density differences, and the density differences are caused by temperature gradients in the fluid, the action is called natural convection. When the currents are due to a mechanical device such as a pump or agitator, the flow is independent of density differences and is called forced convection. Buoyancy forces also exist in forced convection, but usually they have only a small effect.

Radiation. Radiation is a term given to the transfer of energy through space by electromagnetic waves. If radiation is passing through empty space, it is not transformed to heat or any other form of energy, nor is it diverted from its path. If, however, matter appears in its path, the radiation will be transmitted, reflected, or absorbed. It is only the absorbed energy that appears as heat, and this transformation is quantitative. For example, fused quartz transmits practically all the radiation that strikes it; a polished opaque surface or mirror will reflect most of the radiation impinging on it; a black or matte surface will absorb most of the radiation received by it and will transform such absorbed energy quantitatively to heat. The energy emitted by a black body is proportional to the fourth power of the absolute temperature

$$W_b = \sigma T^4 \qquad (4.1\text{-}3)$$

where W_b=rate of radiant energy emission per unit area
σ=Stefan-Boltzmann constant
T=absolute temperature

Monatomic and most diatomic gases are transparent to thermal radiation, and it is quite common to find that heat is flowing through masses of such gases both by radiation and by conduction-convection. Examples are the loss of heat from a radiator or uninsulated steam pipe to the air of a room and heat transfer in furnaces and other high-temperature gas-heating equipment. The two mechanisms are mutually independent and occur in parallel, so that one type of heat flow can be controlled or varied independently of the other. Conduction-convection and radiation can be studied separately and their separate effects added in cases where both are important. In very general terms, radiation becomes important at high temperatures and is independent of the circumstances of the flow of the fluid. Conduction-convection is sensitive to flow conditions and is relatively unaffected by temperature level.

Section 4.2 deals with heat conduction in solids and stationary fluids, sections 4.3 to 4.5 with heat transfer by conduction and convection in fluids in motion, and section 4.6 with heat transfer by radiation. In section 4.7 the principles developed in preceding sections are applied to the design of equipment for heating, cooling and condensing.

4.2 Heat Transfer by Conduction

Conduction is most easily understood by considering heat flow in homogeneous isotropic solids because in these there is no convection and the effect of radiation is negligible unless the solid is translucent to electromagnetic waves. First, the general law of conduction is discussed; second, situations of steady-state heat conduction, where the temperature distribution within the solid does not change with time, are treated; third, the cases of unsteady conduction, where the temperature distribution does change with time, are not considered.

4.2.1 Basic Law of Conduction

The basic relation for heat flow by conduction is the proportionality between heat flux and the temperature gradient. It is known as Fourier's law, which for one-dimensional flow in the x direction has already been given by Eq.(4.1-1). To repeat,

$$\frac{dq}{dA} = -k\frac{dT}{dx} \tag{4.2-1}$$

where q=rate of heat flow in direction normal to surface
A=surface area
T=temperature
x=distance measured normal to surface
k=thermal conductivity

The general expressions of Fourier's law for flow in all three directions in an isotropic material are

$$\frac{dq}{dA} = -k\left(\frac{\partial T}{\partial x} + \frac{\partial T}{\partial y} + \frac{\partial T}{\partial y}\right) = -k\nabla T \tag{4.2-2}$$

Equation (4.2-2) states that the flux vector dq/dA is proportional to the temperature gradient ∇T and is oppositely directed. In an isotropic material, therefore, heat flows by conduction in the direction of steepest temperature descent.

In cylindrical coordinates Eq. (4.2-2) becomes

$$\frac{\partial q}{\partial A} = -k\left(\frac{\partial T}{\partial r} + \frac{1}{r}\frac{\partial T}{\partial \theta} + \frac{\partial T}{\partial z}\right) = -k\nabla T \tag{4.2-3}$$

In spherical coordinates it is

$$\frac{\partial q}{\partial A} = -k\left(\frac{\partial T}{\partial r} + \frac{1}{r}\frac{\partial T}{\partial \theta} + \frac{1}{r\sin\theta}\frac{\partial T}{\partial \phi}\right) = -k\nabla T \tag{4.2-4}$$

Thermal conductivity
The proportionality constant k is a physical property of the substance called the thermal conductivity. It, like the newtonian viscosity μ, is one of the so-called transport properties of the material. This terminology is based on the analogy between Eq.(1.2-3) and Eq. (4.1-1). In

Eq.(1.2-3) the quantity τ is a rate of momentum flow per unit area, the quantity du/dy is the velocity gradient, and μ is the required proportionality factor. In Eq. (4.1-1), q/A is the rate of heat flow per unit area, dT/dx is the temperature gradient, and k is the proportionality factor. The minus sign is omitted in Eq.(1.2-3) because of convention in choosing the direction of the force vector. In SI units, q is measured in watts and dT/dx in ℃/m or K/m. Then the units of k are W/(m·℃) or W/(m·K).

Fourier's law states that k is independent of the temperature gradient but not necessarily of temperature itself. Experiment does confirm the independence of k for a wide range of temperature gradients, except for porous solids, where radiation between particles, which does not follow a linear temperature law, becomes an important part of the total heat flow. On the other hand, k is a function of temperature, but, except for some gases, not a strong one. For small ranges of temperature, k may be considered constant. For larger temperature ranges, the thermal conductivity can usually be approximated by an equation of the form

$$k = a + bT \qquad (4.2\text{-}5)$$

where a and b are empirical constants.

Thermal conductivities of metals cover a wide range of values, from about 17 W/(m·℃) for stainless steel and 45 W/(m·℃) for mild steel, to 380 W/(m·℃) for copper and 415 W/(m·℃), for silver. The thermal conductivity of metals is generally nearly constant or decreases slightly as the temperature is increased, and the conductivity of alloys is less than that of pure metals. For glass and most nonporous materials, the thermal conductivities are much lower, from about 0.35 to 3.5 W/(m·℃); for these materials k may either increase or decrease as the temperature rises.

For most liquids k is lower than that for solids, with typical values of about 0.17 W/(m·℃), and k decreases by 3 to 4 percent for a 10℃ rise in temperature. Water is an exception, with k=0.5 to 0.7 W/(m·℃), and k goes through a maximum as the temperature is raised.

Gases have thermal conductivities an order of magnitude lower than those for liquids. For an ideal gas, k is proportional to the average molecular velocity, the mean free path, and the molar heat capacity. For monatomic gases, a hard-sphere model gives the theoretical equation:

$$k = \frac{0.0832}{\sigma^2}\left(\frac{T}{M}\right)^{1/2} \qquad (4.2\text{-}6)$$

where T = temperature, K
M = molecular weight
σ = effective collision diameter, Å
k = thermal conductivity, W/(m·K)

Note the similarity of Eq. (4.2-6) to Eq.(1.2-4) for estimating the viscosity of simple gases. Both equations contain the term $T^{1/2}/\sigma^2$, but momentum transfer, as shown by Eq.(1.2-4), varies with $M^{1/2}$ whereas the thermal conductivity depends on $M^{-1/2}$.

Equation (4.2-6) generally underestimates the thermal conductivity of polyatomic gases, which have higher heat capacities than monatomic gases because of the rotational and vibrational degrees of freedom. The higher heat capacities can also make k increase quite rapidly with temperature. A change from 300K to 600K may increase the thermal conductivity 3- to 4-fold. Several methods of predicting k for gases and gas mixtures are reviewed by Reid et al. The thermal conductivity of gases is nearly independent of pressure up to about 10 atm; at higher pressures k increases slightly with pressure. Values of k for some solids, liquids, and gases are given in Apps. 5 and 9 through 12. More complete tables are available in the literature.

Solids having low thermal conductivities are used for insulation on pipes, vessels, and buildings. Porous materials such as fiberglass pads or polymer foams act by entrapping air and eliminating convection. Their k values may be nearly as low as that of air itself, and if a high-molecular-weight gas is trapped in a closed-cell foam, k can be less than that for air.

4.2.2 Steady-State Conduction

Simple examples of steady-state conduction are shown in Fig. 4.1. In Fig. 4.1(a) a flat-walled insulated tank contains a refrigerant at perhaps $-10\,^\circ\mathrm{C}$, while the air outside the tank is at $28\,^\circ\mathrm{C}$. The temperature falls linearly with distance across the layer of insulation as heat flows from the air to the refrigerant. As we will see in a later section, there may actually be a temperature drop between the bulk of the air and the outside surface of the insulation, but it is assumed to be negligible in Fig. 4.1(a). Figure 4.1(b) shows a similar tank containing boiling water at $100\,^\circ\mathrm{C}$, losing heat to air at $20\,^\circ\mathrm{C}$. As before, the temperature profile in the insulation is linear, but heat flows in the opposite direction and x in Eq. (4.1-1) must be measured outward from the inside surface. Again, there may be a temperature change in the air near the tank wall; again it is assumed to be negligible.

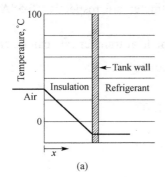
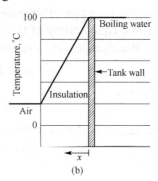

Fig. 4.1 Temperature gradients outside insulated tanks: (a) heat flow into the tank; (b) heat flow from the tank.

The rate of heat flow is found as follows, assuming that k is independent of temperature. Since in steady state there can be neither accumulation nor depletion of heat within the slab, q is constant along the path of heat flow. If x is the distance from the hot side, Eq. (4.1-1) can be written

$$dT = -\frac{q}{kA} dx \tag{4.2-7}$$

Since the only variables in this equation are x and T, direct integration gives

$$\frac{q}{A} = k\frac{T_1 - T_2}{x_2 - x_1} = k\frac{\Delta T}{B} \tag{4.2-8}$$

where $x_2 - x_1 = B =$ thickness of layer of insulation
  $T_1 - T_2 = \Delta T =$ temperature drop across layer

When the thermal conductivity k varies linearly with temperature, in accordance with Eq. (4.2-5), Eq. (4.2-7) still can be used rigorously by taking an average value $\bar{k}$ for k, which may be found either by using the arithmetic average of individual values of k for the two surface temperatures T_1 and T_2 or by calculating the arithmetic average of the temperatures and using the value of k at that temperature.

Equation (4.2-8) can be written in the form

$$\frac{q}{A} = \frac{\Delta T}{R} \qquad (4.2\text{-}9)$$

where R is the thermal resistance of the solid between points 1 and 2. Equation (4.2-9) is an instance of the general rate principle, which equates a rate to the ratio of a driving force to a resistance. In heat conduction, q is the rate and ΔT is the driving force. The resistance R, as shown by Eq. (4.2-9) and using $\bar{k}$ for k to account for a linear variation of k with temperature, is $B/\bar{k}$. The reciprocal of a resistance is a heat-transfer coefficient h, as in Newton's law [Eq. (4.1-2)]. For heat conduction, then, $h = \bar{k}/B$. Both R and h depend on the dimensions of the solid as well as on the thermal conductivity k, which is a property of the material.

EXAMPLE 4.1 An experiment is carried out to measure the thermal conductivity of flat plate with surface area 0.01m^2 and thickness 20mm. The surface temperature of flat plate is measured by a thermocouple. Heat quantity input into the flat plate is measured by ampere meter voltmeter. The temperatures of both sides of flat plate are 180℃ and 30℃, respectively. If the readings of ampere meter voltmeter are respective 0.58A and 100V, what is the thermal conductivity of flat plate?
Solution. From the electrical current and voltage, the heat transfer rate can be calculated

$$q = IV = 0.58 \times 100 = 58 \text{W}$$

Temperature drop is

$$\Delta T = 180 - 30 = 150 ℃$$

Heat transfer area and thickness of flat plate are

$$A = 0.01 \text{ m}^2, \quad B = 0.02 \text{m}$$

From Eq. (4.2-8)

$$\frac{q}{A} = \frac{\Delta T}{B/k} = \frac{150}{0.02/k} = 58/0.01$$

Hence

$$k = 0.773 \text{ W/(m·℃)}$$

Compound resistances in series

Consider a flat wall constructed of a series of layers, as shown in Fig. 4.2. Let the thicknesses of the layers be B_A, B_B, and B_C and the average conductivities of the materials of which the layers are made be k_A, k_B, and k_C, respectively. Also let the area of the compound wall, at right angles to the plane of the illustration, be A. Let ΔT_A, ΔT_B and ΔT_C be the temperature drops across layers A, B, and C, respectively. Assume, further, that the layers are in excellent thermal contact, so that no temperature difference exists across the interfaces between the layers. Then if ΔT is the total temperature drop across the entire wall,

$$\Delta T = \Delta T_A + \Delta T_B + \Delta T_C \qquad (4.2\text{-}10)$$

It is desired, first, to derive an equation for calculating the rate of heat flow through the series of resistances and, second, to show how the rate can be calculated as the ratio of the overall temperature drop ΔT to the overall resistance of the wall. Equation (4.2-8) can be written for each layer, using $\bar{k}$ in place of k,

$$\Delta T_A = q_A \frac{B_A}{k_A A} \qquad \Delta T_B = q_B \frac{B_B}{k_B A} \qquad \Delta T_C = q_C \frac{B_C}{k_C A}$$

Adding the parts of equations above gives

$$\Delta T_A + \Delta T_B + \Delta T_C = \frac{q_A B_A}{A \overline{k}_A} + \frac{q_B B_B}{A \overline{k}_B} + \frac{q_C B_C}{A \overline{k}_C} = \Delta T$$

Since, in steady heat flow, all the heat that passes through the first resistance must pass through the second and in turn pass through the third, q_A, q_B, and q_C are equal and all can be denoted by q. Using this fact and solving for q/A give

$$\frac{q}{A} = \frac{\Delta T}{B_A/\overline{k}_A + B_B/\overline{k}_B + B_C/\overline{k}_C}$$

$$= \frac{\Delta T}{R_A + R_B + R_C} = \frac{\Delta T}{R} \quad (4.2\text{-}11)$$

where R_A, R_B and R_C = resistance of individual layers
R = overall resistance

Equation (4.2-11) shows that in heat flow through a series of layers the overall thermal resistance equals the sum of the individual resistances.

The rate of flow of heat through several resistances in series clearly is analogous to the current flowing through several electric resistances in series. In an electric circuit the potential drop over any one of several resistances is to the total potential drop in the circuit as the individual resistances are to the total resistance. In the same way the potential drops in a thermal circuit, which are the temperature differences, are to the total temperature drop as the individual thermal resistances are to the total thermal resistance. This can be expressed mathematically as

$$\frac{\Delta T}{R} = \frac{\Delta T_A}{R_A} = \frac{\Delta T_B}{R_B} = \frac{\Delta T_C}{R_C} \quad (4.2\text{-}12)$$

Figure 4.2 also shows the pattern of temperatures and the temperature gradients. Depending on the thickness and thermal conductivity of the layer, the temperature drop in that layer may be a large or small fraction of the total temperature drop; a thin layer of low conductivity may well cause a much larger temperature drop and a steeper thermal gradient than a thick layer of high conductivity.

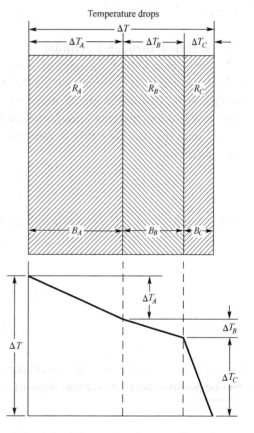

Fig. 4.2 Thermal resistances in series

EXAMPLE 4.2 A flat furnace wall is constructed of 120-mm layer of sil-o-cel brick, with a thermal conductivity 0.08 W/(m·℃), backed by a 150-mm of common brick, of conductivity 0.8 W/(m·℃), the temperature of inner face of the wall is 1400℃, and that of the outer face is 200℃.

(a) What is the heat loss through the wall, in W per square meter.
(b) To reduce the heat loss to 600 W/m^2 by adding a layer of cork with k of 0.2 W/(m·°C) on the outside of common brick, how many meters of cork are required?

Solution.

(a) Given: B_1=0.12m, k_1=0.08 W/(m·°C), B_2=0.15m, k_2=0.8 W/(m·°C)

From Eq. (4.2-11), $\dfrac{q}{A} = \dfrac{\sum \Delta T_i}{\sum R_i} = \dfrac{\Delta T}{B_1/k_1 + B_2/k_2} = \dfrac{1400-200}{0.12/0.08 + 0.15/0.8} = 711 \text{W}/\text{m}^2$

(b) Given: B_1=0.12m, k_1=0.08 W/(m·°C), B_2=0.15m, k_2=0.8 W/(m·°C), k_3=0.2 W/(m·°C), q/A=600W/m^2, suppose the thickness of a layer of cork is x, substituting in Eq.(4.2-11) gives

$\dfrac{q}{A} = \dfrac{\sum \Delta T_i}{\sum R_i} = \dfrac{\Delta T}{B_1/k_1 + B_2/k_2 + x/k_3} = \dfrac{1400-200}{0.12/0.08 + 0.15/0.8 + x/0.2} = 600 \text{W}/\text{m}^2$

Hence x=0.0625m

Heat flow through a cylinder

Consider a hollow cylinder of length L with an inside radius r_i and an outside r_o. The cylinder is made of material with a thermal conductivity k. The temperature of the outside surface is T_o; that of the inside surface is T_i, with $T_i>T_o$. At radius r from the center the heat flow rate is q and the area through which it flows is A. The area is a function of the radius; at steady state the heat flow rate is constant. Equation (4.2-3) becomes, since heat flows only in the r direction,

$$\dfrac{q}{A} = \dfrac{q}{2\pi r L} = -k\dfrac{dT}{dr} \qquad (4.2\text{-}13)$$

Rearranging Eq. (4.2-13) and integrating between limits gives

$$\int_{r_i}^{r_o} \dfrac{dr}{r} = \dfrac{2\pi L k}{q} \int_{T_o}^{T_i} dT$$

$$\ln r_o - \ln r_i = \dfrac{2\pi L k}{q}(T_i - T_o)$$

$$q = \dfrac{k(2\pi L)(T_i - T_o)}{\ln(r_o/r_i)} \qquad (4.2\text{-}14)$$

Equation (4.2-14) can be used to calculate the flow of heat through a thick-walled cylinder. It can be put in a more convenient form by expressing the rate of flow of heat as

$$q = \dfrac{k \overline{A}_L (T_i - T_o)}{r_o - r_i} \qquad (4.2\text{-}15)$$

This is of the same general form as Eq. (4.2-8) for heat flow through a flat wall with the exception of A_L, which must be so chosen that the equation is correct. The term A_L can be determined by equating the right-hand sides of Eqs. (4.2-14) and (4.2-15) and solving for A_L

$$\overline{A}_L = \dfrac{2\pi L (r_o - r_i)}{\ln(r_o/r_i)} \qquad (4.2\text{-}16)$$

Note from Eq. (4.2-16) that A_L is the area of a cylinder of length L and radius r_L, where

$$\bar{r}_L = \frac{r_o - r_i}{\ln(r_o / r_i)} \tag{4.2-17}$$

The form of the right-hand side of Eq. (4.2-17) is important enough to repay memorizing. It is known as the logarithmic mean, and in the particular case of $\bar{r}_L$ is called the logarithmic mean radius. It is the radius that, when applied to the integrated equation for a flat wall, will give the correct rate of heat flow through a thick-walled cylinder.

The logarithmic mean is less convenient than the arithmetic mean, and the latter can be used without appreciable error for thin-walled tubes, where r_o/r_i is nearly 1. The ratio of the logarithmic mean $\bar{r}_L$ to arithmetic mean $\bar{r}_a$ is a function of r_o/r_i, as shown in Fig. 4.3. Thus, when $r_o/r_i = 2$, the logarithmic mean is 0.96 $\bar{r}_a$ and the error in the use of the arithmetic mean is 4 percent. The error is 1 percent where $r_o/r_i = 1.4$.

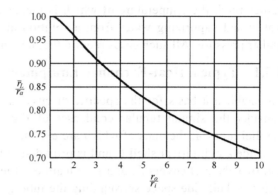

Fig. 4.3 Relation between logarithmic and arithmetic means.

EXAMPLE 4.3 A steel tube with dimension $\Phi 60 \times 3$ mm is backed by a 30-mm thick cork, with a thermal conductivity of 0.043 W/(m·℃), and is followed by a 40-mm layer of some insulating material, of 0.07 W/(m·℃). The temperature of the outer face of steel tube is -110℃, and that of the outer face of insulating material is 10℃. What is the heat loss per meter tube through the wall?

Solution. Similar to compound resistances in series through the flat wall, the total resistance through two layers of cylinders is as follows

$$R = R_1 + R_2 = \frac{B_1}{k_1 \bar{A}_{L1}} + \frac{B_2}{k_2 \bar{A}_{L2}}$$

The thermal resistance of the cork layer per meter tube is

$$R_1 = \frac{B_1}{k_1 \bar{A}_{L1}} = \frac{0.03}{0.043 \times 2\pi(0.06 - 0.03)/\ln(0.06/0.03)} = 2.57$$

and that of the insulating material per meter tube is

$$R_2 = \frac{B_2}{k_2 \bar{A}_{L2}} = \frac{0.04}{0.07 \times 2\pi(0.1 - 0.06)/\ln(0.1/0.06)} = 1.17$$

The overall temperature drop is

$$\Delta T = 10 - (-110) = 120 \text{ ℃}$$

Heat loss per meter tube through the wall is

$$\frac{q}{L} = \frac{\Delta T}{R} = \frac{120}{2.57 + 1.17} = 32 \text{ W/m}$$

4.3 Principles of Heat Flow in Fluids

Heat transfer from a warmer fluid to a cooler fluid, usually through a solid wall separating the two fluids, is common in chemical engineering practice. The heat transferred may be latent heat accompanying a phase change such as condensation or vaporization, or it may be sensible heat from the rise or fall in the temperature of a fluid without any phase change. Typical examples are reducing the temperature of a fluid by transfer of sensible heat to a cooler fluid, the temperature of which is increased thereby; condensing steam using cooling water; and vaporizing water from a solution at a given pressure by condensing steam at a higher pressure. All such cases require that heat be transferred by conduction and convection.

4.3.1 Typical Heat-Exchange Equipment

To establish a basis for the specific discussion of heat transfer to and from flowing fluids, consider the simple tubular condenser of Fig. 4.4. It consists essentially of a bundle of parallel tubes A, the ends of which are expanded into tube sheets B_1 and B_2. The tube bundle is inside a cylindrical shell C and is provided with two channels D_1 and D_2, one at each end, and two channel covers E_1 and E_2. Steam or other pure vapor is introduced through nozzle F into the shell-side space surrounding the tubes, condensate is withdrawn through connection G, and any noncondensable gas that might be present at start-up or that might enter with the inlet vapor is removed through vent K. Connection G leads to a trap, which is a device that allows liquid to flow but holds back vapor. The fluid to be heated is pumped through connection H into channel D_2. It flows through the tubes into the other channel D_1 and is discharged through connection J. The two fluids are physically separated but are in thermal contact with the thin metal tube walls separating them. Heat flows through the tube walls from the condensing vapor to the cooler fluid in the tubes.

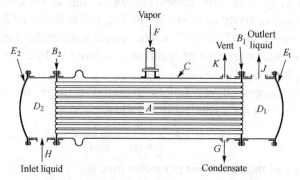

Fig. 4.4 Single-pass tubular condenser: A, tubes; B_1, B_2, tube sheets; C, shell; D_1, D_2, channels; E_1, E_2, channel covers; F, vapor inlet; G, condensate outlet; H, cold-liquid inlet; J, warm-liquid outlet; K, noncondensed gas vent.

If the vapor entering the condenser is a single component, not a mixture, and is not superheated, and if the condensate is not subcooled below its condensing temperature, then the temperature throughout the shell side of the condenser is constant. The reason for this is that the temperature of the condensing vapor is fixed by the pressure of the shell-side space, and the pressure in that space is constant. The temperature of the fluid in the tubes increases continuously as the fluid flows through the tubes.

The temperatures of the condensing vapor and of the liquid are plotted against the tube length in Fig. 4.5. The horizontal line represents the temperature of the condensing vapor, and the curved line below it represents the rising temperature of the tube-side fluid. In Fig. 4.5, the inlet and outlet fluid temperatures are T_{ca} and T_{cb}, respectively, and the constant temperature of the vapor is T_h. At a length L from the entrance end of the tubes, the fluid temperature is T_c, and the local difference between the temperatures of vapor and fluid is $T_h - T_c$. This temperature difference is called a *point temperature difference* and is denoted by ΔT. The point temperature difference at the inlet of the tubes is $T_h - T_{ca}$, denoted by ΔT_1, and that at the exit end is $T_h - T_{cb}$, denoted by ΔT_2. The terminal point temperature differences ΔT_1 and ΔT_2 are called the *approaches*.

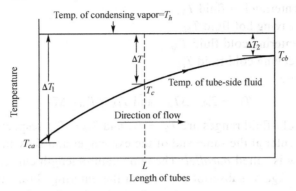

Fig. 4.5 Temperature–length curves for condenser.

The change in temperature of the fluid $T_{cb} - T_{ca}$ is called the *temperature range* or, simply, the *range*. In a condenser there is but one range, that of the cold fluid being heated.

In this text the symbol ΔT is used exclusively to signify a temperature *difference* between two objects or two fluids. It does *not* denote the temperature change in a given fluid.

A second example of simple heat-transfer equipment is the double-pipe exchanger shown in Fig. 4.6. It is assembled of standard metal pipe and standardized return bends and return heads, the latter equipped with stuffing boxes. One fluid flows through the inside pipe, and the second fluid flows through the annular space between the outside and the inside pipe. The function of a heat exchanger is to increase the temperature of a cooler fluid and decrease that of a hotter fluid. In a typical exchanger, the inner pipe may be $1\frac{1}{4}$ in. and the outer pipe $2\frac{1}{2}$ in., both IPS. Such an exchanger may consist of several passes arranged in a vertical stack. Double-pipe exchangers are useful when not more than 9.29 to 13.9 m² (100 to 150 ft²) of surface is required. For larger capacities, more elaborate shell-and-tube exchangers, containing up to hundreds of square meters of area, and described later, are used.

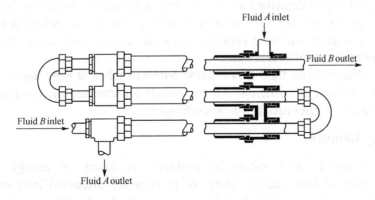

Fig. 4.6 Double-pipe heat exchanger.

Countercurrent and parallel flows

The two fluids enter at different ends of the exchanger, shown in Fig. 4.6, and pass in opposite directions through the unit. This type of flow is that commonly used and is called *counterflow* or *countercurrent flow*. The temperature–length curves for this case are shown in Fig. 4.7(*a*). The four terminal temperatures are denoted as follows:

Temperature of entering hot fluid T_{ha}
Temperature of leaving hot fluid T_{hb}
Temperature of entering cold fluid T_{ca}
Temperature of leaving cold fluid T_{cb}
The approaches are

$$T_{ha} - T_{cb} = \Delta T_2 \quad \text{and} \quad T_{hb} - T_{ca} = \Delta T_1 \tag{4.3-1}$$

The warm-fluid and cold-fluid ranges are $T_{ha} - T_{hb}$ and $T_{cb} - T_{ca}$, respectively.

If the two fluids enter at the same end of the exchanger and flow in the same direction to the other end, the flow is called *parallel*. The temperature–length curves for parallel flow are shown in Fig. 4.7(*b*). Again, subscript *a* refers to the entering fluids and subscript *b* to the leaving fluids. The approaches are $\Delta T_1 = T_{ha} - T_{ca}$ and $\Delta T_2 = T_{hb} - T_{cb}$.

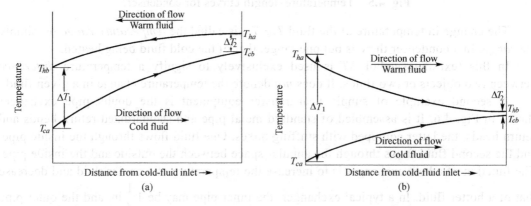

Fig. 4.7 Temperatures in (*a*) countercurrent flow and (*b*) parallel flow.

Parallel flow is rarely used in a single-pass exchanger such as that shown in Fig. 4.6 because, as inspection of Fig. 4.7(*b*) will show, it is not possible with this method of flow to bring the exit temperature of one fluid nearly to the entrance temperature of the other, and the heat that can be transferred is less than that possible in countercurrent flow. In the multipass shell-and-tube exchangers described later, counterflow is usually desirable, but parallel flow is used in some passes. Parallel flow is used in special situations in which it is important to change the temperature of one fluid very rapidly, such as when quenching a hot fluid from a chemical reactor to stop further reactions.

In some exchangers one fluid flows across banks of tubes at right angles to the axis of the tubes. This is known as *crossflow*. An automobile radiator and the condenser in a home refrigerator are examples of crossflow heat exchangers.

4.3.2 Energy Balances

Quantitative attack on heat-transfer problems is based on energy balances and estimations of rates of heat transfer. Rates of transfer are discussed later in this chapter. Many, perhaps most, heat-transfer devices operate under steady-state conditions, and only steady-state operation is considered here.

Enthalpy balances in heat exchangers

In heat exchangers there is no shaft work, and mechanical, potential, and kinetic energies are small in comparison with the other terms in the energy-balance equation. Thus, for one stream through the exchanger

$$\dot{m}(H_b - H_a) = q \tag{4.3-2}$$

where $\dot{m}$ = flow rate of stream

$q = Q/t$ = rate of heat transfer into stream

H_a, H_b = enthalpies per unit mass of stream at entrance and exit, respectively. Equation (4.3-2) can be written for each stream flowing through the exchanger.

A further simplification in the use of the heat-transfer rate q is justified. One of the two fluid streams, that outside the tubes, can gain or lose heat by transfer with the ambient air if the fluid is colder or hotter than the ambient. Heat transfer to or from the ambient is not usually desired in practice, and it is usually reduced to a small magnitude by suitable insulation. It is customary to neglect it in comparison with the heat transfer through the walls of the tubes from the warm fluid to the cold fluid, and q is interpreted accordingly.

Accepting these assumptions, Eq. (4.3-2) can be written for the warm fluid as

$$\dot{m}_h(H_{hb} - H_{ha}) = q_h \tag{4.3-3}$$

and for the cold fluid as

$$\dot{m}_c(H_{cb} - H_{ca}) = q_c \tag{4.3-4}$$

where $\dot{m}_c, \dot{m}_h$ = mass flow rates of cold fluid and warm fluid, respectively

H_{ca}, H_{ha} = enthalpy per unit mass of entering cold fluid and entering warm fluid, respectively

H_{cb}, H_{hb} = enthalpy per unit mass of leaving cold fluid and leaving hot fluid, respectively

q_c, q_h = rates of heat addition to cold fluid and warm fluid, respectively

The sign of q_c is positive, but that of q_h is negative because the warm fluid loses, rather than gains, heat. The heat lost by the warm fluid is gained by the cold fluid, and

$$q_c = -q_h$$

Therefore, from Eqs. (4.3-3) and (4.3-4),

$$\dot{m}_h(H_{ha} - H_{hb}) = \dot{m}_c(H_{cb} - H_{ca}) = q \tag{4.3-5}$$

Equation (4.3-5) is called the *overall enthalpy balance*.

If only sensible heat is transferred and constant specific heats are assumed, the overall enthalpy balance for a heat exchanger becomes

$$\dot{m}_h c_{ph}(T_{ha} - T_{hb}) = \dot{m}_c c_{pc}(T_{cb} - T_{ca}) = q \tag{4.3-6}$$

where c_{pc} = specific heat of cold fluid
c_{ph} = specific heat of warm fluid

Enthalpy balances in total condensers

For a condenser

$$\dot{m}_h \lambda = \dot{m}_c c_{pc}(T_{cb} - T_{ca}) = q \tag{4.3-7}$$

where $\dot{m}_h$ = rate of condensation of vapor

λ = latent heat of vaporization of vapor

Equation (4.3-7) is based on the assumption that the vapor enters the condenser as saturated vapor (no superheat) and the condensate leaves at condensing temperature without being further cooled. If either of these sensible-heat effects is important, it must be accounted for by an added term in the left-hand side of Eq. (4.3-7). For example, if the condensate leaves at a temperature T_{hb} that is less than T_h, the condensing temperature of the vapor, Eq. (4.3-7) must be written

$$\dot{m}_h [\lambda + c_{ph}(T_h - T_{hb})] = \dot{m}_c c_{pc}(T_{cb} - T_{ca}) \tag{4.3-8}$$

where c_{ph} is now the specific heat of the condensate.

4.3.3 Heat Flux and Heat-Transfer Coefficients

Heat-transfer calculations are based on the area of the heating surface and are expressed in watts per square meter of surface through which the heat flows. The rate of heat transfer per unit area is called the *heat flux*. In many types of heat-transfer equipment the transfer surfaces are constructed from tubes or pipe. Heat fluxes may then be based on either the inside area or the outside area of the tubes. Although the choice is arbitrary, it must be clearly stated, because the numerical magnitude of the heat fluxes will not be the same for both.

Average temperature of fluid stream. When a fluid is being heated or cooled, the temperature will vary throughout the cross section of the stream. If the fluid is being heated, the temperature of the fluid is a maximum at the wall of the heating surface and decreases toward the center of the stream. If the fluid is being cooled, the temperature is a minimum at the wall and increases toward the center. Because of these temperature gradients throughout the cross section of the stream, it is necessary, for definiteness, to state what is meant by the temperature of the stream. It is agreed that it is the temperature that would be attained if the entire fluid stream flowing across the section in question were withdrawn and mixed adiabatically to a uniform temperature. The temperature so defined is called the *average* or *mixing-cup stream temperature*. The temperatures plotted in Fig. 4.7 are all average stream temperatures.

Overall Heat-Transfer Coefficient

As shown before, Eqs. (4.2-8) and (4.2-11), the heat flux through layers of solids in series is proportional to a driving force, the overall temperature difference ΔT. This also applies to heat flow through liquid layers and solids in series. In a heat exchanger the driving force is taken as $T_h - T_c$, where T_h is the average temperature of the hot fluid and T_c is that of the cold fluid. The quantity $T_h - T_c$ is the *overall local temperature difference* ΔT. It is clear from Fig. 4.7 that ΔT can vary considerably from point to point along the tube; and therefore, since the heat flux is proportional to ΔT, the flux also varies with tube length. It is necessary to start with a differential equation by focusing attention on a differential area dA through which a differential heat flow dq occurs under the driving force of a local value of ΔT. The local flux is then dq/dA and is related to the local value of ΔT by the equation

$$\frac{dq}{dA} = U \Delta T = U(T_h - T_c) \tag{4.3-9}$$

The quantity U, defined by Eq. (4.3-9) as a proportionality factor between dq/dA and ΔT, is called the *local overall heat-transfer coefficient*.

To complete the definition of U for a tubular exchanger, it is necessary to specify the area. If A is taken as the outside tube area A_o, then U becomes a coefficient based on that area and is written U_o. Likewise, if the inside area A_i is chosen, the coefficient is also based on that area and is denoted by U_i. Since ΔT and dq are independent of the choice of area, it follows that

$$\frac{U_o}{U_i} = \frac{dA_i}{dA_o} = \frac{D_i}{D_o} \qquad (4.3\text{-}10)$$

where D_i and D_o are the inside and outside tube diameters, respectively.

In a plate-type heat exchanger the areas on both sides are the same, and there is only one value of U.

Integration over total surface; logarithmic mean temperature difference. To apply Eq. (4.3-9) to the entire area of a heat exchanger, the equation must be integrated. This can be done formally where certain simplifying assumptions are accepted. The assumptions are that (1) the overall coefficient U is constant, (2) the specific heats of the hot and cold fluids are constant, (3) heat exchange with the ambient air is negligible, and (4) the flow is steady and either parallel or countercurrent, as shown in Fig. 4.7.

The most questionable of these assumptions is that of a constant overall coefficient. The coefficient does in fact vary with the temperatures of the fluids, but its change with temperature is gradual, so that when the temperature ranges are moderate, the assumption of constant U is not seriously in error.

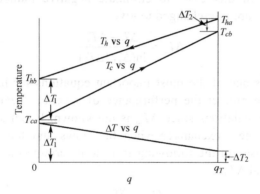

Fig. 4.8 Temperature versus heat flow rate in countercurrent flow

Assumptions 2 and 4 imply that if T_c and T_h are plotted against q, as shown in Fig. 4.8, straight lines are obtained. Since T_c and T_h vary linearly with q, ΔT does likewise and $d(\Delta T)/dq$, the slope of the graph of ΔT versus q, is constant. Therefore

$$\frac{d\Delta T}{dq} = \frac{\Delta T_2 - \Delta T_1}{q_T} \qquad (4.3\text{-}11)$$

where $\Delta T_1, \Delta T_2$ = approaches.
$\qquad q_T$ = rate of heat transfer in entire exchanger.
Elimination of dq from Eqs. (4.3-9) and (4.3-11) gives

$$\frac{d(\Delta T)}{U\Delta T dA} = \frac{\Delta T_2 - \Delta T_1}{q_T} \qquad (4.3\text{-}12)$$

The variables ΔT and A can be separated, and if U is constant, the equation can be integrated over the limits A_T and 0 for A and ΔT_2 and ΔT_1, where A_T is the total area of the heat-transfer surface. Thus

$$\int_{\Delta T_1}^{\Delta T_2} \frac{d(\Delta T)}{\Delta T} = \frac{U(\Delta T_2 - \Delta T_1)}{q_T} \int_0^{A_T} dA$$

or
$$\ln\frac{\Delta T_2}{\Delta T_1} = \frac{U(\Delta T_2 - \Delta T_1)}{q_T} A_T \tag{4.3-13}$$

Equation (4.3-13) can be written

$$q_T = UA_T \frac{\Delta T_2 - \Delta T_1}{\ln(\Delta T_2/\Delta T_1)} = UA_T \Delta T_m \tag{4.3-14}$$

where
$$\Delta T_m = \frac{\Delta T_2 - \Delta T_1}{\ln(\Delta T_2/\Delta T_1)} \tag{4.3-15}$$

Equation (4.3-15) defines the *logarithmic mean temperature difference* (LMTD). It is of the same form as Eq. (4.2-17) for the logarithmic mean radius of a thick walled tube. When ΔT_1 and ΔT_2 are nearly equal, their arithmetic average can be used for ΔT_m within the same limits of accuracy given for Eq. (4.2-17), as shown in Fig. 4.3. If one of the fluids is at constant temperature, as in a condenser, no difference exists among countercurrent flow, parallel flow, or multipass flow, and Eq. (4.3-15) applies to them all. In countercurrent flow, with temperature changes on both sides, ΔT_2, the warm-end approach, may be less than ΔT_1, the cold-end approach. In this case, to eliminate negative numbers and logarithms, the subscripts in Eq. (4.3-15) are interchanged to give

$$\Delta T_m = \frac{\Delta T_1 - \Delta T_2}{\ln(\Delta T_1/\Delta T_2)} \tag{4.3-15a}$$

Equation (4.3-14) is one of the most important equations for heat transfer to and from fluids. It can be used to predict the performance of a certain heat exchanger, which may require trial-and-error calculations since ΔT_m is not known. It can also be used to calculate the area required for a new exchanger when the flows and temperature approaches are specified. Finally, it is used in the following form to determine the overall coefficient from measured values of q_T and ΔT_m:

$$U = \frac{q_T}{A_T \Delta T_m} \tag{4.3-16}$$

Variable overall coefficient. When the overall coefficient varies regularly, the rate of heat transfer may be predicted from Eq. (4.3-17), which is based on the assumption that U varies linearly with the temperature drop over the entire heating surface1:

$$q_T = A_T \frac{U_2 \Delta T_1 - U_1 \Delta T_2}{\ln[(U_2 \Delta T_1)/(U_1/\Delta T_2)]} \tag{4.3-17}$$

where U_1, U_2 = local overall coefficients at ends of exchanger.
$\Delta T_1, \Delta T_2$ = temperature approaches at corresponding ends of exchanger.

Equation (4.3-17) calls for use of a logarithmic mean value of the $U\Delta T$ cross product, where the overall coefficient at one end of the exchanger is multiplied by the temperature approach at the other. The derivation of this equation requires that assumptions 2 to 4 given earlier be accepted.

LMTD not always valid. The LMTD is not always the correct mean temperature difference to use. It should *not* be used when U changes appreciably or when ΔT is not a linear function of q. As an example, consider an exchanger used to cool and condense a superheated vapor, with the temperature diagram shown in Fig. 4.9. The ΔT driving force is a linear function of q while the vapor is being cooled, but ΔT is a different linear function of q in the condensing section of the exchanger. Furthermore, U is not the same in the two parts of the exchanger. The cooling and condensing sections must be sized separately by using the appropriate values of q, U, and LMTD rather than some kind of average U and an overall LMTD.

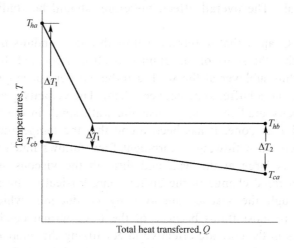

Fig. 4.9 Temperature profiles in cooling and condensing superheated vapor

Multipass exchangers. In multipass shell-and-tube exchangers the flow pattern is complex, with parallel, countercurrent, and crossflow all present. Under these conditions, even when the overall coefficient U is constant, the LMTD cannot be used. Calculation procedures for multipass exchangers are given later.

Individual Heat-Transfer Coefficients

The overall coefficient depends on many variables including the physical properties of the fluids and of the solid wall, the flow rates, and the exchanger dimensions. The only logical way to predict the overall coefficient is to use correlations for the individual resistances of the solid and the fluid layers and to add these resistances to find the overall resistance, which is the reciprocal of the overall coefficient. Consider the local overall coefficient at a specific point in the double-pipe exchanger shown in Fig. 4.6. For definiteness, assume that the warm fluid is flowing through the inside pipe and that the cold fluid is flowing through the annular space. Assume also that the Reynolds numbers of the two fluids are sufficiently large to ensure turbulent flow and that both surfaces of the inside tube are clear of dirt or scale. If now a plot is prepared, as shown in Fig. 4.10, with temperature as the ordinate and distance perpendicular to the wall as the abscissa, several important facts become evident. In the figure, the metal wall of the

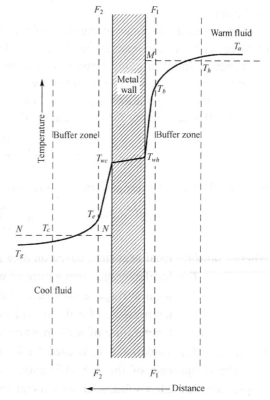

Fig. 4.10 Temperature gradients in forced convection.

tube separates the warm fluid on the right from the cold fluid on the left. The change in temperature with distance is shown by the line T_a T_b T_{wh} T_{wc} T_e T_g. The temperature profile is thus divided into three separate parts, one through each of the two fluids and the other through the metal wall. The overall effect, therefore, should be studied in terms of these individual parts.

It was shown in Chap. 1 that in turbulent flow through conduits three zones exist, even in a single fluid, so that the study of one fluid is itself complicated. In each fluid shown in Fig. 4.10 there is a thin sublayer at the wall, a turbulent core occupying most of the cross section of the stream, and a buffer zone between them. The velocity gradients were described in Chap. 1. The velocity gradient is large near the wall, small in the turbulent core, and in rapid change in the buffer zone. It has been found that the temperature gradient in a fluid being heated or cooled when flowing in turbulent flow follows much the same course. The temperature gradient is large at the wall and through the viscous sublayer, small in the turbulent core, and in rapid change in the buffer zone. Basically, the reason for this is that heat must flow through the viscous sublayer by conduction, which calls for a steep temperature gradient in most fluids because of the low thermal conductivity, whereas the rapidly moving eddies in the core are effective in equalizing the temperature in the turbulent zone. In the buffer zone there are fewer eddies than in the turbulent zone, but they add significantly to the conductive heat transfer. In Fig. 4.10 the dashed lines F_1F_1 and F_2F_2 represent the boundaries of the viscous sublayer and of the buffer zone.

The average temperature of the warm stream is somewhat less than the maximum temperature T_a and is represented by the horizontal line MM, which is drawn at temperature T_h. Likewise, line NN, drawn at temperature T_c, represents the average temperature of the cold fluid.

The overall resistance to the flow of heat from the warm fluid to the cold fluid is a result of three separate resistances operating in series. In general, as shown in Fig. 4.10, the wall resistance is small in comparison with that of the fluids. The fluid resistances are generally computed by using correlations for individual heat-transfer coefficients or film coefficients, which are the reciprocals of the resistances.

The film coefficient for the warm fluid is defined by the equation

$$h = \frac{dq/dA}{T_h - T_{wh}} \tag{4.3-18}$$

For the cold fluid the terms in the denominator are reversed to make h positive

$$h = \frac{dq/dA}{T_{wc} - T_c} \tag{4.3-19}$$

where dq/dA = local heat flux, based on area in contact with fluid
 T_h = local average temperature of warm fluid
 T_c = local average temperature of cold fluid
 T_{wc} = temperature of wall on cold side
 T_{wh} = temperature of wall on warm side

Often $T_{wh} - T_{wc}$ is small and T_w is used for the wall temperature on both sides.

The reciprocals of these coefficients, $1/h_h$ and $1/h_c$, are the thermal resistances. For conduction through a solid, such as a metal wall of thickness x_w and thermal conductivity k, the thermal resistance equals x_w/k. Appropriately corrected for changes in area, the individual resistances may be added to give the overall resistance $1/U$.

A second expression for h is derived from the assumption that heat transfer very near the wall occurs only by conduction, and the heat flux is given by Eq. (4.2-1), noting that the normal distance x may be replaced by y, the distance normal to the wall in Fig. 4.10. Thus

$$\frac{dq}{dA} = -k\left(\frac{dT}{dy}\right)_w \tag{4.3-20}$$

The subscript w calls attention to the fact that the gradient must be evaluated at the wall. Eliminating dq/dA from Eqs. (4.3-19) and (4.3-20) gives

$$h = -k\frac{(dT/dy)_w}{T - T_w} \tag{4.3-21}$$

In Eq. (4.3-21), T is the average fluid temperature, which is T_h for the warm side and T_c for the cool side. The denominator is changed to $T_w - T$ for the cool side to make h positive.

Equation (4.3-21) can be put into a dimensionless form by multiplying by the ratio of an arbitrary length to the thermal conductivity. The choice of length depends on the situation. For heat transfer at the inner surface of a tube, the tube diameter D is the usual choice. Multiplying Eq. (4.3-21) by D/k gives

$$Nu = \frac{hD}{k} = -D\frac{(dT/dy)_w}{T - T_w} \tag{4.3-22}$$

On the cold-fluid side of the tube wall $T < T_w$, and the denominator in Eqs. (4.3-21) and (4.3-22) becomes $T_w - T$. The dimensionless group hD/k is called a *Nusselt number Nu*. That shown in Eq. (4.3-22) is a local Nusselt number based on diameter. The physical meaning of the Nusselt number can be seen by inspection of the right-hand side of Eq. (4.3-22). The numerator $(dT/dy)_w$ is, of course, the gradient at the wall. The factor $(T - T_w)/D$ can be considered the average temperature gradient across the entire pipe, and the Nusselt number is the ratio of these two gradients.

Another interpretation of the Nusselt number can be obtained by considering the gradient that would exist if all the resistance to heat transfer were in a laminar layer of thickness x in which heat transfer was only by conduction. The heat transfer rate and coefficient follow from Eqs. (4.2-1) and (4.10)

$$\frac{dq}{dA} = \frac{k(T - T_w)}{x} \tag{4.3-23}$$

$$h = \frac{k}{x} \tag{4.3-24}$$

From the definition of the Nusselt number,

$$\frac{hD}{k} = Nu = \frac{k}{x}\frac{D}{k} = \frac{D}{x} \tag{4.3-25}$$

The Nusselt number is the ratio of the tube diameter to the equivalent thickness of the laminar layer. Sometimes x is called the film thickness, and it is generally slightly greater than the thickness of the laminar boundary layer because there is some resistance to heat transfer in the buffer zone.

Equation (4.3-18), when applied to the two fluids of Fig. 4.10, becomes, for the inside of the tube (the warm side in Fig. 4.10),

$$h_i = \frac{dq/dA_i}{T_h - T_{wh}} \tag{4.3-26}$$

and for the outside of the tube (the cold side), from Eq. (4.3-19)

$$h_o = \frac{dq/dA_o}{T_{wc} - T_c} \tag{4.3-27}$$

where A_i, and A_o are the inside and outside areas of the tube, respectively.

The cold fluid could, of course, be inside the tubes and the warm fluid outside. Coefficients h_i and h_o refer to the *inside* and the *outside* of the tube, respectively, and not to a specific fluid.

Calculation of overall coefficients from individual coefficients. The overall coefficient is constructed from the individual coefficients and the resistance of the tube wall in the following manner. The rate of heat transfer through the tube wall is given by the differential form of Eq. (4.2-15),

$$\frac{dq}{d\bar{A}_L} = \frac{k_m(T_{wh} - T_{wc})}{B} \tag{4.3-28}$$

where $T_{wh} - T_{wc}$ = temperature difference through tube wall
k_m = thermal conductivity of wall
B = tube wall thickness
$dq/d\bar{A}_L$ = local heat flux, based on logarithmic mean of inside and outside areas of tube

If Eqs. (4.3-26) to (4.3-28) are solved for the temperature differences and the temperature differences added, the result is

$$(T_h - T_{wh}) + (T_{wh} - T_{wc}) + (T_{wc} - T_c) = T_h - T_c = \Delta T$$

$$= dq\left(\frac{1}{dA_i h_i} + \frac{B}{d\bar{A}_L k_m} + \frac{1}{dA_o h_o}\right) \tag{4.3-29}$$

Assume that the heat-transfer rate is arbitrarily based on the outside area. If Eq. (4.3-29) is solved for dq, and if both sides of the resulting equation are divided by dA_o, the result is

$$\frac{dq}{dA_o} = \frac{T_h - T_c}{\frac{1}{h_i}\left(\frac{dA_o}{dA_i}\right) + \frac{B}{k_m}\left(\frac{dA_o}{d\bar{A}_L}\right) + \frac{1}{h_o}} \tag{4.3-30}$$

Now

$$\frac{dA_o}{dA_i} = \frac{D_o}{D_i} \quad \text{and} \quad \frac{dA_o}{d\bar{A}_L} = \frac{D_o}{\bar{D}_L}$$

where D_o, D_i, and $\bar{D}_L$ are the outside, inside, and logarithmic mean diameters of the tube, respectively. Therefore

$$\frac{dq}{dA_o} = \frac{T_h - T_c}{\frac{1}{h_i}\left(\frac{D_o}{D_i}\right) + \frac{B}{k_m}\left(\frac{D_o}{\bar{D}_L}\right) + \frac{1}{h_o}} \tag{4.3-31}$$

Comparing Eq. (4.3-9) with Eq. (4.3-31) shows that

$$U_o = \cfrac{1}{\cfrac{1}{h_i}\left(\cfrac{D_o}{D_i}\right) + \cfrac{B}{k_m}\left(\cfrac{D_o}{\bar{D}_L}\right) + \cfrac{1}{h_o}} \qquad (4.3\text{-}32)$$

If the inside area A_i is chosen as the base area, division of Eq. (4.3-29) by dA_i gives for the overall coefficient

$$U_i = \cfrac{1}{\cfrac{1}{h_i} + \cfrac{B}{k_m}\left(\cfrac{D_i}{\bar{D}_L}\right) + \cfrac{1}{h_o}\left(\cfrac{D_i}{D_o}\right)} \qquad (4.3\text{-}33)$$

Resistance form of overall coefficient. A comparison of Eqs. (4.2-12) and (4.3-32) suggests that the reciprocal of an overall coefficient can be considered to be an overall resistance composed of three resistances in series. The total, or overall, resistance is given by the equation

$$R_o = \frac{1}{U_o} = \frac{D_o}{D_i h_i} + \frac{B}{k_m}\frac{D_o}{\bar{D}_L} + \frac{1}{h_o} \qquad (4.3\text{-}34)$$

The individual terms on the right-hand side of Eq. (4.3-34) represent the individual resistances of the two fluids and of the metal wall. The overall temperature drop is proportional to $1/U$, and the temperature drops in the two fluids and the wall are proportional to the individual resistances, or, for the case of Eq. (4.3-34),

$$\frac{\Delta T}{1/U_o} = \frac{\Delta T_i}{D_o/D_i h_i} = \frac{\Delta T_w}{(B/k_m)(D_o/\bar{D}_L)} = \frac{\Delta T_o}{1/h_o} \qquad (4.3\text{-}35)$$

where ΔT = overall temperature drop
ΔT_i = temperature drop through inside fluid
ΔT_w = temperature drop through metal wall
ΔT_o = temperature drop through outside fluid

Equation (4.3-35) can be used to determine the wall temperatures T_{wh} and T_{wc} by calculating ΔT_i or ΔT_o and adding these values to, or subtracting them from, T_i or T_o.

Fouling factors. In actual service, heat-transfer surfaces do not remain clean. Scale, dirt, and other solid deposits form on one or both sides of the tubes, provide additional resistances to heat flow, and reduce the overall coefficient. The effect of such deposits is taken into account by adding a term $1/(dA\ h_d)$ to the term in parentheses in Eq. (4.3-29) for each scale deposit. Thus, assuming that scale is deposited on both the inside and the outside surface of the tubes, Eq. (4.3-29) becomes, after correction for the effects of scale,

$$\Delta T = dq\left(\frac{1}{dA_i h_{di}} + \frac{1}{dA_i h_i} + \frac{B}{d\bar{A}_L k_m} + \frac{1}{dA_o h_o} + \frac{1}{dA_o h_{do}}\right) \qquad (4.3\text{-}36)$$

where h_{di} and h_{do} are the *fouling factors* for the scale deposits on the inside and outside tube surfaces, respectively. The following equations for the overall coefficients based on outside and inside areas, respectively, follow from Eq. (4.3-36):

$$U_o = \cfrac{1}{D_o/D_i h_{di} + D_o/D_i h_i + (B/k_w)(D_o/\bar{D}_L) + 1/h_o + 1/h_{do}} \qquad (4.3\text{-}37)$$

and

$$U_i = \frac{1}{1/h_{di} + 1/h_i + (B/k_m)(D_i/\overline{D}_L) + D_i/D_o h_o + D_i/D_o h_{do}}$$ (4.3-38)

The actual thicknesses of the deposits are neglected in Eqs. (4.3-37) and (4.3-38).

Numerical values of fouling factors are given in References corresponding to satisfactory performance in normal operation, with reasonable service time between cleanings. They cover a range of approximately 600 to 11000 W/(m² · ℃). Fouling factors for ordinary industrial liquids fall in the range of 1700 to 6000 W/(m² · ℃). Fouling factors are usually set at values that also provide a safety factor for design.

Special cases of the overall coefficient. Although the choice of area to be used as the basis of an overall coefficient is arbitrary, sometimes one particular area is more convenient than others. Suppose, for example, that one individual coefficient h_i is large numerically in comparison with the other h_o, and that fouling effects are negligible. Also, assuming the term representing the resistance of the metal wall is small in comparison with $1/h_o$, the ratios D_o/D_i and $D_o/\overline{D}_L$ have so little significance that they can be disregarded, and Eq. (4.3-32) can be replaced by the simpler form

$$U_o = \frac{1}{1/h_o + B/k_m + 1/h_i}$$ (4.3-39)

In such a case, it is advantageous to base the overall coefficient on that area that corresponds to the largest resistance, or the lowest value of h.

For large-diameter thin-walled tubes, flat plates or any other case in which a negligible error is caused by using a common area for A_i, $\overline{A}_L$, and A_o, Eq. (4.3-39) can be used for the overall coefficient, and U_i and U_o are identical.

Sometimes one coefficient, say, h_o, is so very small in comparison with both x_w/k and the other coefficient h_i that the term $1/h_o$ is very large compared with the other terms in the resistance sum. When this is true, the larger resistance is called the *controlling resistance*, and it is sufficiently accurate to equate the overall coefficient to the small individual coefficient, or in this case, $h_o = U_o$.

Classification of individual heat-transfer coefficients. The problem of predicting the rate of heat flow from one fluid to another through a retaining wall reduces to the problem of predicting the numerical values of the individual coefficients of the fluids concerned in the overall process. A wide variety of individual cases are met in practice, and each type of phenomenon must be considered separately. The following classification is followed in this text:

(1) Heat flow to or from fluids inside tubes, without phase change
(2) Heat flow to or from fluids outside tubes, without phase change
(3) Heat flow from condensing fluids
(4) Heat flow to boiling liquids

Magnitude of heat-transfer coefficients. The ranges of values covered by the coefficient h vary greatly, depending upon the character of the process. Some typical ranges are shown in Table 4.1.

Table 4.1 Magnitude of heat-transfer coefficients

Type of processes	Range of values of h	
	W/(m²·°C)	Btu/(ft²·h·°F)
Steam (dropwise condensation)	30 000–100 000	5 000–20 000
Steam (film-type condensation)	6 000–20 000	1 000–3 000
Boiling water	1 700–50 000	300–9 000
Condensing organic vapors	1 000–2 000	200–400
Water (heating or cooling)	300–20 000	50–3 000
Oils (heating or cooling)	50–1 500	10–300
Steam (superheating)	30–100	5–20
Air (heating or cooling)	1–50	0.2–10
To convert from Btu/(ft²·h·°F) to W/(m²·°C), multiply by 5.6783.		

Source: By permission of author and publisher from W. H. McAdams. *Heat Transmission*, 3rd ed., p.5. Copyright by author, 1954, McGraw-Hill Book Company.

EXAMPLE 4.4 The butyl alcohol with flow rate 1930kg/h and specific heat c_p 2.98 kJ/(kg·°C) is cooled from 90°C to 50°C in a 6m² exchanger having overall heat transfer coefficient 230w/(m²·°C). Water at 18°C is used as a cooling medium. If the exchanger is operated with counter courrent flow and heat loss can be ignored, what are the outlet temperature and flow rate of cooling water?

Solution. From the heat transfer rate equation and energy balance equation,

$$q = UA\Delta T_m = \dot{m}_h c_{ph}(T_{ha} - T_{hb})$$

so $\Delta T_m = \dot{m}_h c_{ph}(T_{ha} - T_{hb})/UA = \dfrac{1930}{3600} \times 2.98 \times 1000(90-50)/(230 \times 6) = 46.3$ °C

Suppose the logarithmic mean temperature difference can be replaced by the arithmetic mean temperature difference, that is

$$\Delta T_m = [(T_{ha} - T_{cb}) + (T_{hb} - T_{ca})]/2 = [(90 - T_{cb}) + (50 - 18)]/2 = 46.3 \text{ °C}$$

so $T_{cb} = 29.4$°C

Check:
$$\Delta T_1 / \Delta T_2 = (90 - 29.4)/(50 - 18) = 1.89 < 2$$

Outlet temperature of cooling water is 29.4°C.

From the heat balance,

$$\dot{m}_c c_{pc}(T_{cb} - T_{ca}) = \dot{m}_h c_{ph}(T_{ha} - T_{hb})$$

from which,

$$\dot{m}_c = 1930 \times 2.98 \times (90-50)/[(4.187 \times (29.8-18)] = 4820 \text{kg}/\text{h}$$

EXAMPLE 4.5 Heated oil is cooled by water in a double pipe exchanger with inner pipe dimension $\Phi 19 \times 2$mm. Cooling water flows through the pipe. The individual heat transfer coefficients of water-side h_i and oil-side h_o are 3490W/(m²·K) and 258 W/(m²·K), respectively. The thermal conductivity k of pipe wall is 45 W/(m·K). What is overall heat transfer coefficient U_o based on the outer surface area of tube? After the exchanger is used for some time, scales are formed on both sides of pipe wall and fouling resistances in water and oil sides are 0.00026m²·K/W and 0.000176m²·K/W, respectively, what is overall heat transfer coefficient U_o based on the outer surface area of tube? What is the increased

percentage of heat resistance?
Solution. From Eq.(4.3-32), overall heat transfer coefficient U_o based on the outer surface area of tube is

$$U_o = \cfrac{1}{\cfrac{D_o}{h_i D_i} + \cfrac{BD_o}{k_m \overline{D}_L} + \cfrac{1}{h_o}} = \cfrac{1}{\cfrac{19}{3490 \times 15} + \cfrac{0.002 \times 19}{45 \times 17} + \cfrac{1}{258}} = 235.6 \text{ W/(m}^2 \cdot \text{K)}$$

After the exchanger is used for some time, scales are formed on both sides of pipe wall, overall heat transfer coefficient U_o based on the outer surface area of tube is

$$U_o = \cfrac{1}{\cfrac{D_o}{h_i D_i} + \cfrac{D_o}{D_i} R_{si} + \cfrac{BD_o}{k_m \overline{D}_L} + \cfrac{1}{h_o} + R_{so}} = \cfrac{1}{\cfrac{19}{3490 \times 15} + \cfrac{19}{15} \times 0.00026 + \cfrac{0.002 \times 19}{45 \times 17} + \cfrac{1}{258} + 0.000176}$$

$= 210 \text{ W/(m}^2 \cdot \text{K)}$

The increased percentage of heat resistance after forming scale is

$$\cfrac{0.00026 \times \cfrac{19}{15} + 0.000176}{\cfrac{1}{235.6} - (0.00026 \times \cfrac{19}{15} + 0.000176)} \times 100\% = 11.9\%$$

4.4 Heat Transfer to Fluids without Phase Change

In a great many applications of heat exchange, heat is transferred between fluid streams without any phase change in the fluids. This is especially important in heat recovery operations, such as when the hot effluent from an exothermic reactor is used to preheat the incoming cooler feed. Other examples include the transfer of heat from a stream of hot gas to cooling water, and the cooling of a hot liquid stream by air. In such situations the two streams are separated by a metal wall, which constitutes the heat-transfer surface. The surface may consist of tubes or other channels of constant cross section, of flat plates, or in such devices as jet engines and advanced power machinery, of special shapes designed to pack a maximum area of transfer surface into a small volume.

Most fluid-to-fluid heat transfer is accomplished in steady-state equipment, but thermal regenerators, in which a bed of solid shapes is alternately heated by a hot fluid and the hot shapes then used to warm a colder fluid, are also used, especially in high-temperature heat transfer.

4.4.1 Boundary Layers

Regimes of heat transfer

A fluid being heated or cooled may be flowing in laminar flow, in turbulent flow, or in the transition range between laminar and turbulent flow. Also, the fluid may be flowing in forced or natural convection. In some instances more than one flow type may occur in the same stream; for instance, in laminar flow at low velocities, natural convection may be superimposed on forced laminar flow.

The direction of flow of the fluid may be parallel to that of the heating surface, so that boundary layer separation does not occur; or the direction of flow may be perpendicular or at an angle to the heating surface, and then boundary layer separation often occurs.

At ordinary velocities the heat generated from fluid friction is usually negligible in comparison with the heat transferred between the fluids. In most cases friction heating may be neglected. It may be important, however, in operations involving very viscous fluids such as the injection molding of polymers. Friction heating of crude oil in the Alaska pipeline helps keep the oil above the ambient temperature. This decreases the viscosity and lowers the pumping cost. In gas flow at high velocities, at Mach numbers approaching 1.0, friction heat becomes appreciable and cannot be ignored. At very high velocities it may be of controlling importance, as it is for spacecraft reentering the earth's atmosphere.

Because the conditions of flow at the entrance to a tube differ from those well downstream from the entrance, the velocity field and the associated temperature field may depend on the distance from the tube entrance. Also, in some situations the fluid flows through a preliminary length of unheated or uncooled pipe, so that the fully developed velocity field is established before heat is transferred to the fluid, and the temperature field is created within an existing velocity field.

Finally, the properties of the fluid—viscosity, thermal conductivity, specific heat, and density—are important parameters in heat transfer. Each of these, especially viscosity, is temperature-dependent. Since the temperature varies from point to point in a flowing stream undergoing heat transfer, a problem appears in the choice of temperature at which the properties should be evaluated. For small temperature differences between fluid and wall and for fluids with weak dependence of viscosity on temperature, the problem is not acute. But for highly viscous fluids such as heavy petroleum oils or where the temperature difference between the tube wall and the fluid is large, the variations in fluid properties within the stream become large, and the difficulty of calculating the heat-transfer rate is increased.

Because of the various effects noted above, the entire subject of heat transfer to fluids without phase change is complex and in practice is treated as a series of special cases rather than as a general theory. All cases considered in this chapter do, however, have a phenomenon in common: In all the formation of a thermal boundary layer, analogous to the hydrodynamic Prandtl boundary layer described before, takes place; it profoundly influences the temperature field and so controls the rate of heat flow.

Thermal boundary layer

Consider a flat plate immersed in a stream of fluid that is in steady flow parallel to the plate, as shown in Fig. 4.11(a). Assume that the stream approaching the plate does so at velocity u_0 and temperature T_∞ and that the surface of the plate is maintained at a constant temperature T_w. Assume that T_w is greater than T_∞, so that the fluid is heated by the plate. As described hydrodynamic theory, a boundary layer develops within which the velocity varies from $u = 0$ at the wall to $u = u_0$ at the outer boundary of the layer. This boundary layer, called the *hydrodynamic boundary layer*, is shown by line OA in Fig. 4.11(a). The penetration of heat by transfer from the plate to the fluid changes the temperature of the fluid near the surface of the plate, and a temperature gradient is generated. The temperature gradient also is confined to a layer next to the wall, and within the layer the temperature varies from T_w at the wall to T_∞ at its outside boundary. This layer, called the *thermal boundary layer,* is shown as line OB in Fig. 4.11(a). As drawn, lines OA and OB show that the thermal boundary layer is thinner than the hydrodynamic layer at all values of x, where x is the distance from the leading edge of the plate.

Prandtl number

The relationship between the thicknesses of the two boundary layers at a given point along the plate depends on the dimensionless *Prandtl number*, which is the ratio of the diffusivity of momentum ν or μ/ρ to the thermal diffusivity α or $k/\rho c_p$. Thus

$$Pr \equiv \frac{\nu}{\alpha} = \frac{c_p \mu}{k} \qquad (4.4\text{-}1)$$

When the Prandtl number is greater than unity, which is true for most liquids, the thermal boundary layer is thinner than the hydrodynamic layer because of the relatively low rate of heat conduction. This situation is illustrated in Fig. 4.11(a). The Prandtl number for water at 70°C is about 2.5; for viscous liquids and concentrated solutions it may be as large as 600. Prandtl numbers for most liquids decrease as the temperature rises because of the decrease in viscosity. With a high viscosity fluid the hydrodynamic boundary layer extends farther from the surface of the plate, which can perhaps be understood intuitively. Imagine moving a flat plate through a very viscous liquid such as glycerol: Fluid at a considerable distance from the plate will be set in motion, which means a thick boundary layer.

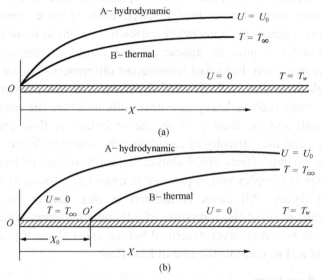

Fig. 4.11 Thermal and hydrodynamic boundary layers on flat plate: (a) entire plate heated; (b) unheated length = x_0.

The Prandtl number of a gas is usually close to 1.0 (0.69 for air, 1.06 for steam), and the two boundary layers have about the same thickness. The Prandtl number for gases is almost independent of temperature because the viscosity and thermal conductivity both increase with temperature at about the same rate. Prandtl numbers for gases and liquids are given in Apps. 15 and 16.

Liquid metals have very low Prandtl numbers, ranging from about 0.01 to 0.04, because of the high thermal conductivity. The temperature gradients extend well beyond the hydrodynamic boundary layer, and special correlations are needed to predict the rate of heat transfer.

In Fig. 4.11(a) it is assumed that the entire plate is heated and that both boundary layers start at the leading edge of the plate. If the first section of the plate is not heated and if the heat-transfer area begins at a definite distance x_0 from the leading edge, as shown by line $O'B$ in Fig. 4.11(b), a hydrodynamic boundary layer already exists at x_0, where the thermal boundary layer begins to form.

The sketches in Fig. 4.11 exaggerate the thickness of the boundary layers for clarity. The actual thicknesses are usually less than 1 percent of the distance from the leading edge of the plate.

In flow through a tube, it has been shown in Chap.1 that the hydrodynamic boundary layer thickens as the distance from the tube entrance increases, and finally the layer reaches the center of the tube. The velocity profile so developed, called *fully developed flow*, establishes a velocity distribution that is unchanged with additional pipe length. The thermal boundary layer in a heated or cooled tube also reaches the center of the tube at a definite length from the entrance of the heated length of the tube, and the temperature profile is considered fully developed at this point since it covers the entire pipe. Unlike the velocity profile, however, the temperature profile flattens as the length of the tube increases; and in very long pipes the entire fluid stream reaches the temperature of the tube wall, the temperature gradients disappear, and heat transfer ceases.

4.4.2 Heat Transfer by Forced Convection in Laminar Flow

In laminar flow, heat transfer occurs only by conduction, as there are no eddies to carry heat by convection across an isothermal surface. The problem is amenable to mathematical analysis based on the partial differential equations for continuity, momentum, and energy. Mathematical solutions depend on the boundary conditions established to define the conditions of fluid flow and heat transfer. When the fluid approaches the heating surface, it may have an already completed hydrodynamic boundary layer or a partially developed one. Or the fluid may approach the heating surface at a uniform velocity, and both boundary layers may be initiated at the same time. A simple flow situation where the velocity is assumed constant in all cross sections and tube lengths is called *plug* or *rodlike flow*. Independent of the conditions of flow, (1) the heating surface may be isothermal; or (2) the heat flux may be equal at all points on the heating surface, in which case the average temperature of the fluid varies linearly with tube length. Other combinations of boundary conditions are possible. The basic differential equation for the several special cases is the same, but the final integrated relationships differ.

Most of the simpler mathematical derivations are based on the assumptions that the fluid properties are constant and temperature-independent and that flow is truly laminar with no crosscurrents or eddies. These assumptions are valid when temperature changes and gradients are small, but with large temperature changes the simple model is not in accord with physical reality for two reasons. First, variations in viscosity across the tube distort the usual parabolic velocity distribution profile of laminar flow. Thus, if the fluid is a liquid and is being heated, the layer near the wall has a lower viscosity than the layers near the center and the velocity gradient at the wall increases. A crossflow of liquid toward the wall is generated. If the liquid is being cooled, the reverse effect occurs. Second, since the temperature field generates density gradients, natural convection may set in, which further distorts the flow lines of the fluid. The effect of natural convection may be small or large, depending on a number of factors to be discussed in the section on natural convection.

In this section three types of heat transfer in laminar flow are considered: (1) heat transfer to a fluid flowing along a flat plate, (2) heat transfer in plug flow in tubes, and (3) heat transfer to a fluid stream that is in fully developed flow at the entrance to the tube. In all cases, the temperature of the heated length of the plate or tube is assumed to be constant, and the effect of natural convection is ignored.

4.4.2.1 Laminar flow heat transfer to flat plate

Consider heat flow to the flat plate shown in Fig. 4.11(b). The conditions are assumed to be as follows:

Velocity of fluid approaching plate and at and beyond the edge of the boundary layer OA: u_0.

Temperature of fluid approaching plate and at and beyond the edge of the thermal boundary layer $O'B$: T_∞.

Temperature of plate: from $x = 0$ to $x = x_0$, $T = \infty$; for $x > x_0$, $T = T_w$, where $T_w > T_\infty$.

The following properties of the fluid are constant and temperature-independent: density ρ, conductivity k, specific heat c_p, and viscosity μ.

Detailed analysis of the situation yields the equation

$$\left(\frac{dT}{dy}\right)_w = \frac{0.332(T_w - T_\infty)}{\sqrt[3]{1-(x_0/x)^{3/4}}} \sqrt[3]{\frac{c_p\mu}{k}} \sqrt{\frac{u_0\rho}{\mu x}} \tag{4.4-2}$$

where $(dT/dy)_w$ is the temperature gradient at the wall. From Eq. (4.3-21), the relation between the local heat-transfer coefficient h_x at any distance x from the leading edge and the temperature gradient at the wall is

$$h_x = \frac{k}{T_w - T_\infty}\left(\frac{dT}{dy}\right)_w \tag{4.4-3}$$

$(dT/dy)_w$ is eliminated and this equation can be put into a dimensionless form by multiplying by x/k, giving

$$\frac{h_x x}{k} = \frac{0.332}{\sqrt[3]{1-(x_0/x)^{3/4}}} \sqrt[3]{\frac{c_p\mu}{k}} \sqrt{\frac{u_0 x\rho}{\mu}} \tag{4.4-4}$$

The left-hand side of this equation is a Nusselt number corresponding to the distance x, or Nu_x. The second group is the Prandtl number Pr, and the third group is a Reynolds number corresponding to distance x, denoted by Re_x. Equation (4.4-4) then can be written

$$Nu_x = \frac{0.332}{\sqrt[3]{1-(x_0/x)^{3/4}}} \sqrt[3]{Pr}\sqrt{Re_x} \tag{4.4-5}$$

When the plate is heated over its entire length, as shown in Fig. 4.11(a), $x_0 = 0$ and Eq. (4.4-5) becomes

$$Nu_x = 0.332\sqrt[3]{Pr}\sqrt{Re_x} \tag{4.4-6}$$

Equation (4.4-6) gives the local value of the Nusselt number at distance x from the leading edge. More important in practice is the average value of N_u over the entire length of the plate x_1, defined as

$$Nu = \frac{hx_1}{k} \tag{4.4-7}$$

where

$$h = \frac{1}{x_1}\int_0^{x_1} h_x dx$$

Equation (4.4-4) can be written for a plate heated over its entire length, since $x_0 = 0$, as

$$h_x = \frac{C}{\sqrt{x}}$$

where C is a constant containing all factors other than hx and x. Then

$$h = \frac{C}{x_1}\int_0^{x_1} \frac{dx}{\sqrt{x}} = \frac{2C}{x_1}\sqrt{x_1} = \frac{2C}{\sqrt{x_1}} = 2h_{x1} \qquad (4.4\text{-}8)$$

The average coefficient is clearly twice the local coefficient at the end of the plate, and Eq. (4.4-6) gives

$$Nu = 0.664\sqrt[3]{Pr}\sqrt{Re_{x_1}} \qquad (4.4\text{-}9)$$

These equations are valid only for Prandtl numbers of 1.0 or greater, since the derivation assumes a thermal boundary layer no thicker than the hydrodynamic layer. However, they can be used for gases with $Pr \approx 0.7$ with little error. The equations are also restricted to cases where the Nusselt number is fairly large, say, 10 or higher, since axial conduction, which was neglected in the derivation, has a significant effect for thick boundary layers.

4.4.2.2 Laminar flow heat transfer in tubes

Graetz and Peclet numbers

The *Graetz number* is commonly used in treating heat transfer to fluids and is defined by the equation

$$Gz \equiv \frac{\dot{m}c_p}{kL} \qquad (4.4\text{-}10)$$

where $\dot{m}$ is the mass flow rate. Since $\dot{m} = \frac{\pi}{4}D^2 V\rho$,

$$Gz = \frac{\pi}{4} \times \frac{\rho V c_p D^2}{kL} \qquad (4.4\text{-}11)$$

The Graetz number can also be calculated from the Reynolds and Prandtl numbers and the D/L ratio

$$Gz = \frac{\pi}{4}\rho V D^2 \frac{c_p}{kL} \times \frac{\mu}{\mu} = \frac{\pi}{4} RePr\frac{D}{L} \qquad (4.4\text{-}12)$$

The *Peclet number* Pe is defined as the product of the Reynolds number and the Prandtl number, or

$$Pe = RePr = \frac{DV\rho}{\mu} \times \frac{c_p\mu}{k} = \frac{\rho V c_p D}{k} = \frac{DV}{\alpha} \qquad (4.4\text{-}13)$$

Fully developed flow

With a newtonian fluid in fully developed laminar flow, the actual velocity distribution at the entrance to the heated section and the theoretical distribution throughout the tube are both parabolic. The empirical correlations have been developed for design purposes. These correlations are based on the Graetz number, but they give the film coefficient or the Nusselt number rather than the change in temperature, since this permits the fluid resistance to be combined with other resistances in determining an overall heat-transfer coefficient. The Nusselt number for heat transfer to a fluid inside a pipe is the film coefficient multiplied by D/k:

$$Nu \equiv \frac{h_i D}{k} \tag{4.4-14}$$

The film coefficient h_i is the average value over the length of the pipe and is calculated as follows for the case of constant wall temperature:

$$h_i = \frac{mc_p(\bar{T}_b - T_a)}{\pi D L \Delta T_m} \tag{4.4-15}$$

Since

$$\Delta T_m = \frac{(T_w - T_a) - (T_w - \bar{T}_b)}{\ln[(T_w - T_a)/(T_w - \bar{T}_b)]} \tag{4.4-16}$$

so

$$Nu = \frac{\dot{m}c_p}{\pi k L} \ln[(T_w - T_a)/(T_w - \bar{T}_b)] \tag{4.4-17}$$

or

$$Nu = \frac{Gz}{\pi} \ln[(T_w - T_a)/(T_w - \bar{T}_b)] \tag{4.4-18}$$

For Graetz numbers greater than 20, the theoretical Nusselt number increases with about the one-third power of Gz. Data for air and for moderate-viscosity liquids follow a similar trend, but the coefficients are about 15 percent greater than predicted from theory. An empirical equation for moderate Graetz numbers (greater than 20) is

$$Nu \cong 2.0 Gz^{1/3} \text{ or } 1.85(Re \times Pr \times D/L)^{1/3} \tag{4.4-19}$$

The increase in film coefficient with increasing Graetz number or decreasing length is a result of the change in shape of the temperature profile. For short lengths, the thermal boundary layer is very thin, and the steep temperature gradient gives a high local coefficient. With increasing distance from the entrance, the boundary layer becomes thicker and eventually reaches the center of the pipe, giving a nearly parabolic temperature profile. The local coefficient is approximately constant from that point on, but the average coefficient continues to decrease with increasing length until the effect of the high initial coefficient is negligible. In practice, the change in local coefficient with length is usually not calculated, and the length average film coefficient is used in obtaining the overall coefficient.

Correction for heating or cooling

For viscous liquids with large temperature drops, a modification of Eq. (4.4-19) is required to account for differences between heating and cooling. When a liquid is being heated, the lower viscosity near the wall makes the velocity profile more like that for plug flow, with a very steep gradient near the wall and little gradient near the center. This leads to a higher rate of heat transfer. When a viscous liquid is cooled, the velocity gradient at the wall is decreased, giving a lower rate of accounts for the heat transfer. A dimensionless, but empirical, correction factor ϕ_v accounts for the difference between heating and cooling:

$$\phi_v \equiv \left(\frac{\mu}{\mu_w}\right)^{0.14} \tag{4.4-20}$$

This factor is added to Eq. (4.4-19) to give the final equation for laminar flow heat transfer:

$$Nu = 2\left(\frac{\dot{m}c_p}{kL}\right)^{1/3}\left(\frac{\mu}{\mu_w}\right)^{0.14} = 2Gz^{1/3}\phi_v \qquad (4.4\text{-}21)$$

In Eqs. (4.4-20) and (4.4-21), μ is the viscosity at the arithmetic mean temperature of the fluid $(T_a+T_b)/2$, and, μ_w is the viscosity at the wall temperature T_w. For liquids $\mu_w<\mu$ and $\phi_v >1.0$ when the liquid is being heated,, and $\mu_w>\mu$ and $\phi_v <1.0$ when the liquid is being cooled.

4.4.3 Heat Transfer by Forced Convection in Turbulent Flow

Perhaps the most important situation in heat transfer is the heat flow in a stream of fluid in turbulent flow in a closed channel, especially in tubes. Turbulence is encountered at Reynolds numbers greater than about 2100, and since the rate of heat transfer is greater in turbulent flow than in laminar flow, most equipment is operated in the turbulent range.

The earliest approach to this case was based on empirical correlations of test data guided by dimensional analysis. The equations so obtained still are much used in design. Subsequently, theoretical study has been given to the problem. A deeper understanding of the mechanism of turbulent flow heat transfer has been achieved, and improved equations applicable over wider ranges of conditions have been obtained.

Dimensional analysis method

Dimensional analysis of the heat flow to a fluid in turbulent flow in a long, straight pipe yields the dimensionless relationship

$$\frac{hD}{k} = \Phi\left(\frac{DV\rho}{\mu}, \frac{c_p\mu}{k}\right) \qquad (4.4\text{-}22)$$

Here the mass velocity G is used in place of its equal $V\rho$. Dividing hD/k by the product $(DG/\mu)(c_p\mu/k)$ gives an alternate relationship

$$\frac{h}{c_p G} = \Phi_1\left(\frac{DG}{\mu}, \frac{c_p\mu}{k}\right) \qquad (4.4\text{-}23)$$

The three groups in Eq. (4.4-22) are the Nusselt, Reynolds, and Prandtl numbers, respectively. The left-hand group in Eq. (4.4-23) is called the *Stanton number St*. The four groups are related by the equation

$$St\ Re\ Pr = Nu \qquad (4.4\text{-}24)$$

Thus, only three of the four are independent.

Empirical equations

To use Eq. (4.4-22) or (4.4-23), the function Φ or Φ_1 must be known. One empirical correlation for long tubes with sharp-edged entrances is the *Dittus-Boelter equation*

$$Nu = \frac{h_i D}{k} = 0.023 Re^{0.8} Pr^n \qquad (4.4\text{-}25)$$

where n is 0.4 when the fluid is being heated and 0.3 when it is being cooled.

Using different values of n in Eq. (4.4-25) is one way of allowing for the higher coefficients found when liquids are heated than when they are cooled. However, the ratio of the coefficients for heating and cooling, according to Eq. (4.4-25), equals $Pr^{0.1}$ and does not depend on conditions at the wall of the pipe. A better relationship for turbulent flow is known as the *Sieder-Tate equation;* it uses the same correction factor ϕ_v as for laminar flow

$$Nu = \frac{h_i D}{k} = 0.023 Re^{0.8} Pr^{1/3} \phi_v \qquad (4.4\text{-}26)$$

The viscosity ratio term ϕ_v is not very important for low-viscosity liquids such as water, but for viscous oils, where the wall and bulk viscosities may differ by 10-fold, the coefficient for heating may be twice that for cooling.

An alternate form of Eq. (4.4-26) is obtained by dividing both sides by Re Pr and transposing to give what is called the Colburn equation.

$$StPr^{2/3} = \frac{0.023 \phi_v}{Re^{0.2}} \qquad (4.4\text{-}27)$$

In using these equations the physical properties of the fluid, except for μ_w, are evaluated at the bulk fluid temperature T. For gases, the viscosity increases with temperature, but the change is small and the viscosity term in Eqs. (4.4-26) and (4.4-27) is usually ignored. These equations should not be used when the Reynolds number is below 6000 or for molten metals, which have very low Prandtl numbers.

Effect of tube length. Near the tube entrance, where the temperature gradients are still forming, the local coefficient h_x is greater than h_∞ for fully developed flow. At the entrance itself, where there is no previously established temperature gradient, h_x is infinite. Its value drops rapidly toward h_∞ in a comparatively short length of tube. Dimensionally, the effect of tube length is accounted for by another dimensionless group x/D, where x is the distance from the tube entrance. The local coefficient approaches h_∞ asymptotically with an increase in x, but it is practically equal to h_∞ when x/D is about 50. The average value of h_x over the tube length is denoted by h_i. The value of h_i is found by integrating h_x over the length of the tube. Since $h_x \to h_\infty$ as $x \to \infty$, the relation between h_i and h_∞ is of the form

$$\frac{h_i}{h_\infty} = 1 + \psi\left(\frac{L}{D}\right) \qquad (4.4\text{-}28)$$

An equation for short tubes with sharp-edged entrances, where the velocity at the entrance is uniform over the cross section, is

$$\frac{h_i}{h_\infty} = 1 + \left(\frac{D}{L}\right)^{0.7} \qquad (4.4\text{-}29)$$

The effect of tube length on h_i fades out when L/D becomes greater than about 50.

Average value of h_i in turbulent flow

Since the temperature of the fluid changes from one end of the tube to the other and the fluid properties μ, k, and c_p are all functions of temperature, the local value of h_i also varies from point to point along the tube. This variation is independent of the effect of tube length.

The effect of fluid properties can be shown by condensing Eq. (4.4-26) to read, assuming $\mu/\mu_w = 1$,

$$h_i = 0.023 \frac{G^{0.8} k^{2/3} c_p^{1/3}}{D^{0.2} \mu^{0.47}} \qquad (4.4\text{-}30)$$

For gases the effect of temperature on h_i is small. At constant mass velocity in a given tube, h_i varies with $k^{2/3} c_p^{1/3} \mu^{-0.47}$. The increase in thermal conductivity and heat capacity with temperature offsets the rise in viscosity, giving a slight increase in h_i. For example, for air h_i increases about 6 percent when the temperature changes from 50 to 100 °C.

For liquids the effect of temperature on h_i is much greater than for gases because of the rapid decrease in viscosity with rising temperature. The effects of k, cp, and μ in Eq. (4.4-30) all act in the same direction, but the increase in h_i with temperature is due mainly to the effect of temperature on viscosity. For water, for example, h_i increases about 50 percent over a temperature range from 50 to 100°C. For viscous oils the change in h_i may be two- or three fold for a 50°C increase in temperature.

In practice, unless the variation in h_i over the length of the tube is more than about 2:1, an average value of h_i is calculated and used as a constant in calculating the overall coefficient U. This procedure neglects the variation of U over the tube length and allows the use of the LMTD in calculating the area of the heating surface. The average value of h_i is computed by evaluating the fluid properties cp, k, and μ at the average fluid temperature, defined as the arithmetic mean between the inlet and outlet temperatures. The value of h_i calculated from Eq. (4.4-26), using these property values, is called the *average coefficient*. For example, assume that the fluid enters at 30°C and leaves at 90°C. The average fluid temperature is $(30 + 90)/2 = 60°C$, and the values of the properties used to calculate the average value of h_i are those at 60°C.

For larger changes in h_i, two procedures can be used: (1) The values of h_i at the inlet and outlet can be calculated, corresponding values of U_1 and U_2 found, and Eq. (4.3-17) used. Here the effect of L/D on the entrance value of h_i is ignored. (2) For even larger variations in h_i, and therefore in U, the tube can be divided into sections and an average U used for each section. Then the lengths of the individual sections can be added to account for the total length of the tube.

Estimation of wall temperature T_w

To evaluate μ_w, the viscosity of the fluid at the wall, temperature T_w must be found. The estimation of T_w requires an iterative calculation based on the resistance equation (4.3-35). If the individual resistances can be estimated, the total temperature drop ΔT can be split into the individual temperature drops by the use of this equation and an approximate value for the wall temperature found. To determine T_w in this way, the wall resistance $(x_w/k_m)(D^o/\bar{D}_L)$ can usually be neglected, and Eq. (4.3-35) used as follows.

From the first two members of Eq. (4.3-35)

$$\Delta T_i = \frac{D_o/D_i h_i}{1/U_o} \Delta T \tag{4.4-31}$$

Substituting $1/U_o$ from Eq. (4.3-32) and neglecting the wall resistance term give

$$\Delta T_i = \frac{1/h_i}{1/h_i + D_i/D_o h_o} \Delta T \tag{4.4-32}$$

In qualitative terms Eq. (4.4-32) may be written

$$\Delta T_i = \frac{\text{inside resistance}}{\text{overall resistance}} \Delta T$$

Use of Eq. (4.4-32) requires preliminary estimates of the coefficients h_i and h_o. To estimate h_i, Eq. (4.4-26) can be used, neglecting ϕ_v. The calculation of h_o will be described later. The wall temperature T_w is then obtained from the following equations:

For heating: $\qquad\qquad T_w = T + \Delta T_i \qquad\qquad (4.4\text{-}33)$

For cooling: $\qquad\qquad T_w = T - \Delta T_i \qquad\qquad (4.4\text{-}34)$

where T is the average fluid temperature.

If the first approximation is not sufficiently accurate, a second calculation of T_w based on the results of the first can be made. Unless the factor ϕ_v is quite different from unity, however, the second approximation is unnecessary.

EXAMPLE 4.6 Toluene is being condensed at 110 °C on the outside of copper condenser tubes with inner and outer diameters of 20 and 25-mm through which cooling water is flowing at an average temperature of 25 °C. Individual heat-transfer coefficients for toluene and water are 2840 W/(m²·°C) and 3970 W/(m²·°C), respectively. Neglecting the resistance of the tube wall, what is the tube wall temperature?
Solution. From Eq.(4.4-32),

$$\Delta T_i = \frac{1/3970}{1/3970 + 0.02/(0.025 \times 2840)}(110-25) = 40 \text{ °C}$$

Since water is being heated by the condensing toluene, the wall temperature is found from Eq.(4.4-33) as follows:

$$T_w = 25 + 40 = 65 \text{ °C}$$

Cross sections other than circular

To use Eq. (4.4-26) or (4.4-27) for cross sections other than circular, it is only necessary to replace the diameter D in both Reynolds and Nusselt numbers by the equivalent diameter D_e, defined as 4 times the hydraulic radius r_H. The method is the same as that used in calculating friction loss.

Effect of roughness

For equal Reynolds numbers the heat-transfer coefficient in turbulent flow is somewhat greater for a rough tube than for a smooth one. The effect of roughness on heat transfer is much less than on fluid friction, and economically it is usually more important to use a smooth tube for minimum friction loss than to rely on roughness to yield a larger heat-transfer coefficient. The effect of roughness on h_i is neglected in practical calculations.

EXAMPLE 4.7 Benzene at 20 °C flows through the pipe of a 1-2 pass shell-tube exchanger consisting of 96 steel tubes ($\phi25\times2.5$mm) with flow rate of 9.5 kg/s and is heated to 80 °C by saturated vapor which goes through shell-side.
(a) What is the heat transfer coefficient h_i between benzene and pipe wall?
(b) If benzene flow rate is double and other conditions are unchanged, what is the heat transfer coefficient between benzene and pipe wall?
(c) If the pipe diameter is decreased to the half of origin and other conditions are similar to case (1), what is the heat transfer coefficient between benzene and pipe wall?
Solution. The arithmetic average temperature of benzene $\bar{t} = (20+80)/2 = 50$ °C. The physical properties of benzene at 50 °C are obtained from App. 8, 14

$$c_p = 1.9 \text{ kJ/(kg·K)}, \quad \mu = 0.37 \times 10^{-3} \text{Pa·s}, \quad \rho = 840 \text{ kg/m}^3 \quad k = 0.138 \text{W/(m·K)}$$

(a) Benzene velocity in a pipe

$$V_1 = \frac{\dot{m}_1}{A_1\rho} = \frac{9.5}{840 \times \frac{\pi}{4} 0.02^2 \times 48} = 0.75 \text{ m/s}$$

$$Re = \frac{V_1 d_1 \rho}{\mu} = \frac{0.75 \times 840 \times 0.02}{0.37 \times 10^{-3}} = 3.4 \times 10^4 > 10^4$$

$$Pr = \frac{c_p \mu}{k} = \frac{1.9 \times 10^3 \times 0.37 \times 10^{-3}}{0.138} = 5.09$$

For fully developed turbulent flow, the Dittus-Boelter equation can be used.

$$Nu = 0.023 Re^{0.8} Pr^{0.4} = 0.023 \times (3.4 \times 10^4)^{0.8} \times 5.09^{0.4} = 186$$

so
$$h_i = \frac{Nu \, k}{d} = 1283.7 \text{ W/(m}^2 \cdot \text{K)}$$

(b) When the flow rate of benzene is double, it is also fully developed turbulent flow and the Dittus-Boelter equation can also be satisfied.

$$\dot{m}_2 = 2\dot{m}_1, \qquad V_2 = 2V_1$$

Hence,
$$\frac{h_{i2}}{h_{i1}} = \left(\frac{V_2}{V_1}\right)^{0.8} = 2^{0.8} = 1.74$$

$$h_{i2} = 1.74 \times 1283.7 = 2233.6 \text{ W/(m}^2 \cdot \text{K)}$$

(c) The diameter of pipe is decreased to the half of origin and the flow rate keeps the same, velocity ratio is

$$\frac{V_3}{V_1} = \frac{\dot{m}/\rho A_3}{\dot{m}/\rho A_1} = \frac{A_1}{A_3} = \left(\frac{d_1}{d_3}\right)^2$$

$$2 d_3 = d_1 \qquad V_3 = 4V_1$$

$$\frac{h_{i3}}{h_{i1}} = \left(\frac{V_3}{V_1}\right)^{0.8} \left(\frac{d_1}{d_3}\right)^{0.2} = 2.64$$

$$h_{i3} = 1283.7 \times 2.64 = 3388.9 \text{ W/(m}^2 \cdot \text{K)}$$

EXAMPLE 4.8 Cold fluid at 25℃ and with specific heat c_p 1.76kJ/(kg·℃) and mass flow rate 715.4kg/h flows turbulently through the inner pipe (ϕ25×2.5mm) of a double pipe exchanger with 2 meter long, and is heated. Saturated vapor is used as a heating medium, entering the jacket at 119.6℃ and leaving at 119.6℃ as saturated liquid. Individual heat transfer coefficient of vapor side h_o is 11000W/(m²·℃) and heat resistance of cold fluid side R_i is 5 times more than that of vapor side. If heat loss and heat resistance of pipe wall can be ignored, what are the overall heat transfer coefficient U_i(based on the inner surface area of pipe) and outlet temperature of fluid? If one more same exchanger is provided and the outlet temperature of cold fluid needs to be above 85℃, how are two exchangers set up?

Solution. From the definition of overall heat transfer coefficient,

$$U_i = \frac{1}{1/h_i + d_i/d_o h_o} = 1/(\frac{6}{h_o} + 0.02/(0.025 \times h_o)) = h_o/6.8 = 11000/6.8 = 1617.6 \text{ W/(m}^2 \cdot ℃)$$

From heat transfer rate equation and heat balance equation,

$$q = U_i A_i \Delta T_m = \dot{m}_c c_{pc}(T_{cb} - T_{ca})$$

Since A_i=0.02×π×2=0.126m², $\dot{m}_c$ =715.4kg/h, c_{pc}=1.76kJ/(kg·℃), T_{ca}=25℃, T_{ha}=T_{hb}=119.6℃

$$\Delta T_1 = T_{ha} - T_{ca} = 119.6 - 25 = 94.6, \quad \Delta T_2 = T_{hb} - T_{cb} = 119.6 - T_{cb}$$

Hence outlet temperature of cold fluid is $T_{cb} = 66.6\,°C$.

There are two methods to set up two same exchangers.

(a) Two same exchangers operate in series.

All above conditions are the same except for heat transfer area which is double. From heat transfer rate equation and heat balance equation,

$$q = U_i A'_i \Delta T_m = \dot{m}_c c_{pc}(T_{cb} - T_{ca})$$

Since $A'_i = 2A_i = 0.252\,m^2$, $\dot{m}_c = 715.4\,kg/h$, $c_{pc} = 1.76\,kJ/(kg·°C)$, $T_{ca} = 25\,°C$, $T_{ha} = T_{hb} = 119.6\,°C$

$$\Delta T_1 = T_{ha}/T_{ca}, \quad \Delta T_2 = T_{hb} - T'_{cb}$$

Hence outlet temperature of cold fluid is $T'_{cb} = 89.9\,°C > 85\,°C$.

This arrange can be satisfied.

(b) Two same exchangers operate in parallel flow.

The flow rate or velocity for each exchanger is decreased to 50% of origin. For turbulent flow of cold fluid,

$$\frac{h'_i}{h_i} = \left(\frac{V'}{V}\right)^{0.8} = \left(\frac{1}{2}\right)^{0.8} = 0.574$$

$$U'_i = \frac{1}{1/h'_i + d_i/d_o h_o} = 1/\left[\frac{6}{0.574 h_o} + 0.02/(0.025 \times h_o)\right]$$

$$= h_o/11.25 = 11000/11.25 = 978\,W/(m^2·°C)$$

From heat transfer rate and heat balance,

$$q = U'_i A'_i \Delta T_m = \dot{m}_c c_{pc}(T_{cb} - T_{ca})$$

Since $A'_i = 2A_i = 0.252\,m^2$, $\dot{m}_c = 715.4\,kg/h$, $c_{pc} = 1.76\,kJ/(kg·°C)$, $T_{ca} = 25\,°C$, $T_{ha} = T_{hb} = 119.6\,°C$

$$\Delta T_1 = T_{ha}/T_{ca}, \quad \Delta T_2 = T_{hb} - T'_{cb}$$

Hence $T'_{cb} = 72.8\,°C < 85\,°C$.

Therefore, two same exchangers should operate in series to enable outlet temperature of cold fluid to be above $85\,°C$.

4.4.4 Heat Transfer in Transition Region between Laminar and Turbulent Flow

Equation (4.4-27) applies only for Reynolds numbers greater than 6000 and Eq. (4.4-21) only for Reynolds numbers less than 2100. The range of Reynolds numbers between 2100 and 6000 is called the *transition region,* and no simple equation applies here. A graphical method therefore is used. The method is based on graphs of Eqs. (4.4-21) and (4.4-27). To obtain an equation for the laminar flow range, it is necessary to transform Eq. (4.4-21) in the following manner. Substituted for the Graetz number, using Eq. (4.4-13), is the quantity $(\pi D/4L)RePr$. The result is

$$Nu = 2\left(\frac{\pi D}{4L}RePr\right)^{1/3}\left(\frac{\mu}{\mu_w}\right)^{0.14} \qquad (4.4-35)$$

4.4.5 Heating and Cooling of Fluid in Forced Convection Outside Tubes

The mechanism of heat flow in forced convection outside tubes differs from that of flow inside tubes, because of differences in the fluid flow mechanism. As has been shown before, no form drag exists inside tubes except perhaps for a short distance at the entrance end, and all friction is wall friction. Because of the lack of form friction, there is no variation in the local heat transfer at different points in a given circumference, and a close analogy exists between friction and heat transfer. An increase in heat transfer is obtainable at the expense of added friction simply by increasing the fluid velocity. Also, a sharp distinction exists between laminar and turbulent flow, which calls for different treatment of heat-transfer relations for the two flow regimes.

On the other hand, in the flow of fluids across a cylindrical shape, boundary layer separation occurs, and a wake develops that causes form friction. No sharp distinction is found between laminar and turbulent flow, and a common correlation can be used for both low and high Reynolds numbers. Also, the local value of the heat-transfer coefficient varies from point to point around a circumference. In Fig. 4.12 the local value of the Nusselt number is plotted radially for all points around the circumference of the tube. At low Reynolds numbers, Nu_θ is a maximum at the front and back of the tube and a minimum at the sides. In practice, the variations in the local coefficient h_θ are often of no importance, and average values based on the entire circumference are used.

Radiation may be important in heat transfer to outside tube surfaces. Inside tubes, the surface cannot see surfaces other than the inside wall of the same tube, and heat flow by radiation does not occur. Outside tube surfaces, however, are necessarily in sight of external surfaces, if not nearby, at least at a distance, and the surrounding surfaces may be appreciably hotter or cooler than the tube wall. Heat flow by radiation, especially when the fluid is a gas, is appreciable in comparison with heat flow by conduction and convection. The total heat flow is then a sum of two independent flows, one by radiation and the other by conduction and convection. The relations given in the remainder of this section have to do with conduction and convection only. Radiation, as such and in combination with conduction and convection, is discussed later.

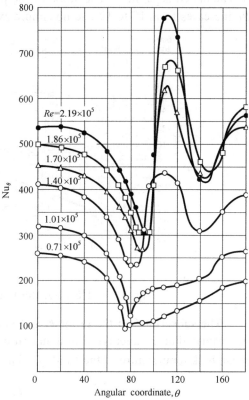

Fig. 4.12 Local Nusselt number for airflow normal to a circular cylinder.

Fluids flowing normal to a single tube

The variables affecting the coefficient of heat transfer to a fluid in forced convection outside a tube are D_o, the outside diameter of the tube; c_p, μ, and k, the specific heat at constant pressure, the viscosity, and the thermal conductivity, respectively, of the fluid; and G, the mass velocity of the fluid approaching the tube. Dimensional analysis gives, then, an equation of the type of Eq. (4.4-22):

$$\frac{h_o D_o}{k} = \psi_0 \left(\frac{D_o G}{\mu}, \frac{c_p \mu}{k} \right) \qquad (4.4\text{-}36)$$

Here, however, ends the similarity between the two types of process—the flow of heat to fluids inside tubes and the flow of heat to fluids outside tubes—and the functional relationships in the two cases differ.

For any one gas for which the Prandtl number is nearly independent of temperature, the Nusselt number is a function only of the Reynolds number. Experimental data for air are plotted in this way in Fig. 4.13. The effect of radiation is not included in this curve and must be calculated separately. The slope of the plot in Fig. 4.13 increases from about 0.4 to 0.7 as Re increases from 10 to 10^5, so no simple exponential equation fits the data.

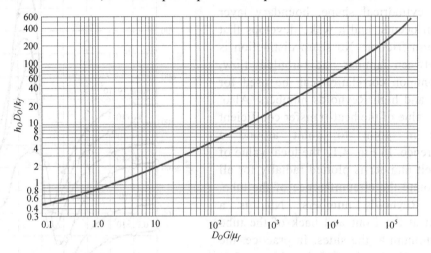

Fig. 4.13 Heat transfer to air flowing normal to a single tube.

The subscript f on the terms k_f and μ_f indicates that in using Fig. 4.13 these terms must be evaluated at the average film temperature T_f midway between the wall temperature and the mean bulk temperature of the fluid $\overline{T}$. Therefore, T_f is given by the equation

$$T_f = \frac{1}{2}(T_w + \overline{T}) \qquad (4.4\text{-}37)$$

Figure 4.13 can be used for both heating and cooling.

For heating and cooling liquids flowing normal to single cylinders the following equation is used:

$$\frac{h_o D_o}{k_f} \left(\frac{c_p \mu f}{k_f} \right)^{-0.3} = 0.35 + 0.56 \left(\frac{D_o G}{\mu_f} \right)^{0.52} \qquad (4.4\text{-}38)$$

This equation can also be used for gases from $Re = 1$ to $Re = 10^4$, but it gives lower values of the Nusselt number than Fig. 4.13 at higher Reynolds numbers.

4.4.6 Natural Convection

As an example of natural convection, consider a hot, vertical plate in contact with the air in a room. The temperature of the air in contact with the plate will be that of the surface of the plate, and a temperature gradient will exist from the plate out into the room. At the bottom of the plate, the temperature gradient is steep, as shown by the full line marked "Z = 10 mm" in Fig. 4.14. At distances above the bottom of the plate, the gradient becomes less steep, as shown by the full curve marked "Z = 240 mm" of Fig. 4.14. At a height of about 600 mm from the bottom of the plate, the temperature–distance curves approach an asymptotic condition and do not change with further increase in height.

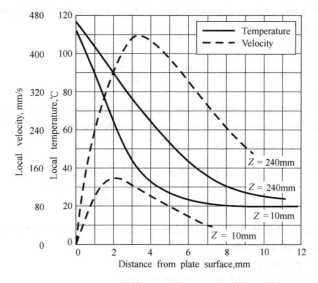

Fig. 4.14 Velocity and temperature gradients, natural convection from heated vertical plate.

The density of the heated air immediately adjacent to the plate is less than that of the unheated air at a distance from the plate, and the buoyancy of the hot air causes an unbalance between the vertical layers of air of differing density. The unbalanced forces generate a circulation by which hot air near the plate rises and cold air flows toward the plate from the room to replenish the rising air stream. A velocity gradient near the plate is formed. Since the velocities of the air in contact with the plate and out in the room are both zero, the velocity is a maximum at a definite distance from the wall. The velocity reaches its maximum a few millimeters from the surface of the plate. The dashed curves in Fig. 4.14 show the velocity gradients for heights of 10 and 240 mm above the bottom of the plate.

The temperature difference between the surface of the plate and the air in the room at a distance from the plate causes a transfer of heat by conduction into the current of gas next to the wall, and the stream carries the heat away by convection in a direction parallel to the plate.

The natural convection currents surrounding a hot, horizontal pipe are more complicated than those adjacent to a vertical heated plate, but the mechanism of the process is similar. The layers of air immediately next to the bottom and sides of the pipe are heated and tend to rise. The rising layers of hot air, one on each side of the pipe, separate from the pipe at points short of the top center of the pipe and form two independent rising currents with a zone of relatively stagnant and unheated air between them.

Natural convection in liquids follows the same pattern, because liquids are also less dense hot than cold. The buoyancy of heated liquid layers near a hot surface generates convection currents just as in gases.

On the assumption that h depends upon pipe diameter, specific heat, thermal conductivity, viscosity, coefficient of thermal expansion, acceleration of gravity, density, and temperature difference, dimensional analysis gives

$$\frac{hD_o}{k} = \Phi\left(\frac{c_p\mu}{k}, \frac{D_o^3\rho^2 g}{\mu^2}, \beta\Delta T\right) \quad (4.4\text{-}39)$$

Since the effect of β is through buoyancy in a gravitational field, the product $g\beta\Delta T$ acts as a single factor, and the last two groups fuse into a dimensionless group called the *Grashof number Gr*, defined by

$$Gr = \frac{D_o^3\rho_f^2\beta g\Delta T_o}{\mu_f^2} \quad (4.4\text{-}40)$$

For single horizontal cylinders, the heat-transfer coefficient can be correlated by an equation containing three dimensionless groups, the Nusselt number, the Prandtl number, and the Grashof number, or specifically,

$$\frac{hD_o}{k_f} = \Phi\left(\frac{c_p\mu_f}{k_f}, \frac{D_o^3\rho_f^2\beta g\Delta T_o}{\mu_f^2}\right) \quad (4.4\text{-}41)$$

where h =average heat-transfer coefficient, based on entire pipe surface
  D_o=outside pipe diameter
  k_f =thermal conductivity of fluid
  c_p =specific heat of fluid at constant pressure
  ρ_f=density of fluid
  β=coefficient of thermal expansion of fluid, is a property of the fluid, equals the reciprocal of the absolute temperature ($1/T$) for ideal gas and can be found from the data of thermodynamics of fluids for the actual fluids.
  g =acceleration of gravity
  ΔT_o =average difference in temperature between outside of pipe and fluid distant from wall
  μ_f=viscosity of fluid

Unlike in the equations for laminar or turbulent flow in pipes, where μ, ρ, and k are evaluated at the bulk temperature, the fluid properties μ_f, ρ_f, and k_f are evaluated at the mean film temperature [Eq. (4.4-37)]. Radiation is not accounted for in this equation.

In Fig. 4.15 is shown a relationship, based on Eq. (4.4-41), which satisfactorily correlates experimental data for heat transfer from a single horizontal cylinder to liquids or gases. The range of variables covered by the single line of Fig. 4.15 is very great.

For magnitudes of log $Gr\,Pr$ of 4 or more, the line of Fig. 4.15 follows closely the empirical equation

$$Nu = 0.53(GrPr)_f^{0.25} \quad (4.4\text{-}42)$$

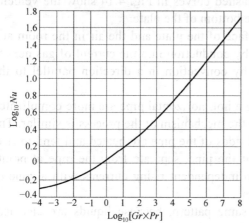

Fig. 4.15 Heat transfer between single horizontal cylinders and fluids in natural convection.

Natural convection to air from vertical shapes and horizontal planes

Equations for heat transfer in natural convection between fluids and solids of definite geometric shape are of the form

$$\frac{hL}{k_f} = b\left[\frac{L^3 \rho_f^2 g \beta_f \Delta T}{\mu_f^2}\left(\frac{c_p \mu}{k}\right)_f\right]^n \quad (4.4\text{-}43)$$

where b, n = constants
L = height of vertical surface or length of horizontal square surface

Properties are taken at the mean film temperature. Equation (4.4-43) can be written

$$Nu_f = b(Gr\ Pr)_f^n \quad (4.4\text{-}44)$$

Values of the constants b and n for various conditions are given in Table 4.2.

Table 4.2   Values of constants in Eq. (4.4-44)

System	Range of $Gr\ Pr$	b	a
Vertical plates, vertical cylinders	10^4–10^9	0.59	0.25
	10^9–10^{12}	0.13	0.333
Horizontal plates:			
Heated, facing upward or cooled, facing down	10^5–2×10^7	0.54	0.25
	2×10^7–3×10^{10}	0.14	0.333
Cooled, facing upward or heated, facing down	3×10^5–3×10^{10}	0.27	0.25

Source: By permission of author and publishers, from W. H. McAdams, *Heat Transmission*, 3rd ed., pp. 172. 180. Copyright by author, 1954, McGraw-Hill Book Company.

4.5   Heat Transfer to Fluids with Phase Change

Processes of heat transfer accompanied by phase change are more complex than simple heat exchange between fluids. A phase change involves the addition or subtraction of considerable quantities of heat at constant or nearly constant temperature. The rate of phase change may be governed by the rate of heat transfer, but it is often influenced by the rate of nucleation of bubbles, drops, or crystals and by the behavior of the new phase after it is formed. This part covers condensation of vapors and boiling of liquids.

4.5.1   Heat Transfer from Condensing Vapors

The condensation of vapors on the surfaces of tubes cooler than the condensing temperature of the vapor is important when vapors such as those of water, hydrocarbons, and other volatile substances are processed. Some examples will be met later in this text, in discussing the unit operations of evaporation, distillation, and drying.

The condensing vapor may consist of a single substance, a mixture of condensable and noncondensable substances, or a mixture of two or more condensable vapors. Friction losses in a condenser are normally small, so that condensation is essentially a constant-pressure process. The condensing temperature of a single pure substance depends only on the pressure, and therefore the process of condensation of a pure substance is isothermal. Also, the condensate is a pure liquid. Mixed vapors, condensing at constant pressure, condense over a temperature range and yield a condensate of variable composition until the entire vapor stream is condensed, when the composition of the condensate equals that of the original uncondensed vapor.

Common examples of condensation in the presence of an inert gas are the condensation of water from a mixture of steam and air and the recovery of hydrocarbon solvents from air streams leaving extraction or drying operations.

Condensation of mixed vapors and condensation in the presence of noncondensing gases are discussed briefly later in this chapter. The following discussion is limited to the condensation of a single volatile substance on a cold tube.

4.5.1.1 Dropwise and film-type condensation

A vapor may condense on a cold surface in one of two ways, which are well described by the terms *dropwise* and *film type*. In film condensation, which is more common than dropwise condensation, the liquid condensate forms a film, or continuous layer, of liquid that flows over the surface of the tube under the action of gravity. It is the layer of liquid interposed between the vapor and the wall of the tube that provides the resistance to heat flow and therefore fixes the magnitude of the heat transfer coefficient.

In dropwise condensation the condensate begins to form at microscopic nucleation sites. Typical sites are tiny pits, scratches, and dust specks. The drops grow and coalesce with their neighbors to form visible fine drops like those often seen on the outside of a cold-water pitcher in a humid room. The fine drops, in turn, coalesce into rivulets, which flow down the tube under the action of gravity, sweep away condensate, and clear the surface for more droplets. During dropwise condensation, large areas of the tube surface are covered with an extremely thin film of liquid of negligible thermal resistance. Because of this the heat-transfer coefficient at these areas is very high; the average coefficient for dropwise condensation may be 5 to 10 times that for film-type condensation. On long tubes, condensation on some of the surface may be film condensation and the remainder dropwise condensation.

The most important and extensive observations of dropwise condensation have been made on steam, but it has also been observed in ethylene glycol, glycerin, nitrobenzene, isoheptane, and some other organic vapors. Liquid metals usually condense in the dropwise manner. The appearance of dropwise condensation depends upon the wetting or nonwetting of the surface by the liquid, and fundamentally, the phenomenon lies in the field of surface chemistry. Much of the experimental work on the dropwise condensation of steam is summarized in the following paragraphs.

(1) Film-type condensation of water occurs on tubes of the common metals if both the steam and the tube are clean, in the presence or absence of air, on rough or on polished surfaces.

(2) Dropwise condensation is obtainable only when the cooling surface is not wetted by the liquid. In the condensation of steam it is often induced by contamination of the vapor with droplets of oil. It is more easily maintained on a smooth surface than on a rough surface.

(3) The quantity of contaminant or promoter required to cause dropwise condensation is minute, and apparently only a monomolecular film is necessary.

(4) Effective drop promoters are strongly adsorbed by the surface, and substances that merely prevent wetting are ineffective. Some promoters are especially effective on certain metals, for example, mercaptans on copper alloys; other promoters, such as oleic acid, are quite generally effective. Some metals, such as steel and aluminum, are difficult to treat to give dropwise condensation.

(5) The average coefficient obtainable in pure dropwise condensation may be as high as 115 kW/(m^2 · ℃).

Although attempts are sometimes made to realize practical benefits from these large coefficients by artificially inducing dropwise condensation, this type of condensation is so unstable and the difficulty of maintaining it so great that the method is not common. Also the resistance of the layer of steam condensate even in filmtype condensation is ordinarily small in comparison with the resistance inside the condenser tube, and the increase in the overall coefficient is relatively small when dropwise condensation is achieved. For normal design, therefore, film-type condensation is assumed.

4.5.1.2 Coefficients for film-type condensation

The basic equations for the rate of heat transfer in film-type condensation were first derived by Nusselt. The *Nusselt equations* are based on the assumption that the vapor and liquid at the outside boundary of the liquid layer are in thermodynamic equilibrium, so that the only resistance to the flow of heat is that offered by the layer of condensate flowing downward in laminar flow under the action of gravity. It is also assumed that the velocity of the liquid at the wall is zero, that the velocity of the liquid at the outside of the film is not influenced by the velocity of the vapor, and that the temperatures of the wall and the vapor are constant. Superheat in the vapor is neglected, the condensate is assumed to leave the tube at the condensing temperature, and the physical properties of the liquid are taken at the mean film temperature.

Vertical tubes. In film-type condensation, the Nusselt theory shows that the condensate film starts to form at the top of the tube and that the thickness of the film increases rapidly near the top of the tube and then more and more slowly in the remaining length. The heat is assumed to flow through the condensate film solely by conduction, and the local coefficient h_x is therefore given by

$$h_x = \frac{k_f}{\delta} \tag{4.5-1}$$

where δ is the local film thickness.

The local coefficient therefore changes inversely with the film thickness. Film thickness δ is typically two or three orders of magnitude smaller than the tube diameter; it can therefore be found, for flow either inside or outside a tube, from the equation for a flat plate, Eq.(1.3-43). Since there is a temperature gradient in the film, the properties of the liquid are evaluated at the average film temperature T_f, given later by Eq. (4.5-11). For condensation on a vertical surface, for which $\cos \varphi = 1$, Eq. (1.3-43) becomes

$$\delta = \left(\frac{3\Gamma \mu_f}{\rho_f^2 g} \right)^{1/3} \tag{4.5-2}$$

where Γ is the condensate loading, the mass rate per unit length of periphery.

Substitution for δ in Eq. (4.5-1) gives for the local heat-transfer coefficient, at a distance L from the top of the vertical surface, the equation

$$h_x = k_f \left(\frac{\rho_f^2 g}{3\Gamma \mu_f} \right)^{1/3} \tag{4.5-3}$$

Equation (4.5-3) applies to condensation either inside or outside a tube. Pure vapors are usually condensed on the outside of tubes; for this situation, for vertical tubes, the local coefficient is given by the relations

$$h_x = \frac{dq}{\Delta T_o dA_o} = \frac{\lambda d\dot{m}}{\Delta T_o \pi D_o dL} \tag{4.5-4}$$

where λ = heat of vaporization

$\dot{m}$ = local flow rate of condensate

Since $\dot{m}/\pi D_o = \Gamma$, Eq. (4.5-4) may be written

$$h_x = \frac{\lambda d\Gamma}{\Delta T_o dL} \qquad (4.5\text{-}5)$$

The *average* coefficient h for the entire tube is defined by

$$h \equiv \frac{q_T}{A_o \Delta T_o} = \frac{\dot{m}_T \lambda}{\pi D_o L_T \Delta T_o} = \frac{\Gamma_b \lambda}{L_T \Delta T_o} \qquad (4.5\text{-}6)$$

where q_T = total rate of heat transfer

$\dot{m}_T$ = total rate of condensation

L_T = total tube length

Γ_b = condensate loading at bottom of tube

Eliminating h_x from Eqs. (4.5-3) and (4.5-5) and solving for ΔT_o give

$$\Delta T_o = \left(\frac{3\Gamma \mu_f}{\rho_f^2 g}\right)^{1/3} \frac{\lambda d\Gamma}{k_f dL} \qquad (4.5\text{-}7)$$

Substituting ΔT_o from Eq. (4.5-7) into Eq. (4.5-6) gives

$$h = \frac{\Gamma_b k_f}{L_T} \left(\frac{\rho_f^2 g}{3\mu_f}\right)^{\frac{1}{3}} \frac{dL}{\Gamma^{1/3} d\Gamma} \qquad (4.5\text{-}8)$$

Rearranging Eq. (4.5-8) and integrating between limits lead to

$$h \int_0^{\Gamma_b} \Gamma^{\frac{1}{3}} d\Gamma = \frac{\Gamma_b k_f}{L_T} \left(\frac{\rho_f^2 g}{3\mu_f}\right)^{\frac{1}{3}} \int_0^{L_T} dL$$

from which

$$h = \frac{4k_f}{3} \left(\frac{\rho_f^2 g}{3\Gamma_b \mu_f}\right)^{\frac{1}{3}} \qquad (4.5\text{-}9)$$

Comparing Eqs. (4.5-9) and (4.5-3) shows that the average coefficient for a vertical tube, provided flow is laminar, is 4/3 times the local coefficient at the bottom of the tube.

Equation (4.5-9) can be rearranged to include the exit Reynolds number for the condensate film $4\Gamma_b/\mu_f$.

$$h \left(\frac{u_f^2}{k_f^3 \rho_f^2 g}\right)^{1/3} = 1.47 \left(\frac{4\Gamma_b}{\mu_f}\right)^{-1/3} = 1.47 Re^{-1/3} \qquad (4.5\text{-}10)$$

On the assumption that the temperature gradient is constant across the film and that $1/\mu$ varies linearly with temperature, the reference temperature for evaluating μ_f, k_f, and ρ_f is given by the equation

$$T_f = T_h - \frac{3(T_h - T_w)}{4} = T_h - \frac{3\Delta T_o}{4} \qquad (4.5\text{-}11)$$

where T_f = reference temperature
 T_h = temperature of condensing vapor
 T_w = temperature of outside surface of tube wall

Equation (4.5-10) is often used in an equivalent form, in which the term Γ_b has been eliminated by combining Eqs. (4.5-6) and (4.5-10) to give

$$h = 0.943 \left(\frac{k_f^3 \rho_f^2 g \lambda}{\Delta T_o L \mu_f} \right)^{1/4} \tag{4.5-12}$$

Equations (4.5-10) and (4.5-11) were derived on the assumption of laminar flow of the condensate, and experimental data agree with the theory, but only for low Reynolds numbers. When the Reynolds number $4\Gamma_b/\mu_f$ is greater than about 30, ripples or waves appear on the surface of the condensate film and increase the rate of heat transfer. The average film thickness is still in reasonable agreement with the laminar flow theory, up to $Re \cong 1200$, but as Re is increased, the heat transfer coefficient becomes considerably greater than predicted. When Re is greater than 1800, the film becomes turbulent and local heat-transfer rates are much higher than predicted from Eq. (4.5-3).

A correlation for the average coefficient as a function of the Reynolds number at the bottom of the tube is given in Fig. 4.16.

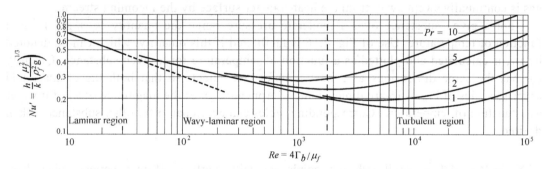

Fig. 4.16 Film coefficient for condensation on vertical surfaces

Horizontal tubes. Corresponding to Eqs. (4.5-10) and (4.5-12) for vertical tubes, the following equations apply to single horizontal tubes:

$$h \left(\frac{\mu_f^2}{k_f^3 \rho_f^2 g} \right)^{1/3} = 1.51 \left(\frac{4\Gamma'}{\mu_f} \right)^{-1/3} \tag{4.5-13}$$

and

$$h = 0.729 \left(\frac{k_f^3 \rho_f^2 g \lambda}{\Delta T_o D_o \mu_f} \right)^{1/4} \tag{4.5-14}$$

where Γ' is the condensate loading per unit *length* of tube $\dot{m}/L$ and all the other symbols have the usual meaning.

For horizontal tubes, Γ' is typically only about $0.1\Gamma_b$ for vertical tubes, and the flow of the condensate is usually laminar. Some authors, however, recommend increasing h found from Eq. (4.5-13) by 20 percent to allow for rippling when $Re > 40$.

Practical use of Nusselt equations

In the absence of high vapor velocities, experimental data check Eqs. (4.5-13) and (4.5-14) well, and these equations can be used as they stand for calculating heat transfer coefficients for film-type condensation on a single horizontal tube. Also, Eq. (4.5-14) can be used for film-type condensation on a vertical stack of horizontal tubes, where the condensate falls cumulatively from tube to tube and the total condensate from the entire stack finally drops from the bottom tube. The average coefficient h_N for the stack of tubes is less than that for one tube; it is given by the equation

$$h_N = h_1 N^{-1/4} \tag{4.5-15}$$

where h_N = average coefficient for entire stack
h_1 = condensing coefficient for top tube in stack
N = number of tubes in stack

For a stack of tubes Eq. (4.5-14) becomes

$$h_N = 0.729 \left(\frac{k_f^3 \rho_f^2 g \lambda}{N \Delta T_o D_o \mu_f} \right)^{1/4} \tag{4.5-16}$$

In general, the coefficient of a film condensing on a horizontal tube is larger than that on a vertical tube under otherwise similar conditions unless the tubes are very short or there are very many horizontal tubes in each stack. Vertical tubes are preferred when the condensate must be appreciably subcooled below its condensation temperature. Mixtures of vapors and noncondensing gases are usually cooled and condensed *inside* vertical tubes, so that the inert gas is continually swept away from the heat-transfer surface by the incoming stream.

For a given substance over a moderate pressure range the quantity $(k_f^3 \rho_f^2 g/\mu^2)^{1/3}$ is a function of temperature. Use of Fig. 4.16 is facilitated if this quantity, which can be denoted by ψ_f, is calculated and plotted as a function of temperature for a given substance. Quantity ψ_f has the same dimensions as a heat transfer coefficient, so that both the ordinate and abscissa scales of Fig. 4.16 are dimensionless. Appendix gives the magnitude of ψ_f for water as a function of temperature. Corresponding tables can be prepared for other substances when desired.

EXAMPLE 4.9 A shell-and-tube condenser with vertical 19-mm copper tubes has chlorobenzene condensing at atmospheric pressure in the shell. The latent heat of condensation of chlorobenzene is 324.9 J/g. The tubes are 1.52 m long. Cooling water at an average temperature of 79°C is flowing in the tubes. The water-side coefficient is 4540 W/(m²·°C). (*a*) What is the coefficient of the condensing chlorobenzene? (*b*) What would the coefficient be in a horizontal condenser with the same number of tubes if the average number of tubes in a vertical stack were 6? Neglect fouling factors and tube wall resistance.
Solution.
(*a*) A trial-and-error solution may be necessary since the condensate film coefficient depends on the Reynolds number, which in turn depends on the rate of condensation. Also, the wall temperature must be estimated to determine the physical properties of the condensate.
The quantities that may be specified directly are

$\lambda = 324.9$ J/g $g = 9.8$ m/s² $L = 1.52$ m $h_i = 4540$ W/(m²·°C)

The condensing temperature T_f is 132°C, so the wall temperature T_w must lie between 79 and 132°C. It is probably closer to 79°C because the resistance of the organic condensate is usually greater than that of flowing water. As a first approximation, T_w is taken as 90°C. From Eq. (4.5-11),

$$T_f = 132 - 3/4(132 - 90) = 100\,°C$$
$$\rho = 1106 \text{ kg/m}^3 \text{ at } 20\,°C$$
$$\rho_f = 1018 \text{ kg/m}^3 \text{ at } 100\,°C \quad \text{(assumes 1 percent decrease per } 10\,°C)$$
$$k_f = 0.144 \text{ W/(m} \cdot °C) \text{ at } 10\,°C \quad \text{(App.12)}$$
$$k_f = 0.11 \text{ W/(m} \cdot °C) \text{ at } 100\,°C \quad \text{(assumes 3 percent decrease per } 10\,°C)$$
$$\mu_f = 0.32 \text{ cP at } 100\,°C = 3.2 \times 10^{-4} \text{ Pa} \cdot \text{s (App.8)}$$
$$c_p = 1550 \text{ J/(kg} \cdot °C) \quad \text{(App.14)}$$

Then
$$\Pr = 1550(3.2 \times 10^{-4})/0.11 = 4.51$$

For a first estimate, assume $Re = 2000$. From Fig.4.16,

$$Nu' = \frac{h}{k_f}\left(\frac{\mu_f^2}{\rho_f^2 g}\right)^{1/3} = 0.23$$

[Note: This result may be checked by substituting the values for Re and Pr into Eq. (4.5-13). This gives $Nu' = 0.231$.] Then

$$h = 0.23 \times 0.11\left[\frac{1018^2 \times 9.8}{(3.2 \times 10^{-4})^2}\right]^{1/3} = 1171 \text{ W/(m}^2 \cdot °C)$$

$$1/U = 1/1171 + 1/4540 = 1.07 \times 10^{-3}$$
$$U = 931 \text{ W/(m}^2 \cdot °C)$$

Check the wall temperature, using Eq.(4.4-31).

$$\Delta T_i = \frac{1/4540}{1/931} \times 53 = 11\,°C$$

$$T_w = 79 + 11 = 90\,°C \quad \text{(as assumed)}$$

From App.4, the area per tube is
$$A = 0.0909 \text{ m}^2$$

Condensate per tube:
$$\frac{931 \times 0.0909(132 - 79)}{324.9} = 13.8 \text{ g/s}$$

$$\Gamma = \frac{13.8 \times 10^{-3}}{\pi(0.019)} = 0.231 \text{ kg/(s·m)}$$

$$Re = \frac{4\Gamma}{\mu_f} = \frac{4 \times 0.231}{3.2 \times 10^{-4}} = 2890$$

From Fig. 4.16, Nu' is slightly more than 0.23. [Solving Eq. (4.5-13) with $Re = 2890$ gives $Nu' = 0.233$] Either way, the first estimate of Re is close enough.
$$h = 1171 \text{ W/(m}^2 \cdot °C)$$

(b) For a horizontal condenser, use Eq. (4.5-17) and assume the wall temperature to be about $90\,°C$. For $N = 6$,

$$h = 0.729 \times \left(\frac{0.144^4 \times 1018^2 \times 9.8 \times 324900}{6 \times 53 \times 0.019 \times 3.2 \times 10^{-4}}\right)^{1/4} = 1095 \text{ W/(m}^2 \cdot °C)$$

This is only slightly less than the value for a vertical tube, so no adjustment of the wall temperature is needed.

4.5.1.3 Effect of Different Factors on Condensation Heat Transfer

Condensation of superheated vapors

If the vapor entering a condenser is superheated, both the sensible heat of superheat and the latent heat of condensation must be transferred through the cooling surface. For steam, because of the low specific heat of the superheated vapor and the large latent heat of condensation, the heat of superheat is usually small in comparison with the latent heat. For example, 50 °C of superheat represents only 100 J/g, as compared with approximately 2300 J/g for latent heat. In the condensation of organic vapors, such as petroleum fractions, the superheat may be appreciable in comparison with the latent heat. When the heat of superheat is important, either it can be calculated from the degrees of superheat and the specific heat of the vapor and added to the latent heat; or if tables of thermal properties are available, the total heat transferred per kilogram of vapor can be calculated by subtracting the enthalpy of the condensate from that of the superheated vapor.

The effect of superheat on the rate of heat transfer depends upon whether the temperature of the surface of the tube is higher or lower than the condensation temperature of the vapor. If the temperature of the tube is lower than the temperature of condensation, the tube is wet with condensate, just as in the condensation of saturated vapor, and the temperature of the outside boundary of the condensate layer equals the saturation temperature of the vapor at the pressure existing in the equipment. The situation is complicated by the presence of a thermal resistance between the bulk of the superheated vapor and the outside of the condensate film and by the existence of a temperature drop, which is equal to the degrees of superheat in the vapor, across that resistance. Practically, however, the net effect of these complications is small, and it is satisfactory to assume that the entire heat load, which consists of the heats of superheat and of condensation, is transferred through the condensate film; that the temperature drop is that across the condensate film; and that the coefficient is the average coefficient for condensing vapor as read from Fig. 4.16. The procedure is summarized by the equation

$$q = hA(T_h - T_w) \qquad (4.5\text{-}17)$$

where q = total heat transferred, including latent heat and superheat
A = area of heat-transfer surface in contact with vapor
h = coefficient of heat transfer, from Fig. 4.16
T_h = saturation temperature of vapor
T_w = temperature of tube wall

When the vapor is highly superheated and the exit temperature of the cooling fluid is close to that of condensation, the temperature of the tube wall may be greater than the saturation temperature of the vapor, condensation cannot occur, and the tube wall will be dry. The tube wall remains dry until the superheat has been reduced to a point where the tube wall becomes cooler than the condensing temperature of the vapor, and condensation takes place. The equipment can be considered as two sections, one a desuperheater and the other a condenser. In calculations, the two sections must be considered separately. The desuperheater is essentially a gas cooler. The logarithmic mean temperature difference applies, and the heat-transfer coefficient is that for cooling a fixed gas. The condensing section is treated by the methods described in the previous paragraphs.

Because of the low individual coefficient on the gas side, the overall coefficient in the desuperheater section is small, and the area of the heating surface in that section is large in comparison with the amount of heat removed. This situation should be avoided in practice. Superheat can be eliminated more economically by injection of a spray of liquid directly into the superheated vapor, since small drops evaporate very rapidly, cooling the vapor to the saturation temperature. The desuperheating section is thereby eliminated, and condensation occurs with high coefficients.

Effect of noncondensables

When a multicomponent mixture contains a noncondensing gas, the rate of condensation is seriously reduced. As in the condensation of a mixture of condensable vapors, there is mass transfer of one or more components in the vapor phase; but here the condensing molecules must diffuse through a film of noncondensing gas which does not move toward the condensate surface. As condensation proceeds, the relative amount of this inert gas in the vapor phase increases significantly.

The partial pressure of the condensing vapor is less than the total pressure, which lowers the equilibrium condensation temperature. In addition, the partial pressure of the condensing vapor at the condensate surface must be less than it is in the bulk vapor-gas phase, to provide the driving force for mass transfer through the gas film. This further lowers the condensing temperature, and usually the change in temperature due to mass transfer is greater than the change in the equilibrium temperature.

Even a small amount of gas can have a large effect on the rate of condensation. Less than 1 percent air in steam can reduce the condensation rate by more than one-half, and 5 percent of inert gas can decrease the steam condensation rate by a factor of 5. Whenever air or other noncondensable gas is present in the feed, a fraction of the incoming gas must be vented from the condenser. If the vapor fed to a nonvented condenser, for instance, contains 0.1 percent air and 99 percent of the vapor is condensed, then the remaining vapor will contain 10 percent air and the condensation rate will be quite low in the last part of the condenser.

Figure 4.17 shows the profiles of temperature and partial pressure in such a condenser. The condensing temperature drops as the composition of the gas-vapor mixture, and hence its dew point, changes as condensation proceeds. Rigorous methods of solving the general problem are based on equating the heat flow to the condensate surface at any point to the heat flow away from the surface. This involves trial-and-error solutions for the point temperature of the condensate surface and from these an estimation of the point values of the heat flux $U\Delta T$. The values of $1/(U\Delta T)$ for each point are plotted against the heat transferred to that point and the area of the condenser surface found by numerical integration.

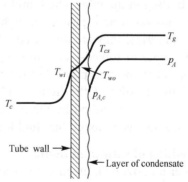

Fig. 4.17 Profiles of partial pressure and temperature in a condenser when noncondensables are present: p_A, partial pressure of condensable vapor in bulk gas phase; $p_{A,c}$, partial pressure at surface of condensate layer; T_g, temperature of bulk gas phase; T_{cs}, temperature at condensate surface; T_{wi}, temperature inside tube wall; T_{wo}, temperature outside tube wall; T_c, temperature of coolant.

4.5.2 Heat Transfer to Boiling Liquids

Heat transfer to a boiling liquid is a necessary step in evaporation, distillation, and steam generation, and it may also be used to control the temperature of a chemical reactor. The boiling liquid may be contained in a vessel equipped with a heating surface fabricated from horizontal or vertical tubes, in which steam or other vapor is condensed or hot fluid is circulated to supply the heat needed to boil the liquid. Alternatively, the liquid to be boiled may flow, under either natural or forced convection, inside heated tubes. An important application of boiling in tubes is the evaporation of water from solution, as discussed later.

When boiling is accomplished by a hot immersed surface, the temperature of the mass of the liquid is the same as the boiling point of the liquid under the pressure existing in the equipment. Bubbles of vapor are generated at the heating surface, rise through the mass of liquid, and disengage from the surface of the liquid. Vapor accumulates in a vapor space over the liquid; a vapor outlet from the vapor space removes the vapor as fast as it is formed. This type of boiling can be described as *pool boiling of saturated liquid* since the vapor leaves the liquid in equilibrium with the liquid at its boiling temperature. When a liquid is boiled under natural circulation inside a vertical tube, relatively cool liquid enters the bottom of the tube and is heated as it flows upward at a low velocity. The liquid temperature rises to the boiling point under the pressure prevailing at that particular level in the tube. Vaporization begins, and the upward velocity of the two-phase liquid-vapor mixture increases enormously. The resulting pressure drop causes the boiling point to fall as the mixture proceeds up the tube and vaporization continues. Liquid and vapor emerge from the top of the tubes at very high velocity.

With forced circulation through horizontal or vertical tubes, the liquid may also enter at a fairly low temperature and be heated to its boiling point, changing into vapor near the discharge end of the tube. Sometimes a flow control valve is placed in the discharge line beyond the tube so that the liquid in the tube may be heated to a temperature considerably above the boiling point corresponding to the downstream pressure. Under these conditions there is no boiling in the tube: The liquid is merely heated, as a liquid, to a high temperature, and flashes into vapor as it passes through the valve. Natural- and forced-circulation boilers are called *calandrias;* they are discussed later.

In some types of forced-circulation equipment, the temperature of the mass of the liquid is below that of its boiling point, but the temperature of the heating surface is considerably above the boiling point of the liquid. Bubbles form on the heating surface, but on release from the surface are absorbed by the mass of the liquid. This type of heat transfer is called *subcooled boiling,* even though the fluid leaving the heat exchanger is entirely liquid.

Pool boiling of saturated liquid

Consider a horizontal, electrically heated wire immersed in a vessel containing a boiling liquid. Assume that q/A, the heat flux, and ΔT, the difference between the temperature of the wire surface T_w and that of the boiling liquid T, are measured. Start with a very low temperature drop ΔT. Now raise T_w and increase the temperature drop by steps, measuring q/A and ΔT at each step, until very large values of ΔT are reached. A plot of q/A versus ΔT on logarithmic coordinates will give a curve of the type shown in Fig. 4.18. This curve can be divided into four segments. In the first segment, at low temperature drops, the line AB is straight and has a slope of 1.25. This is consistent with the correlation for natural convection [Eq. (4.4-46)] and corresponds to the equation

$$\frac{q}{A} = a\Delta T^{1.25} \tag{4.5-18}$$

where a is a constant. The second segment, line BC, is also approximately straight, but its slope is greater than that of line AB. The slope of line BC depends upon the specific experiment; it usually lies between 3 and 4. The second segment terminates at a definite point of maximum flux, which is point C in Fig. 4.18. The temperature drop corresponding to point C is called the *critical temperature drop,* and the flux at point C is the *peak flux.* In the third segment, line CD in Fig. 4.18, the flux decreases as the temperature drop rises and reaches a minimum at point D. Point D is called the *Leidenfrost point.* In the last segment, line DE, the flux again increases with ΔT and, at large temperature drops, surpasses the previous maximum reached at point C.

Because, by definition, $h = (q/A)/\Delta T$, the plot of Fig. 4.18 is readily convertible into a plot of h versus ΔT. This curve is shown in Fig. 4.19. A maximum and a minimum coefficient are evident in Fig. 4.19. They do not, however, occur at the same values of the temperature drop as the maximum and minimum fluxes indicated in Fig. 4.18. The coefficient is normally a maximum at a temperature drop slightly lower than that at the peak flux; the minimum coefficient occurs at a much higher temperature drop than that at the Leidenfrost point. The coefficient is proportional to $\Delta T^{0.25}$ in the first segment of the line in Fig. 4.18 and to between ΔT^2 and ΔT^3 in the second segment.

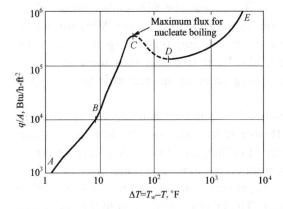

Fig. 4.18 Heat flux versus temperature drop, boiling water at 212°F on an electrically heated wire: *AB,* natural convection; *BC,* nucleate boiling; *CD,* transition boiling; *DE,* film boiling. *(After McAdams et al)*

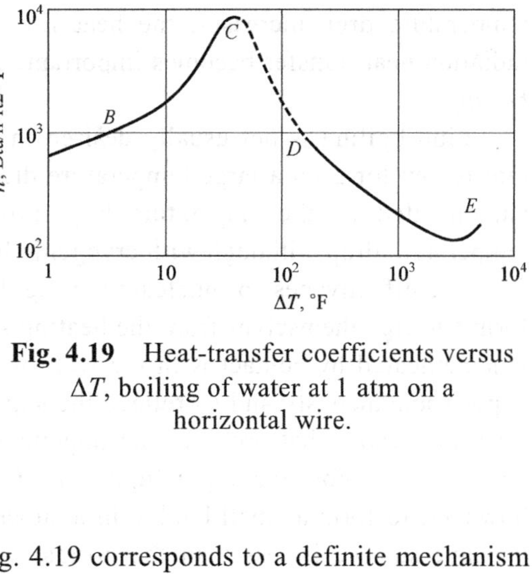

Fig. 4.19 Heat-transfer coefficients versus ΔT, boiling of water at 1 atm on a horizontal wire.

Each of the four segments of the graph in Fig. 4.19 corresponds to a definite mechanism of boiling. In the first section, at low temperature drops, the mechanism is that of heat transfer to a liquid in natural convection, and the variation of h with ΔT agrees with that given by Eq. (4.4-46). Bubbles form on the surface of the heater, are released from it, rise to the surface of the liquid, and are disengaged into the vapor space; but they are too few to disturb appreciably the normal currents of free convection. At larger temperature drops, lying between 5 and 25°C in the case shown in Fig. 4.19, the rate of bubble production is large enough for the stream of bubbles moving up through the liquid to increase the velocity of the circulation currents in the mass of liquid, and the coefficient of heat transfer becomes greater than that in undisturbed natural convection. As ΔT is increased, the rate of bubble formation increases and the coefficient increases rapidly.

The action occurring at temperature drops below the critical temperature drop is called *nucleate boiling,* in reference to the formation of tiny bubbles, or vaporization nuclei, on the heating surface. During nucleate boiling, the bubbles occupy but a small portion of the heating surface at a time, and most of the surface is in direct contact with liquid. The bubbles are generated at localized active sites, usually small pits or scratches on the heating surface.

As the temperature drop is raised, more sites become active, improving the agitation of the liquid and increasing the heat flux and the heat-transfer coefficient.

Eventually, however, so many bubbles are present that they tend to coalesce and cover portions of the heating surface with a layer of insulating vapor. This layer has a highly unstable surface, from which miniature "explosions" send jets of vapor away from the heating element into the bulk of the liquid. This type of action is called *transition boiling*. In this region, corresponding to segment *CD* in Fig. 4.18, increasing the temperature drop increases the thickness of the layer of vapor and reduces the number of explosions that occur in a given time. The heat flux and the heat-transfer coefficient both fall as the temperature drop is raised.

Near the Leidenfrost point another distinct change in mechanism occurs. The hot surface becomes covered with a quiescent film of vapor, through which heat is transferred by conduction and (at very high temperature drops) by radiation. The random explosions characteristic of transition boiling disappear and are replaced by the slow and orderly formation of bubbles at the interface between the liquid and the film of hot vapor. These bubbles detach themselves from the interface and rise through the liquid. Virtually all the resistance to heat transfer is offered by the vapor sheath covering the heating element. As the temperature drop increases, the heat flux rises, slowly at first and then more rapidly as radiation heat transfer becomes important. The boiling action in this region is known as *film boiling*.

Film boiling is not usually desired in commercial equipment because the heat transfer rate is low for such a large temperature drop. Heat-transfer apparatus should be so designed and operated that the temperature drop in the film of boiling liquid is smaller than the critical temperature drop, although with cryogenic liquids this is not always feasible.

The effectiveness of nucleate boiling depends primarily on the ease with which bubbles form and free themselves from the heating surface. The layer of liquid next to the hot surface is superheated by contact with the wall of the heater. The superheated liquid tends to form vapor spontaneously and so relieve the superheat. It is the tendency of superheated liquid to flash into vapor that provides the impetus for the boiling process. Physically, the flash can occur only by forming vapor-liquid interfaces in the form of small bubbles. It is not easy, however, to form a small bubble in a superheated liquid, because the vapor pressure over a concave surface is less than the normal value, and the smaller the bubble the greater the effect. A very small bubble can exist in equilibrium with superheated liquid, and the smaller the bubble, the greater the equilibrium superheat and the smaller the tendency to flash. By taking elaborate precautions to eliminate all gas and other impurities from the liquid and to prevent shock, it is possible to superheat water by several hundred degrees Fahrenheit without formation of bubbles.

A second difficulty appears if the bubble does not readily leave the surface once it is formed. The important factor in controlling the rate of bubble detachment is the interfacial tension between the liquid and the heating surface. If this interfacial tension is large, the bubble tends to spread along the surface and blanket the heat transfer area, as shown in Fig. 4.20(*c*), instead of leaving the surface

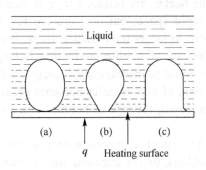

Fig. 4.20 Effect on bubble formation of interfacial tension between liquid and heating surface.

to make room for other bubbles. If the interfacial tension between liquid and solid is low, the bubble will pinch off easily, in the manner shown in Fig. 4.20(a). An example of intermediate interfacial tension is shown in Fig. 4.5(b).

The high rate of heat transfer in nucleate boiling is primarily the result of the turbulence generated in the liquid by the dynamic action of the bubbles.

The coefficient obtained during nucleate boiling is sensitive to a number of variables, including the nature of the liquid, the type and condition of the heating surface, the composition and purity of the liquid, the presence or absence of agitation, and the temperature or pressure. Minor changes in some variables cause major changes in the coefficient. The reproducibility of check experiments is poor.

Qualitatively, the effects of some variables can be predicted from a consideration of the mechanisms of boiling. A roughened surface provides centers for nucleation that are not present on a polished surface. Thus roughened surfaces usually give larger coefficients than smooth surfaces. This effect, however, is due in part to the fact that the total surface of a rough tube is larger than that of a smooth surface of the same projected area. A very thin layer of scale may increase the coefficient of the boiling liquid, but even a thin scale will reduce the overall coefficient by adding a resistance that reduces the overall coefficient more than the improved boiling liquid coefficient increases it. Gas or air adsorbed on the surface of the heater or contaminants on the surface often facilitate boiling by either the formation or the disengaging of bubbles. A freshly cleaned surface may give a higher or lower coefficient than the same surface after it has been stabilized by a previous period of operation. This effect is associated with a change in the condition of the heating surface. Agitation increases the coefficient by increasing the velocity of the liquid across the surface, which helps to sweep away bubbles.

Curves similar to those in Figs. 4.18 and 4.19 would be obtained when studying pool boiling on a horizontal tube or pipe, although the critical temperature drop and the Leidenfrost point would probably not be the same as for a small-diameter wire.

4.6 Radiation Heat Transfer

Radiation, which may be considered to be energy streaming through space at the speed of light, may originate in various ways. Some types of material will emit radiation when they are treated by external agencies, such as electron bombardment, electric discharge, or radiation of definite wavelengths. Radiation due to these effects will not be discussed here. All substances at temperatures above absolute zero emit radiation that is independent of external agencies. Radiation that is the result of temperature only is called thermal radiation, and this discussion is restricted to radiation of this type.

Fundamental facts concerning radiation

Radiation moves through space in straight lines, or beams, and only substances in sight of a radiating body can intercept radiation from that body. The fraction of the radiation falling on a body that is reflected is called the reflectivity ρ. The fraction that is absorbed is called the absorptivity α. The fraction that is transmitted is called the transmissivity τ. The sum of these fractions must be unity, or

$$\alpha + \rho + \tau = 1 \tag{4.6-1}$$

Radiation as such is not heat, and when transformed to heat on absorption, it is no longer radiation. In practice, however, reflected or transmitted radiation usually falls on other absorptive bodies and is eventually converted to heat, perhaps after many successive reflections.

The maximum possible absorptivity is unity, attained only if the body absorbs all radiation incident upon it and reflects or transmits none. A body that absorbs all incident radiation is called a blackbody.

The complex subject of thermal radiation transfer has received much study in recent years and is covered in a number of texts. The following introductory treatment discusses the following topics: emission of radiation, absorption by opaque solids, radiation between surfaces, radiation to and from semitransparent materials.

4.6.1 Emission of Radiation

The radiation emitted by any given body of material is independent of that being emitted by other material in sight of, or in contact with, the body. The net energy gained or lost by a body is the difference between the energy emitted by the body and that absorbed by it from the radiation reaching it from other bodies. Heat flow by conduction and convection may also be taking place independently of the radiation.

When bodies at different temperatures are placed in sight of one another inside an enclosure, the hotter bodies lose energy by emission of radiation faster than they receive energy by absorption of radiation from the cooler bodies, and the temperatures of the hotter bodies decrease. Simultaneously the cooler bodies absorb energy from the hotter ones faster than they emit energy, and the temperatures of the cooler bodies increase. The process reaches equilibrium when all the bodies reach the same temperature, just as in heat flow by conduction and convection. The conversion of radiation to heat on absorption and the attainment of temperature equilibrium through the net transfer of radiation justify the usual practice of calling radiation heat.

Wavelength of radiation

Known electromagnetic radiation covers an enormous range of wavelengths, from the short cosmic rays having wavelengths of about 10^{-11}cm to long wave broadcasting waves having lengths of 1000m or more.

Radiation of a single wavelength is called monochromatic. A beam of thermal radiation is not monochromatic but has a wide distribution of wavelengths, as shown in Fig. 4.21. The range between 0.1 and 100μm is important in heat flow by radiation, while visible light covers a narrow range of wavelengths from about 0.39 to 0.78 μm. Thermal radiation at ordinary industrial temperatures has wavelengths in the infrared spectrum, which includes waves just longer than the longest visible waves. At temperatures above about 500℃ heat radiation in the visible spectrum becomes significant, and the phrases red heat and white heat refer to this fact. The higher the temperature of the radiating body, the shorter the predominant wave-length of the thermal radiation

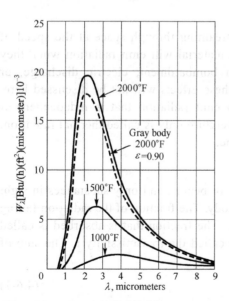

Fig. 4.21 Energy distribution in spectra of blackbodies and gray bodies

emitted by it.

For a given temperature, the rate of thermal radiation varies with the state of aggregation and the molecular structure of the substance. Monatomic and diatomic gases such as oxygen, argon, and nitrogen radiate weakly, even at high temperatures. Under industrial conditions, these gases neither emit nor absorb appreciable amounts of radiation. Polyatomic gases, including water vapor, carbon dioxide, ammonia, sulfur dioxide, and hydrocarbons emit and absorb radiation at several wavelengths, and infrared absorption at selected wavelengths can be used to analyze mixtures of such gases. Absorption of infrared radiation by polyatomic gases has a major effect on our climate, since a considerable fraction of the energy radiated from the earth's surface is absorbed by these gases in the lower atmosphere. The gas concentrations are low, making the absorption per unit volume quite small, but there are several kilometers of atmosphere in which absorption occurs. At very high temperatures the absorption of radiation by CO_2 and water vapor is strong enough, even over short distances, that it must be considered in the design of furnaces.

Solids and liquids, except in thin layers, absorb and emit radiation over the entire spectrum. In contrast to gases, which are semitransparent volume emitters and absorbers, most solids and liquids are opaque to radiation, and emission and absorption are correlated with the exposed surface area. The following discussion deals mainly with emission and absorption by surfaces and radiation between surfaces.

Emissive power

The monochromatic energy emitted by a radiating surface depends on the temperature of the surface and on the wavelength of the radiation. At constant surface temperature, a curve can be plotted showing the rate of energy emission as a function of the wavelength. Typical curves of this type are shown in Fig. 4.21. Each curve rises steeply to a maximum and decreases asymptotically to zero emission at very large wavelengths. The unit chosen for measuring the monochromatic radiation is based on the fact that, from a small area of a radiating surface, the energy emitted is "broadcast" in all directions through any hemisphere centered on the radiation area. The monochromatic radiation emitted in this manner from unit area in unit time divided by the wavelength is called the monochromatic radiating power W_λ. The ordinates in Fig. 4.21 are values of W_λ.

For the entire spectrum of the radiation from a surface, the total radiating power W is the sum of all the monochromatic radiations from the surface, or mathematically,

$$W = \int_0^\infty W_\lambda \, d\lambda \tag{4.6-2}$$

Graphically, W is the entire area under any of the curves in Fig. 4.21 from wavelengths of zero to infinity. Physically, the total radiating power is the total radiation of all wavelengths emitted by unit area in unit time in all directions through a hemisphere centered on the area.

Blackbody radiation; emissivity

A blackbody is defined as an ideal emitter that has the maximum attainable emissive power at any given temperature. It also absorbs all incoming radiation, and is the standard to which all other radiators are referred. The ratio of the total emissive power W of a body to that of a blackbody W_b is, by definition, the *emissivity* ε of the body. Thus,

$$\varepsilon \equiv \frac{W}{W_b} \tag{4.6-3}$$

The *monochromatic emissivity* ε_λ is the ratio of the monochromatic emissive power to that of a blackbody at the same wavelength, or

$$\varepsilon_\lambda \equiv \frac{W_\lambda}{W_{b,\lambda}} \qquad (4.6\text{-}4)$$

If the monochromatic emissivity of a body is the same for all wavelengths, the body is called a *gray body*.

Emissivities of solids

Emissivities of solids are tabulated in standard references. Emissivities of polished metals are low, in the range of 0.02 to 0.10, and they usually increase with temperature. Emissivities of nonconductors generally decrease with increasing temperature. Those of most oxidized metals range from 0.6 to 0.85; those of nonmetals such as refractories, paper, boards, and building materials, from 0.65 to 0.95; and those of paints, other than aluminum paint, from 0.80 to 0.96.

Practical source of blackbody radiation

No actual substance is a blackbody, although some materials, such as certain grades of carbon black, do approach blackness. An experimental equivalent of a blackbody is an isothermal enclosure containing a small peephole. If a sight is taken through the peephole on the interior wall of the enclosure, the effect is the same as viewing a blackbody. The radiation emitted by the interior of the walls or admitted from outside through the peephole is completely absorbed after successive reflections, and the overall absorptivity of the interior surface is unity.

Laws of blackbody radiation

A basic relationship for blackbody radiation is the *Stefan-Boltzmann law,* which states that the total emissive power of a blackbody is proportional to the fourth power of the absolute temperature, or

$$W_b = \sigma T^4 \qquad (4.6\text{-}5)$$

where σ is a universal constant depending only upon the units used to measure T and W_b. The Stefan-Boltzmann law is an exact consequence of the laws of thermodynamics and electromagnetism.

The distribution of energy in the spectrum of a blackbody is known accurately. It is given by *Planck's law*

$$W_{b,\lambda} = \frac{2\pi h c^2 \lambda^{-5}}{e^{hc/k\lambda T} - 1} \qquad (4.6\text{-}6)$$

where $W_{b,\lambda}$ = monochromatic emissive power of blackbody
h = Planck's constant
c = speed of light
λ = wavelength of radiation
k = Boltzmann's constant
T = absolute temperature

Equation (4.6-6) can be written

$$W_{b,\lambda} = \frac{C_1 \lambda^{-5}}{e^{C_2/\lambda T} - 1} \qquad (4.6\text{-}7)$$

where C_1 and C_2 are constants with 3.742×10^{-16} W·m² and 1.439 cm·K.

Plots of $W_{b,\lambda}$ versus λ from Eq. (4.6-6) are shown as solid lines in Fig. 4.21 for blackbody radiation at temperatures of 1000, 1500, and 2000°F. The dotted line shows the monochromatic radiating power of a gray body of emissivity 0.9 at 2000°F.

Planck's law can be shown to be consistent with the Stefan-Boltzmann law by substituting $W_{b,\lambda}$ from Eq. (4.6-6) into Eq. (4.6-2) and integrating.

At any given temperature, the maximum monochromatic radiating power is attained at a definite wavelength, denoted by λ_{max}. *Wien's displacement law* states that λ_{max} is inversely proportional to the absolute temperature, or

$$T\lambda_{max} = C \tag{4.6-8}$$

The constant C is 2890 when λ_{max} is in micrometers and T is in Kelvins, or 5200 when T is in degrees Rankine.

Wien's law also can be derived from Planck's law [Eq. (4.6-6)] by differentiating with respect to λ, equating the derivative to zero, and solving for λ_{max}.

4.6.2 Absorption of Radiation by Opaque Solids

When radiation falls on a solid body, a definite fraction ρ may be reflected, and the remaining fraction $1-\rho$ enters the solid to be either transmitted or absorbed. Most solids (other than glasses, certain plastics, quartz, and some minerals) absorb radiation of all wavelengths so readily that, except in thin sheets, the transmissivity τ is zero, and all nonreflected radiation is completely absorbed in a thin surface layer of the solid. The absorption of radiation by an opaque solid is therefore a surface phenomenon, not a volume phenomenon, and the interior of the solid is not of interest in the absorption of radiation. The heat generated by the absorption can flow into or through the mass of an opaque solid only by conduction.

Reflectivity and absorptivity of opaque solids

Since the transmissivity of an opaque solid is zero, the sum of the reflectivity and the absorptivity is unity, and the factors that influence reflectivity affect absorptivity in the opposite sense. In general, the reflectivity of an opaque solid depends on the temperature and character of the surface, the material of which the surface is made, the wavelength of the incident radiation, and the angle of incidence. Two main types of reflection are encountered, specular and diffuse. The first is characteristic of smooth surfaces such as polished metals; the second is found in reflection from rough surfaces or from dull, or matte, surfaces. In specular reflection, the reflected beam makes a definite angle with the surface, and the angle of incidence equals the angle of reflection. The reflectivity from these surfaces approaches unity, and the absorptivity approaches zero. Matte, or dull, surfaces reflect diffusely in all directions, there is no definite angle of reflection, and the absorptivity can approach unity. Rough surfaces, in which the scale of roughness is large in comparison with the wavelength of the incident radiation, will reflect diffusely even if the radiation from the individual units of roughness is specular. Reflectivities of rough surfaces may be either large or small, depending upon the reflective characteristic of the material itself. Most industrial surfaces of interest to the chemical engineer give diffuse reflection, and in treating practical cases, the important simplifying assumption can usually be made that reflectivity and absorptivity are independent of the angle of incidence. This assumption is equivalent to the *cosine law* [see Eq. (4.6-14)], which states that for a perfectly diffusing surface the intensity (or brightness, in the case of visible light) of the radiation leaving the surface is independent of the angle from which the surface is viewed. This is true whether the radiation is emitted by the surface, giving *diffuse radiation,* or is reflected by it, giving *diffuse reflection.*

The reflectivity may vary with the wavelength of the incident radiation, and the absorptivity of the entire beam is then a weighted average of the monochromatic absorptivities and depends upon the entire spectrum of the incident radiation.

The absorptivity of a gray body, like the emissivity, is the same for all wavelengths. If the surface of the gray body gives diffuse radiation or reflection, its monochromatic absorptivity is also independent of the angle of incidence of the radiant beam. The total absorptivity equals the monochromatic absorptivity and is also independent of the angle of incidence.

Kirchhoff's law

An important generalization concerning the radiating power of a substance is Kirchhoff's law, which states that, at temperature equilibrium, the ratio of the total radiating power of any body to its absorptivity depends only upon the temperature of the body. Thus, consider any two bodies in temperature equilibrium with common surroundings. Kirchhoff's law states that

$$\frac{W_1}{\alpha_1} = \frac{W_2}{\alpha_2} \tag{4.6-9}$$

where W_1, W_2 = total radiating powers of two bodies
α_1, α_2 = absorptivities of two bodies
This law applies to both monochromatic and total radiation.

If the first body referred to in Eq. (4.6-9) is a blackbody, $\alpha_1=1$, and

$$W_1 = W_b = \frac{W_2}{\alpha_2} \tag{4.6-10}$$

where W_b denotes the total radiating power of a blackbody. Thus

$$\alpha_2 = \frac{W_2}{W_b} \tag{4.6-11}$$

But, by definition, the emissivity of the second body ε_2 is

$$\varepsilon_2 = \frac{W_2}{W_b} = \alpha_2 \tag{4.6-12}$$

Thus, when any body is at temperature equilibrium with its surroundings, its emissivity and absorptivity are equal. This relationship may be taken as another statement of Kirchhoff's law. In general, except for blackbodies or gray bodies, absorptivity and emissivity are not equal if the body is not in thermal equilibrium with its surroundings.

The absorptivity and emissivity, monochromatic or total, of a blackbody are both unity. The cosine law also applies exactly to a blackbody, as the reflectivity is zero for all wavelengths and all angles of incidence.

Kirchhoff's law applies to volumes as well as to surfaces. Since absorption by an opaque solid is effectively confined to a thin layer at the surface, the radiation emitted from the surface of the body originates in this same surface layer. Radiating substances absorb their own radiation, and radiation emitted by the material in the interior of the solid is also absorbed in the interior and does not reach the surface.

Because the energy distribution in the incident radiation depends upon the temperature and character of the originating surface, the absorptivity of the receiving surface may also depend upon these properties of the originating surface. Kirchhoff's law does not, therefore, always apply to nonequilibrium radiation. If, however, the receiving surface is gray, a constant fraction, independent of wavelength, of the incident radiation is absorbed by the receiving surface, and Kirchhoff's law applies whether or not the two surfaces are at the same temperature.

The majority of industrial surfaces, unfortunately, are not gray, and their absorptivities vary strongly with the nature of the incident radiation. Figure 4.22 shows how the absorptivity of various solids varies with the peak wavelength of the incident radiation and thus with the temperature of the source. A few solids, such as slate, are almost truly gray, and their absorptivities are almost constant. For polished metallic surfaces the absorptivity α_2 rises with the absolute temperature of the source T_1 and also that of surface T_2

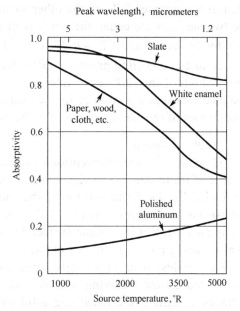

Fig. 4.22 Absorptivities of various solids versus source temperature and peak wavelength of incident radiation.

4.6.3 Radiation between Surfaces

The total radiation for a unit area of an opaque body of area A_1, emissivity ε_1, and absolute temperature T_1 is

$$\frac{q}{A_1} = \sigma \varepsilon_1 T_1^4 \qquad (4.6\text{-}13)$$

Most surfaces, however, that emit radiation also receive radiation from other surfaces at different temperatures. Some of this incoming radiation is absorbed and must be allowed for in determining the total flux of radiant energy. For example, a steam line in a room is surrounded by the walls, floor, and ceiling of the room, all of which are radiating to the pipe, and although the pipe loses more energy than it absorbs from its surroundings, the net loss by radiation is less than that calculated from Eq. (4.6-13). Even when a surface is radiating to a clear night sky, the radiated energy is partially absorbed by the water and carbon dioxide in the atmosphere, and part of this absorbed energy is radiated back to the surface.

In furnaces and other high-temperature equipment, where radiation is particularly important, the usual objective is to obtain a controlled rate of net heat exchange between one or more hot surfaces, called *sources,* and one or more cold surfaces, called *sinks.* In many cases the hot surface is a flame, but exchange of energy between surfaces is common, and a flame can be considered to be a special form of translucent surface. The following treatment is limited to the radiant energy transfer between opaque surfaces in the absence of any absorbing medium between them.

The simplest type of radiation between two surfaces occurs where each surface can see only the other, for example, where the surfaces are very large parallel planes, as shown in Fig. 4.23(*a*), and where both surfaces are black. The energy emitted per unit area by the first plane is σT_1^4; that emitted by the second plane is σT_2^4. Assume that $T_1 > T_2$. All the radiation from each of the surfaces falls on the other surface and is completely absorbed. Since the areas of the two surfaces are equal, the net loss of energy per unit area by the first plane and the net gain by the second are $\sigma T_1^4 - \sigma T_2^4$, or $\sigma (T_1^4 - T_2^4)$.

Actual engineering problems differ from this simple situation in the following ways: (1) One or both of the surfaces of interest see other surfaces. In fact, an element of surface in a concave area sees a portion of its own surface. (2) No actual surface is exactly black, and the emissivities of the surfaces must often be considered.

Angle of vision

Qualitatively, the interception of radiation from an area element of a surface by another surface of finite size can be visualized in terms of the angle of vision, which is the solid angle subtended by the finite surface at the radiating element. The solid angle subtended by a hemisphere is 2π steradians (sr). This is the maximum angle of vision that can be subtended at any area element by a plane surface in sight of the element. It will be remembered that the total radiating power of an area element is defined to take this fact into account. If the angle of vision is less than 2π sr, only a fraction of the radiation from the area element will be intercepted by the receiving area and the remainder will pass on to be absorbed by other surfaces in sight of the remaining solid angle. Some of the hemispherical angle of vision of an element of a concave surface is subtended by the originating surface itself.

Figure 4.23 shows several typical radiating surfaces. Figure 4.23(a) shows how, in two large parallel planes, an area element on either plane is subtended by a solid angle of 2π sr by the other. The radiation from either plane cannot escape being intercepted by the other. A point on the hot body of Fig. 4.23(b) sees only the cold surface, and the angle of vision is again 2π sr. Elements of the cold surface, however, see, for the most part, other portions of the cold surface, and the angle of vision for the hot body is small. This effect of self-absorption is also shown in Fig. 4.23(c), where the angle of vision of an element of the hot surface subtended by the cold surface is relatively small. In Fig. 4.23(d), the cold surface subtends a small angle at the hot surface, and the bulk of the radiation from the hot surface passes on to some undetermined background. Figure 4.23(e) shows a simple furnace with insulated refractory walls. Radiation from the hot floor, or source, is intercepted partly by the row of tubes across the top of the furnace, which form the sink, and partly by the refractory walls and the refractory ceiling behind the tubes. The refractory in such assemblies is assumed to absorb and emit energy at the same rate, so the net energy effect at the refractory is zero. The refractory ceiling absorbs the energy

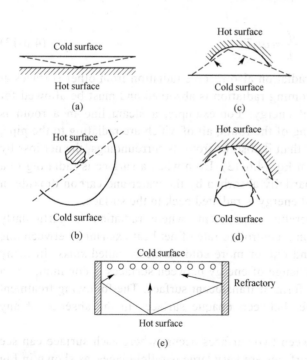

Fig. 4.23 Angle of vision in radiant heat flow.

that passes between the tubes and reradiates it to the backs of the tubes.

Quantitative calculation of radiation between black surfaces

For a given combination of finite surfaces,the net rate of transfer dq_{12} between the two area elements is found in the form

$$q_{12} = \sigma AF(T_1^4 - T_2^4) \tag{4.6-14}$$

where q_{12} = net radiation between two surfaces
A = area of either of two surfaces, chosen arbitrarily
F = dimensionless geometric factor

The factor F is called the *view factor* or *angle factor;* it depends upon the geometry of the two surfaces, their spatial relationship with each other, and the surface chosen for A. If surface A_1 is chosen for A, Eq. (4.6-14) can be written

$$q_{12} = \sigma A_1 F_{12}(T_1^4 - T_2^4) \tag{4.6-15}$$

If surface A_2 is chosen,

$$q_{12} = \sigma A_2 F_{21}(T_1^4 - T_2^4) \tag{4.6-16}$$

Comparing Eqs. (4.6-15) and (4.6-16) gives

$$A_1 F_{12} = A_2 F_{21} \tag{4.6-17}$$

Factor F_{12} may be regarded as the fraction of the radiation leaving area A_1 that is intercepted by area A_2. If surface A_1 can see only surface A_2, the view factor F_{12} is unity. If surface A_1 sees a number of other surfaces and if its entire hemispherical angle of vision is filled by these surfaces, then

$$F_{11} + F_{12} + F_{13} + \cdots = 1.0 \tag{4.6-18}$$

The factor F_{11} covers the portion of the angle of vision subtended by other portions of body A_1. If the surface of A_1 cannot see any portion of itself, F_{11} is zero. The net radiation associated with an F_{11} factor is, of course, zero.

In some situations the view factor may be calculated simply. For example, consider a small blackbody of area A_2 having no concavities and surrounded by a large black surface of area A_1. The factor F_{21} is unity, as area A_2 can see nothing but area A_1. The factor F_{12} is, by Eq. (4.6-27),

$$F_{12} = \frac{F_{21} A_2}{A_1} = \frac{A_2}{A_1} \tag{4.6-19}$$

By Eq. (4.6-18),

$$F_{11} = 1 - F_{12} = 1 - \frac{A_2}{A_1} \tag{4.6-20}$$

The factor F has been determined by Hottel for a number of important special cases. Figure 4.24 shows the F factor for equal parallel planes directly opposed. Line 1 is for disks, line 2 for squares, line 3 for rectangles having a ratio of length to width of 2 : 1, and line 4 for long, narrow rectangles. In all cases, the factor F is a function of the ratio of the side or diameter of the planes to the distance between them.

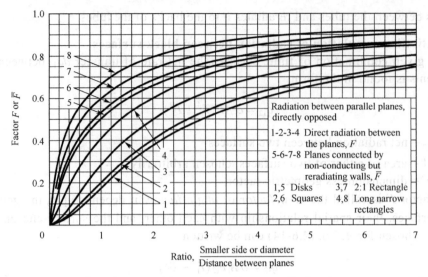

Fig. 4.24 View factor and interchange factor, radiation between opposed parallel disks, rectangles, and squares.

Allowance for refractory surfaces

When the source and sink are connected by refractory walls in the manner shown in Fig. 4.23(*e*), factor F can be replaced by an analogous factor, called the *interchange factor* $\overline{F}$, and Eqs. (4.6-15) and (4.6-16) written as

$$q_{12} = \sigma A_1 \overline{F}_{12}(T_1^4 - T_2^4) = \sigma A_2 \overline{F}_{21}(T_1^4 - T_2^4) \tag{4.6-21}$$

The interchange factor $\overline{F}$ has been determined accurately for some simple situations. Lines 5 to 8 of Fig. 4.24 give values of $\overline{F}$ for directly opposed parallel planes connected by refractory walls. Line 5 applies to disks, line 6 to squares, line 7 to 2:1 rectangles, and line 8 to long, narrow rectangles.

An approximate equation for $\overline{F}$ in terms of F is

$$\overline{F}_{12} = \frac{A_2 - A_1 F_{12}^2}{A_1 + A_2 - 2A_1 F_{12}} \tag{4.6-22}$$

Equation (4.6-22) applies where there is but one source and one sink, where neither area A_1 nor A_2 can see any part of itself. It is based on the assumption that the temperature of the refractory surface is constant. This last is a simplifying assumption, as the local temperature of the refractory usually varies between those of the source and the sink.

Nonblack surfaces

The treatment of radiation between nonblack surfaces, in the general case where absorptivity and emissivity are unequal and both depend upon wavelength and angle of incidence, is obviously complicated. Several important special cases can, however, be treated simply.

A simple example is a small body that is not black surrounded by a black surface. Let the areas of the enclosed and surrounding surfaces be A_1 and A_2, respectively, and let their temperatures be T_1 and T_2, respectively. The radiation from surface A_2 falling on surface A_1 is $\sigma A_2 F_{21} T_2^4$. Of this, the fraction α_1, the absorptivity of area A_1 for radiation from surface A_2, is absorbed by surface A_1. The remainder is reflected to the black surroundings and completely reabsorbed by area A_2. Surface A_1 emits radiation in amount $\sigma A_1 \varepsilon_1 T_1^4$, where ε_1 is the emissivity of surface A_1. All this radiation is absorbed by surface A_2, and none is returned by another reflection. The emissivity ε_1 and absorptivity α_1 are not in general equal, because the two surfaces are not at the same temperature. The net energy loss by surface A_1 is

$$q_{12} = \sigma \varepsilon_1 A_1 T_1^4 - \sigma A_2 F_{21} \alpha_1 T_2^4 \tag{4.6-23}$$

But by Eq. (4.6-17), $A_2 F_{21} = A_1$, and after elimination of $A_2 F_{21}$, Eq. (4.6-23) becomes

$$q_{12} = \sigma A_1 (\varepsilon_1 T_1^4 - \alpha_1 T_2^4) \tag{4.6-24}$$

If surface A_1 is gray, $\varepsilon_1 = \alpha_1$ and

$$q_{12} = \sigma A_1 \varepsilon_1 (T_1^4 - T_2^4) \tag{4.6-25}$$

In general, for gray surfaces, Eqs. (4.6-15) and (4.6-16) can be written

$$q_{12} = \sigma A_1 \mathscr{F}_{12} (T_1^4 - T_2^4) = \sigma A_2 \mathscr{F}_{21} (T_1^4 - T_2^4) \tag{4.6-26}$$

where $\mathscr{F}_{12}$ and $\mathscr{F}_{21}$ are the *overall interchange factors* and are functions of ε_1 and ε_2. The value of σ is 5.672×10^{-8} W/(m$^2 \cdot$K^4).

Two large parallel planes. In simple cases factor $\mathscr{F}$ can be calculated directly by considering the paths of reflected beams that are successively reflected and reabsorbed. For two large parallel plates that have unequal emissivities ε_1 and ε_2, the overall interchange factor is

$$\mathscr{F}_{12} = \frac{1}{1/\varepsilon_1 + 1/\varepsilon_2 - 1} \tag{4.6-27}$$

One gray surface completely surrounded by another. Let the area of the enclosed body be A_1 and that of the enclosure be A_2. The overall interchange factor for this case is given by

$$\mathscr{F}_{12} = \frac{1}{1/\varepsilon_1 + (A_1/A_2)[(1/\varepsilon_2) - 1]} \tag{4.6-28}$$

Equation (4.6-28) applies strictly to concentric spheres or concentric cylinders, but it can be used without serious error for other shapes. The case of a gray body surrounded by a black one can be treated as a special case of Eq. (4.6-28) by setting $\varepsilon_2 = 1.0$. Under these conditions $F_{12} = \varepsilon_1$.

For gray surfaces in general the following approximate equation may be used to calculate the overall interchange factor:

$$\mathscr{F}_{12} = \frac{1}{1/\overline{F}_{12} + [(1/\varepsilon_1) - 1] + (A_1/A_2)[(1/\varepsilon_2) - 1]} \tag{4.6-29}$$

where ε_1 and ε_2 are the emissivities of source and sink, respectively. If no refractory is present, $\mathscr{F}$ is used in place of $\overline{F}$.

Gebhart describes a direct method for calculating $\mathscr{F}$ in enclosures of gray surfaces where more than two radiating surfaces are present. Problems involving nongray surfaces are discussed in other literatures.

EXAMPLE 4.10 A chamber for heat-curing large aluminum sheets, lacquered black on both sides, operates by passing the sheets vertically between two steel plates 150 mm apart. One of the plates is at 300°C, and the other, exposed to the atmosphere, is at 25°C. (*a*) What is the temperature of the lacquered sheet? (*b*) What is the heat transferred between the walls when equilibrium has been reached? Neglect convection effects. Emissivity of steel is 0.56; emissivity of lacquered sheets is 1.0.

Solution.

(a) Let subscript 1 refer to hot plate, 2 to lacquered sheets, and 3 to cold plate:
$$\varepsilon_1, \varepsilon_3 = 0.56 \quad \varepsilon_2 = 1.0 \quad T_1 = 573K \quad T_3 = 298K$$

From Eq. (4.6-26)
$$q_{12} = \sigma A_1 \mathscr{F}_{12}(T_1^4 - T_2^4)$$

$$q_{23} = \sigma A_2 \mathscr{F}_{23}(T_2^4 - T_3^4)$$

At equilibrium $q_{12} = q_{23}$. From Eq. (4.6-27)
$$\mathscr{F}_{12} = \frac{1}{1/0.56 + 1/1.0 - 1} = 0.56 = \mathscr{F}_{23}$$

Since $A_1 = A_2$,
$$\left(\frac{T_1}{100}\right)^4 - \left(\frac{T_2}{100}\right)^4 = \left(\frac{T_2}{100}\right)^4 - \left(\frac{T_3}{100}\right)^4$$

$$5.73^4 - \left(\frac{T_2}{100}\right)^4 = \left(\frac{T_2}{100}\right)^4 - 2.98^4$$

$$T_2 = 490.4 \text{ K} = 217.4 ^\circ\text{C}$$

(b) From Eq. (4.6-27) the heat flux is
$$\frac{q_{12}}{A} = 5.672 \times 0.56 \times (5.73^4 - 4.904^4) = 1587 \text{W/m}^2$$

Check:
$$\frac{q_{23}}{A} = 5.672 \times 0.56 \times (4.904^4 - 2.98^4)$$
$$= 1,587 \text{W/m}^2$$

Note: If the lacquered sheet is removed, $q_{13} = 3174 \text{W/m}^2$

4.7 Heat-Exchange Equipment

In industrial processes heat energy is transferred by a variety of methods, including conduction in electric-resistance heaters; conduction-convection in exchangers, boilers, and condensers; radiation in furnaces and radiant heat dryers; and by special methods such as dielectric heating. Often the equipment operates under steady-state conditions, but in many processes it operates cyclically, as in regenerative furnaces and agitated process vessels.

This part deals with equipment types that are of greatest interest to a process engineer: tubular and plate exchangers; extended-surface equipment; mechanically aided heat-transfer devices; condensers and vaporizers; and packed-bed reactors or regenerators. Evaporators are described in next Chapter. Information on all types of heat-exchange equipment is given in engineering texts and handbooks.

General design of heat-exchange equipment

The design and testing of practical heat-exchange equipment are based on the general principles given before. From material and energy balances, the required heat-transfer rate is calculated. Then, using the overall coefficient and the average ΔT, the required heat-transfer area is determined, and in cyclic equipment, the cycle time. In simple devices these quantities can be evaluated easily and with considerable accuracy, but in complex processing units the evaluation may be difficult and subject to considerable uncertainty. The final design is nearly always a compromise, based on engineering judgment, to give the best overall performance in light of the service requirements.

Sometimes the design is governed by considerations that have little to do with heat transfer, such as the space available for the equipment or the pressure drop that can be tolerated in the fluid streams. Tubular exchangers are, in general, designed in accordance with various standards and codes, such as the Standards of the Tubular Exchanger Manufacturers Association (TEMA) and the ASME-API Unfired Pressure Vessel Code*.

In designing an exchanger many decisions—some of them arbitrary—must be made to specify the materials of construction, tube diameter, tube length, baffle spacing, number of passes, and so forth. Compromises must also be made. For example, a high fluid velocity inside small tubes leads to improved heat-transfer coefficients and a small required area, but increases the friction losses and pumping costs. The design of an individual exchanger may be optimized by a formal procedure to balance the heat-transfer area, and hence the price of the equipment and the fixed costs, against the cost of energy to pump the fluids. In processing plants, however, the exchangers are components of a complex network of heat-transfer equipment, and it is the network, not the individual units, that is optimized to give minimum investment and operating costs.

4.7.1 Shell-and-Tube Heat Exchangers

Tubular heat exchangers are so important and so widely used in the process industries that their design has been highly developed. Standards devised and accepted by TEMA are available covering in detail the materials, methods of construction, technique of design, and dimensions for exchangers. The following sections describe the more important types of exchanger and cover the fundamentals of their engineering, design, and operation.

Single-pass 1-1 exchanger

The simple double-pipe exchanger shown in Fig. 4.6 is inadequate for flow rates that cannot readily be handled in a few tubes. If several double pipes are used in parallel, the weight of metal required for the outer tubes becomes so large that the shell-and-tube construction, such as that shown in Fig. 4.25, where one shell serves for many tubes, is more economical. This exchanger, because it has one shell-side pass and one tube-side pass, is a 1-1 exchanger.

In an exchanger the shell-side and tube-side heat-transfer coefficients are of comparable importance, and both must be large if a satisfactory overall coefficient is to be attained. The velocity and turbulence of the shell-side liquid are as important as those of the tube-side fluid. To promote crossflow and raise the average velocity of the shell-side fluid, baffles are installed in the shell. In the construction shown in Fig. 4.25, baffles A consist of circular disks of sheet metal with one side cut away. Common practice is to cut away a segment having a height equal to one-fourth the inside diameter of the shell. Such baffles are called 25 *percent baffles.* The cut edges of the baffles may be horizontal for up-and-down flow or rotated 90° to provide side-to-side flow. The baffles are perforated to receive the tubes. To minimize leakage, the clearances between baffles and shell and tubes should be small. The baffles are supported by one or more guide rods C, which are fastened between the tube sheets D and D' by setscrews. To fix the baffles in place, short sections of tube E are slipped over rod C between the baffles. In assembling such an exchanger, it is necessary to do the tube sheets, support rods, spacers, and baffles first and then to install the tubes.

The stuffing box shown at the right-hand end of Fig. 4.25 provides for expansion. This construction is practicable only for small shells.

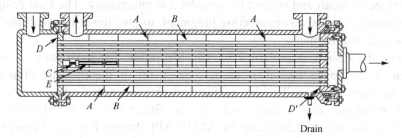

Fig. 4.25 Single-pass 1-1 counterflow heat exchanger: A, baffles; B, tubes; C, guide rods; D, D', tube sheets; E, baffle spacers (only two shown).

Tubes and tube sheets. Tubes are drawn to definite wall thickness, and they are available in all common metals. Tables of dimensions of standard tubes are given in App.4. Standard lengths of tubes for heat-exchanger construction are 8, 12, 16, and 20 ft. Tubes are arranged in a triangular or square layout, known as *triangular pitch* or *square pitch* (pitch is the distance between centers of adjacent tubes). Triangular pitch is used unless the shell side tends to foul badly, because more heat-transfer area can be packed into a shell of given diameter than with square pitch. If the center-to-center distance between tubes is too small, tubes in triangular pitch cannot be cleaned by running a brush between rows, whereas tubes in square pitch are readily cleaned. Also, square pitch gives a lower shell-side pressure drop than triangular pitch.

TEMA standards specify a minimum pitch of 1.25 times the outside diameter of the tubes for triangular pitch and a minimum cleaning lane of ¼ in. for square pitch.

Shell and baffles. Shell diameters are standardized. For shells up to and including 23 in. the diameters are fixed in accordance with American Society for Testing and Materials (ASTM) pipe standards. Standard inside diameters are 8, 10, 12, 13¼, 15¼, 17¼, 19¼, 21¼, and 23¼ in., then 25, 27 in., and so on in 2-in. increments. These shells are constructed of rolled plate.

The distance between baffles (center to center) is the *baffle pitch*, or baffle spacing. It should not be less than one-fifth the diameter of the shell or more than the inside diameter of the shell.

Tubes are usually attached to the tube sheets by grooving the holes circumferentially and rolling the tube ends into the holes by means of a rotating tapered mandrel, which stresses the metal of the tube beyond the elastic limit, so the metal flows into the grooves. In high-pressure exchangers, the tubes are welded or brazed to the tube sheet after rolling.

Alternate designs

Shell-and-tube exchangers with segmented plate baffles may have vibration problems caused by the fluid flowing at high velocity across the tubes. In the *ROD baffle exchanger* developed by Phillips Petroleum Company, metal rods rather than sheet-metal baffles are used to support the tubes, and flow in the shell is mainly parallel to the tube axis. The tubes are arranged in square pitch, and rods with a diameter equal to the clearance between tube rows are attached to ring supports and placed between alternate tubes in both horizontal and vertical directions. The normal rod diameter is ¼ in., and each tube is supported on all four sides at several points along the exchanger, as shown in Fig. 4.26.

Correlations for the outside film coefficient have been developed using the hydraulic diameter for the Reynolds and Nusselt numbers and allowing for the effects of baffle spacing and leakage around the tube bundle. Flow across the rods leads to vortex formation, and the coefficients for turbulent flow are about 1.5 times those predicted for the same Reynolds number using the Dittus-Boelter equation [Eq. (4.4-25)]. The coefficients are not as high as those for a segmentally baffled exchanger with close baffle spacing, but the lower pressure drop and reduced vibration failure make the ROD baffle exchanger preferred for many applications.

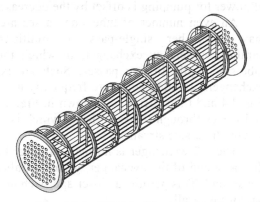

Fig. 4.26 ROD baffle exchanger, cutaway view.

Another design that requires no baffles uses tubes that are twisted into a helical shape with an oval cross section, so that each tube is supported over its entire length by multiple contact points with adjacent tubes. The end sections are kept circular to permit mounting in standard tube sheets. The twisted tubes give improved heat transfer coefficients inside and outside because of greater turbulence, and the decrease in required surface area may more than offset the higher cost per square foot. This design eliminates tube vibration and may also reduce the rate of fouling.

With both the twisted tube and ROD baffle exchangers, flow distribution on the shell side is a problem in large-diameter units. With a single inlet pipe, tubes near the inlet would get more than the average flow, and those opposite the inlet would get little flow for an appreciable distance down the exchanger. Flow distribution is improved by enlarging the shell at the ends of the exchanger to make annular zones where fluid enters or leaves radially at lower velocity.

Multipass exchangers

The 1-1 exchanger has limitations, because when the tube-side flow is divided evenly among all the tubes, the velocity may be quite low, giving a low heat-transfer coefficient. If the number of tubes is reduced and the length increased so that the velocity is sufficiently high, the tube length required may be impractical. Using multipass construction with two, four, or more tube passes permits the use of standard tube lengths while ensuring a high velocity and a high tube-side coefficient. The disadvantages are that (1) the construction of the exchanger is slightly more complicated, (2) some sections in the exchanger have parallel flow, which limits the temperature approach, and (3) the friction loss is greatly increased. For example, the average velocity in the tubes of a four-pass exchanger is 4 times that in a single-pass exchanger having the same number and size of tubes and operated at the same liquid flow rate. The tube-side coefficient of the four-pass exchanger is approximately $4^{0.8} = 3.03$ times that for the single-pass exchanger, or even more if the velocity in the single-pass unit is sufficiently low to give laminar flow. The pressure drop per unit length is $4^{1.8}$ times greater, and the length is increased by 4 times; consequently the total friction loss is $4^{2.8} = 48.5$ times that in the single-pass unit, not including the additional expansion and contraction losses. The most economic design calls for such a velocity in the tubes that the increased cost of power for pumping is offset by the decreased cost of the apparatus.

An even number of tube-side passes are used in multipass exchangers. The shell side may be either single-pass or multipass. A common construction is the 1-2 parallel-counterflow exchanger, in which the shell-side liquid flows in one pass and the tube-side liquid in two passes. Such an exchanger is shown in Fig. 4.27. In multipass exchangers, floating heads are frequently used, and the bulge in the shell of the condenser in Fig. 4.4 and the stuffing box shown in Fig. 4.25 are unnecessary. The tube-side liquid enters and leaves through the same head, which is divided by a baffle to separate the entering and leaving tube-side streams.

The 1-2 exchanger is normally arranged so that the cold fluid and the hot fluid enter at the same end of the exchanger, giving parallel flow in the first tube pass and counterflow in the second. This permits a closer approach at the exit end of the exchanger than if the second pass were parallel.

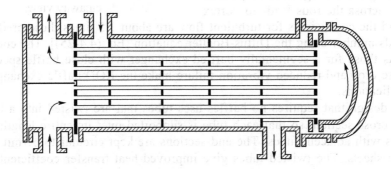

Fig. 4.27 A 1-2 parallel-counterflow exchanger.

2-4 Exchanger

The 1-2 exchanger has an important limitation. Because of the parallel-flow pass, the exchanger is unable to bring the exit temperature of one fluid very near to the entrance temperature of the other. Another way of stating the same limitation is that the heat recovery of a 1-2 exchanger is inherently poor.

A better recovery can be obtained by adding a longitudinal baffle to give two shell passes. A 2-2 exchanger of this kind closely approximates the performance of a double-pipe exchanger, but even with two tube-side passes the total tube length may be insufficient for good heat transfer. More common is the 2-4 exchanger, which has two shell-side and four tube-side passes. This type of exchanger also gives higher velocities and a larger overall heat-transfer coefficient than a 1-2 exchanger having two tube-side passes and operating with the same flow rates. An example of a 2-4 exchanger is shown in Fig. 4.28.

Temperature patterns in multipass exchangers

Temperature-length curves for a 1-2 exchanger are shown in Fig. 4.29(a) using the following temperature designations:

Inlet temperature of hot fluid T_{ha}
Outlet temperature of hot fluid T_{hb}
Inlet temperature of cold fluid T_{ca}
Outlet temperature of cold fluid T_{cb}
Intermediate temperature of cold fluid T_{ci}

Curve T_{ha}-T_{hb} applies to the shell-side fluid, which is assumed to be the hot fluid. Curve

T_{ca}-T_{ci} applies to the first pass of the tube-side liquid, and curve T_{ci}-T_{cb} to the second pass of the tube-side liquid. In Fig. 4.29(a) curves $T_{ha} - T_{hb}$ and $T_{ca} - T_{ci}$ taken together are those of a parallel-flow exchanger, and curves $T_{ha} - T_{hb}$ and $T_{ci} - T_{cb}$ taken together correspond to a countercurrent exchanger. The curves for a 2-4 exchanger are given in Fig. 4.29(b). The dotted lines refer to the shell-side fluid and the solid lines to the tube-side fluid. Again it is assumed that the hotter fluid is in the shell. The hotter pass of the shell-side fluid is in thermal contact with the two hottest tube-side passes and the cooler shell-side pass with the two coolest tube-side passes. The exchanger as a whole approximates a true countercurrent unit more closely than is possible with a 1-2 exchanger.

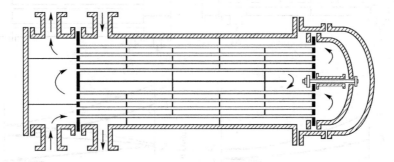

Fig. 4.28 A 2-4 exchanger.

Correction of LMTD in multipass exchangers

The LMTD as given by Eq. (4.3-15) does not apply to the exchanger as a whole, nor to the individual tube passes using ($T_{hb} - T_{ci}$) as one of the driving forces in the LMTD. The reason is that the ΔT for each tube pass is not a linear function of the heat transferred, as was assumed in the derivation of Eq. (4.3-15). It is customary to define a correction factor F_G by which the LMTD must be multiplied to get a correct average driving force. If the overall heat transfer coefficient and the specific heats are constant, and all elements of a given fluid have the same thermal history in passing through the exchanger, F_G can be calculated from the equation.

$$F_G = \frac{(Z^2+1)^{1/2} \ln\left(\dfrac{1-\eta_H}{1-Z\eta_H}\right)}{(Z-1)\ln\left(\dfrac{2-\eta_H(Z+1-(Z^2+1)^{1/2})}{2-\eta_H(Z+1+(Z^2+1)^{1/2})}\right)} \tag{4.7-1}$$

where

$$Z = \frac{T_{ha}-T_{hb}}{T_{cb}-T_{ca}}, \quad \eta_H = \frac{T_{cb}-T_{ca}}{T_{ha}-T_{ca}}$$

The factor Z is the ratio of the true drop in temperature of the hot fluid to the rise in temperature of the cold fluid. It is also equal to the ratio of the flow rates times the heat capacities of the streams, or

$$Z = \frac{\dot{m}_c c_{pc}}{\dot{m}_h c_{ph}} \tag{4.7-2}$$

The factor η_H is the *heating effectiveness,* or the ratio of the actual temperature rise of the cold fluid to the maximum possible rise if the warm end approach, based on countercurrent flow, were zero. The effects of η_H and Z on F_G are shown in Fig. 4.30(a). For $Z = 1.0$ and $\eta_H = 0.5$, $F_G = 0.80$, and F_G drops rapidly with further increase in η_H. When F_G is less than 0.8, the exchanger should be redesigned with two shell passes or larger temperature differences; otherwise the heat transfer surface is inefficiently used.

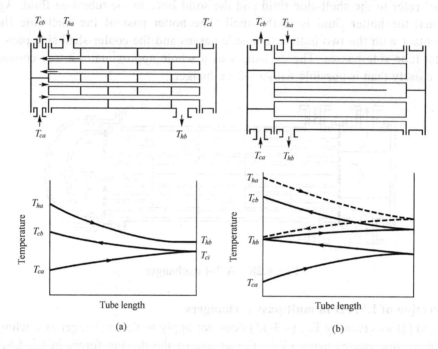

Fig. 4.29 Temperature–length curves: (a) 1-2 exchanger; (b) 2-4 exchanger.

Although Eq. (4.7-1) was derived for two tube passes, the factor F_G is almost the same for one shell pass and four, six, or any even number of tube passes, so Eq. (4.7-1) can be used for these cases. When there are two shell passes and four tube passes, as shown in Fig. 4.30(b), the exchanger operates closer to countercurrent flow, as illustrated by the F_G factors in Fig. Fig. 4.30(b). For example, with $Z = 1.0$ and $\eta_H = 0.5$, $F_G = 0.95$ (compared with $F_G = 0.80$ for the 1-2 exchanger), and for $\eta_H = 0.67$, $F_G = 0.80$. Other combinations of shell-side passes and tube-side passes are sometimes used, and plots for these cases are available, but the 1-2 and 2-4 are the most common arrangements.

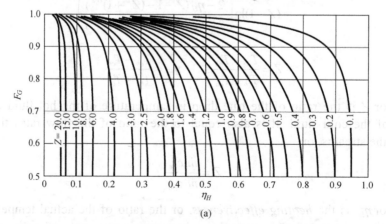

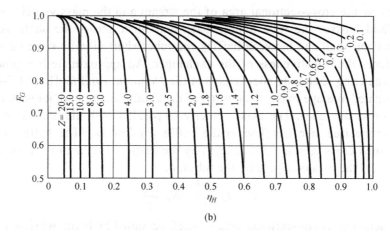

Fig. 4.30 Correction of LMTD: (*a*) 1-2 exchangers; (*b*) 2-4 exchangers.

EXAMPLE 4.11 In the 1-2 exchanger sketched in Fig. 4.29(*a*), the values of the temperatures are $T_{ca} = 70°C$; $T_{cb} = 120°C$; $T_{ha} = 240°C$; $T_{hb} = 120°C$. What is the correct mean temperature drop in this exchanger?

Solution. The correction factor F_G is found from Eq. (4.7-1) or Fig. 4.30(*a*). For this case,

$$\eta_H = \frac{120-70}{240-70} = 0.294 \qquad Z = \frac{240-120}{120-70} = 2.4$$

$$(Z^2+1)^{1/2} = (2.4^2+1)^{1/2} = 2.60$$

$$F_G = \frac{2.6\ln\left(\dfrac{1-0.294}{1-(2.4\times 0.294)}\right)}{1.4\ln\left(\dfrac{2-0.294(3.4-2.6)}{2-0.294(3.4+2.6)}\right)} = 0.807$$

The value from Fig. 4.30(*a*) is $F_G = 0.82 \pm 0.01$. The temperature drops are
At shell inlet: $\Delta T = 240-120 = 120°C$
At shell outlet: $\Delta T = 120-70 = 50°C$

$$\Delta T_m = \frac{120-50}{\ln(120/50)} = 80°C$$

The correct mean is $\Delta T_m = 0.81 \times 80 = 65°C$.

EXAMPLE 4.12 What is the correct mean temperature difference in a 2-4 exchanger operating with the same inlet and outlet temperatures as in the exchanger in Example 4.11?

Solution. For a 2-4 exchanger, when $\eta_H = 0.294$ and $Z = 2.4$, the correction factor from Fig.4.30(*a*) is $F_G = 0.96$. The logarithmic mean temperature difference ΔT_m is the same as in Example 4.11. The correct mean $= 0.96 \times 80 = 77°C$

Heat-transfer coefficients in shell-and-tube exchangers

The heat-transfer coefficient h_i for the tube-side fluid in a shell-and-tube exchanger can be calculated from Eq. (4.4-26) or (4.4-27). The coefficient for the shell-side h_o cannot be so easily calculated because the direction of flow is partly parallel to the tubes and partly across them, and because the cross-sectional area of the stream and the mass velocity of the stream vary as the fluid crosses the tube bundle back and forth across the shell. Also, leakage between baffles and shell and between baffles and tubes short-circuits some of the shell-side liquid and reduces the effectiveness of the exchanger. An approximate but generally useful equation for predicting shell-side coefficients is the *Donohue equation*[Eq. (4.7-6)], which is based on a weighted average mass velocity G_e of the fluid flowing parallel with the tubes and that flowing across the tubes. The mass velocity G_b parallel with the tubes is the mass flow rate divided by the free area for flow in the baffle window S_b. (The baffle window is the portion of the shell cross section not occupied by the baffle.) This area is the total area of the baffle window less the area occupied by the tubes, or

$$S_b = f_b \frac{\pi D_s^2}{4} - N_b \frac{\pi D_o^2}{4} \tag{4.7-3}$$

where f_b = fraction of cross-sectional area of shell occupied by baffle window (0.1955 for 25 percent baffles)
D_s = inside diameter of shell
N_b = number of tubes in baffle window
D_o = outside diameter of tubes

In crossflow the mass velocity passes through a local maximum each time the fluid passes a row of tubes. For correlating purposes the mass velocity G_c for crossflow is based on the area S_c for transverse flow between the tubes in the row at or closest to the centerline of the exchanger. In a large exchanger S_c can be estimated from the equation

$$S_c = PD_s\left(1 - \frac{D_o}{p}\right) \tag{4.7-4}$$

where p = center-to-center distance between tubes
P = baffle pitch

The mass velocities are then

$$G_b = \frac{\dot{m}}{S_b} \quad \text{and} \quad G_c = \frac{\dot{m}}{S_c} \tag{4.7-5}$$

The Donohue equation is

$$\frac{h_o D_o}{k} = 0.2\left(\frac{D_o G_e}{\mu}\right)^{0.6}\left(\frac{c_p \mu}{k}\right)^{0.33}\left(\frac{\mu}{\mu_w}\right)^{0.14} \tag{4.7-6}$$

where $G_e = \sqrt{G_b G_c}$. This equation tends to give conservatively low values of h_o, especially at low Reynolds numbers. More elaborate methods of estimating shell side coefficients are available in other references.

After the individual coefficients are known, the total area required is found in the usual way from the overall coefficient using an equation similar to Eq. (4.3-14). As discussed previously, the LMTD must often be corrected for the departure from true counterflow.

EXAMPLE 4.13 Methane at 1 atm flows through the shell-side of 1-1 pass shell-tube exchanger, leaves the exchanger at 10 m/s, and is cooled from 120℃ to 30℃. The inner diameter of shell is 190mm, and there are 37 steel tubes (ϕ 19×2mm) in the exchanger. What is the individual heat transfer coefficient of methane to the pipe wall?
Solution. The arithmetic average temperature of methane is (120+30)/2=75℃. At this temperature, the physical properties of methane are as follows from App. 8, 12, and 14.

$\mu=1.15\times10^{-5}$ Pa·s, $k=0.0407$ W/(m·℃), $c_p=2.5$ kJ/(kg·℃)

$$\rho=\frac{PM}{RT}=\frac{1.013\times10^5\times16}{8314\times348}=0.56\text{kg/m}^3$$

For the convective heat transfer in the non-circular tube, the equivalent diameter is

$$D_e=\frac{4\times(\frac{\pi}{4}D_i^2-n\frac{\pi}{4}d_o^2)}{\pi(D_i+nd_o)}=\frac{D_i^2-37d_i^2}{D_i+37d_i}=\frac{0.19^2-37\times0.019^2}{0.19+37\times0.019}=0.0255\text{m}$$

Hence,

$$Re=\frac{D_e V\rho}{\mu}=\frac{0.0255\times10\times0.56}{1.15\times10^{-5}}=12420$$

$$Pr=\frac{c_p\mu}{k}=\frac{2.5\times10^3\times1.15\times10^{-5}}{0.0407}=0.707$$

So, the individual heat transfer coefficient for the shell side is

$$h_o=0.023\frac{k}{D_e}Re^{0.8}Pr^{0.3}=0.023\frac{0.0407}{0.0255}\times12420^{0.8}\times0.707^{0.3}=62.4\text{W/(m}^2\cdot\text{℃)}$$

EXAMPLE 4.14 Some oil flows through the shell-side of shell-tube exchanger at the mass flow rate of 30kg/s, and is used to heat the reactant of mass flow rate 38kg/s. Inlet and outlet temperatures of reactant are 25℃ and 60℃, and those of oil are 150℃ and 110℃, respectively. There is a shell-tube exchanger in shock, whose dimensions are: the inside diameter of shell is 0.6m; the shell-side is a single pass while the tube-side is two pass; there are 324 tubes (ϕ 19×2mm, tube length 3m) placed in square pitch arrangement and the tube pitch is 25.4mm. In the shell-side, there is 25% arcuate baffle and the distance between two baffles is 230mm. Check whether this exchanger can meet the above requirement of heat transfer. It is known that the physical properties at the average temperature of fluids are as follows:

fluid	Specific heat, c_p kJ/(kg·℃)	Viscosity, μPa·s	$(\mu/\mu_w)^{0.14}$	Thermal conductivity, k W/(m·℃)
reactant	1.986	0.0029		0.136
oil	2.2	0.0052	0.95	0.119

Soluiton. Calculate the convection heat transfer coefficients h_i and h_o. The tube number for each pass is 324/2=162. The flow area for each pass is

$$162\times0.785\times0.015^2=0.0286\text{m}^2$$

Mass flux of reactant

$$G=\frac{38}{0.0286}=1328\text{kg/(m}^2\cdot\text{s)}$$

$$Re=\frac{dG}{\mu}=\frac{0.015\times1328}{0.0029}=6869$$

Hence, h_i in the tube side is

$$h_i=0.023\frac{k}{D_i}Re^{0.8}Pr^{0.4}=0.023\frac{0.136}{0.15}\times6869^{0.8}\times(\frac{1986\times0.029}{0.136})^{0.4}=1093\text{W/(m}^2\cdot\text{℃)}$$

The flow area of shell side is

$$hD_i(1-\frac{d_o}{t}) = 0.23 \times 0.6 \times (1-\frac{0.019}{0.0254}) = 0.0348 \text{m}^2$$

Equivalent diameter of shell d_e is

$$d_e = \frac{4(t^2 - \pi d_o^2/4)}{\pi d_o} = \frac{4(0.0254^2 - 0.785 \times 0.019^2)}{\pi \times 0.019} = 0.0243 \text{m}$$

$$Re = \frac{0.0243 \times 30}{0.0052 \times 0.0348} = 4021$$

hence, h_o in the shell side is

$$h_o = 0.2 \frac{k}{D_o} Re^{0.6} Pr^{1/3} (\frac{\mu}{\mu_w})^{0.14} = 0.2 \times \frac{0.119}{0.019} \times 4021^{0.6} \times (\frac{2200 \times 0.0052}{0.119})^{1/3} \times 0.95 = 781 \text{W/(m}^2 \cdot \text{°C)}$$

Overall heat transfer coefficient based on the outer surface area of tube is

$$\frac{1}{U_o} = \frac{1}{h_o} + \frac{d_o}{h_i d_i} = \frac{1}{781} + \frac{0.019}{1093 \times 0.015} = 0.00244$$

$$U_o = 410 \text{W/(m}^2 \cdot \text{°C)}$$

The heat transfer area A_o to be required is :

$$A_o = \frac{q}{U_o \Delta T_m}$$

where

$$q = \dot{m}_h c_{ph}(T_{ha} - T_{hb}) = 30 \times 2200 \times (150 - 110) = 2.64 \times 10^6 \text{ W}$$

$$\Delta T_m = \frac{\Delta T_1 + \Delta T_2}{2} = \frac{(150-60)+(150-25)}{2} = 87.5 \text{ °C}$$

Hence

$$A_o = \frac{q}{U_o \Delta T_m} = \frac{2.64 \times 10^6}{410 \times 87.5} = 73.6 \text{m}^2$$

The heat transfer area of exchanger in stock A'_o is

$$A'_o = n\pi d_o L = 324 \times \pi \times 0.019 \times 3 = 58 \text{m}^2$$

Since the heat transfer area of exchanger in stock A'_o is less than the heat transfer area A_o to be required, it is not feasible.

Choice of tube-side fluid

Several factors must be considered in deciding which fluid to put in the tubes and which to put in the shell of a shell-and-tube heat exchanger. If one of the fluids is quite corrosive it should be put in the tubes, which can be made of a corrosion-resistant metal or alloy, rather than in the shell which would require that both the shell and tubes be made of the more expensive material. If corrosion is not a problem but one of the fluids is dirty and likely to form deposits on the wall, that fluid should be inside the tubes since it is much easier to clean the inside of the tubes than the outside. Very hot fluids are placed in the tubes for reasons of safety and heat economy. Finally, the decision might be based on which arrangement gives higher overall heat transfer coefficients or lower pressure drop. Very viscous liquids are often placed on the shell side, because flow across the tubes promotes some turbulence and gives better heat transfer than would laminar flow in the tubes.

Crossflow exchangers

In some exchangers, such as air heaters, the shell is rectangular and the number of tubes in each row is the same. Flow is directly across the tubes, and baffles are not needed. Figure 4.31 shows the factor F_G for crossflow exchangers, derived on the assumption that neither stream mixes with itself during flow through the exchanger. The quantities Z and η_H are given by Eqs. (4.7-1) and (4.7-2), and as before, F_G is so defined that when it is multiplied by the counterflow LMTD, the product is the correct mean temperature drop.

For the shell-side heat-transfer coefficient in a crossflow exchanger, the following equation is recommended.

$$\frac{h_o D_o}{k} = 0.287 \left(\frac{D_o G}{\mu}\right)^{0.61} \left(\frac{c_p \mu}{k}\right)^{0.33} F_a \qquad (4.7\text{-}7)$$

where G is the mass velocity outside the tubes, based on the minimum area for flow in any tube row, and F_a is an "arrangement factor" that depends on Re and the tube spacing p. The other symbols are the same as in Eq. (4.7-6). Typical values of F_a are given in Table 4.3.

Table 4.3   Arrangement factor F_a for crossflow with square pitch

p/D_o	F_a			
	$Re=2000$	$Re=8000$	$Re=20000$	$Re=40000$
1.25	0.85	0.92	1.03	1.02
1.5	0.94	0.90	1.06	1.04
2.0	0.95	0.85	1.05	1.02

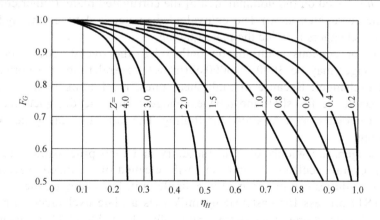

Fig. 4.31   Correction of LMTD for crossflow.

4.7.2   Plate-Type Exchangers

For many applications at moderate temperature and pressure, an alternative to the shell-and-tube exchanger is the gasketed plate exchanger, which consists of many corrugated stainless-steel sheets separated by polymer gaskets and clamped in a steel frame. Inlet portals and slots in the gaskets direct the hot and cold fluid to alternate spaces between the plates. The corrugations induce turbulence for improved heat transfer, and each plate is supported by multiple contacts with adjoining plates, which have a different pattern or angle of corrugation. The space between plates is equal to the depth of the corrugations and is usually 2 to 5mm. A typical plate design is shown in Fig. 4.32.

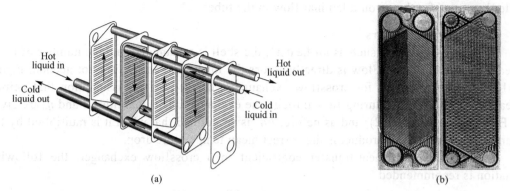

Fig. 4.32   Plate heat exchanger: (*a*) general layout, (*b*) detail of plate design.

For a liquid-liquid exchanger, the usual fluid velocity is 0.2 to 1.0m/s, and because of the small spacing, the Reynolds number is often less than 2100. However, the corrugations give the flow turbulent characteristics at Reynolds numbers of 100 to 400, depending on the plate design. Evidence for turbulent flow is that the heat transfer coefficient varies with the 0.6 to 0.8 power of the flow rate, and the pressure drop depends on the 1.7 to 2.0 power of the flow rate. The heat-transfer correlation for a common plate design is

$$Nu = \frac{hD_e}{k} = 0.37 Re^{0.67} Pr^{0.33} \qquad (4.7\text{-}8)$$

The pressure drop is given by the Fanning equation with the following friction factor:

$$f = 2.5 Re^{-0.3} \qquad (4.7\text{-}9)$$

In Eq. (4.7-8), h is based on the nominal area of the corrugated plate. (Other correlations may be based on the corrugated area.) The equivalent diameter is 4 times the hydraulic radius, which for most exchangers is twice the plate spacing.

With water or aqueous solutions on both sides, the overall coefficient for a clean plate-type exchanger may be 3000 to 6000W/(m²·K), several times the normal value for a shell-and-tube exchanger. Because of high shear rates, the fouling factors experienced are much lower than those for shell-and-tube exchangers, and the designer may just add 10 percent to the calculated area to allow for fouling. The units can easily be taken apart for thorough cleaning.

Plate exchangers are widely used in the dairy and food processing industries because they have high overall coefficients and are easily cleaned or sanitized. Several exchangers with different heat duties can be grouped in a single unit. For example, the high-temperature short-time (HTST) process for pasteurizing milk uses a plate exchanger with three or four sections. In the first or regeneration section, raw milk is heated to 68℃ by exchange with hot pasteurized milk. In the next section, hot water raises the milk temperature to 72℃, the pasteurization temperature. The hot milk is held for at least 15 seconds in an external coil or zigzag holding tube and then returned to the regenerator. In the last section chilled brine rapidly cools the product to 4℃. In some plants the regenerator has two sections, and milk heated to 55℃ is sent to a centrifuge to remove some of the fat before returning to the second part of the regenerator. The pressure of the pasteurized milk is kept higher than the pressure of the other fluids to prevent contamination if a pinhole leak develops.

Compact exchangers have found many other applications in the chemical industry and heat recovery networks. Since nearly ideal countercurrent flow can be achieved, the heating effectiveness may be higher than is possible in multipass shell-and-tube exchangers. New designs and better gaskets permit operation at up to 200℃ and 25atm. Exchangers with plate areas of 2m² and a total area of 1500m² are available.

4.7.3 Extended-Surface Equipment

Difficult heat-exchange problems arise when one of two fluid streams has a much lower heat transfer coefficient than the other. A typical case is heating a fixed gas, such as air, by means of condensing steam. The individual coefficient for the steam is typically 100 to 200 times that for the air stream; consequently, the overall coefficient is essentially equal to the individual coefficient for the air, the capacity of a unit area of heating surface will be low, and many meters of tube will be required to provide reasonable capacity. Other variations of the same problem are found in heating or cooling viscous liquids or in treating a stream of fluid at low flow rate, because of the low rate of heat transfer in laminar flow.

To conserve space and to reduce the cost of the equipment in these cases, certain types of heat exchange surfaces, called *extended surfaces,* have been developed in which the outside area of the tube is multiplied, or extended, by fins, pegs, disks, and other appendages and the outside area in contact with the fluid thereby made much larger than the inside area. The fluid stream having the lower coefficient is brought into contact with the extended surface and flows outside the tubes, while the other fluid, having the high coefficient, flows through the tubes. The quantitative effect of extending the outside surface can be seen from the overall coefficient, written in the following form, in which the resistance of the tube wall is neglected:

$$U_i = \frac{1}{1/h_i + A_i / A_o h_o} \quad (4.7\text{-}10)$$

Equation (4.7-10) shows that if h_o is small and h_i large, the value of U_i will be small; but if the area A_o is made much larger than A_i, the resistance $A_i / A_o h_o$ becomes small and U_i increases just as if h_o were increased, with a corresponding increase in capacity per unit length of tube or unit of inside area.

Types of extended surface

Two common types of extended surfaces are available, examples of which are shown in Fig. 4.33 Longitudinal fins are used when the direction of flow of the fluid is parallel to the axis of the tube; transverse fins are used when the direction of flow of the fluid is across the tubes. Spikes, pins, studs, or spines are also used to extend surfaces, and tubes carrying these can be used for either direction of flow. In all types, it is important that the fins be in tight contact with the tube, both for structural reasons and to ensure good thermal contact between the base of the fin and the wall.

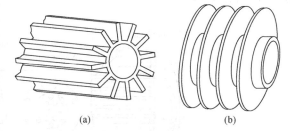

Fig. 4.33 Types of extended surface: (*a*) longitudinal fins; (*b*) transverse fins.

Air-cooled exchangers

As cooling water has become scarcer and pollution controls more stringent, the use of air-cooled exchangers has increased. These consist of bundles of horizontal finned tubes, typically 25 mm in diameter and 2.4 to 9 m long, through which air is circulated by a large fan. Hot process fluids in the tubes, at temperatures from 100 to 400 ℃ or more, can be cooled to about 20 ℃ above the dry-bulb temperature of the air. Heat-transfer areas, based on the outside surface of the tubes, range from 50 to 500 m²; the fins multiply this by a factor of

7 to 20. Air flows between the tubes at velocities of 3 to 6 m/s. The pressure drop and power consumption are low, but sometimes to reduce the fan noise to an acceptable level, the fan speed must be lower than that for maximum efficiency. In air-cooled condensers the tubes are usually inclined. Detailed design procedures are given in the literatures, but the coefficient can be approximated using Eq. (4.7-7) for h_o.

4.7.4 Condensers

As discussed before, heat-transfer devices used to liquefy vapors by removing their latent heat are called *condensers*. The latent heat is removed by absorbing it in a cooler liquid, called the *coolant*. Since the temperature of the coolant obviously is increased in a condenser, the unit also acts as a heater; but functionally it is the condensing action that is important, and the name reflects this fact. Condensers fall into two classes. In the first, called shell-and-tube condensers, the condensing vapor and coolant are separated by a tubular heat-transfer surface. In the second, called contact condensers, the coolant and vapor streams, both of which are usually water, are physically mixed and leave the condenser as a single stream.

Shell-and-tube condensers

The condenser shown in Fig. 4.4 is a single-pass unit, since the entire stream of cooling liquid flows through all the tubes in parallel. In large condensers, this type of flow has a serious limitation. The number of tubes is so large that in single pass flow the velocity through the tubes is too small to yield an adequate heat transfer coefficient, and the unit is uneconomically large. Also, because of the low coefficient, long tubes are needed if the cooling fluid is to be heated through a reasonably large temperature range, and such long tubes are not practicable. To obtain larger velocities, higher heat transfer coefficients, and shorter tubes, the multipass principle used in heat exchangers may also be used for the coolant in a condenser. In a multipass condenser it is not necessary to use the correction factor F_G since the jacket temperature is constant (corresponding to $Z = 0$) and the direction of flow in the tubes has no effect on the driving force. An example of a two-pass condenser is shown in Fig.4.34.

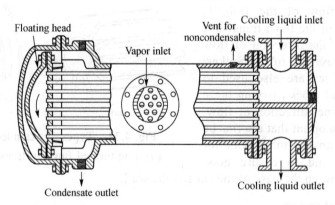

Fig. 4.34 Two-pass floating-head condenser.

Provision for thermal expansion. Because of the differences in temperature existing in condensers, expansion strains may be set up that are sufficiently severe to buckle the tubes or pull them loose from the tube sheets. The most common method of avoiding damage from expansion is the use of the floating-head construction, in which one of the tube sheets (and therefore one end of the tubes) is structurally independent of the shell. This principle is used in the condenser of Fig. 4.34. The figure shows how the tubes may expand or contract, independent of the shell. A perforated plate is set over the vapor inlet to prevent cutting of the tubes by drops of liquid that may be carried by the vapor.

Dehumidifying condensers

A condenser for mixtures of vapors and noncondensable gases is shown in Fig. 4.35. It is set vertically, not horizontally as are most condensers for vapor containing no noncondensable gas; also, vapor is condensed inside the tubes, not outside, and the coolant flows through the shell. This provides a positive sweep of the vapor-gas mixture through the tubes and avoids the formation of any stagnant pockets of inert gas that might blanket the heat-transfer surface. The modified lower head acts to separate the condensate from the uncondensed vapor and gas.

Contact condensers

An example of a contact condenser is shown in Fig. 4.36. Contact condensers are much smaller and cheaper than surface condensers. In the design shown in Fig. 4.36, part of the cooling water is sprayed into the vapor stream near the vapor inlet, and the remainder is directed into a discharge throat to complete the condensation. When a shell-and-tube condenser is operated under vacuum, the condensate is usually pumped out, but it may be removed by a barometric leg. This is a vertical tube about 10 m long, sealed at the bottom by a condensate-receiving tank. In operation the level of liquid in the leg automatically adjusts itself so that the difference in head between levels in leg and tank corresponds to the difference in pressure between the atmosphere and the vapor space in the condenser. Then the liquid flows down the leg as fast as it is condensed, without breaking the vacuum. In a direct-contact condenser the pressure regain downstream from the water jet inlet is often sufficient to eliminate the need for a barometric leg.

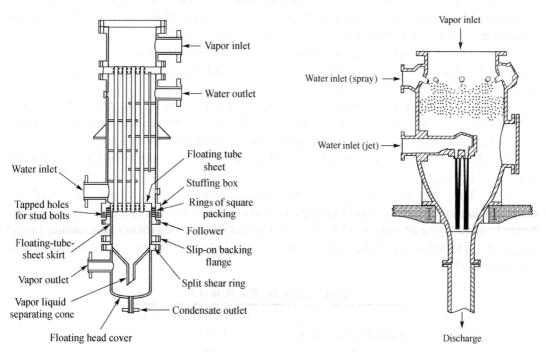

Fig. 4.35 Outside packed dehumidifying cooler-condenser.

Fig. 4.36 Contact condenser.

PROBLEMS

4.1 A layer of pulverized cork 152-mm thick is used as a layer of thermal insulation in a flat wall. The temperature of the cold side of the cork is 4.4℃, and that of the warm side is 82.2℃. The thermal conductivity of the cork at 0℃ is 0.036W/(m·℃), and that at 93.3℃ is 0.055W/(m·℃). The area of the wall is 2.32m². What is the rate of heat flow through the wall in watts?

4.2 A flat furnace wall is constructed of a 114-mm layer of Sil-o-cel brick, with a thermal conductivity of 0.138W/(m·℃) backed by a 229-mm layer of common brick, of conductivity 1.38W/(m·℃). The temperature of the inner face of the wall is 760℃, and that of the outer face is 76.6℃. (a) What is the heat loss through the wall? (b) What is the temperature of the interface between the refractory brick and the common brick? (c) Supposing that the contact between the two brick layers is poor and that a "contact resistance" of 0.088℃·m²/W is present, what would be the heat loss?

4.3 A tube of 60-mm outer diameter (OD) is insulated with a 50-mm layer of silica foam, for which the conductivity is 0.055W/(m·℃), followed with a 40-mm layer of cork with a conductivity of 0.05W/(m·℃). If the temperature of the outer surface of the pipe is 150℃ and the temperature of the outer surface of the cork is 30℃, calculate the heat loss in watts per meter of pipe.

4.4 The vapor pipe (d_o=426mm) is covered by a 426mm insulating layer [k =0.615 W/(m·℃)]. If the temperature of outer surface of pipe is 177℃ and the temperature outside the insulating layer is 38℃, what are the heat loss per meter pipe and the temperature profile within the insulating layer?

4.5 The outer diameter of a steel tube is 150mm. The tube wall is backed by two insulating layers to reduce the heat loss. The ratio of thermal conductivity of two insulating materials is $k_2/k_1 = 2$ and both insulating materials have the same thickness 30mm. If the temperature difference between pipe wall and outer surface of insulating material is constant, which insulating material should be packed inside to enable less heat loss?

4.6 A steel spherical shell has inside radius r_i and outside radius r_o. The temperatures inside and outside walls are t_i and t_o, respectively, and the conductivity is k. Derive the equation for heat transfer by conduction.

4.7 Air at the normal pressure passes through the pipe (d_i 20mm) and is heated from 20℃ to 100℃. What is the film heat transfer coefficient h_i between the air and pipe wall if the average velocity of air is 10 m/s? The properties of air at 60℃ are as follows.

Air: density ρ 1.06kg/m³, viscosity μ 0.02 cP, conductivity k 0.0289W/(m·℃), and heat capacity c_p 1kJ/(kg·K).

4.8 Methyl alcohol flowing in the inner pipe of a double-pipe exchanger is cooled with water flowing in the jacket. The inner pipe is made from 25-mm Schedule 40 steel pipe. The thermal conductivity of steel is 45W/(m·℃). The individual coefficients and fouling factors are given in the following Table. What is the overall coefficient, based on the outside area of the inner pipe?

Table Data for Problem 4.8

	Coefficient, W/(m²·℃)
Alcohol coefficient, h_i	1,020
Water coefficient, h_o	1,700
Inside fouling factor, h_{di}	5,680
Outside fouling factor, h_{do}	2,840

4.9 Benzene with mass flow 1800kg·h^{-1} flows through the annulus of double pipe exchanger. Its temperature changes from 20℃ to 80℃. The dimension of inner tube is ϕ19×2.5mm, and the dimension of outer tube is ϕ38×3mm. What is individual heat transfer coefficient of benzene?

4.10 Air at 2 atm and 20℃ flows through in the pipe of tubular exchanger at 60m^3·h^{-1}, and is heated to 80℃. The dimension of pipe is ϕ57×3.5mm and its length is 3m long. What is individual heat transfer coefficient of air side?

4.11 The dimensional analysis method is used to derive the Dimensionless Number Equation of natural convection heat transfer coefficient h between the wall and liquid.

4.12 A hot fluid with a mass flow rate 2250kg/h passes through a ϕ25×2.5mm tube. The physical properties of fluid are as follows:
k=0.5W/(m·℃), c_p =4kJ/(kg·K), viscosity 10^{-3} N·s/m^2, density 1000kg/m^3
Find:
(a) Heat transfer film coefficient h_i, in W/(m^2·K).
(b) If the flow rate decreases to 1125 kg/h and other conditions are the same, what is the h_i?
(c) If the diameter of tube (inside diameter) decreases to 10 mm, and the velocity u keeps the same as that of case (a), calculate h_i.
(d) When the average temperature of fluid and quantity of heat flow per meter of tube are 40℃ and 400W/m, respectively, what is the average temperature of pipe wall for case (a)?
(e) From this problem, in order to increase the heat transfer film coefficient and enhance heat transfer, what kinds of methods can you use and which is better, explain?
Hint: for laminar flow, Nu=1.86[$Re\ Pr$]$^{1/3}$
    for turbulent flow, Nu=0.023$Re^{0.8}\ Pr^{1/3}$

4.13 The saturated vapor at temperature of 100℃ is condensed to saturated liquid on the surface of a single vertical pipe with length 2.5m and diameter 38mm (outside). The temperature of pipe wall is 92℃. What is the quantity of condensed vapor per hour? If the pipe is placed horizontally, what is the quantity of condensed vapor per hour?

4.14 In a double pipe exchange (ϕ23×2mm), the cold fluid [c_p=1kJ/(kg·K), flow rate 500 kg/h] passes through the pipe and the hot fluid goes through the annular. The inlet and outlet temperatures of cold fluid are 20℃ and 80℃, and the inlet and outlet temperatures of hot fluid are 150℃ and 90℃, respectively. The h_i (heat transfer coefficient inside a pipe) is 700W/(m^2·℃) and overall heat transfer coefficient U_o (based on the outside surface of pipe) is 300W/(m^2·℃), respectively. If the heat loss is ignored and the conductivity of pipe wall (steel) is taken as 45W/(m·℃), find:
(a) Heat transfer coefficient outside the pipe h_o?
(b) The pipe length required for counter flow, in m?
(c) What is the pipe length required if the heating medium changes to saturated vapor (140℃) which condenses to saturated liquid and other conditions keep unchanged?
(d) When the exchanger is used for a year, it is found that it cannot meet the need of production (the outlet temperature of cold fluid cannot reach 80℃), explain why?

4.15 Water flows turbulently in the pipe of ϕ25×2.5mm shell tube exchanger. When the velocity of water u is 1m/s, overall heat transfer coefficient U_o (based on the outer surface area of pipe) is 2115W/(m^2·℃). If the u becomes 1.5m/s and other conditions keep unchanged, U_o is 2660W/(m^2·℃). What is the film coefficient h_o outside the pipe? (Heat resistances of pipe wall and scale are ignored)

4.16 The vapor at saturation temperature t_s 100℃ is condensed on the surface of a single vertical pipe with pipe length 2m and outer diameter 0.04m. The outside surface temperature of pipe t_w is 94℃. Calculate the heat transfer coefficient for vapor condensation and the quantity of the condensed vapor per hour. If the pipe is placed horizontally, what are the heat transfer coefficient for vapor condensation and the quantity of the condensed vapor per hour?

4.17 Water and oil pass parallelly through an exchanger which is 1 m long. The inlet and outlet temperatures of water are 15℃ and 40℃, and those of oil are 150℃ and 100℃, respectively. If the outlet temperature of oil decreases to 80℃, and the flow rates and physical properties and inlet temperatures of water and oil maintain the same, what is the pipe length of new exchanger? (Heat loss and pipe wall resistance are neglected)

4.18 Air which passes through the pipe in turbulent flow is heated from 20℃ to 80℃. The saturated vapor at 116.3℃ condenses to saturated water outside the pipe. If air flow rate increases to 120% of the origin and inlet and outlet temperatures of air stay constant, what kind of method can you employ in order to do that? (Heat resistance of pipe wall and scale can be ignored.)

4.19 Water flows through the pipe of a $\phi 25 \times 2.5$mm shell-tube exchanger from 20℃ to 50℃. The hot fluid (c_p 1.9kJ/(kg·℃), flow rate 1.25kg/s) goes counter currently along the shell and the temperatures change from 80℃ to 30℃. Heat transfer coefficients of water and hot fluid are 0.85kW/(m²·℃) and 1.7kW/(m²·℃). What is the overall heat transfer coefficient U_o (based on outer surface area of tube) and heat transfer area if the scale resistance can be ignored? (The conductivity of steel is 45W/(m·℃)).

4.20 Heated oil (heat capacity c_p=3.35kJ/(kg·℃)) is cooled by water (heat capacity c_p=4.187kJ/(kg·℃)) and they flow countercurrently through a double pipe exchanger with inner pipe dimension $\phi 180 \times 10$mm. Water at 15℃ of inlet temperature goes through the pipe and leaves at 55℃. Oil passes through the annulus with mass flow rate 500kg/h and inlet and outlet temperatures of oil are 90℃ and 40℃, respectively. The heat transfer coefficients h_i and h_o for water and oil are 1000W/(m²·℃) and 299W/(m²·℃), and heat resistances of pipe wall and fouling as well as heat loss can be ignored. Find:
(a) Water flow rate, in kg/h.
(b) Overall heat transfer coefficient U_o based on outer surface of pipe.
(c) LMTD Δt_m and pipe length of pipe, in m.
(d) If inlet temperature of water becomes 20℃ and oil inlet temperature keeps the same, what happens to that?
(e) In order to enhance heat transfer, what ways can be employed?

4.21 A single pass (1-1) shell-tube exchanger is made of many $\phi 25 \times 2.5$ mm tubes. Organic solution with u=0.5m/s, mass flow rate=15000kg/h, c_p=1.76kJ/(kg·℃), ρ=858kg/m³, passes through the tube, its temperature changes from 20℃ to 50℃. The saturated vapor at 130℃ condenses to the saturated water, which goes through the shell. The individual heat transfer coefficients h_i and h_o in the pipe and shell are 700W/(m²·℃) and 1.0×10^4W/(m²·℃), respectively. The thermal conductivity k of pipe wall is 45W/(m·℃). If the heat loss and resistances of fouling can be ignored, find:
(a) Overall heat transfer coefficient U_o (based on outside tube area).
(b) Number of pipes and total length of pipes.
(c) If the inlet temperature of organic solution keeps the same, what can you do in order to increase the heat transfer rate of the exchanger? Explain simply?

4.22 There are two same shell-tube exchangers (1-1 pass) in stock. Each one consists of 64 tubes with dimension $\phi 27 \times 3.5$mm and heat transfer area 20m². The saturated vapor with condensation latent heat 2054kJ/kg at 170℃ goes through the shell and condenses to saturated liquid. Air at 30℃ passes fully turbulently through the tubes with mass flow rate 2.5kg/s.

(a) If two exchangers are installed parallelly, mass flow rate of air (c_p= 1kJ/(kg·K)) is 1.25kg/s for each exchanger and heat transfer coefficient of air is 38W/(m²·K). If the flow pattern of air is fully turbulent, what are outlet temperature of air and vapor flow rate?
(b) If two exchangers work in series with air mass flow rate 2.5kg/s, what are outlet temperature of air and vapor flow rate now? The heat resistances of pipe wall and vapor and the heat loss are ignored.

    4.23 The saturated vapor at 120℃ condenses on the outside surface of a double-pipe exchanger. The air goes fully turbulently through the tube. Vapor-side and air-side coefficients are 10000W/(m²·K)and 50W/(m²·K), inlet and outlet temperatures of air are 30℃ and 80℃, respectively. The heat resistance of pipe wall, heat loss, and air property change induced due to temperature variation can be ignored. In order to enhance heat transfer, two following ways are proposed.
(a) The enhanced surface (tube) will be employed but the heat transfer area keeps the same, so the heat transfer coefficient h_i in air side will be double. If the inlet temperature and mass flow rate of air, and saturated vapor temperature do not change, find:
(1) Outlet temperature of air.
(2) What is ratio of vapor flow rate to origin. $\dot{m}'/\dot{m}$ =? ($\dot{m}'$: vapor flow rate in a new exchanger, $\dot{m}$: vapor flow rate in an old exchanger)
(b) If mass flow rate of air is double, what are the outlet temperature of air and $\dot{m}'/\dot{m}$ =?
(c) Discuss and state simply the same and different points for two enhanced methods.
Hint: the heat resistance of vapor side can be neglected.

    4.24 Two parallel flat plates 5 mm apart are exposed to air. Emissivity and temperature of the first plate are 0.1 and 350K, while those of the second one are 0.05 and 300K, respectively. If the first plate is coated with some material and its emissivity changes from 0.1 to 0.025. What is the variation percentage of heat transfer rate? Suppose the convection heat transfer can be neglected.

    4.25 Hot air flows through the steel pipe(ϕ 426×9mm), in which the thermocouple is set up to measure the temperature of air. In order to reduce the reading error, a tube is used to shelter the thermocouple. The emissivity of the tube is 0.3 and its area is 90 times as much as that of contact point of the thermocouple. The temperature of pipe wall measured is 110℃ and the reading of the thermocouple is 220 ℃. If the convective heat transfer coefficient between air and the tube is 10W/(m²·℃), and the convective heat transfer coefficient between air and thermocouple contact point is 12W/(m²·℃).The emissivity of thermocouple joint is 0.8. Calculate:(a) The real temperature of air t_a; (b) The temperature of sheltering tube t_i; (c) The reading error of thermocouple.

名 词 术 语

英文 / 中文	英文 / 中文
absorptivity / 吸收率	boiling point / 沸点
air-cooled exchanger / 空冷器	boiling temperature / 沸腾温度
arithmetic means / 算术平均	condensate / 冷凝液
baffle / 挡板	condensate loading / 冷凝液负荷
baffle pitch / 挡板间距	condensation / 冷凝
barometric leg / 大气腿	condenser / 冷凝器
black body / 黑体	conduction / 传导
boiling / 沸腾	conductor / 导体

英文 / 中文	英文 / 中文
contact condenser / 接触式冷凝器	natural (free) convection / 自然对流
controlling heat resistance / 控制热阻	Newton's law of cooling / 牛顿冷却定律
convection / 对流	noncondensed gas / 不凝性气体
countercurrent flow / 逆流	nucleate boiling / 核沸腾
critical temperature drop / 临界温度降	numerical integration / 数值积分
crossflow / 错流	Nusselt number / 努塞尔数
dehumidifying condenser / 除湿冷凝器	one-dimensional heat transfer / 一维传热
dimensional analysis / 无量纲分析	opaque solid / 不透射体
dimensionless geometric factor / 无量纲几何因子	overall heat-transfer coefficient / 总传热系数
	overall temperature drop / 总温降
double-pipe heat exchanger / 套管换热器	parallel flow / 并流
dropwise condensation / 滴状冷凝	partial pressure / 分压
electromagnetic radiation / 电磁辐射	phase change / 相变
emission / 发射	plate heat exchanger / 板式换热器
emissive power / 发射能力	pool boiling / 池沸腾
emissivity / 发射率	Prandtl number / 普朗特数
enthalpy balance / 焓衡算(平衡)	provision for thermal expansion / 热膨胀补偿
evaporator / 蒸发器	quantitative calculation / 定量计算
expansion joint / 膨胀节	radiation / 辐射
extended-surface / 扩展表面	radiation heat transfer / 辐射传热
film boiling / 膜沸腾	rate of heat flow / 热流率
film-type condensation / 膜状冷凝	reflectivity / 反射率
fin tube exchanger / 翅片管换热器	roughness / 粗糙度
fixed-sheet exchanger / 固定板式换热器	saturated liquid / 饱和液体
floating-head exchanger / 浮头式换热器	sensible heat / 显热
forced convection / 强制对流	shell-and-tube exchanger / 列管换热器
fouling factor / 污垢因子	single-pass 1-1 exchanger / 单管程单壳程换热器
Fourier's law / 傅立叶定律	
Grashof number / 格拉晓夫数	single-pass / 单程
gray body / 灰体	specific heat / 比热
guide rod / 定距杆	steady-state conduction / 稳态热传导
heat flow / 热流	Stefan-boltzmann constant / 斯蒂芬波尔茨曼常数
heat flux / 热通量	
heat loss / 热损失	subcooled liquid / 过冷液体
heat transfer equipment / 换热设备	subheated liquid / 过热液体
heat transfer / 传热	temperature difference / 温差
heating effectiveness / 热效率	temperature gradient / 温度梯度
heat-transfer coefficient / 传热系数	thermal boundary layer / 热边界层
individual heat-transfer coefficient / 传热分(膜)系数	thermal conductivity / 导热系数
	thermal radiation / 热辐射
insulator / 绝缘体	thermal resistance / 热阻
Kirchhoff's law / 柯斯克夫定律	transition boiling / 过渡沸腾
latent heat of vaporization / 汽化潜热	transmissivity / 透射率
logarithmic mean temperature difference (LMTD) / 对数平均温度差	tube pitch / 管间距
	tube sheet / 管板
logarithmic mean / 对数平均	wavelength / 波长
multipass / 多程	white body / 白体

Chapter 5

Evaporation

5.1 Summarize

Evaporation is one of the oldest unit operations, it is also an area in which much has changed in the last quarter century. Heat transfer to a boiling liquid has been discussed generally in Chap.4. Especial case occurs so often that it is considered an individual operation. It is called evaporation and is the subject of this chapter.

The objective of evaporation is to concentrate a solution consisting of a nonvolatile solute and a volatile solvent. In the overwhelming majority of evaporations the solvent is water. Evaporation is conducted by vaporizing a portion of the solvent to produce a concentrated solution of thick liquor. Evaporation differs from drying in that the residue is a liquid—sometimes a highly viscous one—rather than a solid; it differs from distillation in that the vapor usually is a single component, and even when the vapor is a mixture, no attempt is made in the evaporation step to separate the vapor into fractions; it differs from crystallization in that emphasis is placed on concentrating a solution rather than forming and building crystals. In certain situations, for example, in the evaporation of brine to produce common salt, the line between evaporation and crystallization is far from sharp. Evaporation sometimes produces a slurry of crystals in a saturated mother liquor.

Normally, in evaporation the thick liquor is the valuable product, and the vapor is condensed and discarded. In one specific situation, however, the reverse is true. Mineral-bearing water often is evaporated to give a solid-free product for boiler feed, for special process requirements, or for human consumption. This technique is often called water distillation, but technically it is evaporation. Large-scale evaporation processes have been developed and used for recovering potable water from seawater. Here the condensed water is the desired product. Only a fraction of the total water in the feed is recovered, and the remainder is returned to the sea.

5.1.1 Single-and Multiple-Effect Operation

Most evaporators are heated by steam condensing on metal tubes. Except in some special horizontal-tube evaporators, the material to be evaporated flows inside the tubes. Usually the steam is at a low pressure, below 3 atm abs; often the boiling liquid is under moderate vacuum, at pressures down to about 0.05 atm abs. Reducing the boiling temperature of the liquid increases the temperature difference between the steam and the boiling liquid and thus increases the heat-transfer rate in the evaporator.

As shown in Fig. 5.1, when a single evaporator is used, the vapor from the boiling liquid is condensed and discarded. This method is called *single-effect evaporation*, and although it is simple, it utilizes steam ineffectively. To evaporate 1 kg of water from a solution calls for from 1 to 1.3 kg of steam. If the vapor from one evaporator is fed into the steam chest of a second evaporator and the vapor from the second is then sent to a condenser, the operation becomes double-effect. The heat in the original steam reused in the second effect, and the evaporation achieved by a unit mass of steam fed to the first effect is approximately doubled. Additional effects can be added in the same manner. The general method of increasing the evaporation per unit kg of steam by using a series of evaporators between the steam supply and the condenser is called *multiple-effect evaporation* which as shown in Fig. 5.2.

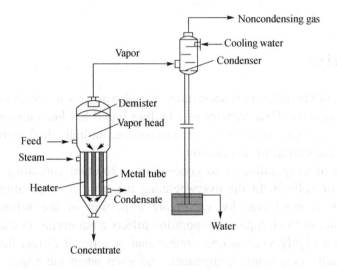

Fig. 5.1 Simplified diagram of single-effect evaporation.

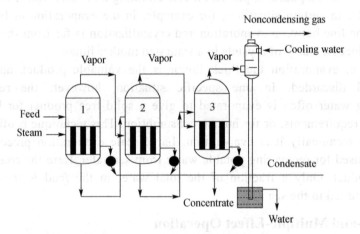

Fig. 5.2 Simplified diagram of multiple-effect evaporation.

5.1.2 Vacuum Evaporation

For the evaporation of liquids that are adversely affected by high temperatures, it may be necessary to reduce the temperature of boiling by operating under reduced pressure. There are relationship between vapor pressure and boiling temperature, for water. When the vapor pressure of the liquid reaches the pressure of its surroundings, the liquid boils. The reduced pressures required to boil the liquor at lower temperatures are obtained by mechanical or steam jet ejector vacuum pumps, combined generally with condensers for the vapors from the evaporator. Mechanical vacuum pumps are generally cheaper in running costs but more expensive in terms of capital than are steam jet ejectors. The condensed liquid can either be pumped from the system or discharged through a tall barometric column in which a static column of liquid balances the atmospheric pressure. Vacuum pumps are then left to deal with the non-condensibles, which of course are much less in volume but still have to be discharged to the atmosphere.

5.1.3 Liquid Characteristics

The practical solution of an evaporation problem is profoundly affected by the character of the liquor to be concentrated. It is the wide variation in liquor characteristics (which demands judgment and experience in designing and operating evaporators) that broadens this operation from simple heat transfer to a separate art. Some of the most important properties of evaporating liquids are as follows.

Concentration. Although the thin liquor fed to an evaporator may be sufficiently dilute to have many of the physical properties of water, as the concentration increases, the solution becomes more and more individualistic. The density and viscosity increase with solid content until either the solution becomes saturated or the liquor becomes too viscous for adequate heat transfer. Continued boiling of a saturated solution causes crystals to form; these must be removed or the tubes clog. The boiling point of the solution may also rise considerably as the solid content increases, so that the boiling temperature of a concentrated solution may be much higher than that of water at the same pressure.

Foaming. Some materials, especially organic substances, foam during vaporization. A stable foam accompanies the vapor out of the evaporator, causing heavy entrainment.

Temperature sensitivity. Many fine chemicals, pharmaceutical products, and foods are damaged when heated to moderate temperatures for relatively short times. In concentrating such materials special techniques are needed to reduce both the temperature of the liquid and the time of heating.

Scale. Some solutions deposit scale on the heating surface. The overall coefficient then steadily diminishes, until the evaporator must be shut down and the tubes cleaned.

Materials of construction. Whenever possible, evaporators are made of some kind of steel. Many solutions, however, attack ferrous metals or are contaminated by them. Special materials such as copper, nickel, stainless steel, aluminum, impervious graphite, and lead are then used. Since these materials are expensive, high heat-transfer rates become especially desirable to minimize the first cost of the equipment.

Many other liquid characteristics must be considered by the designer of an evaporator. Some of these are the specific heat, heat of concentration, freezing point, gas liberation on boiling, toxicity, explosion hazards, radioactivity, and necessity for sterile operation. Because of the variation in liquor properties, many different evaporator designs have been developed. The choice for any specific problem depends primarily on the characteristics of the liquid.

5.2 Evaporation Equipment

The evaporator has two principal functions, to exchange heat and to separate the vapor that is formed from the liquid. In evaporation, heat is added to a solution to vaporize the solvent which is usually water. The heat is generally provided by the condensation of a vapor such as steam on one side of a metal surface with the evaporating liquid on the other side. The type of equipment used depends primarily on the configuration of the heat-transfer surface and on the means employed to provide agitation or circulation of the liquid.

The typical evaporator is made up of three functional sections: the heat exchanger, the evaporating section, where the liquid boils and evaporates, and the separator in which the vapor leaves the liquid and passes off to the condenser or to other equipment. Considered as a piece of process plant, the evaporators may be operated either as *circulation units* or as *once-through units*. These general types are discussed below.

5.2.1 Circulation Evaporators

In circulation evaporators a pool of liquid is held within the equipment. Incoming feed mixes with the liquid from the pool, and the mixture passes through the tubes. Unevaporated liquid discharged from the tubes returns to the pool, so that only part of the total evaporation occurs in one pass. All forced-circulation evaporators are operated in this way; climbing-film evaporators are usually circulation units.

The thick liquor from a circulation evaporator is withdrawn from the pool. All the liquor in the pool must therefore be at the maximum concentration. Since the liquid entering the tubes may contain several parts of thick liquor for each part of feed, its viscosity is high and the heat-transfer coefficient tends to be low.

Circulation evaporators are not well suited to concentrating heat-sensitive liquids. With a reasonably good vacuum the temperature of the bulk of the liquid may be nondestructive, but the liquid is repeatedly exposed to contact with hot tubes. Some of the liquid, therefore, may be heated to an excessively high temperature. Although the average residence time of the liquid in the heating zone may be short, part of the liquid is retained in the evaporator for a considerable time. Prolonged heating of even a small part of a heat-sensitive material such as a food can ruin the entire product.

Circulation evaporators, however, can operate over a wide range of concentration between feed and thick liquor in a single unit, and are well adapted to single effect evaporation. They may operate either with *natural circulation*, with the flow through the tubes induced by density differences, or with *forced circulation*, with flow provided by a pump.

Vertical type evaporator

By using vertical, rather than horizontal tubes, the natural circulation of the heated liquid can be made to give good heat transfer. The standard vertical natural circulation evaporator, shown in Fig. 5.3 is an example of this type. Recirculation of the liquid is through a large "downcomer" so that the liquors rise through the vertical tubes about 5-8 cm diameter, boil in the space just above the upper tube plate and recirculate through the downcomers. The hydrostatic head reduces boiling on the lower tubes, which are covered by the circulating liquid. The length to diameter ratio of the tubes is of the order of 15:1. This natural circulation increases the heat transfer coefficient. It is not used with viscous liquids. A variation of this is the basket type, which shown in Fig. 5.4 is a variant of the calandria evaporator in which the steam chest is contained in a basket suspended in the lower part of the evaporator, and recirculation occurs through the annular space round the basket. This type is widely used in the sugar, salt, and caustic soda industries.

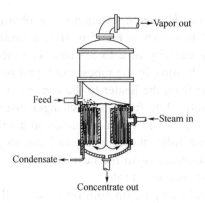
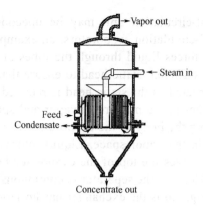

Fig. 5.3   Vertical type evaporator.   Fig. 5.4   Basket type evaporator.

Long-tube vertical type evaporator

A typical long-tube vertical evaporator with upward flow of the liquid is shown in Fig. 5.5. The essential parts are (1) a tubular exchanger with steam in the shell and liquid to be concentrated in the tubes, (2) a separator or vapor space for removing entrained liquid from the vapor, and (3) when operated as a circulation unit, a return leg for the liquid from the separator to the bottom of the exchanger. Since the heat transfer coefficient on the steam side is very high compared to that on the evaporating liquid side, high liquid velocities are desirable. The tubes are 3 to 10 m long and the formation of vapor bubbles inside the tubes causes a pumping action giving quite high liquid velocities. Generally, the liquid passes through the tubes only once and is not recirculated. Contact times can be quite low in this type. Long-tube vertical evaporators are especially effective in concentrating liquids that tend to foam. Foam is broken when the high-velocity mixture of liquid and vapor impinges against the vapor-head baffle.

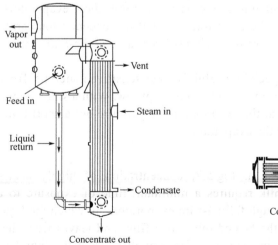
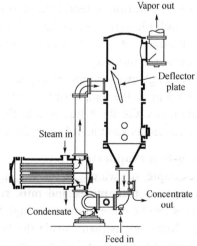

Fig. 5.5   Long-tube vertical evaporator.   Fig. 5.6   Forced-circulation evaporator.

Forced-circulation evaporator

In a natural-circulation evaporator the liquid enters the tubes at 0.3 to 1.2 m/s. The linear velocity increases greatly as vapor is formed in the tubes, so that in general the rates of heat transfer are satisfactory. With viscous liquids, however, the overall coefficient in a natural-circulation unit may be uneconomically low. Higher coefficients are obtained in forced-circulation evaporators, an example of which is shown in Fig. 5.6. Here a centrifugal pump forces liquid through the tubes at an entering velocity of 2 to 5.5 m/s. The tubes are under sufficient static head to ensure that there is no boiling in the tubes; the liquid becomes superheated as the static head is reduced during flow from the heater to the vapor space, and it flashes into a mixture of vapor and spray in the outlet line from the exchanger just before entering the body of the evaporator. The mixture of liquid and vapor impinges on a deflector plate in the vapor space. Liquid returns to the pump inlet, where it meets incoming feed; vapor leaves the top of the evaporator body to a condenser or to the next effect. Part of the liquid leaving the separator is continuously withdrawn as concentrate.

Fig. 5.6 is the exchanger has horizontal tubes and two-pass on both tube and shell sides. In others, vertical single-pass exchangers are used. In both types the heat-transfer coefficients are high, especially with thin liquids, but the greatest improvement over natural-circulation evaporation is seen with viscous liquids. With thin liquids the improvement with forced circulation does not warrant the added pumping costs over natural circulation; but with viscous material the added costs are justified, especially when expensive metals must be used. An example is caustic soda concentration, which must be done in nickel equipment. In multiple-effect evaporators producing a viscous final concentrate, the first effects may be natural-circulation units and the later ones, handling viscous liquid, forced-circulation units. Because of the high velocities in a forced-circulation evaporator, the residence time of the liquid in the tubes is short—about 1 to 3s—so that moderately heat-sensitive liquids can be concentrated in them. They are also effective in evaporating salting liquors or those that tend to foam.

5.2.2 Once-Through Evaporators

In once-through operation, the feed liquor passes through the tubes only once, releases the vapor, and leaves the unit as thick liquor. All the evaporation is accomplished in a single pass. The ratio of evaporation to feed is limited in single-pass units; thus these evaporators are well adapted to multiple-effect operation, where the total amount of concentration can be spread over several effects. *Agitated-film evaporators, falling-film* and *climbing-film evaporators* are always operated in this way.

Once-through evaporators are especially useful for heat-sensitive materials. By operating under high vacuum, the temperature of the liquid can be kept low. With a single rapid passage through the tubes, the thick liquor is at the evaporation temperature but a short time and can be quickly cooled as soon as it leaves the evaporator.

Falling-film evaporator

An example of which is shown in Fig.5.7, concentration of highly heat-sensitive materials such as fruit juices and milk requires a minimum time of exposure to a heated surface. This can be done in once-through falling-film evaporators, in which the liquid enters at the top, flows downstream inside the heated tubes as a film, and leaves from the bottom. The tubes are large—50 to 250 mm in diameter. Vapor evolved from the liquid is usually carried downward with the liquid and leaves from the bottom of the unit. In appearance these evaporators resemble long, vertical, tubular exchangers with a liquid-vapor separator at the bottom and a distributor for the liquid at the top.

The chief problem in a falling-film evaporator is that of distributing the liquid uniformly as a film inside the tubes. This is done by a set of perforated metal plates above a carefully leveled tube sheet, by inserts in the tube ends to cause the liquid to flow evenly into each tube, or by "spider" distributors with radial arms from which the feed is sprayed at a steady rate on the inside surface of each tube. Still another way is to use an individual spray nozzle inside each tube.

When recirculation is allowable without damaging the liquid, distribution of liquid to the tubes is facilitated by a moderate recycling of liquid to the tops of the tubes. This provides a larger volume of flow through the tubes than is possible in once-through operation.

During evaporation the amount of liquid is continuously reduced as it flows downward, and too great a reduction can lead to dry spots near the bottom of the tube. Thus the amount of concentration that can be done in a single pass is limited.

Falling-film evaporators, with no recirculation and short residence times, handle sensitive products that can be concentrated in no other way. They are also well adapted to concentrating viscous liquids.

Climbing film evaporator

As shown in Fig. 5.8, the climbing film evaporator is designed to operate as either a batch or continuous evaporator. It can be operated under free or forced circulation and demonstrates the basic principle employed by an evaporator of this kind. The apparatus consists of a steam heated, long tube Calandria evaporator. Feed material enters at the base of the Calandria and is partially evaporated. Vapor and liquid pass into the cyclone where the phases are separated, the vapor passing through a water-cooled condenser and into a receiver while the liquid passes down into the concentrate receiver.

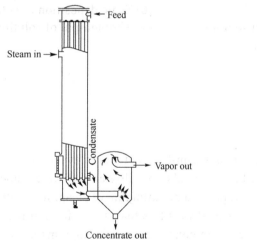

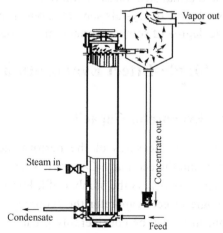

Fig. 5.7 Falling-film evaporator.   **Fig. 5.8** Climbing film evaporator.

Agitated-film evaporator

The principal resistance to overall heat transfer from the steam to the boiling liquid in an evaporator is on the liquid side. One way of reducing this resistance, especially with viscous liquids, is by mechanical agitation of the liquid film, as in the evaporator shown in Fig. 5.9. This is a modified falling-film evaporator with a single jacketed tube containing an internal agitator. Feed enters at the top of the jacketed section and is spread out into a thin, highly turbulent film by the vertical blades of the agitator. Concentrate leaves from the bottom of the jacketed section; vapor rises from the vaporizing zone into an unjacketed separator, which is somewhat larger in diameter than the evaporating tube. In the separator the agitator blades throw entrained liquid outward against stationary vertical plates. The droplets coalesce on

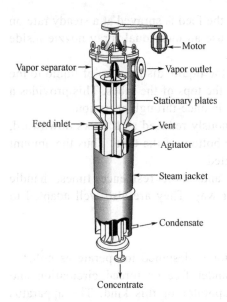

Fig. 5.9 Agitated-film evaporator.

these plates and return to the evaporating section. Liquid-free vapor escapes through outlets at the top of the unit.

The chief advantage of an agitated-film evaporator is its ability to give high rates of heat transfer with viscous liquids. The product may have a viscosity as high as 1000 P at the evaporation temperature. With highly viscous materials the coefficient is appreciably greater than in forced circulation evaporators and much greater than in natural-circulation units. The agitated-film evaporator is particularly effective with such viscous heat-sensitive products as gelatin, rubber latex, antibiotics, and fruit juices. Its disadvantages are the high cost; the internal moving parts, which may need considerable maintenance; and the small capacity of single units, which is far below that of multitubular evaporators.

5.2.3 Choice of the Evaporators

(1) Viscosity of solution
(2) The thermal steadiness of solution
(3) The degree of crystals separated out from solution
(4) The degree of some materials form a foam
(5) Corrosively of solution
(6) Scale deposition of solution
(7) Through-put of solution

5.3 Single-Effect Evaporation

5.3.1 Evaporator Capacity

The principal measures of the performance of a steam-heated tubular evaporator are the capacity and the economy. *Capacity* is defined as the number of kilograms of water vaporized per hour. *Economy* is the number of kilograms vaporized per kilogram of steam fed to the unit. In a single-effect evaporator the economy is nearly always less than 1, but in multiple-effect equipment it may be considerably greater. The steam consumption, in kilograms per hour, is also important. It equals the capacity divided by the economy.

The rate of heat transfer is the product of three factors: the area of the heat-transfer surface, the overall heat-transfer coefficient and the overall temperature drop.

If the feed to the evaporator is at the boiling temperature corresponding to the absolute pressure in the vapor space, almost all the heat transferred through the heating surface is available for evaporation, and the capacity is nearly proportional to the rate of heat transfer. As evaporation proceeds the boiling point rises and some sensible heat is required, but the gain in sensible heat is usually small in comparison with the heat of vaporization. If the feed is cold, the heat required to heat it to its boiling point may be quite large and the capacity for a given value of the rate of heat transfer is reduced accordingly, as heat used to heat the feed is not available for evaporation. Conversely, if the feed is at a temperature above the boiling point in the vapor space, a portion of the feed evaporates spontaneously by adiabatic equilibration with the vapor-space pressure and the capacity is greater than that corresponding to the rate of heat transfer. This process is called *flash evaporation*.

The actual temperature drop across the heating surface depends on the solution being evaporated, the difference in pressure between the steam chest and the vapor space above the boiling liquid, and the depth of liquid over the heating surface. In some evaporators the velocity of the liquid in the tubes also influences the temperature drop because the frictional loss in the tubes increases the effective pressure of the liquid. When the solution has the characteristics of pure water, its boiling point can be read from steam tables if the pressure is known, as can the temperature of the condensing steam. In actual evaporators, however, the boiling point of a solution is affected by two factors, *boiling point elevation and liquid head*.

5.3.2 Boiling-Point Elevation and Dühring's Rule

The vapor pressure of aqueous solutions is less than that of water at the same temperature. Consequently, for a given pressure the boiling point of the solutions is higher than that of pure water. The increase in boiling point over that of water is known as the *boiling-point elevation* of the solution.

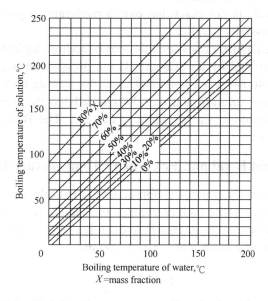

Fig. 5.10 Dühring lines, system sodium hydroxide–water.

For a given pressure, BPE is boiling-point elevation of the solution which is represented by the Δ'. Thus an empirical equation can be written

$$\Delta' = f \Delta'_a \qquad (5.3\text{-}1)$$

$$f = \frac{0.0162(T'+273)^2}{r'} \qquad (5.3\text{-}2)$$

where Δ' = BPE, boiling-point elevation, ℃
 Δ'_a = the boiling-point elevation of the solution at pressure p =1atm , ℃
 f = emendations coefficient, dimensionless
 T' = boiling temperature of water at pressure of vapor space

r' = latent heat of vaporization from thick liquor

It is small for dilute solutions and for solutions of organic colloids but may be as large as 80 ℃ for concentrated solutions of inorganic salts. The BPE must be subtracted from the temperature drop that is predicted from the steam tables.

For strong solutions the BPE is best found from an empirical rule known as Dühring's rule, which states that the boiling point of a given solution is a linear function of the boiling point of pure water at the same pressure.

Dühring's rule can be written

$$\frac{T_A - T_A'}{t_w - t_w'} = k \tag{5.3-3}$$

where k = the slope of Dühring's line, dimensionless
T_A = the boiling point of the solution at pressure p_M, ℃
T_A' = the boiling point of the solution at pressure p_N, ℃
t_w = the boiling point of water at pressure p_M, ℃
t_w' = the boiling point of water at pressure p_N, ℃

Thus if the boiling point of the solution is plotted against that of water at the same pressure, a straight line results. Different lines are obtained for different concentrations. Over wide ranges of pressure the rule is not exact, but over a moderate range the lines are very nearly straight, though not necessarily parallel.

Fig. 5.10 is a set of Dühring lines for solutions of sodium hydroxide in water. The use of this figure may be illustrated by an example. If the pressure over a 40 percent solution of sodium hydroxide is such that water boils at 93.3 ℃, by reading up from the x axis at 93.3 ℃ to the line for 40 percent solution and then horizontally to the y axis, it is found that the boiling point of the solution at this pressure is 121.1 ℃. The BPE for this solution at this pressure is therefore 27.8 ℃.

5.3.3   Effect of Liquid Head and Friction on Temperature Drop

If the depth of liquid in an evaporator is appreciable, the boiling point corresponding to the pressure in the vapor space is the boiling point of the surface layer of liquid only. A mass of liquid at a distance z m below the surface is under a pressure of the vapor space plus a head of z m of liquid and therefore has a higher boiling point.

$$p_m = p' + \frac{z\rho g}{2} \tag{5.3-4}$$

where p' = the pressure in the vapor space, Pa
z = distance below liquid surface, m
p_m = the average pressure of liquid, Pa

The average boiling point of the liquid in the tubes is higher than the boiling point corresponding to the pressure in the vapor space. This increase in boiling point lowers the average temperature drop between the steam and the liquid and reduces the capacity. The amount of reduction cannot be estimated quantitatively with precision, but the qualitative effect of liquid head, especially with high liquor levels should not be ignored. If the static head is represented by the Δ'', then

$$\Delta'' = t_{pm} - t_p' \tag{5.3-5}$$

where t_{pm} = the boiling point of pure water at average pressure p_m, ℃
t_p' = the boiling point of water at pressure in the vapor space p', ℃

In general, in any actual evaporator, the friction on temperature drop is very small, can be negligible.

Although some correlations are available for determining the true temperature drop from the operating conditions, usually this quantity is not available to the designer, and the net temperature drop, corrected for BPE and static head only, are used.

Therefore, the true temperature drop corrected for both boiling elevation and static head, the liquor boiling point is

$$t = T' + \Delta' + \Delta'' \tag{5.3-6}$$

where T' = the temperature of the vapor
       t = the boiling point of the solution

EXAMPLE 5.1 In the single-effective evaporator, the pressure in an evaporator is given as 1atm and a solution of 40.8% $CaCl_2$ liquor is being boiled. The height of the liquor in the heating room is 1m, and the density of $CaCl_2$ solution is 1340 kg/m³. Calculate:
(a) The boiling point rise BPR of solution over that of water at the same pressure.
(b) The boiling temperature of the $CaCl_2$ solution.
Solution.
(a) From the Appendix 22, we know the boiling point of the 40.8% $CaCl_2$ solution under the average press is 120℃. So the boiling point rise

$$BPR = \Delta' = 120 - 100 = 20 \text{ ℃}$$

(b) The average pressure of liquid

$$p_m = p' + \frac{\rho g Z}{2} = 103.3 \times 10^3 + \frac{1340 \times 9.81 \times 1}{2} = 107.88 \text{kPa}$$

From the steam tables in Appendix 6. the boiling point of pure water at 107.88kPa is t_{pm} = 102℃, the boiling point of pure water at 101.325kPa(1 atm abs) is t_p' =100℃, the static head $\Delta'' = 102 - 100 = 2$ ℃. Therefore, the true temperature drop

$$\Delta = \Delta' + \Delta'' = 20 + 2 = 22 \text{ ℃}$$

Then the boiling temperature of $CaCl_2$ solution

$$t = \Delta + T_w' = 22 + 100 = 122 \text{ ℃}$$

5.4 Calculation Methods for Single-Effect Evaporator

The chief factor influencing the economy of an evaporator system is the number of effects. By proper design the enthalpy of vaporization of the steam to the first effect can be used one or more times, depending on the number of effects. The economy also is influenced by the temperature of the feed. If the temperature is below the boiling point in the first effect, the heating load uses a part of the enthalpy of vaporization of the steam and only a fraction is left for evaporation; if the feed is at a temperature above the boiling point, the accompanying flash contributes some evaporation over and above that generated by the enthalpy of vaporization in the steam. Quantitatively, evaporator economy is entirely a matter of enthalpy balances.

5.4.1 Methods of Operation of Evaporators

A simplified diagram of a vertical-tube, single-effect evaporator is given in Figure 5.11. The

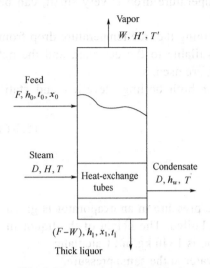

Fig.5.11 Material and enthalpy balances in evaporator.

rate of steam flow and of condensate is D, that of the thin liquor, or feed, is F, and that of the thick liquor is L. The rate of vapor flow to the condenser, assuming that no solids precipitate from the liquor, is W. Also, let T be the condensing temperature of the steam, t_1 the boiling temperature of the liquid in the evaporator, and t_0 the temperature of the feed.

If the solution to be evaporated is assumed to be dilute and like water, then 1 kg of steam condensing will evaporate approximately 1 kg of vapor. This will hold if the feed entering has a temperature t_0 near the boiling point.

The basic equation for solving for the capacity of single-effect evaporator is Eq. (5.4-1), which can be written as

$$q = UA\Delta T = UA(T - t_1) \qquad (5.4\text{-}1)$$

where q = the rate of heat transfer
U = the overall heat transfer coefficient, W/(m²·K)
A = the heat transfer area, m²
T = the temperature of the condensing steam, K
t_1 = the boiling point of the liquid, K
ΔT = the difference in temperature between the condensing steam and the boiling liquid in the evaporator

In order to solve Eq. (5.4-1) the value of q must be determined by making a heat and material balance on the evaporator shown in Figure 5.11.

5.4.2 Material Balances for Single-Effect Evaporator

Single-effect evaporators are often used when the required capacity of operation is relatively small and/or the cost of steam is relatively cheap compared to the evaporator cost. However, for large-capacity operation, using more than one effect will markedly reduce steam costs.

For the material balance since we are at steady state, the rate of mass in = rate of mass out .Then, for total balance,

$$F = W + L \qquad (5.4\text{-}2)$$

For the solute balance alone,

$$Fx_0 = (F - W)x_1 \qquad (5.4\text{-}3)$$

where W = the rate of vapor flow to the condenser
F = the rate of thin liquor, or feed, kg/h
L = the rate of thick liquor, kg/h
x_0 = a solids content of mass fraction of feed
x_1 = a solids content of mass fraction of thick liquor

5.4.3 Enthalpy (Heat) Balances for Single-Effect Evaporator

In a single-effect evaporator, the latent heat of condensation of the steam is transferred through a heating surface to vaporize water from a boiling solution. Two enthalpy balances are needed, one for the steam and one for the vapor or liquid side.

It is assumed that there is no leakage or entrainment, that the flow of noncondensables is negligible, and that heat losses from the evaporator need not be considered. The steam entering the steam chest may be superheated, and the condensate usually leaves the steam chest somewhat subcooled below its boiling point. Both the superheat and the subcooling of the condensate are small, however, and it is acceptable to neglect them in making an enthalpy balance. The small errors made in neglecting them are approximately compensated by neglecting heat losses from the steam chest.

Under these assumptions the difference between the enthalpy of the steam and that of the condensate is simply r, the latent heat of condensation of the steam. The enthalpy balance for the steam side is

$$q_s = D(H - h_w) = Dr \tag{5.4-4}$$

where q_s = rate of heat transfer through heating surface from steam
  H = specific enthalpy of steam
  h_w = specific enthalpy of condensate
  r = latent heat of condensation of steam
  D = rate of flow of steam

The enthalpy balance for the liquor side is

$$q = WH' - Fh_0 + (F - W)h_1 \tag{5.4-5}$$

where q = rate of heat transfer from heating surface to liquid
  H' = specific enthalpy of vapor
  h_0 = specific enthalpy of thin liquor
  h_1 = specific enthalpy of thick liquor
  W = the rate of vapor flow to the condenser

In the absence of heat losses, the heat transferred from the steam to the tubes equals that transferred from the tubes to the liquor, and $q_s = q$. Thus, by combining Eqs. (5.4-4) and (5.4-5)

$$q = Dr = WH' - Fh_0 + (F - W)h_1 \tag{5.4-6}$$

$$D = \frac{WH' - Fh_0 + (F - W)h_1}{r} \tag{5.4-7}$$

The liquor-side enthalpies H', h_0, and h_1 depend upon the characteristics of the solution being concentrated. Most solutions when mixed or diluted at constant temperature do not give much heat effect. This is true of solutions of organic substances and of moderately concentrated solutions of many inorganic substances. Thus sugar, salt, and paper mill liquors do not possess appreciable heats of dilution or mixing. Sulfuric acid, sodium hydroxide, and calcium chloride, on the other hand, especially in concentrated solutions, evolve considerable heat when diluted and so possess appreciable heats of dilution. An equivalent amount of heat is required, in addition to the latent heat of vaporization, when dilute solutions of these substances are concentrated to high densities.

Enthalpy balance with appreciable heat of dilution

If the heat of dilution of the liquor being concentrated is too large to be neglected, an enthalpy–concentration diagram is used for the values of h_0 and h_1 in Eq. (5.4-6). In an enthalpy–concentration diagram the enthalpy, in joules per gram of solution, is plotted against concentration, in mass fraction or weight percentage of solute. Isotherms drawn on the diagrams show the enthalpy as a function of concentration at constant temperature. Figure 5.12 is an enthalpy–concentration diagram for solutions of sodium hydroxide and water.

The enthalpy of water is referred to the same datum as in the steam tables, namely, liquid water at 0 ℃, so enthalpies from the figure can be used with those from the steam tables when liquid water or steam is involved in the calculations. In finding data for substitution into Eq. (5.4-6), values of h_0 and h_1 are taken from Fig. 5.12, and the enthalpy H' of the vapor leaving the evaporator is obtained from the steam tables.

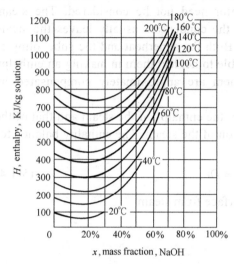

Fig.5.12 Enthalpy–concentration diagram, system sodium hydroxide-water.

The curved boundary lines on which the isotherms of Fig. 5.12 terminate represent conditions of temperature and concentration under which solid phases form. These are various solid hydrates of sodium hydroxide. The enthalpies of all single-phase solutions lie above this boundary line. The enthalpy–concentration diagram can also be extended to include solid phases.

The isotherms on an enthalpy–concentration diagram for a system with no heat of dilution are straight lines. Enthalpy–concentration diagrams can be constructed, of course, for solutions having negligible heats of dilution, but they are unnecessary in view of the simplicity of the specific-heat method described in the last section.

Enthalpy balance with negligible heat of dilution

For solutions having negligible heats of dilution, the enthalpy balances over a single-effect evaporator can be calculated from the specific heats and temperatures of the solutions. The heat-transfer rate q on the liquor side includes q_f, the heat transferred to the thin liquor to change its temperature from t_0 to the boiling temperature t_1, and q_v, the heat to accomplish the evaporation. That is,

$$q = q_f + q_v \tag{5.4-8}$$

If the specific heat of the thin liquor is assumed constant over the temperature range from t_0 to t_1, then

$$q_f = Fc_{p0}(t_1 - t_0) \tag{5.4-9}$$

Also
$$q_v = Wr' \tag{5.4-10}$$

where t_1 = the boiling temperature of the liquid in the evaporator
       t_0 = the temperature of the feed, K or ℃
       c_{p0} = specific heat of thin liquor(feed)
       r' = latent heat of vaporization from thick liquor

If the boiling-point elevation of the thick liquor is negligible, $r' = r$, the latent heat of vaporization of water at the pressure in the vapor space. When the boiling-point elevation is appreciable, the vapor leaving the solution is superheated by an amount, in degrees, equal to the boiling-point elevation, and r' differs slightly from r. In practice, however, it is nearly always sufficiently accurate to use r, which may be read directly from steam tables (see Appendix 6).

Substitution from Eqs. (5.4-9) and (5.4-10) into Eq. (5.4-8) gives the final equation for the enthalpy balance over a single-effect evaporator when the heat of dilution is negligible:

$$q = Fc_{p0}(t_1 - t_0) + Wr' \qquad (5.4-11)$$

$$D = \frac{Fc_{p0}(t_1 - t_0) + Wr'}{r} \qquad (5.4-12)$$

If the temperature t_0 of the thin liquor is greater than t_1, the term $Fc_{p0}(t_1 - t_0)$ is negative and is the net enthalpy brought into the evaporator by the thin liquor. This item is the *flash evaporation*. If the temperature t_0 of the thin liquor fed to the evaporator is less than t_1, the term $Fc_{p0}(t_1 - t_0)$ is positive and for a given evaporation additional steam will be required to provide this enthalpy. The term $Fc_{p0}(t_1 - t_0)$ is therefore the heating load. In words, Eq. (5.4-11) states that the heat from the condensing steam is utilized:

(1) to vaporize water from the solution and;

(2) to heat the feed to the boiling point; if the feed enters above the boiling point in the evaporator, part of the evaporation is from flash.

5.4.4 Heat-Transfer Coefficients

As shown by Eq. (5.4-1), the heat flux and the evaporator capacity are affected by changes both in the temperature drop and in the overall heat-transfer coefficient. The temperature drop is fixed by the properties of the steam and the boiling liquid and except for the effect of hydrostatic head is not a function of the evaporator construction. The overall coefficient, on the other hand, is strongly influenced by the design and method of operation of the evaporator.

The concept of an overall heat transfer coefficient is used in the calculation of the rate of heat transfer in an evaporator. The general equation can be written

$$\frac{1}{U_o} = \frac{d_o}{h_i d_i} + R_i \frac{d_o}{d_i} + \frac{1}{h_o} + R_o + \frac{b d_o}{\lambda d_m} \qquad (5.4-13)$$

where U_o = overall coefficient
h_o = a film coefficient of shell side
h_i = a film coefficient of tube side
d_i, d_o = diameter of inside and outside the tubes
R_i = fouling factor inside wall
R_o = fouling factor outside wall

As Eq. (5.4-13) show, the overall resistance to heat transfer between the steam and the boiling liquid is the sum of five individual resistances: the steam-film resistance; the two scale resistances, inside and outside the tubes; the tube wall resistance; and the resistance from the boiling liquid. The overall coefficient is the reciprocal of the overall resistance. In most evaporators the fouling factor of the condensing steam and the resistance of the tube wall are very small, and they are usually neglected in evaporator calculations. In an agitated-film evaporator the tube wall is fairly thick, so that its resistance may be a significant part of the total.

Because of the difficulty of measuring the high individual film coefficients in an evaporator, experimental results are usually expressed in terms of overall coefficients. These are based on the net temperature drop corrected for boiling-point elevation. The overall coefficient, of course, is influenced by the same factors influencing individual coefficients; but if one resistance (say, that of the liquid film) is controlling, large changes in the other resistances have almost no effect on the overall coefficient.

Typical overall coefficients for various types of evaporators, are given in Table 5.1. These coefficients apply to conditions under which the various evaporators are ordinarily used. A small accumulation of scale reduces the coefficients to a small fraction of the clean-tube values.

An agitated-film evaporator gives a seemingly low coefficient with a liquid having a viscosity of 100 P, but this coefficient is much larger than would be obtained in any other type of evaporator which could handle such a viscous material at all.

Table 5.1 Typical overall coefficients in evaporators

Type	Overall coefficient U, W/(m$^2 \cdot$ ℃)
Long-tube vertical evaporators	
Natural circulation	1000–2500
Forced circulation	2000–5000
Agitated-film evaporator, Newtonian liquid, viscosity	
1 cP	2000
1 P	1500
100 P	600

5.4.5 Single-Effect Calculations

The use of material balances, enthalpy balances, and the capacity equation (5.4-1) in the design of single-effect evaporators is shown in Example 5.2.

EXAMPLE 5.2 A continuous single-effect evaporator is to concentrates 9072 kg/h of a 1.0wt% salt solution entering at 311.0 K(37.8℃) to a final concentration of 1.5wt%. The vapor space of the evaporator is at 101.325kPa(1 atm abs) and the steam supplied is saturated at 143.3 kPa. The overall coefficient U=1704W/(m$^2 \cdot$ ℃). Calculate :
(a) The amounts of vapor and liquid product;
(b) The heat-transfer area required.
Assume that, since it is dilute, the solution has the same boiling point as water.
Solution. The heat capacity of the feed is assumed to be c_{p0}=4.14 kJ/(kg·K). (Often, for feeds of inorganic salts in water, the c_p can be assumed approximately as that of water alone.)
    From the steam tables in Appendix 6.
at 143.3 kPa   specific enthalpy of vapor H = 2693 kJ/kg
                 specific enthalpy of water h_w= 460 kJ/kg
                 the latent heat of the steam r = 2230 kJ/kg
                 the saturation water temperature T=383.2 K
at 101.32 kPa   specific enthalpy of vapor H'= 2677 kJ/kg
                 the latent heat of the steam r' = 2258 kJ/kg
                 the boiling point of the dilute solution in the evaporator t_1=373.2 K
(a) The amounts of vapor and liquid product is found from a material balance.
  For the material balance, substituting into Eq.(5.4-2)

$$F = W + L$$
$$9072 = W + L$$

Substituting into Eq.(5.4-3) and solving,

$$Fx_0 = (F - W)x_1$$
$$9072 \times 0.01 = L \times 0.15$$

The amounts of liquid product is $L = 6048$ kg/h of liquid.
Substituting into Eq.(5.4-3) and solving, The amounts of vapor is $W = 3024$ kg/h of vapor.
(b) To make a heat balance using Eq. (5.4-12)

$$D = \frac{Fc_{p0}(t_1 - t_0) + Wr'}{r}$$

Substituting into Eq.(5.4-12), the rate of vapor flow to the condenser $W = 3024$ kg/h of vapor.

$$D = \frac{9072 \times 4.14 \times (373.3 - 311.0) + 3024 \times 2258}{2230} = 4108 \text{ kg steam/h}$$

The heat q transferred through the heating surface area A is from Eq. (5.4-4)

$$q_s = D(H - h_w) = Dr \qquad (5.4\text{-}4)$$

Eq. (5.4-4) substituting into Eq. (5.4-1)

$$q = UA\Delta T = UA(T - t_1) \qquad (5.4\text{-}1)$$

The heat-transfer area

$$A = \frac{Dr}{U(T - t_1)} = \frac{4108 \times 2230 \times 1000}{3600 \times 1704 \times (383.2 - 373.2)} = 149.3 \text{m}^2$$

5.5 Multiple-Effect Evaporators

5.5.1 Methods of Feeding

Forward feed multiple-effect evaporator

A simplified diagram of a *forward-feed multiple-effect evaporation* system is shown in Fig. 5.13. The usual method of feeding a multiple-effect evaporator is to pump the thin liquid into the first effect and send it in turn through the other effects, the concentration of the liquid increases from the first effect to the last. This pattern of liquid flow is the simplest. It requires a pump for feeding dilute solution to the first effect, since this effect is often at about atmospheric pressure, and a pump to remove thick liquor from the last effect. The transfer from effect to effect, however, can be done without pumps, since the flow is in the direction of decreasing pressure, and control valves in the transfer line are all that is required.

Backward-feed multiple-effect evaporators

Another common method is *backward-feed*, in which dilute liquid is fed to the last effect and then pumped through the successive effects to the first, as shown in Fig. 5.14. This method requires a pump between each pair of effects in addition to the thick-liquor pump, since the flow is from low pressure to high pressure. Backward feed often gives a higher capacity than forward feed when the thick liquor is viscous, but it may give a lower economy than forward feed when the feed liquor is cold.

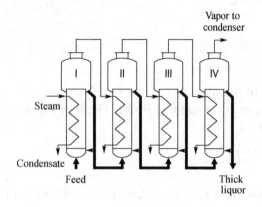

Fig. 5.13 Forward-feed multiple-effect evaporators.

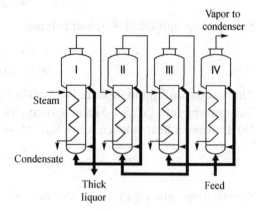

Fig. 5.14 Backward-feed multiple-effect evaporators.

Parallel-feed multiple-effect evaporators

As shown in Fig. 5.15, parallel feed in multiple-effect evaporators involves the adding of fresh feed and the withdrawal of concentrated product from each effect. The vapor from each effect is still used to heat the next effect. This method of operation is mainly used when the feed is almost saturated and solid crystals are the product, as in the evaporation of brine to make salt. In parallel feed there is no transfer of liquid from one effect to another.

Mixed-feed multiple-effect evaporators

Other patterns of feed are sometimes used. In *mixed-feed* the dilute liquid enters an intermediate effect, flows in forward feed to the end of the series, and is then pumped back to the first effects for final concentration, as shown in Fig. 5.16. This eliminates some of the pumps needed in backward feed and yet permits the final evaporation to be done at the highest temperature.

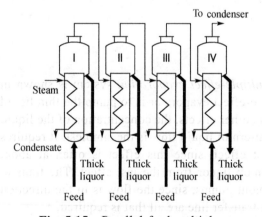

Fig. 5.15 Parallel-feed multiple-effect evaporators.

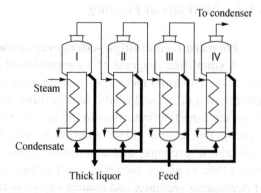

Fig. 5.16 Mixed-feed multiple-effect evaporators.

5.5.2 Calculation Methods for Multiple-Effect Evaporators

Figure 5.17 shows three long-tube natural-circulation evaporators connected to form a triple-effect system. Connections are made so that the vapor from one effect serves as the heating medium for the next. A condenser and air ejector establish a vacuum in the third effect in the series and withdraw noncondensables from the system. The first effect of a multiple-effect evaporator is the effect to which the raw steam is fed and in which the pressure in the vapor space is the highest. The last effect is that in which the vapor-space pressure is a minimum. In this manner the pressure difference between the steam and the condenser is spread across two or more effects in the multiple-effect system. The pressure in each effect is lower than that in the effect from which it receives steam and higher than that of the effect to which it supplies vapor. Each effect, in itself, acts as a single-effect evaporator, and each has a temperature drop across its heating surface corresponding to the pressure drop in that effect. Every statement that has so far been made about a single-effect evaporator applies to each effect of a multiple-effect system. Arranging a series of evaporator bodies into a multiple-effect system is a matter of interconnecting piping, and not of the structure of the individual units. The numbering of the effects is independent of the order in which liquor is fed to them—they are always numbered in the direction of decreasing pressure. In Figure 5.17 dilute feed enters the first effect, where it is partly concentrated; it flows to the second effect for additional concentration and then to the third effect for final concentration. Thick liquor is pumped out of the third effect.

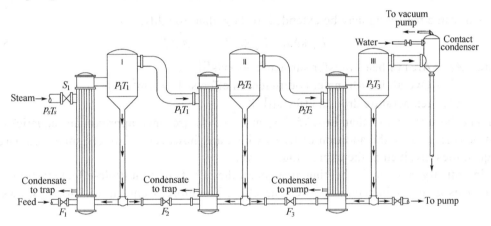

Fig.5.17 Triple-effect evaporator: I, II, III, first, second, and third effects; F_1, F_2, F_3, feed or liquor control valves; S_1 steam valves; p_s, p_1, p_2, p_3, pressures; T_s, T_1, T_2, T_3, temperatures.

In steady operation the flow rates and evaporation rates are such that neither solvent nor solute accumulates or depletes in any of the effects. The temperature, concentration, and flow rate of the feed are fixed, the pressures in steam inlet and condenser established, and all liquor levels in the separate effects maintained. Then all internal concentrations, flow rates, pressures, and temperatures are automatically kept constant by the operation of the process itself. The concentration of the thick liquor can be changed only by changing the rate of flow of the feed. If the thick liquor is too dilute, the feed rate to the first effect is reduced; and if the thick liquor is too concentrated, the feed rate is increased. Eventually the concentration in the last effect and in the thick-liquor discharge will reach a new steady state at the desired level.

The heating surface in the first effect will transmit per hour an amount of heat given by the equation

$$q_1 = U_1 A_1 \Delta T_1 \qquad (5.5\text{-}1)$$

where q_1 = rate of heat transfer in effects I, W
A_1 = area of heat-transfer surface in effects I, m^2

U_1 = overall heat-transfer coefficient in effects I, W/(m²·℃)
ΔT_1 = temperature drop in effects I, K

If the part of this heat that goes to heat the feed to the boiling point is neglected for the moment, it follows that practically all this heat must appear as latent heat in the vapor that leaves the first effect. The temperature of the condensate leaving the second effect is very near the temperature T_1 of the vapors from the boiling liquid in the first effect. Therefore, in steady operation practically all the heat that was expended in creating vapor in the first effect must be given up when this same vapor condenses in the second effect. The heat transmitted in the second effect, however, is given by the equation

$$q_2 = U_2 A_2 \Delta T_2 \tag{5.5-2}$$

where q_2 = Rate of heat transfer in effects II, W
A_2 = Area of heat-transfer surface in effects II, m²
U_2 = Overall heat-transfer coefficient in effects II, W/(m²·℃)
ΔT_2 = temperature drop in effects II, K

As has just been shown, q_1 and q_2 are nearly equal, and therefore

$$U_1 A_1 \Delta T_1 = U_2 A_2 \Delta T_2 \tag{5.5-3}$$

This same reasoning may be extended to show that, roughly,

$$U_1 A_1 \Delta T_1 = U_2 A_2 \Delta T_2 = U_3 A_3 \Delta T_3 \tag{5.5-4}$$

where A_3 = Area of heat-transfer surface in effects III, m²
U_3 = Overall heat-transfer coefficient in effects III, W/(m²·℃)
ΔT_3 = temperature drop in effects III, K

It should be understood that Eqs. (5.5-3) and (5.5-4) are only approximate equations that must be corrected by the addition of terms which are, however, relatively small compared to the quantities involved in the expressions above.

In ordinary practice the heating areas in all the effects of a multiple-effect evaporator are equal. This is to obtain economy of construction. Therefore, from Eq. (5.5-4) it follows that since

$$q_1 = q_2 = q_3 = q \tag{5.5-5}$$

$$\frac{q}{A} = U_1 \Delta T_1 = U_2 \Delta T_2 = U_3 \Delta T_3 \tag{5.5-6}$$

where q_3 = Rate of heat transfer in effects III, W

From this it follows that the temperature drops in a multiple-effect evaporator are approximately inversely proportional to the heat-transfer coefficients. Calling ΔT as follows for no boiling-point rise.

$$\Delta T = \Delta T_1 + \Delta T_2 + \Delta T_3 = T - t_3 \tag{5.5-7}$$

Since ΔT_1 is proportional to $1/U_1$, then

$$\Delta T_1 = \Delta T \times \frac{1/U_1}{1/U_1 + 1/U_2 + 1/U_3} \tag{5.5-8}$$

Similar equations can be written for ΔT_2 and ΔT_3.

EXAMPLE 5.3 A triple-effect evaporator is concentrating a liquid that has no appreciable elevation in boiling point. The temperature of the steam to the first effect is 108℃, and the boiling point of the solution in the last effect is 52℃. The overall heat transfer coefficients, in W/(m² · °C), are 2500 in the first effect, 2000 in the second effect, and 1500 in the third effect. (As the solution becomes more concentrated, the viscosity increases and the overall coefficient is reduced.) At what temperatures will the liquid boil in the first and second effects?

Solution. The total temperature drop is 108–52 = 56°C. As shown by Eq. (5.5-6), the temperature drops in the several effects will be approximately inversely proportional to the coefficients. Thus, from the Eq. (5.5-8),

$$\Delta T_1 = \Delta T \times \frac{1/U_1}{1/U_1 + 1/U_2 + 1/U_3} = 56 \times \frac{1/2500}{1/2500 + 1/2000 + 1/1500} = 14.3 \text{ °C}$$

In the same manner $\Delta T_2 = 17.9$°C and $\Delta T_3 = 23.8$°C. Consequently the boiling point in the first effect will be 108–14.3=93.7°C, and that in the second effect will be 93.7–17.9 =75.8°C.

The increase in economy through the use of multiple-effect evaporation is obtained at the cost of reduced capacity. It might be thought that by providing several times as much heating surface the evaporating capacity would be increased, but this is not the case. The total capacity of a multiple-effect evaporator is usually no greater than that of a single-effect evaporator having a heating surface equal to one of the effects and operating under the same terminal conditions, and, when there is an appreciable boiling-point elevation, is often considerably smaller. When the boiling-point elevation is negligible, the effective overall ΔT equals the sum of the ΔT's in each effect, and the amount of water evaporated *per unit area of surface* in an N-effect multiple-effect evaporator is approximately $(1/N)$th that in the single effect. This can be shown by the following analysis.

If the heating load and the heat of dilution are neglected, the capacity of an evaporator is directly proportional to the rate of heat transfer. The heat transferred in the three effects in Fig. 5.17 is given by the Eqs. (5.5-1), (5.5-2), and (5.5-9)

$$q_3 = U_3 A_3 \Delta T_3 \tag{5.5-9}$$

The total capacity is proportional to the total rate of heat transfer q_T, found by adding these equations.

$$q_T = q_1 + q_2 + q_3 = U_1 A_1 \Delta T_1 + U_2 A_2 \Delta T_2 + U_3 A_3 \Delta T_3 \tag{5.5-10}$$

Assume that the surface area S and the overall coefficient U are the same in each effect and that the boiling-point elevation is negligible. Then Eq. (5.5-10) can be written

$$q_T = UA(\Delta T_1 + \Delta T_2 + \Delta T_3) = UA\Delta T \tag{5.5-11}$$

where ΔT is the total temperature drop between the steam in the first effect and the vapor in the last effect. Suppose now that a single-effect evaporator with a surface area A is operating with the same total temperature drop. If the overall coefficient is the same as in each effect of the triple-effect evaporator, the rate of heat transfer in the single effect is

$$q_T = UA\Delta T \tag{5.5-12}$$

This is exactly the same equation as that for the multiple-effect evaporator. No matter how many effects are used, provided the overall coefficients are the same, the capacity will be no greater than that of a single effect having an area equal to that of each effect in the multiple unit.

The boiling-point elevation tends to make the capacity of a multiple-effect evaporator less than that of the corresponding single effect. Offsetting this are the changes in overall coefficients in a multiple-effect evaporator. In a single-effect unit producing 50 percent NaOH, for example, the overall coefficient U for this viscous liquid would be small. In a triple-effect unit, the coefficient in the final effect would be the same as that in the single effect; but in the other effects, where the NaOH concentration is much lower than 50 percent, the coefficients would be greater. Thus the average coefficient for the triple-effect evaporator would be greater than that for the single effect. In some cases this overshadows the effect of boiling-point elevation, and the capacity of a multiple effect unit is actually greater than that of a single effect.

5.5.3 Effect of Liquid Head and Boiling-Point Elevation

The liquid head and the boiling-point elevation influence the capacity of a multiple-effect evaporator even more than they do that of a single effect. The reduction in capacity caused by the liquid head, as before, cannot be estimated quantitatively. The liquid head reduces the temperature drop available in each effect of a multiple-effect evaporator, as does the boiling-point elevation.

Consider an evaporator that is concentrating a solution with a large boiling-point elevation. The vapor coming from this boiling solution is at the solution temperature and is therefore superheated by the amount of the boiling-point elevation. Superheated steam is essentially equivalent to saturated steam at the same pressure when used as a heating medium. The temperature drop in any effect, therefore, is calculated from the temperature of saturated steam at the pressure of the steam chest, and not from the temperature of the boiling liquid in the previous effect. This means that the boiling-point elevation in any effect is lost from the total available temperature drop. This loss occurs in every effect of a multiple-effect evaporator, and the resulting loss of capacity is often important.

The influence of these losses in temperature drop on the capacity of a multiple effect evaporator is shown in Fig.5.18. The three diagrams in this figure represent the temperature drops in a single-effect, double-effect, and triple-effect evaporator. The terminal conditions are the same in all three; that is, the steam pressure in the first effect and the saturation temperature of the vapor evolved from the last effect are identical in all three evaporators. Each effect contains a liquid with a boiling-point elevation. The total height of each column represents the total temperature spread from the steam temperature to the saturation temperature of the vapor from the last effect.

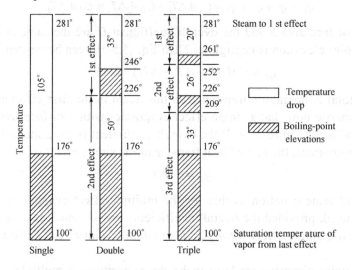

Fig. 5.18 Effect of boiling-point elevation on capacity of evaporators.

Consider the single-effect evaporator. Of the total temperature drop of 181 °C, the shaded part represents the loss in temperature drop due to boiling-point elevation. The remaining temperature drop, 105 °C, the actual driving force for heat transfer, is represented by the unshaded part. The diagram for the double-effect evaporator shows two shaded portions because there is a boiling-point elevation in each of the two effects, and the residual unshaded part, totaling 85 °C, is smaller than in the diagram for the single effect. In the triple-effect evaporator there are three shaded portions since there is a loss of temperature drop in each of three effects, and the total net available temperature drop, 79 °C, is correspondingly smaller.

In extreme cases of a large number of effects or very high boiling-point elevations, the sum of the boiling-point elevations in a proposed evaporator could be greater than the total temperature drop available. Operation under such conditions is impossible. The design or the operating conditions of the evaporator would have to be revised to reduce the number of effects or increase the total temperature drop.

The economy of a multiple-effect evaporator is not influenced by boiling-point elevations if minor factors, such as the temperature of the feed and changes in the heats of vaporization, are neglected. A kilogram of steam condensing in the first effect generates about a kilogram of vapor, which condenses in the second effect, generating another kilogram there, and so on. The economy of a multiple-effect, evaporator depends on heat-balance considerations and not on the rate of heat transfer. The capacity, on the other hand, is reduced by the boiling-point elevation. The capacity of a double-effect evaporator concentrating a solution with a boiling point elevation is generally less than one-half the capacity of two single effects, each operating with the same overall temperature drop. The capacity of a triple effect is generally less than one-third that of three single effects with the same terminal temperatures.

5.5.4 Optimum Number of Effects

The cost of each effect of an evaporator per square meter or square foot of surface is a function of its total area and decreases with area, approaching an asymptote for very large installations. Thus the investment required for an N-effect evaporator is about N times that for a single-effect evaporator of the same capacity. The optimum number of effects must be found from an economic balance between the savings in steam obtained by multiple-effect operation and the added investment required.

5.5.5 Multiple-Effect Calculations

In designing a multiple-effect evaporator the results usually desired are the amount of steam consumed, the area of the heating surface required, the approximate temperatures in the various effects, and the amount of vapor leaving the last effect. As in a single-effect evaporator, these quantities are found from material balances, enthalpy balances, and the capacity equation (5.4-1). In a multiple-effect evaporator, however, a trial-and-error method is used in place of a direct algebraic solution.

Consider, for example, a triple-effect evaporator. There are seven equations which may be written: an enthalpy balance for each effect, a capacity equation for each effect, and the known total evaporation, or the difference between the thin-and thick-liquor rates. If the amount of heating surface in each effect is assumed to be the same, there are seven unknowns in these equations: (1) the rate of steam flow to the first effect, (2) to (4) the rate of flow from each effect, (5) the boiling temperature in the first effect, (6) the boiling temperature in the second effect, and (7) the heating surface per effect. It is possible to solve these equations for the seven unknowns, but the method is tedious and involved.

Another method of calculation is as follows:
(1) Assume values for the boiling temperatures in the first and second effects.
(2) From enthalpy balances find the rates of steam flow and of liquor from effect to effect.
(3) Calculate the heating surface needed in each effect from the capacity equations.
(4) If the heating areas so found are not nearly equal, estimate new values for the boiling temperatures and repeat items 2 and 3 until the heating surfaces are equal.

In practice these calculations are done by computer. Results of a typical calculation are shown in the following example.

EXAMPLE 5.4 A triple-effect evaporator is to be used to concentrate 22700kg/h of a 10% solution to 50%. The feed enters at 40°C. Steam is available at 121°C, and the vapor from the last effect is condensed at 51.7°C. The overall heat-transferring coefficient of every effect are U_1=2840W/(m² · °C), U_2=1700W/(m² · °C), U_3=1135W/(m² · °C). Every effect liquor specific heat and the latent heat of the steam may be taken as c_p=4.186kJ/(kg · °C) and r=2326kJ/kg(not change with temperature and concentration). Radiation and undercooling of condensate may be neglected. Calculate the amount of steam consumed, the per effect liquid product and the area.

Solution.
(1) From the material balance, the total amount vaporized W is

$$W = F\left(1 - \frac{x_0}{x_3}\right) = 22700 \times \left(1 - \frac{0.1}{0.5}\right) = 18160 \text{ kg/h}$$

(2) Estimate per effect liquor boiling point
Suppose $\quad q_1 = q_2 = q_3$
Let $\quad A_1 = A_2 = A_3 = A$
Substituting into Eq. (5.5-6)

$$q = U_1 A \Delta T_1 = U_2 A \Delta T_2 = U_3 A \Delta T_3$$

The BPR in each effect is calculated as follows:

$$\frac{\Delta T_2}{\Delta T_1} = \frac{U_1}{U_2} = \frac{2840}{1700} = 1.67$$

$$\frac{\Delta T_3}{\Delta T_1} = \frac{U_1}{U_3} = \frac{2840}{1135} = 2.5$$

$$\sum \Delta T = 121 - 51.7 = 69.3 \text{ °C}$$

Then $\quad \sum \Delta T = \Delta T_1 + \frac{\Delta T_2}{\Delta T_1} \times \Delta T_1 + \frac{\Delta T_3}{\Delta T_1} \times \Delta T_1 = (1 + 1.67 + 2.5)\Delta T_1$

The boiling point of solution in each effect is

$$\Delta T_1 = \frac{69.3}{1 + 1.67 + 2.5} = 13.4 \text{ °C}$$

$$t_1 = 121 - 13.4 = 107.6 \text{ °C}$$

$$\Delta T_2 = 1.67 \times 13.4 = 22.4 \text{ °C}$$

$$t_2 = 107.6 - 22.4 = 85.2 \text{ °C}$$

$$\Delta T_3 = 2.5 \times 13.4 = 33.5 \ ^\circ C$$
$$t_3 = 85.2 - 33.5 = 51.7 \ ^\circ C$$

(3) The heat balance
Suppose the amounts of liquid product is L, from the Eq. (5.4-12).
First effect:
$$Dr = W_1 r_1 + F c_{Po}(t_1 - t_0)$$
$$D \times 2326 = (F - L_1) \times 2326 + 22700 \times 4.187 \times (107.6 - 40)$$
$$D = 25462 - L_1 \tag{a}$$

Making a heat balance on effects 2 and 3, similar equations can be written:
Second effect:
$$W_1 r_1 = W_2 r_2 + (F - W_1) c_p (t_2 - t_1)$$
$$(F - L_1) r_1 = (L_1 - L_2) r_2 + L_1 c_p (t_2 - t_1)$$
$$22700 - L_1 = L_1 - L_2 + L_1 \times \frac{4.186 \times (85.2 - 107.6)}{2326}$$
$$L_2 = 1.96 L_1 - 22700 \tag{b}$$

Third effect:
$$W_2 r_2 = W_3 r_3 + (F - W_1 - W_2) c_p (t_3 - t_2)$$
$$(L_1 - L_2) \times 2326 = (L_2 - L_3) \times 2326 + L_2 \times 4.187 \times (51.7 - 85.2) \tag{c}$$
$$L_3 = F - W = 22700 - 18160 = 4540 \text{kg/h} \tag{d}$$

Substitutes equation (d) into equation (c), have
$$L_1 = 1.94 L_2 - 4540 \tag{e}$$

Associate with equation (a), (b), and (e). Thus have:
$$L_1 = 17350 \text{kg/h}$$
$$L_2 = 11285 \text{kg/h}$$
$$D = 8109 \text{kg/h}$$

(4) Using Eq. (5.4-1) for each effect, calculate the area A_1, A_2, and A_3:
$$A_1 = \frac{q_1}{U_1 \Delta T_1} = \frac{8110 \times 2326 \times 10^3}{2840 \times 13.4 \times 3600} = 137 \text{m}^2$$
$$A_2 = \frac{(22700 - 17352) \times 2326}{1700 \times 19.3 \times 3.6} = 91 \text{m}^2$$
$$A_3 = \frac{(17352 - 11285) \times 2326}{1135 \times 33.5 \times 3.6} = 103 \text{m}^2$$

Then the average area is
$$A_m = \frac{A_1 + A_2 + A_3}{3} = 110 \text{m}^2$$

These area are not nearly equal, a second trial should be performed as follows.
(5) Obtain new values of $\Delta T'_1$, $\Delta T'_2$, and $\Delta T'_3$, from
$$\Delta T'_1 = \frac{\Delta T_1 A_1}{A_m}, \qquad \Delta T'_2 = \frac{\Delta T_2 A_2}{A_m}, \qquad \Delta T'_3 = \frac{\Delta T_3 A_3}{A_m}$$
$$\Delta T'_1 = \frac{A_1}{A_m} \Delta T_1 = \frac{137}{110} \times 13.4 = 16.6 \ ^\circ C \qquad \text{take } 17.5 \ ^\circ C$$
$$\Delta T'_2 = \frac{91}{110} \times 22.4 = 18.5 \ ^\circ C \qquad \text{take } 19.3 \ ^\circ C$$

$$\Delta T_3' = \frac{103}{110} \times 33.5 = 31.4\,°C \qquad \text{take } 32.5\,°C$$

The actual boiling point of solution in each effect is
$$t_1=103.5\,°C, \qquad t_2=84.2\,°C, \qquad t_1=51.7\,°C$$

(6) Using the new $\Delta T'$ values from step (5), repeat the calculation starting with step (3), solving,

$$D = 25294 - L_1 \qquad (a)$$
$$L_2 = 1.965L_1 - 22700 \qquad (b)$$
$$L_1 = 1.942L_2 - 4540 \qquad (c)$$

The steam consumption, the amounts of liquid product and the area are:
$$D = 8042\,\text{kg/h}$$
$$L_2 = 11233\,\text{kg/h}$$
$$L_1 = 17252\,\text{kg/h}$$

$$A_1 = \frac{8042 \times 2326}{2840 \times 17.5 \times 3.6} = 104\,\text{m}^2$$

$$A_2 = \frac{(22700 - 17252) \times 2326}{1700 \times 19.3 \times 3.6} = 107\,\text{m}^2$$

$$A_3 = \frac{(17252 - 11233) \times 2326}{1135 \times 32.5 \times 3.6} = 105\,\text{m}^2$$

Then the average area is $A_m = 105\,\text{m}^2$, these area are reasonably close to each other, a second trial is not need.

(7) From these results the answers to the problems are found to be
$$D = 8042\,\text{kg/h}$$
$$L_1 = 17252\,\text{kg/h}, \qquad L_2 = 11233\,\text{kg/h}, \qquad L_3 = 4540\,\text{kg/h}$$
$$A = 105\,\text{m}^2$$

5.5.6 Vapor Recompression

The energy in the vapor evolved from a boiling solution can be used to vaporize more water, provided there is a temperature drop for heat transfer in the desired direction. In a multiple-effect evaporator this temperature drop is created by progressively lowering the boiling point of the solution in a series of evaporators through the use of lower absolute pressures. The desired driving force can also be obtained by increasing the pressure (and, therefore, the condensing temperature) of the evolved vapor by mechanical or thermal recompression. The compressed vapor is then condensed in the steam chest of the evaporator from which it came.

Mechanical recompression

The principle of mechanical vapor recompression is illustrated in Fig. 5.19. Cold feed is preheated almost to its boiling point by exchange with hot liquor and is pumped through a heater as in a conventional forced-circulation evaporator. The vapor evolved, however, is not condensed; instead it is compressed to a somewhat higher pressure by a positive-displacement or centrifugal compressor and becomes the "steam" which is fed to the heater. Since the saturation temperature of the compressed vapor is higher than the boiling point of the feed, heat flows from the vapor to the solution, generating more vapor. A small amount of makeup steam may be necessary. The optimum temperature drop for a typical system is about 5 ℃. The energy utilization of such a system is very good: Based on the steam equivalent of the power required to drive the compressor, the economy corresponds to that of an evaporator containing 10 to 15 effects. Important applications of mechanical compression evaporation include production of distilled water from seawater, evaporation of black liquor in the paper industry, evaporation of heat-sensitive materials such as fruit juices, and crystallization of salts.

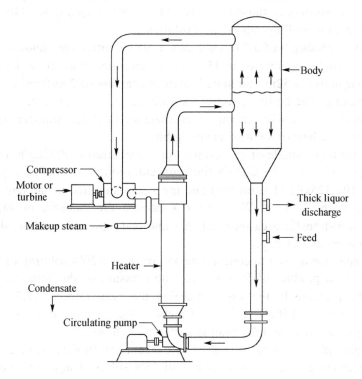

Fig. 5.19 Mechanical recompression applied to forced-circulation evaporator.

Falling-film evaporators are especially well suited for operation with vapor recompression systems.

Thermal recompression

In a thermal recompression system the vapor is compressed by acting on it with high-pressure steam in a jet ejector. This results in more steam than is needed for boiling the solution, so that excess steam must be vented or condensed. The ratio of motive steam to the vapor from the solution depends on the evaporation pressure; for many low-temperature operations, with steam at 8 to 10 atm pressure, the ratio of steam required to the mass of water evaporated is about 0.5.

Since steam jets can handle large volumes of low-density vapor, thermal recompression is better suited than mechanical recompression to vacuum evaporation. Jets are cheaper and easier to maintain than blowers and compressors. The chief disadvantages of thermal recompression are the low efficiency of the jets and lack of flexibility in the system toward changed operating conditions.

PROBLEMS

5.1 Using the empirical formula and Dühring's rule to calculate the boiling point of the 40% aqueous sodium at the pressure of 0.2 kgf/cm².

5.2 As an example of use of the chart, the pressure in an evaporator is given as 25.6kPa and a solution of 30% NaOH is being boiled. Determine the boiling temperature of the NaOH solution and the boiling-point rise BPR of the solution over that of water at the same pressure.

5.3 In a single effective evaporator, $CaCl_2$ solution is evaporated, and the pressure in the vapor space of the evaporator is 1atm. The depth of liquid in an evaporator is 1m. It is known that the concentration of the solution is 40.8% and the density is $1340 kg/m^3$. Estimate the boiling point and the boiling-point rise of solution.

5.4 A feed of 2500kg/h of a 7.0% solution at 95℃ enters continuously a single-effect evaporator and is being concentrated to 45%. The evaporation is at atmospheric pressure and the area of the evaporator is $52 m^2$. Saturated steam at pressure of 2 kgf/cm^2 (gauge pressure) is supplied for heating. The boiling point of the solution in the evaporator is 103℃. The heat loss of the evaporator is 11000W. Attempt to estimate the total heat transfer coefficient of the evaporator. Assume the heat of solution is negligible.

5.5 A continuous single-effect evaporator concentrates 9072kg/h of a 1.0% salt solution entering at 311.0K (37.8℃) to a final concentration of 1.5%. The vapor space of the evaporator is at 101.325kPa (1.0atm abs) and the steam supplied is saturated at 143.3 kPa. The overall coefficient $U=1704$ $W/(m^2 \cdot K)$. Calculate the amounts of vapor and liquid product and the heat-transfer area required. Assume that, since it is dilute, the solution has the same boiling point as water.

5.6 An evaporator is used to concentrate 4536kg/h of a 20% solution of NaOH in water entering at 60℃ to a product of 50% solids. The pressure of the saturated steam used is 172.4kPa and the pressure in the vapor space of the evaporator is 11.7kPa. The overall heat-transfer coefficient is $1560 W/(m^2 \cdot K)$. Calculate the steam used, the steam economy in kg vaporized /kg steam used, and the heating surface area in m^2.

5.7 A single-effect evaporator is used to concentrate 10000kg/h of a 5% solution of $NaNO_3$ in water entering at 40℃ to a product of 25% solids. The pressure of the saturated steam used is 29.4kPa (gauge pressure) and the absolute pressure in the vapor space of the evaporator is 300 mmHg. The overall heat transfer coefficient of the evaporator is $2000 W/(m^2 \cdot ℃)$. The heat loss is 5% of the heat energy supplied to the evaporator. Calculate the heat transfer area of the evaporator and the steam used. Assume that the static head is neglected. The atmosphere is 760mmHg.

5.8 A feed rate of 1000kg/h of 10% aqueous solution is evaporated in a double-effect evaporator with forward feed. In the first effect, the solution is concentrated to 15%, and from 15% to 30% for second effect. The boiling points of the solution in the two effects are 108℃ and 95℃, respectively. Calculate: (a) The amount of water flashed to vapor due to the temperature difference between the two effects; (b) The percentage of the water flashed in the total water is to be evaporated in the second effect.

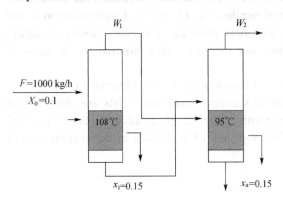

Figure for problem 5.8

5.9 A solution with a negligible boiling- point rise is being evaporated in a triple-effect evaporator using saturated steam at 121.1℃ (394.3K). The pressure in the vapor of the last effect is 25.6kPa abs. The heat-transfer coefficients are $U_1 = 2840$, $U_2 = 1988$, and $U_3 = 1420$ W/(m²·K), and the areas are equal. Estimate the boiling point in each of the evaporators.

5.10 A triple-effect parallel-flow evaporator, if the influences of the boiling-point elevations, the heat of solution, and the heat loss are all neglected. Derive: (a) The amount of solvent vaporized in each effect is equal; (b) The heat transfer rate of each effect is equal.

5.11 A triple-effect forward-feed evaporator is being used to evaporate a sugar solution containing 10*wt*% solids to a concentrated solution of 50%. The boiling-point rise of the solutions (independent of pressure) can be estimated from $\Delta°C = 1.78x + 6.22x^2$, where x is *wt* fraction of sugar in solution. Saturated steam at 205.5kPa (121.1°C saturation temperature) is being used. The pressure in the vapor space of the third effect is 13.4 kPa. The feed rate is 22680 kg/h at 26.7°C. The heat capacity of the liquid solutions is $c_p = 4.19-2.35x$ kJ/(kg·K). The heat of solution is considered to be negligible. The coefficients of heat transfer have been estimated as $U_1 = 3123$, $U_2 = 1987$, and $U_3 = 1136$ W/(m²·K). If each effect has the same surface area, calculate the area, the steam rate used, and the steam economy.

名 词 术 语

英文 / 中文

agitated-film evaporator / 刮板搅拌薄膜蒸发器
backward-feed multiple-effect evaporator / 逆流加料法多效蒸发器
basket type evaporator / 悬筐蒸发器
boiling-point elevation / 溶液沸点升高
circulation evaporator / 循环形蒸发器
climbing-film evaporator / 升膜蒸发器
Dühring's rule / 杜林规则
evaporation / 蒸发
falling-film evaporator / 降膜蒸发器
flash evaporation / 闪蒸发
forced-circulation evaporator / 强制循环蒸发器

英文 / 中文

forward feed multiple-effect evaporator / 顺流加料法多效蒸发器
heat of dilution / 溶液稀释热
liquid head / 液柱静压强
mixed-feed multiple-effect evaporator / 混合加料法多效蒸发器
multiple-effect evaporation / 多效蒸发
parallel-feed multiple-effect evaporator / 平流加料法多效蒸发器
single-effect evaporation / 单效蒸发
temperature drop / 温度差损失
vacuum evaporation / 真空蒸发
vertical type evaporator / 中央循环管式蒸发器

Chapter 6

Principles of Diffusion and Mass Transfer Between Phases

Diffusion is the movement, under the influence of a physical stimulus, of an individual component through a mixture. The most common cause of diffusion is a concentration gradient of the diffusing component. A concentration gradient tends to move the component in such a direction as to equalize concentrations and destroy the gradient. When the gradient is maintained by constantly supplying the diffusing component to the high-concentration end of the gradient and removing it at the low-concentration end, there is a steady-state flux of the diffusing component. This is characteristic of many mass-transfer operations. For example, when ammonia is removed from a gas by absorption in water in a packed column, at each point in the column a concentration gradient in the gas phase causes ammonia to diffuse to the gas-liquid interface, where it dissolves, and a gradient in the liquid phase causes it to diffuse into the bulk liquid. In stripping a solute from a liquid the gradients are reversed; here diffusion brings solute from the bulk liquid to the interface and from there into the gas phase. In some other mass-transfer operations such as leaching and adsorption, unsteady-state diffusion takes place, and the gradients and fluxes decrease with time as equilibrium is approached.

Although the usual cause of diffusion is a concentration gradient, diffusion can also be caused by an activity gradient, as in reverse osmosis, by a pressure gradient, by a temperature gradient, or by the application of an external force field, as in a centrifuge. Molecular diffusion induced by temperature is *thermal diffusion,* and that from an external field is *forced diffusion*. Both are uncommon in chemical engineering. Only diffusion under a concentration gradient is considered in this chapter.

Diffusion is not restricted to molecular transfer through stagnant layers of solid or fluid. It also takes place when fluids of different compositions are mixed. The first step in mixing is often mass transfer caused by the eddy motion characteristic of turbulent flow. This is called *eddy diffusion*. The second step is molecular diffusion between and inside the very small eddies. Sometimes the diffusion process is accompanied by bulk flow of the mixture in a direction parallel to the direction of diffusion.

Role of diffusion in mass transfer

In all the mass-transfer operations, diffusion occurs in at least one phase and often in both phases. In distillation, the low boiler diffuses through the liquid phase to the interface and away from the interface into the vapor. The high boiler diffuses in the reverse direction and passes through the vapor into the liquid. In leaching, diffusion of solute through the solid phase is followed by diffusion into the liquid. In liquid extraction, the solute diffuses through the raffinate phase to the interface and then into the extract phase. In crystallization, solute diffuses through the mother liquor to the crystals and deposits on the solid surfaces. In humidification or dehumidification there is no diffusion through the liquid phase because the liquid phase is pure and no concentration gradient through it can exist; but the vapor diffuses to or from the liquid-gas interface into or out of the gas phase. In membrane separations diffusion occurs in all there phases: in the fluids on either side of the membrane and in the membrane itself.

6.1 Theory of Diffusion

The quantitative relationships for diffusion are discussed in this section, focusing on diffusion in a direction *perpendicular* to the interface between phases and at a definite location in the equipment. The first topic considered is steady-state diffusion, where the concentrations at any point do not change with time. Equations for transient diffusion are presented later. This discussion is restricted to binary or pseudo-binary mixtures.

Fick's first law of diffusion

In the following general equation for one-dimensional diffusion, the molar flux J is similar to the heat flux q/A, and the concentration gradient dc/dz is similar to the temperature gradient dT/dx. Thus

$$J_A = -D_{AB} \frac{dc_A}{dz} \tag{6.1-1}$$

where J_A = molar flux of component A, kmol/(m² · s)
  D_{AB} = volumetric diffusivity, m²/s
  c_A = concentration, kmol/m³
  z = distance in direction of diffusion, m

Equation (6.1-1) is a statement of Fick's first law of diffusion, which may be written in several different forms. As discussed below, it applies whether or not the total mixture is stationary or is in motion with respect to a stationary plane.

For diffusion in three dimensions, Eq. (6.1-1) becomes

$$J_A = -D_{AB} \nabla c_A = \rho_M D_{AB} \nabla x_A \tag{6.1-2}$$

where ρ_M = molar density of the mixture, kmol/m³
  x_A = mol fraction of A in the liquid phase

Analogies with momentum and heat transfer

The similarities among the equations for diffusion, heat transfer, and momentum transfer are evident from the following equations. The dimensions of volumetric diffusivity D_{AB} are area divided by time, the same as those of thermal diffusivity α and kinematic viscosity ν. For one-dimensional systems, from Eqs. (6.1-1), substituting α for k (conductivity), and using ν in place of μ,❶

For mass: $$J_A = -D_{AB} \frac{dc_A}{dz}$$

For heat energy: $$\frac{dq}{dA} = -\alpha \frac{d(\rho c_p T)}{dx}$$

---

❶ *Note* that for momentum transfer, by convention, the negative sign is often omitted.

For momentum: $$\tau = v\frac{d(\rho u)}{dy}$$

where z, x, and y are all measures of distance in the direction of flow. These equations state that mass transport occurs because of a gradient in mass concentration, energy transport because of a gradient in energy concentration, and momentum transport because of a gradient in momentum concentration. These analogies do not apply in two- and three-dimensional systems, because τ is a tensor quantity with nine components, whereas J_A and dq/dA are vectors with three components.

The similarities between diffusion and heat flow permit the solution to the equations for heat conduction to be adapted to problems of diffusion in solids or fluids. This is especially useful for unsteady-state diffusion problems because of the many published solutions for unsteady-state heat transfer (see Chap. 4).

Differences between heat transfer and mass transfer result from the fact that heat is not a substance but energy in transit, whereas diffusion is the physical flow of material. Furthermore, all molecules of a mixture are at the same temperature at a given point in space, so heat transfer in a given direction is based on one temperature gradient and the average thermal conductivity. With mass transfer there are different concentration gradients for each component and often different diffusivities, denoted by using identifying subscripts for c and D_{AB}.

The material nature of diffusion and the resulting flow lead to four types of situations:
(1) Only one component A of the mixture is transferred to or from the interface, and the total flow is the same as the flow of A. Absorption of a single component from a gas into a liquid is an example of this type.
(2) The diffusion of component A in a mixture is balanced by an equal and opposite molar flow of component B, so that there is no net molar flow. This is generally the case in distillation, and it means there is no net volume flow in the gas phase. There is generally a net volume flow or mass flow in the liquid phase because of the difference in the molar densities, but this flow is small and can be neglected.
(3) Diffusion of A and B takes place in opposite directions, but the molar fluxes are unequal. This situation often occurs in diffusion of chemically reacting species to and from a catalyst surface, but the equations are not covered in this text.
(4) Two or more components diffuse in the same direction but at different rates, as in some membrane separations and adsorption processes.

Diffusion quantities

Five interrelated concepts are used in diffusion theory:
(1) Velocity u, defined as usual by length/time.
(2) Flux across a plane N, mol/(area·time).
(3) Flux relative to a plane of zero velocity J, mol/(area·time).
(4) Concentration c and molar density ρ_M, mol/volume (mole fraction may also be used).
(5) Concentration gradient dc/dz, where z is the length of the path perpendicular to the area across which diffusion is occurring.

Appropriate subscripts are used when needed. The equations apply equally well to SI, cgs, and fps units. In some applications, mass, rather than molal units, may be used in flow rates and concentrations.

Velocities in diffusion

Several velocities are needed to describe the movements of individual substances and of the total phase. Since absolute motion has no meaning, any velocity must be based on an arbitrary state of rest. In this discussion *velocity* without qualification refers to the velocity relative to the interface between the phases and is that apparent to an observer at rest with respect to the interface.

The individual molecules of any one component in the mixture are in random motion. If the instantaneous velocities of the components are summed, resolved in the direction perpendicular to the interface, and divided by the number of molecules of the substance, the result is the macroscopic velocity of that component. For component A, for instance, this velocity is denoted by u_A.

Molal flow rate, velocity, and flux

The total molar flux N in a direction perpendicular to a stationary plane is the product of the volumetric average velocity u_0 and the molar density ρ_M. For components A and B crossing a stationary plane, the fluxes are

$$N_A = c_A u_A \qquad (6.1\text{-}3a)$$

$$N_B = c_B u_B \qquad (6.1\text{-}3b)$$

Diffusivities are defined, not with respect to a stationary plane, but relative to a plane moving at the volume-average velocity u_0.[●] By definition there is no net volumetric flow across this reference plane, although in some cases there is a net molar flow or a net mass flow. The molar flux of component A through this reference plane is a diffusion flux designated J_A and is equal to the flux of A for a stationary plane [Eq. (6.1-3a)] minus the flux due to the total flow at velocity u_0 and concentration c_A:

$$J_A = c_A u_A - c_A u_0 = c_A (u_A - u_0) \qquad (6.1\text{-}4)$$

$$J_B = c_B u_B - c_B u_0 = c_B (u_B - u_0) \qquad (6.1\text{-}5)$$

Equation (6.1-1), written in terms of a diffusivity D_{AB} of component A in its mixture with component B, is

$$J_A = -D_{AB} \frac{dc_A}{dz} \qquad (6.1\text{-}6)$$

For component B:
$$J_B = -D_{BA} \frac{dc_B}{dz} \qquad (6.1\text{-}7)$$

Equations (6.1-6) and (6.1-7) are statements of Fick's first law of diffusion for a binary mixture. Note that this law is based on three decisions:
(1) The flux is in moles per unit area per unit time.
(2) The diffusion velocity is relative to the volume-average velocity.
(3) The driving potential is in terms of the molar concentration (moles of component A per unit volume).

Relations between diffusivities

The relationship between D_{AB} and D_{BA} is easily determined for ideal gases, since the molar density does not depend on the composition:

$$c_A + c_B = c_M = \frac{P}{RT} \qquad (6.1\text{-}8)$$

[●] Recall that linear velocity equals volumetric flow rate per unit area. Some authors define diffusivity using molar-average velocity, but the volume-average velocity is more suitable for both gases and liquids.

For diffusion of A and B in a gas at constant temperature and pressure,

$$dc_A + dc_B = dc_M = 0 \qquad (6.1\text{-}9)$$

Choosing the reference plane for which there is zero volume flow, we can set the sum of the molar diffusion fluxes of A and B to zero, since the molar volumes are the same:

$$-D_{AB}\frac{dc_A}{dz} - D_{BA}\frac{dc_B}{dz} = 0 \qquad (6.1\text{-}10)$$

Since $dc_A = -dc_B$, the diffusivities must be equal; that is,

$$D_{AB} = D_{BA} \qquad (6.1\text{-}11)$$

When we are dealing with liquids, the same result is found if all mixtures of A and B have the same mass density.

$$c_A M_A + c_B M_B = \rho = const \qquad (6.1\text{-}12)$$

$$M_A dc_A + M_B dc_B = 0 \qquad (6.1\text{-}13)$$

For no volume flow across the reference plane, the sum of the volumetric flows due to diffusion is zero. The volumetric flow is the molar flow times the molar volume M/ρ and

$$-D_{AB}\frac{dc_A}{dz}\frac{M_A}{\rho} - D_{BA}\frac{dc_B}{dz}\frac{M_B}{\rho} = 0 \qquad (6.1\text{-}14)$$

Substituting Eq. (6.1-13) into Eq. (6.1-14) gives

$$D_{AB} = D_{BA} \qquad (6.1\text{-}15)$$

Other equations can be derived for diffusion in liquids where the density changes, but for most practical applications, equal diffusivities are assumed when we are dealing with binary mixtures. In the following equations a volumetric diffusivity D is used, the driving force for diffusion is based on concentration differences in moles per volume. A common form of the diffusion equation gives the total flux relative to a fixed plane:

$$N_A = c_A u_0 - D\frac{dc_A}{dz} \qquad (6.1\text{-}16)$$

For gases it is often convenient to use mole fraction rather than molar concentrations, and since $c_A = c_M y_A$ and $u_0 = N/c_M$, Eq (6.1-16) becomes:

$$N_A = y_A N - Dc_M \frac{dy_A}{dz} \qquad (6.1\text{-}17)$$

Equation (6.1-17) is sometimes applied to liquids, although it is only approximate if the molar density is not constant.

Interpretation of diffusion equations

Equation (6.1-16) is the basic equation for mass transfer in a nonturbulent fluid phase. It accounts for the amount of component A carried by the convective bulk flow of the fluid and the amount of A being transferred by molecular diffusion. The vector nature of the fluxes and concentration gradients must be understood, since these quantities are characterized by directions and magnitudes. As derived, the positive sense of the vectors is in the direction of increasing z, which may be either toward or away from the interface. As shown in Eq. (6.1-6), the sign of the gradient is opposite to the direction of the diffusion flux, since diffusion is in the direction of lower concentrations, or "downhill," like the flow of heat "down" a temperature gradient.

There are several types of situations covered by Eq. (6.1-16). The simplest case is zero convective flow and equimolal counter diffusion of A and B, as occurs in the diffusive mixing of two gases. This is also the case for the diffusion of A and B in the vapor phase for distillations that have constant molal overflow. The second common case is the diffusion of only one component of the mixture, where the convective flow is caused by the diffusion of that component. Examples include evaporation of a liquid with diffusion of the vapor from the interface into a gas stream and condensation of a vapor in the presence of a noncondensable gas. Many examples of gas absorption also involve diffusion of only one component, which creates a convective flow toward the interface. These two types of mass transfer in gases are treated in the following sections for the simple case of steady-state mass transfer through a stagnant gas layer or film of known thickness. The effects of transient diffusion and laminar or turbulent flow are taken up later.

Equimolal diffusion

For equimolal diffusion in gases, the net volumetric and molar flows are zero, and Eq. (6.1-16) or Eq. (6.1-17) can be used with the convective term set to zero, which makes them equivalent to Eq. (6.1-6). Equation (6.1-17) is integrated over a film thickness Z, assuming a constant flux N_A and zero total flux:

$$-Dc_M \int_{y_{Ai}}^{y_A} dy_A = N_A \int_0^z dz \qquad (6.1\text{-}18)$$

where y_A = mole fraction of A at outer edge of film

y_{Ai} = mole fraction of A at interface or inner edge of film

Integration of Eq. (6.1-18) and rearrangement give

$$N_A = J_A = \frac{Dc_M}{z}(y_{Ai} - y_A) \qquad (6.1\text{-}19)$$

or

$$N_A = J_A = \frac{D}{z}(c_{Ai} - c_A) \qquad (6.1\text{-}20)$$

The concentration gradient for A is linear in the film, and the gradient for B has the same magnitude but the opposite sign, as shown in Fig. 6.1(a). Note that for equimolal diffusion $N_A = J_A$.

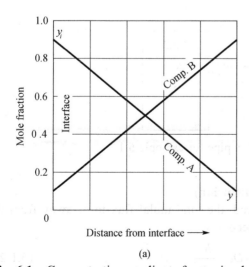
(a)

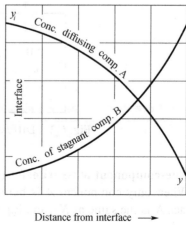
(b)

Fig. 6.1 Concentration gradients for equimolal and one-way diffusion: (a) components A and B diffusing at same molal rates in opposite directions; (b) component A diffusing, component B stationary with respect to interface.

EXAMPLE 6.1 Two large vessels of A and B as shown in Fig. 6.2 are connected by a pipe with an internal diameter of 24.4mm and a length of 0.61m. The total pressure of the system is 101.3kPa, and the temperature is 25℃. Vessel A and B contains homogeneous mixture of ammonia and nitrogen, the partial pressure of ammonia in vessel A and B is 20kPa and 6.67kPa, respectively. The diffusivity of NH_3-N_2 is 2.30×10^{-5} m²/s at 25℃ and 101.3kPa. (a) What is the rate of ammonia in kmol /s transferred from vessel A to vessel B. (b) What is the partial pressure of ammonia in kPa at the place of 0.305m away from the section 1 in pipe.

Solution.

(a) From Eq. (6.1-19) for equimolal diffusion, the flux of ammonia is

$$N_A = J_A = \frac{Dc_M}{z}(y_{A1} - y_{A2}) = \frac{DP}{RTz}(y_{A1} - y_{A2}) = \frac{D}{RTz}(p_{A1} - p_{A2})$$

$$= \frac{2.3 \times 10^{-5}}{8.314 \times 298 \times 0.61} \times (20 - 6.67) = 2.03 \times 10^{-7} \text{ kmol}/(m^2 \cdot s)$$

The area of the pipe is

$$A = \frac{\pi}{4}d^2 = \frac{\pi}{4} \times 0.0244^2 = 4.68 \times 10^{-4} \text{ m}^2$$

The rate of ammonia transferred from section 1 to section 2 is

$$N_A A = 2.03 \times 10^{-7} \times 4.68 \times 10^{-4} = 9.50 \times 10^{-11} \text{ kmol/s}$$

(b) For a steady-state diffusion, $N_A A = const$, so $N_A = const$

Assume $z' = 0.305$m $\qquad p_{A2} = p'_{A2}$
$\qquad z = 0 \qquad p_{A1} = 20$kPa

Thus $\qquad N_A = \dfrac{D}{RTz'}(p_{A1} - p'_{A2})$

$$p'_{A2} = p_{A1} - \frac{N_A RTz}{D} = 20 - \frac{2.03 \times 10^{-7} \times 8.314 \times 298 \times 0.305}{2.30 \times 10^{-5}} = 13.3 \text{kPa}$$

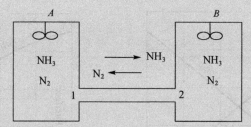

Fig. 6.2 Diffusion through a pipe for example 6.1.

One-component mass transfer (one-way diffusion)

When only component A is being transferred, the total molal flux to or away from the interface N is the same as N_A, and Eq. (6.1-17) becomes

$$N_A = y_A N_A - Dc_M \frac{dy_A}{dz} \qquad (6.1-21)$$

Rearranging and integrating, we have

$$N_A(1-y_A) = -Dc_M \frac{dy_A}{dz} \qquad (6.1\text{-}22)$$

$$\frac{N_A z}{Dc_M} = -\int_{y_{Ai}}^{y_A} \frac{dy_A}{1-y_A} = \ln\frac{1-y_A}{1-y_{Ai}} \qquad (6.1\text{-}23)$$

or

$$N_A = \frac{Dc_M}{z} \ln\frac{1-y_A}{1-y_{Ai}} \qquad (6.1\text{-}24)$$

Equation (6.1-24) can be rearranged by using the logarithmic mean of $1-y_A$ for easier comparison with Eq. (6.1-19) for equimolal diffusion. Since the driving force $y_{Ai} - y_A$ can be written $(1-y_A)-(1-y_{Ai})$, the logarithmic mean becomes

$$(1-y_A)_m = \frac{y_{Ai} - y_A}{\ln\left[(1-y_A)/(1-y_{Ai})\right]} \qquad (6.1\text{-}25)$$

Combining Eqs. (6.1-24) and (6.1-25) gives

$$N_A = \frac{Dc_M}{z} \frac{y_{Ai} - y_A}{(1-y_A)_m} \qquad (6.1\text{-}26)$$

The flux of component A for a given concentration difference is therefore greater for one-way diffusion than for equimolal diffusion, since the term $(1-y_A)_m$ is always less than 1.0.

The concentration gradient for one-way diffusion is not linear but is steeper at low values of y_A, as shown in Fig. 6.1(b). The gradient for component B can be obtained directly from the gradient for A, since $y_A + y_B = 1.0$ or $c_A + c_B = \rho_M$. There is no transfer of B toward the interface in spite of the large concentration gradient shown in Fig. 6.1(b). The explanation is that B tends to diffuse toward the region of lower concentration, but the diffusion flux is just matched by the convective flow carrying B in the opposite direction.

EXAMPLE 6.2 Two large vessels of A and B as shown in Fig. 6.2 are connected by a pipe with an internal diameter of 24.4mm and a length of 0.61m. The total pressure of the system is 101.3kPa, and the temperature is 25℃. Vessel A and B contains uniform mixture of ammonia and nitrogen, the partial pressure of ammonia in vessel A and B is 20kPa and 6.67kPa, respectively. The ammonia is transferred in stagnant nitrogen gas film of the pipe from section 1 to section 2 at the steady-state. The diffusivity of NH_3-N_2 is $2.30\times 10^{-5} m^2/s$ at 25℃ and 101.3kPa. (a) What is the rate of ammonia in kmol /s transferred from section 1 to section 2. (b) What is the partial pressure of ammonia in kPa at the place of 0.305m away from section 1 in pipe.
Solution.
(a) From Eq. (6.1-26) for one-way diffusion, the flux of ammonia in pipe is

$$N_A = \frac{Dc_M}{z} \frac{y_{A1} - y_{A2}}{(1-y_A)_m} = \frac{DP}{RTz} \frac{y_{A1} - y_{A2}}{(1-y_A)_m} = \frac{D}{RTz} \frac{p_{A1} - p_{A2}}{(1-y_A)_m}$$

$$\frac{1}{(1-y_A)_m} = \frac{\ln\left[(1-y_{A2})/(1-y_{A1})\right]}{y_{A1} - y_{A2}} = \frac{1}{y_{B2} - y_{B1}} \ln\left(\frac{y_{B2}}{y_{B1}}\right) = \frac{P}{p_{B2} - p_{B1}} \ln\left(\frac{p_{B2}}{p_{B1}}\right)$$

$$\frac{1}{(1-y_A)_m} = \frac{P}{p_{Bm}} \qquad p_{Bm} = \frac{p_{B2} - p_{B1}}{\ln(p_{B2}/p_{B1})}$$

Thus
$$N_A = \frac{D}{RTz} \times \frac{P}{p_{Bm}}(p_{A1} - p_{A2})$$

$$p_{B2} = P - p_{A2} = 101.3 - 6.67 = 94.6\,\text{kPa}$$

$$p_{B2} = P - p_{A2} = 101.3 - 20 = 81.3\,\text{kPa}$$

$$p_{Bm} = \frac{94.6 - 81.3}{\ln\frac{94.6}{81.3}} = 87.8\,\text{kPa}$$

$$N_A = \frac{D}{RTz} \times \frac{P}{p_{Bm}}(p_{A1} - p_{A2})$$

$$= \frac{2.30 \times 10^{-5}}{8.314 \times 298 \times 0.61} \times \frac{101.3}{87.8} \times (20 - 6.67) = 2.34 \times 10^{-7}\,\text{kmol}/(\text{m}^2\cdot\text{s})$$

The area of the pipe is
$$A = \frac{\pi}{4}d^2 = \frac{\pi}{4} \times 0.0244^2 = 4.68 \times 10^{-4}\,\text{m}^2$$

The rate of ammonia be transferred from section 1 to section 2 is
$$N_A A = 2.34 \times 10^{-7} \times 4.68 \times 10^{-4} = 10.95 \times 10^{-11}\,\text{kmol/s}$$

(b) Assume $z' = 0.305\,\text{m}$ $\quad p_{A2} = p'_{A2} \quad p_{B2} = p'_{B2}$

$z = 0$ $\quad p_{A1} = 20\,\text{kPa} \quad p_{B1} = 81.3\,\text{kPa}$

From Eq. (6.1-24) for one-way diffusion, the flux of ammonia in pipe is
$$N_A = \frac{Dc_M}{z}\ln\frac{1 - y_{A2}}{1 - y_{A1}} = \frac{DP}{RTz}\ln\frac{p_{B2}}{p_{B1}}$$

Thus
$$\ln\frac{p'_{B2}}{p_{B1}} = \frac{N_A RTz'}{DP} = \frac{2.34 \times 10^{-7} \times 8.314 \times 298 \times 0.305}{2.30 \times 10^{-5} \times 101.3} = 7.59 \times 10^{-2}$$

$$\frac{p'_{B2}}{p_{B1}} = e^{0.0759} = 1.08$$

$$p'_{B2} = 1.08 \times 81.3 = 87.8\,\text{kPa}$$

$$p'_{A2} = P - p'_{B2} = 101.3 - 87.8 = 13.5\,\text{kPa}$$

EXAMPLE 6.3 (a) For the diffusion of solute A through a layer of gas to an absorbing liquid, with $y_{Ai} = 0.20$ and $y_A = 0.10$, calculate the transfer rate for one-way diffusion compared to that for equimolal diffusion. (b) What is the value of y_A halfway through the layer for one-way diffusion?

Solution.
(a) From Eq. (6.1-19) for equimolal diffusion,
$$N_A = J_A = \frac{Dc_M}{z}(y_{Ai} - y_A) = \frac{Dc_M}{z}(0.20 - 0.10)$$

From Eq. (6.1-24) for one-way diffusion,
$$N_A = \frac{Dc_M}{z}\ln\frac{1 - y_A}{1 - y_{Ai}} = \frac{Dc_M}{z}\ln\frac{0.9}{0.8} = \frac{Dc_M}{z}(0.1178)$$

(The concentration terms in both equations are reversed to make the flux toward the interface positive.) The ratio of the fluxes is 0.1178/0.10 = 1.18. In this case the transfer rate with one-way diffusion is about 18 percent greater than that with equimolal diffusion.

(b) When $z' = z/2$,

$$\ln\frac{1-y_A}{0.8} = \frac{z}{2} \times \frac{N_A}{Dc_M} = \frac{0.1178}{2} = 0.0589$$

$$1 - y_A = 0.8485 \qquad y_A = 0.1515$$

The concentration at the midpoint is only slightly greater than if the gradient were linear (y_A=0.150).

6.2 Prediction of Diffusivities

Diffusivities are best estimated by experimental measurements, and where such information is available for the system of interest, it should be used directly. Often the desired values are not available, however, and they must be estimated from published correlations. Sometimes a value is available for one set of conditions of temperature and pressure; the correlations are then useful in predicting, from the known value, the desired values for other conditions.

Diffusion in gases

Values of D for some common gases diffusing in air at 0°C and 1 atm are given in App.17. A simple theory for gases shows that D is proportional to the product of the average molecular velocity $\bar{u}$ and the mean free path λ.

$$D \cong \frac{1}{3}\bar{u}\lambda \qquad (6.2\text{-}1)$$

Since the mean free path for ideal gases varies inversely with pressure, D also varies inversely with pressure, and the product DP can be considered constant up to about 10 atm. The mean molecular velocity depends on $T^{0.5}$, and since the mean free path increases with $T^{1.0}$, the simple theory predicts D varies with $T^{1.5}$, a term that appears in some empirical equations for diffusivity. A more rigorous approach based on modern kinetic theory allows for the different sizes and velocities of the molecules and the mutual interactions as they approach one another. Using the *Lennard-Jones* (6-12) *potential* with parameters ε and σ leads to the following equation for binary diffusion.

$$D_{AB} = \frac{0.001858T^{3/2}\left[(M_A + M_B)/M_A M_B\right]^{1/2}}{P\sigma_{AB}^2 \Omega_D} \qquad (6.2\text{-}2)$$

where D_{AB} = diffusivity, cm²/s
T = temperature, K
M_A, M_B = molecular weights of components A and B
P = pressure, atm
$\sigma_{AB} = (\sigma_A + \sigma_B)/2$ = effective collision diameter, Å
Ω_D = collision integral = $f(kT/\varepsilon_{AB})$
k = Boltzmann's constant
ε = Lennard-Jones force constant for common gases

$$\varepsilon_{AB} = \sqrt{\varepsilon_A \varepsilon_B}$$

Tables of σ, ε, and Ω_D for common gases are given in App. 18. Equation (6.2-2) is known as the *Chapman-Enskog equation*.

The collision integral Ω_D decreases with increasing temperature, which makes D_{AB} increase with more than the 1.5 power of the absolute temperature. The change in Ω_D with temperature is not very great, and for diffusion in air at temperatures from 300 to 1000 K, D varies with about $T^{1.7-1.8}$, and $T^{1.75}$ can be used to extrapolate from room temperature data.

Note the similarity of Eq. (6.2-2) to equation for gas viscosity and equation for thermal conductivity of ideal gases.

Diffusion in small pores. When gases diffuse in very small pores of a solid, as may occur during adsorption, drying of porous solids, or some membrane separation processes, the diffusivity is less than the normal value because of molecular collisions with the pore walls. When the pore size is much smaller than the normal mean free path, the diffusion process is called *Knudsen diffusion*, and the diffusivity for a cylindrical pore is

$$D_K = 9700 r \sqrt{\frac{T}{M}} \tag{6.2-3}$$

where D_K = Knudsen diffusivity, cm^2/s
  T = temperature, K
  M = molecular weight
  r = pore radius, cm

For pores of intermediate size, collisions with the wall and with other molecules are both important, and the diffusivity in the pore is predicted by combining the reciprocals of the bulk and Knudsen diffusivities.

$$\frac{1}{D_{pore}} = \frac{1}{D_{AB}} + \frac{1}{D_K} \tag{6.2-4}$$

EXAMPLE 6.4 Predict the volumetric diffusivity for benzene in air at 100°C and 2 atm by using the rigorous equation and by extrapolating from the published value for 0°C and 1 atm.

Solution. From App. 18, the force constants are as follows:

	ε/k	σ	M
Benzene	412.3	5.349	78.1
Air	78.6	3.711	29

Thus
$$\sigma_{AB} = \frac{5.349 + 3.711}{2} = 4.53$$

$$\varepsilon_{AB}/k = (412.3 \times 78.6)^{0.5} = 180$$

$$\frac{kT}{\varepsilon} = \frac{373}{180} = 2.072$$

From App. 18, $\Omega_D = 1.062$. From Eq. (6.2-2),

$$D_{AB} = \frac{0.001858 \times 373^{1.5} \left[(78.1 + 29)/78.1 \times 29 \right]^{0.5}}{2 \times 4.53^2 \times 1.062} = 0.0668 \, cm^2/s$$

From App. 17 at standard temperature and pressure,

$$D_{AB} = 0.299 \, ft^2/h = 0.0722 \, cm^2/s$$

At 373K and 2 atm,

$$D_{AB} \cong 0.0772 \times \frac{1}{2} \times \left(\frac{373}{273}\right)^{1.75} = 0.0666 \, \text{cm}^2/\text{s}$$

Agreement with the value calculated from Eq. (6.2-2) is very good.

Diffusion in liquids

The theory of diffusion in liquids is not as advanced or the experimental data as plentiful as for gas diffusion. Diffusivities in liquids are generally 4 to 5 orders of magnitude smaller than in gases at atmospheric pressure. Diffusion in liquids occurs by random motion of the molecules, but the average distance traveled between collisions is less than the molecular diameter, in contrast to gases, where the mean free path is orders of magnitude greater than the molecular size. As a result, diffusivities in liquids are generally 4 to 5 orders of magnitude smaller than in gases at atmospheric pressure. Because of the much greater liquid densities, however, the fluxes for a given mole fraction gradient in liquid or gas may be nearly the same.

Diffusivities for large spherical molecules in dilute solution can be predicted from the *Stokes-Einstein equation,* which was derived by considering the drag on a sphere moving in a continuous fluid.

$$D = \frac{kT}{6\pi r_0 \mu} \tag{6.2-5}$$

where k is the Boltzmann constant, 1.380×10^{23} J/K.

A convenient form of the equation is

$$D = \frac{7.32 \times 10^{-16} T}{r_0 \mu}$$

where D = diffusivity, cm^2/s
T = absolute temperature, K
r_0 = molecular radius, cm
μ = viscosity, cP

For solutes of small to moderate molecular weight ($M < 400$), the diffusivity in liquids is greater than that calculated from Eq. (6.2-5), because the drag is less than predicted for a continuum fluid. The diffusivity varies with about the 0.6 power of the molar volume rather than the $-\frac{1}{3}$ power derived from the Stokes-Einstein equation. A widely used correlation for the liquid diffusivity of small molecules is the empirical *Wilke-Chang equation*

$$D = 7.4 \times 10^{-8} \times \frac{(\psi_B M_B)^{1/2} T}{\mu V_A^{0.6}} \tag{6.2-6}$$

where D = diffusivity, cm^2/s
T = absolute temperature, K
μ = viscosity of solution, cP
V_A = molar volume of solute as liquid at its normal boiling point, cm^3/g mol
ψ_B = *association parameter* for solvent
M_B = molecular weight of solvent

The recommended values of ψ_B are 2.6 for water, 1.9 for methanol, 1.5 for ethanol, and presumably greater than 1.0 for other polar molecules that can associate by hydrogen bonding. It is 1.0 for benzene, heptane, and other unassociated solvents. When water is the solute, the diffusivities are about 2.3-fold below the values found from Eq. (6.2-6), suggesting an association parameter of 4.0 for water in organic liquids. Equation (6.2-6) is valid only at low solute concentrations and does not apply when the solution has been thickened by addition of high-molecular-weight polymers. Small amounts of polymer can raise the solution viscosity more than 100-fold or even gel the solution, but the diffusivity of small solutes is only slightly reduced, because the polymer chains are too far apart to obstruct the movement of the solute molecules.

For dilute aqueous solutions of nonelectrolytes a simpler equation can be used

$$D = \frac{13.26 \times 10^{-5}}{\mu_B^{1.14} V_A^{0.589}} \tag{6.2-7}$$

where μ_B = viscosity of water, cP

The diffusion coefficient for dilute solutions of completely ionized univalent electrolytes is given by the *Nernst equation*

$$D = \frac{2RT}{\left(1/\lambda_+^0 + 1/\lambda_-^0\right) F_a^2} \tag{6.2-8}$$

where λ_+^0, λ_-^0 = limiting (zero-concentration) ionic conductances,
A/cm^2 · (V/cm) · (g equivalent/cm^3)
R = gas constant, 8.314 J/(K · g mol)
F_a = Faraday constant, = 96500 Coulombs/g equivalent

Table 6.1 Limiting ionic conductances in water at 25℃[21b]

Cation	λ_+^0	Anion	λ_-^0
H$^+$	349.8	OH$^-$	197.6
Li$^+$	38.7	Cl$^-$	76.3
Na$^+$	50.1	Br$^-$	78.3
K$^+$	73.5	I$^-$	76.8
NH$_4^+$	73.4	NO$_3^-$	71.4

Table 6.1 lists values of λ^0 at 25℃. Values for higher temperatures can be estimated from the change in T/μ.

Note that unlike the case for binary gas mixtures the diffusion coefficient for a dilute solution of A in B is not the same as for a dilute solution of B in A, since μ, M_B, and V_A will be different when the solute and solvent are exchanged. For intermediate concentrations, an approximate value of D is sometimes obtained by interpolation between the dilute solution values, but this method can lead to large errors for nonideal solutions.

EXAMPLE 6.5 Winkelmann's Method are used to determine the diffusivity of CCl$_4$ in air, the equipment are showed in Fig. 6.3. The thermostatic vertical fine pipe contains liquid CCl$_4$. The air quickly pass through the horizontal pipe, and the partial pressure of CCl$_4$ in air at the top of the vertical pipe is near zero. Assume that the transport of CCl$_4$ from the surface of the liquid to the top of the vertical pipe is diffusion. The experiment is conducted at the temperature of 321K and the pressure of 101.3kPa. The experimental data are listed in table 6.2. The saturated vapor pressure of CCl$_4$ is 37.6kPa, and the density of liquid CCl$_4$ is 1540 kg/m^3 at 321K and 101.3kPa. Calculate the diffusivity of CCl$_4$ vapor in air at 321K and 101.3kPa.

Table 6.2 The experimental data for example 6.5

Time τ, s	Distance z, mm
0	0
1.6×10^3	2.5
11.1×10^3	12.9
27.4×10^3	23.2
80.2×10^3	43.9
117.5×10^3	54.7
168.6×10^3	67.0
199.7×10^3	73.8
289.3×10^3	90.3
383.1×10^3	104.8

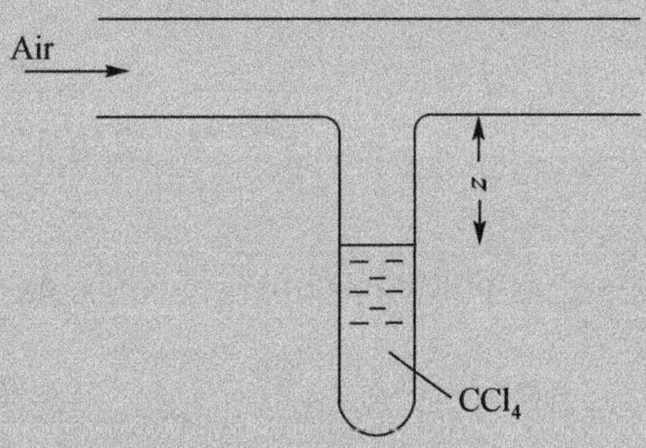

Fig. 6.3 The measurement of diffusivity of CCl_4 in air.

Solution. Because the gas in the vertical pipe is not disturbed by the horizontal gas flow outside the pipe, the transport of CCl_4 vapor from the surface of the liquid to the top of the vertical pipe is the diffusion of CCl_4 through the stagnant air film and keeps equilibrium state. The vaporizing rate of liquid CCl_4 is equal to the transport rate of CCl_4 vapor in the vertical pipe.

From Eq. (6.1-24) for one-way diffusion, the transport rate of CCl_4 vapor in pipe is

$$N_A = \frac{Dc_M}{z} \ln \frac{1-y_{A2}}{1-y_{A1}} = \frac{DP}{RTz} \ln \frac{p_{B2}}{p_{B1}}$$

when $\quad z = z_0 \quad p_{A1} = p^* \quad p_{B1} = P - p^*$
$\quad\quad\quad z = z \quad p_{A2} = 0 \quad p_{B2} = 101.3 \text{kPa} = P$

Thus
$$N_A = \frac{DP}{RTz} \ln \frac{P}{P-p^*} \tag{a}$$

With the vaporizing of liquid CCl_4, the surface of the liquid is decreased and the distance of z is increased gradually. Assume the cross section of the pipe is A and the density of CCl_4 is ρ_L. During the period of $d\tau$, the amount of vaporizing CCl_4 is equal to that of the diffusing CCl_4 that leave the top of pipe. The rate of surface decreasing $\frac{dz}{d\tau}$ has the following relationship with the transport rate of CCl_4 in the vertical pipe.

Thus
$$AN_A d\tau = \frac{\rho_L A dz}{M_A}$$

or
$$N_A = \frac{\rho_L}{M_A} \frac{dz}{d\tau} \tag{b}$$

Substituting the Eq. (b) in Eq. (a) and integrating

$$\int_{z_0}^{z} z dz = \frac{DPM_A}{RT\rho_L} \ln \frac{P}{P-p^*} \int_0^{\tau} d\tau$$

$$\frac{1}{2}(z^2 - z_0^2) = \frac{DPM_A}{RT\rho_L} \ln \frac{P}{P-p^*} \tau \tag{c}$$

where
$$(z^2 - z_0^2) = (z - z_0)(z + z_0) = (z - z_0)\left[(z - z_0) + 2z_0\right] \quad \text{(d)}$$

Substituting the Eq. (d) in Eq. (c)

$$\frac{\tau}{z - z_0} = \frac{RT\rho_L}{2DPM_A \ln\frac{P}{P - p^*}}(z - z_0) + \frac{RT\rho_L}{DPM_A \ln\frac{P}{P - p^*}}z_0$$

$\dfrac{\tau}{z - z_0}$ is plotted against $(z - z_0)$, as shown in table 6.3 and Fig. 6.4. The slope is

$$s = \frac{RT\rho_L}{2DPM_A \ln\frac{P}{P - p^*}} = 3.0 \times 10^7 \, \text{s/m}^2$$

The diffusivity of CCl_4 is

$$D = \frac{RT\rho_L}{2sPM_A \ln\frac{P}{P - p^*}} = \frac{8.314 \times 321 \times 1540}{2 \times 3.0 \times 10^7 \times 101.3 \times 154 \times \ln\frac{101.3}{101.3 - 37.6}} = 9.47 \times 10^{-6} \, \text{m}^2/\text{s}$$

The diffusivity of CCl_4 vapor in air at 321K and 101.3kPa is $9.47 \times 10^{-6} \, \text{m}^2/s$.

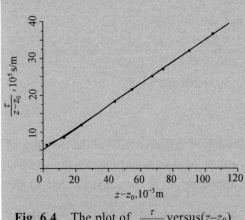

Fig. 6.4 The plot of $\dfrac{\tau}{z - z_0}$ versus $(z - z_0)$.

Table 6.3 The data sheet of $\tau/(z-z_0)$ versus $(z-z_0)$

Time τ, s	$z-z_0$, m	$\dfrac{\tau}{z-z_0}$, s/m
0	0	
1.6×10^3	2.5×10^3	6.4×10^5
11.1×10^3	12.9×10^3	8.6×10^5
27.4×10^3	23.2×10^3	11.8×10^5
80.2×10^3	43.9×10^3	18.3×10^5
117.5×10^3	54.7×10^3	21.5×10^5
168.6×10^3	67.0×10^3	25.2×10^5
199.7×10^3	73.8×10^3	27.1×10^5
289.3×10^3	90.3×10^3	32.0×10^5
383.1×10^3	104.8×10^3	36.6×10^5

Schmidt number

The ratio of the kinematic viscosity to the molecular diffusivity is known as the *Schmidt number*, designated by Sc.

$$Sc = \frac{\upsilon}{D} = \frac{\mu}{\rho D}$$

The Schmidt number is analogous to the Prandtl number, which is the ratio of the kinematic viscosity to the thermal diffusivity (see Chap. 4).

$$Pr = \frac{\upsilon}{\alpha} = \frac{\mu}{\rho\alpha} = \frac{\mu}{\rho\left[k/(\rho c_p)\right]} = \frac{c_p\mu}{k}$$

Schmidt numbers for gases in air at 0℃ and 1 atm are given in App.17. Most of the values are between 0.5 and 2.0. The Schmidt number is independent of pressure when the ideal gas law applies, since the viscosity is independent of pressure, and the effects of pressure on ρ and D cancel. Temperature has only a slight effect on the Schmidt number for gases because μ and ρD both change with about $T^{0.7-0.8}$.

The Schmidt numbers for liquids range from about 10^2 to 10^5 for typical mixtures. For small solutes in water at 20℃ where $D \cong 10^{-5}$ cm²/s, $Sc \cong 10^3$. The Schmidt number decreases markedly with increasing temperature because of the decreasing viscosity and the increase in the diffusivity.

EXAMPLE 6.6 Estimate the diffusivity of benzene in toluene and toluene in benzene at 110℃. The physical properties are as follows:

	M	Boiling point, ℃	V_A at boiling point, cm³/mol	μ at 110℃, cP
Benzene	78.11	80.1	96.5	0.24
Toluene	92.13	110.6	118.3	0.26

Solution. Equation (6.2-6) will be used. For benzene in toluene,

$$D = 7.4 \times 10^{-8} \times \frac{92.13^{1/2} \times 383}{0.26 \times 96.5^{0.6}} = 6.74 \times 10^{-5} \text{ cm}^2/\text{s}$$

For toluene in benzene,

$$D = 7.4 \times 10^{-8} \times \frac{78.11^{1/2} \times 383}{0.24 \times 118.3^{0.6}} = 5.95 \times 10^{-5} \text{ cm}^2/\text{s}$$

Turbulent diffusion

In a turbulent stream the moving eddies transport matter from one location to another, just as they transport momentum and heat energy. By analogy with the equations for momentum transfer and heat transfer in turbulent streams, the equation for mass transfer is

$$J_{A,t} = -D_E \frac{dc_A}{dz} \tag{6.2-9}$$

where $J_{A,t}$ = molal flux of A, *relative to phase as a whole*, caused by turbulent action
D_E = eddy diffusivity

The total molal flux, relative to the entire phase, becomes

$$J_A = -(D + D_E) \frac{dc_A}{dz} \tag{6.2-10}$$

The eddy diffusivity depends on the fluid properties but also on the velocity and position in the flowing stream. Therefore Eq. (6.2-10) cannot be directly integrated to determine the flux for a given concentration difference. This equation is used with theoretical or empirical relationships for D_E in fundamental studies of mass transfer, and similar equations are used for heat or momentum transfer in developing analogies between the transfer processes. Such studies are beyond the scope of this text, but Eq. (6.2-10) is useful in helping to understand the form of some empirical correlations for mass transfer.

6.3 Mass-Transfer Theories

For steady-state mass transfer through a stagnant layer of fluid Eq.(6.1-19) or Eq.(6.1-24) can be used to predict the mass-transfer rate, provided Z is known. However, this is not a common situation, because in most mass-transfer operations turbulent flow is desired to increase the rate of transfer per unit area or to help disperse one fluid in another and create more interfacial area. Furthermore, mass transfer to a fluid interface is often of the unsteady-state type, with continuously changing concentration gradients and mass-transfer rates. In spite of these differences, mass transfer in most cases is treated using the same type of equations, which feature a *mass-transfer coefficient k*. This coefficient is defined as the rate of mass transfer per unit area per unit concentration difference and is usually based on equal molal flows. The concentrations can be expressed in moles per volume or mole fractions, with subscript c indicating concentration and y or x mole fractions in the vapor or liquid phase:

$$k_c = \frac{J_A}{c_{Ai} - c_A} \tag{6.3-1}$$

or

$$k_y = \frac{J_A}{y_{Ai} - y_A} \tag{6.3-2}$$

Since k_c is a molar flux divided by a concentration difference, it has the units of velocity, such as centimeters per second or meters per second:

$$k_c = \frac{\text{mol}}{s \cdot \text{cm}^2 \cdot \text{mol}/\text{cm}^3} = \text{cm}/s$$

For k_y or k_x the units are the same as for J_A, moles per area per time, since the mole fraction driving force is dimensionless. It is apparent that k_c and k_y are related by the molar density as follows:

$$k_y = k_c \rho_M = \frac{k_c p}{RT} \tag{6.3-3}$$

$$k_x = k_c \rho_M = \frac{k_c \rho_s}{M} \tag{6.3-4}$$

Gas-phase coefficients are sometimes based on the partial pressure driving force and are denoted by k_g.

$$k_g = \frac{J_A}{p_{Ai} - p_A} \tag{6.3-5}$$

Then

$$k_g = \frac{k_y}{P} = \frac{k_c}{RT} \tag{6.3-6}$$

The significance of k_c is brought out by combining Eq. (6.3-1) with Eq. (6.1-20) for steady-state equimolal diffusion in a stagnant film. This gives

$$k_c = \frac{J_A}{c_{Ai} - c_A} = \frac{D(c_{Ai} - c_A)}{Z} \times \frac{1}{c_{Ai} - c_A} = \frac{D}{Z} \tag{6.3-7}$$

Thus the coefficient k_c is the molecular diffusivity divided by the thickness of the stagnant layer. When we are dealing with unsteady-state diffusion or diffusion in flowing streams, Eq. (6.3-7) can still be used to give an effective film thickness from known values of k_c and D.

Film Theory

The basic concept of the film theory is that the resistance to diffusion can be considered equivalent to that in a stagnant film of a certain thickness. The implication is that the coefficient k_c varies with the first power of D, which is rarely true, but this does not detract from the value of the theory in many applications. The film theory is often used as a basis for complex problems of multicomponent diffusion or diffusion plus chemical reaction.

For example, consider mass transfer from a turbulent gas stream to the wall of a pipe, with the concentration gradient as shown in Fig. 6.5. There is a laminar layer near the wall, where mass transfer is mainly by molecular diffusion, and the concentration gradient is almost linear. As the distance from the wall increases, turbulence becomes stronger, and the eddy diffusivity increases, which means that a lower gradient is needed for the same flux [see Eq. (6.2-10)]. The value of c_A is a maximum at the center of the pipe, but this value is *not* used in mass-transfer calculations. Instead the driving force is taken as $c_A - c_{Ai}$, where c_A is the concentration reached if the stream were thoroughly mixed. This is the same as a flow-weighted average concentration and is also the concentration to be used in material-balance calculations. (This is analogous to the usage in heat transfer, where the average temperature of a stream is used in defining h.)

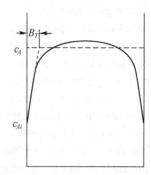

Fig. 6.5 Concentration gradient for mass transfer in a pipe with turbulent flow of gas.

If the gradient near the wall is linear, it can be extrapolated to c_A, and the distance from the wall at this point is the effective film thickness B_T. Generally, the resistance to mass transfer is mainly in the laminar boundary layer very close to the wall, and B_T is only slightly greater than the thickness of the laminar layer. However, as will be brought out later, the value of B_T depends on the diffusivity D and not just on flow parameters, such as the Reynolds number. The concept of an effective film thickness is useful, but values of B_T must not be confused with the actual thickness of the laminar layer.

Effect of one-way diffusion

As shown previously, when only component A is diffusing through a stagnant film, the rate of mass transfer for a given concentration difference is greater than if component B is diffusing in the opposite direction. From Eqs. (6.1-19) and (6.1-26), the ratio of the fluxes is

$$\frac{N_A}{J_A} = \frac{1}{(1-y_A)_m} = \frac{1}{y_{Bm}} \qquad (6.3\text{-}8)$$

This relationship, derived for molecular diffusion in a stagnant film, is assumed to hold reasonably well for unsteady-state diffusion or a combination of molecular and eddy diffusion. Sometimes the mass-transfer coefficient for one-way transfer is denoted by k_c' or k_y', and the coefficients then follow the same relationship as the fluxes in Eq. (6.3-8):

$$\frac{k_c'}{k_c} = \frac{k_y'}{k_y} = \frac{1}{(1-y_A)_m} \qquad (6.3\text{-}9)$$

The rate of one-way mass transfer can be expressed using either type of coefficient:

$$N_A = k_y'(y_{Ai} - y_A) \qquad (6.3\text{-}10)$$

$$N_A = \frac{k_y(y_{Ai} - y_A)}{(1 - y_A)_m} \tag{6.3-11}$$

When the value of y_A is 0.10 or less, the difference between k_y and k'_y is small and often ignored in design calculations. For mass transfer in the liquid phase, the corresponding correction term for one-way diffusion $(1 - x_A)_m$ is usually omitted, because the correction is small compared to the uncertainty in the diffusivity and the mass-transfer coefficient.

Boundary Layer Theory

Although there are few examples of diffusion through a stagnant fluid film, mass transfer often takes place in a thin boundary layer near a surface where the fluid is in laminar flow. If the velocity gradient in the boundary layer is linear and the velocity is zero at the surface, the equations for flow and diffusion can be solved to give the concentration gradient and the average mass-transfer coefficient. The coefficient depends on the two-thirds power of the diffusivity and decreases with increasing distance along the surface in the direction of flow, because an increase either in distance or in D makes the concentration gradient extend farther from the surface, which decreases the gradient dc_A/dz at the surface.

For flow over a flat plate or around a cylinder or sphere, the velocity profile is linear near the surface, but the gradient decreases as the velocity approaches that of the main stream at the outer edge of the boundary layer. Exact calculations show that the mass-transfer coefficient still varies with $D^{2/3}$ if D is low or the Schmidt number $\mu/\rho D$ is 10 or larger. For Schmidt numbers of about 1, typical for gases, the predicted coefficient varies with a slightly lower power of D. For boundary layer flows, no matter what the shape of the velocity profile or value of the physical properties, the transfer rate cannot increase with the 1.0 power of the diffusivity, as implied by the film theory. Boundary layer theory can be used to estimate k_c for some situations, but when the boundary layer becomes turbulent or separation occurs, exact predictions of k_c cannot be made, and the theory serves mainly as a guide in developing empirical correlations. The analogy between heat and mass transfer in boundary layers permits correlations developed for heat transfer to be used for mass transfer. For example, the average coefficient for mass transfer through a laminar boundary layer on a flat plate is obtained from chapter 4 by replacing Nu with Sh and Pr with Sc.

$$Sh = \frac{k_c z}{D} = 0.664(Sc)^{1/3}(Re_z)^{1/2} \tag{6.3-12}$$

where z is the total length of the plate.

Penetration Theory

The penetration theory makes use of the expression for the transient rate of diffusion into a relatively thick mass of fluid with a constant concentration at the surface. The change in concentration with distance and time is governed by Fick's second law:

$$\frac{\partial c_A}{\partial t} = D_{AB}\frac{\partial^2 c_A}{\partial b^2} \tag{6.3-13}$$

The boundary conditions are

$$c_A = \begin{cases} c_{A0} & \text{for } t = 0 \\ c_{Ai} & \text{at } b = 0, t > 0 \end{cases}$$

The particular solution of Eq. (6.3-13) is the same as that for transient heat conduction to a semi-infinite solid.

The instantaneous flux at time t is given by :

$$J_A = \sqrt{\frac{D}{\pi t}}(c_{Ai} - c_A) \qquad (6.3\text{-}14)$$

The average flux over the time interval 0 to t_T is

$$\overline{J}_A = \frac{1}{t_T}\int_0^{t_T} J_A dt = \frac{c_{Ai} - c_A}{t_T}\sqrt{\frac{D}{\pi}}\int_0^{t_T}\frac{dt}{t^{1/2}} = 2\sqrt{\frac{D}{\pi t_T}}(c_{Ai} - c_A) \qquad (6.3\text{-}15)$$

Combining Eqs.(6.3-1) and (6.3-15) gives the average mass-transfer coefficient over time t_T:

$$\overline{k}_c = 2\sqrt{\frac{D}{\pi t_T}} = 1.13\sqrt{\frac{D}{t_T}} \qquad (6.3\text{-}16)$$

Higbie was the first to apply this equation to gas absorption in a liquid, showing that diffusing molecules will not reach the other side of a thin layer if the contact time is short. The depth of penetration, defined as the distance at which the concentration change is 1 percent of the final value, is $3.6\sqrt{Dt_T}$. For $D = 10^{-5}$ cm²/s and $t_T = 10$ s, the depth of penetration is only 0.036 cm. In gas absorption equipment, drops and bubbles often have very short lifetimes because of coalescence, and the penetration theory is likely to apply.

An alternate form of the penetration theory was developed by Danckwerts, who considered the case where elements of fluid at a transfer surface are randomly replaced by fresh fluid from the bulk stream. An exponential distribution of ages or contact times results, and the average transfer coefficient is given by

$$\overline{k}_c = \sqrt{Ds} \qquad (6.3\text{-}17)$$

where s is the fractional rate of surface renewal, in s^{-1}.

Both Eqs.(6.3-16) and (6.3-17) predict that the coefficient varies with the one-half power of the diffusivity and give almost the same value for a given average contact time. A modified version of the penetration theory, which assumes that eddies from a turbulent bulk fluid come to within random distances of the surface, gives slightly higher exponents for the diffusivity, which indicates that this theory might apply for mass transfer to pipe walls or flat surfaces such as a pool of liquid.

The various forms of the penetration theory can be classified as surface renewal models, implying either formation of new surface at frequent intervals or replacement of fluid elements at the surface with fresh fluid from the bulk. The time t_T and its reciprocal, the average rate of renewal, are functions of the fluid velocity, the fluid properties, and the geometry of the system and can be accurately predicted in only a few special cases. However, even if t_T must be determined empirically, the surface renewal models give a sound basis for correlation of mass-transfer data in many situations, particularly for transfer to drops and bubbles. The similarity between Eq.(6.3-16) and equation for the Heat-transfer coefficients in scraped-surface exchanger is an example of the close analogy between heat and mass transfer. It is often reasonable to assume that t_T is the same for both processes and thus to estimate rates of heat transfer from measured mass-transfer rates or vice versa.

Two-Film Theory

In many separation processes, material must diffuse from one phase into another phase, and the rates of diffusion in both phases affect the overall rate of mass transfer. In the two-film theory, proposed by Whitman in 1923, equilibrium is assumed at the interface, and the resistances to mass transfer in the two phases are added to get an overall resistance, just as is done for heat transfer. The reciprocal of the overall resistance is an overall coefficient, which is easier to use for design calculations than the individual coefficients.

What makes mass transfer between phases more complex than heat transfer is the discontinuity at the interface, which occurs because the concentration or mole fraction of diffusing solute is hardly ever the same on opposite sides of the interface. For example, in distillation of a binary mixture, y_A^* is greater than x_A, and the gradients near the surface of a bubble might be as shown in Fig. 6.6(a). For the absorption of a very soluble gas, the mole fraction in the liquid at the interface would be greater than that in the gas, as shown in Fig. 6.6(b).

In the two-film theory, the rate of transfer to the interface is set equal to the rate of the transfer from the interface:

$$N = k_x(x_A - x_{Ai}) \tag{6.3-18}$$

$$N = k_y(y_{Ai} - y_A) \tag{6.3-19}$$

The rate is also set equal to an overall coefficient K_y times an overall driving force $y_A^* - y_A$, where y_A^* is the composition of the vapor that would be in equilibrium with the bulk liquid of composition x_A:

$$N = K_y(y_A^* - y_A) \tag{6.3-20}$$

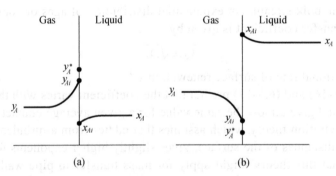

Fig. 6.6 Concentration gradients near a gas-liquid interface: (a) distillation; (b) absorption of a very soluble gas.

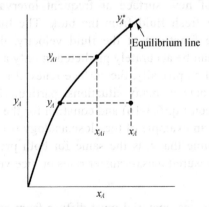

Fig. 6.7 Bulk and interface concentrations typical of distillation.

To get K_y in terms of k_y and k_x, Eq. (6.3-20) is rearranged and the term $y_A^* - y_A$ replaced by $(y_A^* - y_{Ai}) + (y_{Ai} - y_A)$:

$$\frac{1}{K_y} = \frac{y_A^* - y_A}{N} = \frac{y_A^* - y_{Ai}}{N} + \frac{y_{Ai} - y_A}{N} \tag{6.3-21}$$

Equations (6.3-18) and (6.3-19) are now used to replace N in the last two terms of Eq. (6.3-21):

$$\frac{1}{K_y} = \frac{y_A^* - y_{Ai}}{k_x(x_A - x_{Ai})} + \frac{y_{Ai} - y_A}{k_y(y_{Ai} - y_A)}$$

Fig. 6.7 shows typical values of the composition at the interface, and it is apparent that $(y_A^* - y_{Ai})/(x_A - x_{Ai})$ is the local slope of the equilibrium curve. This slope is denoted by m. The equation can then be written

$$\frac{1}{K_y} = \frac{m}{k_x} + \frac{1}{k_y} \tag{6.3-22}$$

The term $1/K_y$ can be considered an overall resistance to mass transfer, and the terms m/k_x and $1/k_y$ are the resistances in the liquid and gas films. These "films" need not be stagnant layers of a certain thickness in order for the two-film theory to apply. Mass transfer in either film may be by diffusion through a laminar boundary layer or by unsteady-state diffusion, as in the penetration theory, and the overall coefficient is still obtained from Eq. (6.3-22). For some problems, such as transfer through a stagnant film into a phase where the penetration theory is thought to apply, the penetration theory coefficient is slightly changed because of the varying concentration at the interface, but this effect is only of academic interest.

The essential part of the two-film theory is the method of allowing for the partition of solute between phases at the interface while combining individual coefficients to get an overall coefficient. This approach is used in the analysis of laboratory data and in equipment design for many types of mass-transfer operations, including absorption, adsorption, extraction, and distillation. The same principle of adding resistances with the appropriate distribution factors also applies to membrane separations, in which there are three resistances in series.

PROBLEMS

6.1 Carbon dioxide is diffusing through nitrogen in one direction at atmospheric pressure and 0°C. The mole fraction of CO_2 at point A is 0.2; at point B, 3 m away, in the direction of diffusion, it is 0.02. Diffusivity D is 0.144 cm²/s. The gas phase as a whole is stationary; that is, nitrogen is diffusing at the same rate as the carbon dioxide, but in the opposite direction. (a) What is the molal flux of CO_2, in kilogram moles per square meter per hour? (b) What is the *net* mass flux, in kilograms per square meter per hour? (c) At what speed, in meters per second, would an observer have to move from one point to the other so that the net mass flux, *relative to him or her,* would be zero? (d) At what speed would the observer have to move so that, relative to him or her, the *nitrogen* is stationary? (e) What would be the molal flux of carbon dioxide relative to the observer under condition (d)?

6.2 An open circular tank 8 m in diameter contains n-propanol at 25°C exposed to the atmosphere in such a manner that the liquid is covered with a stagnant air film estimated to be 5 mm thick. The concentration of propanol beyond the stagnant film is negligible. The vapor pressure of propanol at 25 °C is 20 mm Hg. If propanol is worth $1.20 per liter, what is the value of the loss of propanol from this tank in dollars per day? The specific gravity of propanol is 0.80.

6.3 Ethanol vapor is being absorbed from a mixture of alcohol vapor and water vapor by means of a nonvolatile solvent in which alcohol is soluble but water is not. The temperature is 97 °C, and the total pressure is 760 mm Hg. The alcohol vapor can be considered to be diffusing through a film of alcohol-water vapor mixture 0.1 mm thick. The mole percent of the alcohol in the vapor at the outside of the film is 80 percent, and that on the inside, next to the solvent, is 10 percent. The volumetric diffusivity of alcohol-water vapor mixtures at 25 °C and 1 atm is 0.15 cm^2/s. Calculate the rate of diffusion of alcohol vapor in kilograms per hour if the area of the film is 10 m^2.

6.4 An ethanol-water vapor mixture is being rectified by contact with an alcohol-water liquid solution. Alcohol is being transferred from gas to liquid and water from liquid to gas. The molal flow rates of alcohol and water are equal but in opposite directions. The temperature is 95 °C and the pressure is 1 atm. Both components are diffusing through a gas film 0.1 mm thick. The mole percentage of the alcohol at the outside of the film is 80 percent, and that on the inside is 10 percent. Calculate the rate of diffusion of alcohol and of water in pounds per hour through a film area of 10 m^2.

6.5 A wetted-wall column operating at a total pressure of 518 mm Hg is supplied with water and air, the latter at a rate of 120 g/min. The partial pressure of the water vapor in the air stream is 76 mm Hg, and the vapor pressure of the liquid-water film on the wall of the tower is 138 mm Hg. The observed rate of vaporization of water into the air is 13.1 g/min. The same equipment, now at a total pressure of 820 mm, is supplied with air at the same temperature as before and at a rate of 100 g/min. The liquid vaporized is n-butyl alcohol. The partial pressure of the alcohol is 30.5 mm, and the vapor pressure of the liquid alcohol is 54.5 mm. What rate of vaporization, in grams per minute, may be expected in the experiment with n-butyl alcohol?

6.6 Air at 40 °C and 2.0 atm is passed through a shallow bed of naphthalene spheres 12 mm in diameter at a rate of 2 m/s, based on the empty cross section of the bed. The vapor pressure of naphthalene is 0.35 mm Hg. How many kilograms per hour of naphthalene will evaporate from 1 m^3 of bed, assuming a bed porosity of 40 percent?

6.7 Diffusion coefficients for vapors in air can be determined by measuring the rate of evaporation of a liquid from a vertical glass tube. For a tube 0.2 cm in diameter filled with n-heptane at 21 °C, calculate the expected rate of decrease of the liquid level when the meniscus is 1 cm from the top based on the published diffusivity of 0.071 cm^2/s. At 21 °C the vapor pressure and density of n-heptane are 0.050 atm and 0.66 g/cm^3, respectively. Would there be any advantage in using a larger-diameter tube?

6.8 Estimate the liquid-film mass-transfer coefficient for O_2 diffusing from an air bubble rising through water at 20 °C. Choose a bubble size of 4.0 mm, assume a spherical shape, and assume rapid circulation of gas inside the bubble. Neglecting the change of bubble size with distance traveled, calculate the fraction of oxygen absorbed from the air in 1m of travel if the water contains no dissolved oxygen.

6.9 Small spheres of solid benzoic acid are dissolved in water in an agitated tank. If the Sherwood number is nearly constant at a value of 4.0, show how the time for complete dissolution varies with the initial size of the particle. How much time would be required for

100-μm particles to dissolve completely in pure water at 25°C? Solubility: 0.43 g/100 g H$_2$O. $D = 1.21 \times 10^{-5}$ cm^2/s.

6.10 Water saturated with air at 20°C is passed through hydrophobic hollow fibers at 50cm/s. The fibers are 1m long with an inner diameter of 500μm, and vacuum is applied on the outside to remove oxygen as fast as it diffuses to the fiber wall. Estimate the mass-transfer coefficient for oxygen.

6.11 In the absorption of ammonia from air into water at 20°C, the slope of the equilibrium line is about 1.0. Estimate the fraction of the total resistance in the gas phase, assuming that the penetration theory applied to both phases.

6.12 If the liquid film resistance is 5 times that of the gas film for a gas absorption process, by how much would the rate of absorption change if the liquid film coefficient could be doubled without changing other parameters? What would be the effect of doubling the gas-film coefficient?

6.13 Nitrobenzene was adsorbed from saturated air at 33°C in cylinders of activated carbon 4 mm in diameter and 10mm long. The cylinders were embedded in Teflon with the top circular face exposed to the air so that only axial diffusion took place. The concentration profiles for nitrobenzene were measured at several times using nuclear magnetic resonance. Data taken at 64 hours are as follows:

Distance, mm	2	4	6	8	10
c/c_s	0.78	0.48	0.11	0.01	0

(*a*) Using 8mm as the penetration distance, calculate the average value of the effective diffusivity D_e. (*b*) Using this value of D_e, plot the predicted concentration profile and compare it with the measured profile. What might explain the difference?

6.14 A transdermal patch for drug delivery contains a 0.06M solution of the drug in a capsule 2mm thick and 3.0cm in diameter. Just before application, a 250-μm polymer film is placed between the capsule and the skin. The drug diffusion coefficient in solution is 5.4×10^{-6} cm^2/s; in the polymer it is 1.8×10^{-7} cm^2/s. The equilibrium concentration in the polymer phase is 0.6 times that in the solution. The drug is rapidly absorbed by the skin, so the concentration at the polymer-skin interface is assumed to be zero. (*a*) What fraction of the drug in the capsule is absorbed in the first 2 hours? (*b*) At what time will the rate of drug delivery be only half the initial rate? How many moles will have been delivered at this time? (*c*) Can you suggest a method of getting a more nearly constant rate of drug delivery?

6.15 Set up and solve the differential equation for mass transfer from a small sphere of a pure slightly soluble solid suspended in a large mass of stagnant water.

名 词 术 语

英文 / 中文

activity gradient / 活度梯度
adsorption / 吸收
association parameter for solvent / 溶剂的缔合参数
average molecular velocity / 平均分子速度
boundary layer / 边界层
collision integral / 碰撞积分
concentration gradient / 浓度梯度

英文 / 中文

constant molal overflow / 恒摩尔流
convective bulk flow / 对流总体流动
convective flow / 对流流动
distillation / 蒸馏
downhill / 向下
eddy diffusion / 涡流扩散
eddy diffusivity / 涡流扩散系数
effective collision diameter / 有效碰撞直径

英文 / 中文

effective film thickness / 有效膜厚
equimolal counter diffusion / 等摩尔反向扩散
equimolal diffusion / 等分子扩散
extract phase / 萃取相
flow-weighted average concentration / 流动权重平均浓度
forced diffusion / 强制扩散
gas-liquid interface / 气液界面
high boiler / 高沸物
individual coefficient / 分系数
instantaneous flux / 瞬时通量
leaching / 浸提
Lennard-Jones(6-12)potential / Lennard-Jones 6-12 次方位能
Lennard-Jones force constant / Lennard-Jones 力常数
logarithmic mean / 对数平均
mass-transfer coefficient / 质量传递系数
mean free path / 平均自由程
molar density / 摩尔密度
molar flux / 摩尔通量
molar volume / 摩尔体积
molecular diffusion / 分子扩散
multicomponent diffusion / 多组分扩散

英文 / 中文

one-dimensional diffusion / 一维扩散
one-way diffusion / 单向扩散
one-way mass transfer / 单向质量传递
overall coefficient / 总系数
packed column / 填料塔
partial pressure driving force / 分压推动力
penetration theory / 渗透理论
pressure gradient / 压力梯度
pseudo-binary mixtures / 拟二元混合物
raffinate phase / 萃余相
random motion / 随机运动
reference plane / 参考面
reverse osmosis / 反渗透
Schmidt number / 施密特数
stagnant film / 滞流膜
stagnant gas layer / 气体滞流层
steady-state diffusion / 稳态扩散
temperature gradient / 温度梯度
thermal diffusion / 热扩散
thermal diffusivity / 热扩散系数
transient diffusion / 瞬时扩散
two-film theory / 双膜理论
unsteady-state diffusion / 非稳态扩散
volumetric diffusivity / 体积扩散系数

Chapter 7

Equilibrium Relations and Equilibrium-Stage Operations

7.1 Equilibrium Relations

To evaluate driving forces relating to the mass transfer, a knowledge of equilibria between phases is therefore of basic importance. Several kinds of equilibria are important in mass transfer. In all situations, two phases are involved, and all combinations are found except two gas phases or two solid phases. The controlling variables are temperature, pressure, and concentration. Equilibrium data can be shown in tables, equations, or graphs. For most processes considered in this text, the pertinent equilibrium relationship can be shown in graphically.

7.1.1 Gas-Liquid Equilibrium

As in the gas-liquid systems, the equilibrium in vapor-liquid systems is restricted by the phase rule which we learned from the textbook of Physical Chemistry. As an example we shall use the ammonia-water, vapor-liquid system. For two components and two phases, this means that there are 2 degrees of freedom. The four variables are temperature, pressure, and the composition y_A of NH_3 in the vapor phase and x_A in the liquid phase. The composition of water (B) is fixed if y_A or x_A is specified, since $y_A + y_B = 1.0$ and $x_A + x_B = 1.0$. If the pressure is fixed, only one more variable can be set. If we set the liquid composition, the temperature and vapor composition are automatically set.

Gas–liquid equilibrium data

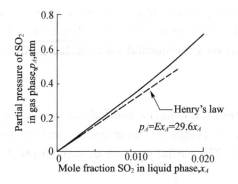

Fig. 7.1 Equilibrium plot for SO_2-water system at 20℃.

At low concentration, the plot of partial pressure versus component passes through the origin and is straight line, this behavior is general and is a basic characteristic of substances in dilute solutions. To illustrate the obtaining of experimental gas-liquid equilibrium data, the system SO_2-air-water will be considered. An amount of gaseous SO_2, air, and water are put in a closed container and shaken repeatedly at a given temperature until equilibrium is reached. Samples of the gas and liquid are analyzed to determine the partial pressure p_A in atm of SO_2 (A) in the gas and mole fraction x_A in the liquid.

Fig.7.1 shows a plot of data from Appendix 21 of the partial pressure p_A of SO_2 in the vapor in equilibrium with the mole fraction x_A of SO_2 in the liquid at 293 K (20℃).

Henry's law

Often the equilibrium relation between p_A in the gas phase and x_A in the liquid phase can be expressed by a straight-line. Henry's law equation at low concentrations:

$$p_A = E x_A \tag{7.1-1}$$

where E is the Henry's law constant in atm/mole fraction for the given system. If both sides of Eq.(7.1-1) are divided by total pressure P in atm,

$$y_A = \frac{E}{P} x_A \tag{7.1-2}$$

and let

$$m = \frac{E}{P} \tag{7.1-3}$$

where m is the Henry's law constant in mole frac gas/mole frac liquid. Note that m depends on total pressure, whereas E does not.

In Fig.7.1 the data follow Henry's law up to a concentration x_A of about 0.005, where $E = 29.6$ atm/mol frac. In general, up to a total pressure of about 5×10^5 Pa (5atm) the value of E is independent of P. Data for some common gases with water are given in Appendix 20.

EXAMPLE 7.1 What will be the concentration of oxygen dissolved in water at 298K when the solution is in equilibrium with air at 1atm total pressure? The Henry's law constant is 4.38×10^4 atm/mol fraction.
Solution. The partial pressure p_A of oxygen (A) in air is 0.21 atm. Using Eq. (7.1-1)
$$0.21 = E x_A = 4.38 \times 10^4 x_A$$
Solving, $x_A = 4.80 \times 10^{-6}$ mol fraction. This means that 4.80×10^{-6} mol O_2 is dissolved in 1.0 mol water plus oxygen, or 0.000853 part O_2/100 parts water.

7.1.2 Vapor-Liquid Equilibrium Relations

The quantity and reliability of data available for equilibrium systems vary widely. Perhaps the most widely and completely investigated type of system is the vapor-liquid system containing paraffin hydrocarbons. For other vapor-liquid systems, such as oxygenated or chlorinated hydrocarbons, the information available is less complete. Generalizations in this area are difficult and the designer is wise to insist on experimental data for the particular system on hand.

Raoult's Law

At low pressure the vapor of mixture approaches ideal behavior and follows the ideal gas law. In some mixtures, Raoult's law applies to each component over entire concentration range from 0 to 1.0, such mixtures are called ideal. The partial pressure of component A in gas is proportional to its concentration in liquid.

$$p_A = P_A x_A \tag{7.1-4}$$

P_A is the vapor pressure of pure A in Pa (or atm), and x_A is the mole fraction of A in the liquid. The same rule, written for component B, is

$$p_B = P_B x_B \tag{7.1-5}$$

This law holds only for ideal solutions. Actually, few solutions are ideal. The assumption of ideal liquid behavior is generally correct only for members of a homologous series that are close together in molecular weight, such as benzene-toluene, hexane-heptane, and methyl alcohol-ethyl alcohol, which are usually substances very similar to each other. Mixtures of polar substances such as water, alcohol, and electrolytes depart greatly from ideal-solution law. If one component of a two-component mixture follows the ideal-solution law, so does the other. However, the assumption of ideal liquid behavior is good from a practically engineering standpoint.

Relative volatility of vapor–liquid systems

The most widely available data are those for vapor-liquid systems and these are frequently referred to as vapor-liquid equilibrium distribution coefficients or K values. Since K values vary considerably with temperature and pressure, the selectivity that is equal to the ratio of the K values is used. For the vapor-liquid systems this is typically referred to as the relative volatility α. For a binary system, this is defined as the ratio of the concentration of A in the vapor to the concentration of A in the liquid divided by the ratio of the concentration of B in the vapor to the concentration of B in the liquid:

$$\alpha_{AB} = \frac{y_A/x_A}{y_B/x_B} = \frac{y_A/x_A}{(1-y_A)/(1-x_A)} \tag{7.1-6}$$

where α_{AB} is the relative volatility of A with respect to B in the binary system. Equation (7.1-6) can be rearranged to give

$$y_A = \frac{\alpha x_A}{1+(\alpha-1)x_A} \tag{7.1-7}$$

where $\alpha = \alpha_{AB}$. For a separation process in which $\alpha = 1$, the compositions of component A would be the same in both phases, so do the component B. obviously, separation is not possible when this occurs since the driving force for mass transfer is zero. When the value of α is above 1.0, a separation is possible. The value of α may change as concentration and pressure change.

If the system obeys Raoult's law, as does the benzene-toluene system,

$$y_A = \frac{P_A}{P} x_A \qquad y_B = \frac{P_B}{P} x_B \tag{7.1-8}$$

Substituting Eq. (7.1-8) into (7.1-6) for an ideal system,

$$\alpha = \frac{P_A}{P_B} \tag{7.1-9}$$

At a given temperature P_A and P_B are fixed, so that for constant-temperature operations with ideal system α is constant. When binary systems follow Raoult's law, the relative volatility often varies only slightly over a large concentration range at constant total pressure.

EXAMPLE7.2 Using the data from table 7.1, calculate the relative volatility for the benzene-toluene system at 85°C and 105°C.

Solution. At 85 ℃, substituting the vapor pressures into Eq. **(7.1-9)** for a system following Raoult's law,

$$\alpha = \frac{P_A}{P_B} = \frac{116.9}{46.0} = 2.54$$

Similarly at 105 ℃,

$$\alpha = \frac{P_A}{P_B} = \frac{204.2}{86.0} = 2.38$$

The variation in α is about 7%.

Table 7.1 Vapor-Pressure and Equilibrium-Mole-Fraction Data for Benzene–Toluene System

Temperature		Benzene vapor pressure		Toluene vapor pressure		Mole Fraction Benzene at 101.325 kPa	
K	℃	kPa	mm Hg	kPa	mm Hg	x_A	y_A
353.3	80.1	101.32	760			1.000	1.000
358.2	85	116.9	877	46.0	345	0.780	0.900
363.2	90	135.5	1016	54.0	405	0.581	0.777
368.2	95	155.7	1168	63.3	475	0.411	0.632
373.2	100	179.2	1344	74.3	557	0.258	0.456
378.2	105	204.2	1532	86.0	645	0.130	0.261
383.8	110.6	240.0	1800	101.32	760	0	0

Boiling-point diagrams and *x-y* plots

Often the vapor-liquid equilibrium relations for a binary mixture of *A* and *B* are given as a boiling-point diagram at constant pressure, shown in Fig. 7.2. The accepted convention is to plot the most volatile or lower boiling constituent increasing in composition in the positive *X* direction. Since only two components are present, the compositions of both components are known if we specify the composition of one. The upper line on Fig. 7.2 represents the saturated vapor line (the *dew-point line*) and the lower line is the saturated liquid line (the *bubble-point line*). The two-phase region is in the region between these two lines.

Figure 7.2 illustrates another use for the temperature-composition diagram. Assume that a mixture of composition x_{A1} = 0.318 exists at the temperature of 75 ℃ represented by point *A* on Fig. 7.2. As this mixture is heated it will remain all liquid until point *B* is reached, where the first small bubble of vapor will form at 98 ℃ and the composition of the first vapor in equilibrium is y_{A1} = 0.532. If

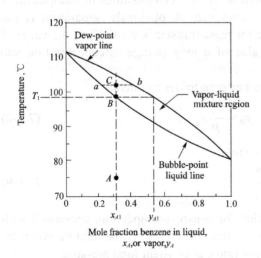

Fig. 7.2 Boiling point diagram for benzene (*A*)-toluene (*B*) at 101.325 kPa total pressure.

heating of the mixture is continued, and the vapor formed is maintained in contact with the liquid, the mixture will eventually reach a point C on the diagram where both vapor and liquid exist. The overall composition of the total mixture remains $x_{A1} = 0.318$, but the liquid portion will have a composition $x_A = 0.22$, moving to the left on the bubble-point line since y_A =4.15 is richer in A. The vapor and liquid in equilibrium (x_A and y_A) are connected by the equilibrium tie line ab. The relative amounts of the vapor and liquid present can be determined by a series of material balances.

The system benzene–toluene follows Raoult's law, so the boiling-point diagram can be calculated by using the pure vapor-pressure data in table 7.1 and the following equations

$$p_A + p_B = P \tag{7.1-10}$$

$$P_A x_A + P_B (1 - x_A) = P \tag{7.1-11}$$

$$y_A = \frac{p_A}{P} = \frac{P_A}{P} x_A \tag{7.1-12}$$

EXAMPLE 7.3 Calculate the vapor and liquid compositions in equilibrium at 95℃ for benzene-toluene using the vapor pressure from table 7.1 at 101.32 kPa.
Solution. At 95℃ from table 7.1 for benzene, P_A = 155.7 kPa and P_B = 63.3 kPa. Substituting into Eq. **(7.1-11)** and solving,

$$155.7 x_A + 63.3(1 - x_A) = 101.32 \text{kPa}$$

Hence, x_A = 0.411 and x_B = 1–x_A = 1–0.411 = 0.589. Substituting into Eq.(7.1-12),

$$y_A = \frac{P_A}{P} x_A = \frac{155.7 \times 0.411}{101.32} = 0.632$$

Another type of equilibrium diagram that is frequently used in vapor-liquid separations is the so-called x-y diagram, as shown in Fig.7.3 where y_A is plotted versus x_A for the benzene-toluene system. This is developed from the temperature-composition diagram by plotting liquid composition on the x axis and composition of the vapor that would be in equilibrium on the y axis. The 45° line is given to show that y_A is richer in component A than x_A.

The boiling-point diagram in Fig.7.2 is typical of an ideal system following Raoult's law. Nonideal systems differ considerably. Vapor-phase nonidealities are fairly regular and are relatively easily predicted. Liquid-phase nonidealities are less regular and are much more difficult to predict. In the extreme cases liquid-phase nonideality can be such as to reverse the volatility of the two components. When this occurs the system is referred to as azeotropic and the composition at which the volatility reversal occurs is referred to as the *azeotrope* or *azeotropic composition*.

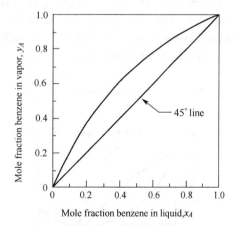

Fig. 7.3 Equilibrium diagram for system benzene (A)–toluene (B) at 101.32 kPa (1atm).

In Fig.7.4 shows the temperature-composition diagram at constant pressure for a binary system which forms a maximum-boiling azeotrope. There are two vapor-liquid envelopes which meet at the azeotropic composition. Within each envelope, vapor-liquid equilibrium can be predicted from a horizontal tie line as in the case for the system exhibiting ideal behavior. However, one terminus of each envelope is at the azeotropic composition. The mixture is called a maximum boiling azeotrope because the boiling point of the constant composition mixture is higher than the boiling point of either pure component. At the azeotrope, or constant-composition point, both the vapor and the liquid in equilibrium have the same composition. Figure 7.5 shows the x-y diagram for the system and clearly indicates the volatility reversal at the azeotropic composition. The maximum temperature T_{max} corresponds to a concentration x_{Az} and $x_{Az} = y_{Az}$ at this point. The plot of y_A versus x_A would show the curve crossing the 45° line at this point. Acetone-chloroform is an example of such a system.

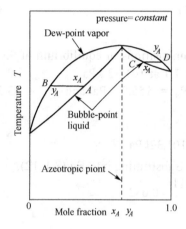 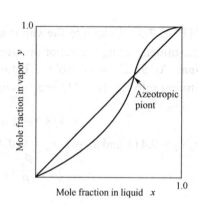

Fig. 7.4 Maximum boiling azeotrope.   **Fig. 7.5** Maximum boiling azeotropic system.

In Figs. 7.6 and 7.7 show the temperature-composition and x-y diagrams, respectively, for a system that exhibits a minimum boiling azeotrope. The behavior of this system is identical with that of the maximum boiling azeotrope except that the temperature of the azeotropic boiling point is below the boiling point of either of the pure component. A minimum-boiling azeotrope is shown with $y_{Az} = x_{Az}$ at T_{min}. Ethanol-water is such a system.

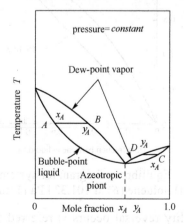 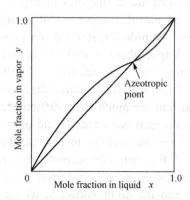

Fig. 7.6 Minimum boiling azeotrope.   **Fig. 7.7** Minimum boiling azeotropic system.

It is impossible to separate the azeotropes by the ordinary distillation because both the vapor and the liquid in equilibrium have the same composition. Separation of the original mixture may be enhanced by adding a solvent that forms an azeotrope with one of the key components. This process is called *azeotropic distillation*, which will be discussed in chapter 9 some more detail.

7.2   Equilibrium-Stage Operations

One class of mass-transfer devices consists of assemblies of individual units, or stages, interconnected so that the materials being processed pass through each stage in turn. The two streams move countercurrently through the assembly; in each stage they are brought into contact, mixed, and then separated. Such multistage systems are called *cascades*. For mass transfer to take place, the streams entering each stage must not be in equilibrium with one another, for it is the departure from equilibrium conditions that provides the driving force for transfer. The leaving streams are usually not in equilibrium either but are much closer to being so than the entering streams are. The closeness of the approach to equilibrium depends on the effectiveness of mixing and mass transfer between the phases. To simplify the design of a cascade, the streams leaving each stage are often assumed to be in equilibrium, which, by definition, makes each stage *ideal*. A correction factor or efficiency is applied later to account for any actual departures from equilibrium.

To illustrate the principle of an equilibrium-stage cascade, two typical countercurrent multistage devices are described here, one for distillation or gas absorption, where the stages are arranged one above the other in a vertical column, and one for solid-liquid contacting as in leaching, where the stages are a series of stirred tanks on the same level. Other types of mass-transfer equipment are discussed in later chapters.

7.2.1   Equipment for Stage Contacts

Typical distillation equipment

Equipment for continuous distillation is shown in Fig.7.8. Column C is fed continuously with the liquid mixture to be distilled, and the liquid in reboiler A is partially converted to vapor by heat transferred from the heating element B. The vapor stream from the still is brought into intimate countercurrent contact with a descending stream of boiling liquid in the column, or tower, C. This liquid must be rich enough in the low boiler that there is mass transfer of the low boiler from the liquid to the vapor at each stage of the column. Such a liquid can be obtained simply by condensing the overhead vapors and returning some of the liquid to the top of the column. This return liquid is called *reflux*. The use of reflux increases the purity of the overhead product, but not without some cost, since the vapor generated in the reboiler must provide both reflux and overhead product, and this energy cost is a large part of the total cost of separation by distillation.

The reflux entering the top of the column is often at the boiling point; but if it is cold, it is almost immediately heated to its boiling point by the vapor. Throughout the rest of the column, the vapor at any stage is at the same temperature as the liquid, which is at its boiling point. The temperature increases on going down the column because of the increase in pressure and the increasing concentration of high-boiling components.

The vapor is enriched at each stage because the vapor coming to a stage contains less low boilers than the vapor that would be in equilibrium with the liquid fed to that stage. If, as is usual, the overhead vapor is totally condensed, it has the same composition as the product and the reflux. The reflux, however, has an equilibrium vapor composition which is richer than the vapor coming up to the top stage. This vapor is therefore enriched in low boilers at the expense of the reflux liquid. This partially depletes the reflux of low boilers, but if the flow rates have been set correctly, the liquid passing down to the second stage is still able to enrich the lower-quality vapor coming up to the second stage. At all stages in the column some low boilers diffuse from the liquid into the vapor, and a corresponding amount of high boilers diffuse from the vapor into the liquid. The heat of vaporization of the low boilers is supplied by the heat of condensation of the high boilers, and the total molal flow of vapor up the column is nearly constant.

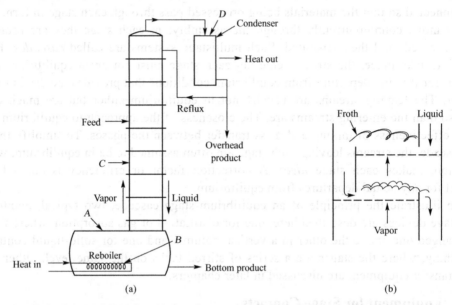

Fig. 7.8 (*a*) Reboiler with fractionating column: *A*, reboiler; *B*, heating element; *C*, column; *D*, condenser. (*b*) Detail of sieve plate.

The upper section of the column, above the feed plate, is known as the *rectifying section*. Here the vapor stream is enriched in low boilers, as it is brought into contact with the reflux. It is immaterial where the reflux originates, provided its concentration in low boilers is sufficiently great to give the desired product. The usual source of reflux is the condensate leaving condenser *D*. Part of the condensate is withdrawn as the product, and the remainder is returned to the top of the column. Reflux is sometimes provided by partial condensation of the overhead vapor; the reflux then differs in composition from the vapor leaving as overhead product. Provided an azeotrope is not formed, the vapor reaching the condenser can be brought as close to complete purity as desired by using a tall tower and a large reflux.

The section of the column below the feed plate is the *stripping* or *enriching section,* in which the liquid is progressively stripped of low boilers and enriched in the high boiling components. If the feed is liquid, as is usually the case, it adds to the liquid flow in the lower section of the column. If in addition the feed is cold, additional vapor must be provided by the reboiler to raise the temperature of the feed to the boiling point. To accomplish this, the additional vapor is condensed when it comes in contact with the feed, adding still more to the liquid flowing down through the stripping section. Details of column performance and the effects of feed condition are discussed in Chap. 9.

From the reboiler, liquid is withdrawn which contains most of the high boiling components and usually only a little of the low boilers. This liquid is called the *bottom product* or *bottoms.*

The column shown in Fig. 7.8(*a*) contains a number of plates, or trays, stacked one above the other. Often these plates are perforated and are known as sieve plates, details of which are shown in Fig. 7.8(*b*). They consist of horizontal trays carrying a number of holes and a vertical plate which acts as a down-comer and a segmental weir. Sometimes, as discussed in Chap.12, the holes contain valves or plugs which are lifted as vapor passes through them. The down-comer from a given plate reaches nearly to the tray below. Liquid flows over the weirs from plate to plate down the column, passing across the plates where the rising vapor causes it to froth. The vapor space above the froth contains a mist of fine droplets formed by the collapsing bubbles. Most of the drops fall back into the liquid, but some are entrained by the vapor and are carried to the plate above. See Chap.12 for a discussion of the effect of this entrainment on column performance.

Typical leaching equipment

In leaching, soluble material is dissolved from its mixture with an inert solid by means of a liquid solvent. A diagrammatic flow sheet of a typical countercurrent leaching plant is shown in Fig. 7.9. It consists of a series of units, in each of which the solid from the previous unit is mixed with the liquid from the succeeding unit, and the mixture is allowed to settle. The solid is then transferred to the next succeeding unit, and the liquid to the previous unit. As the liquid flows from unit to unit, it becomes enriched in solute, and as the solid flows from unit to unit in the reverse direction, it becomes impoverished in solute. The solid discharged from one end of the system is well extracted, and the solution leaving at the other end is strong in solute. The thoroughness of the extraction depends on the amount of solvent and the number of units. In principle, the unextracted solute can be reduced to any desired amount if enough solvent and a sufficient number of units are used.

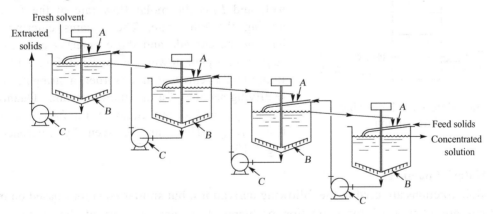

Fig. 7.9 Countercurrent leaching plant: *A*, launder; *B*, rake; *C*, slurry pump.

Multistage leaching can also be carried out in a single piece of equipment in which the solid phase is moved mechanically to achieve countercurrent flow. Two examples are shown in Fig.10.1.

7.2.2 Principles of Stage Processes

In the sieve-plate tower and the countercurrent leaching plant shown in Figs. 7.8 and 7.9, the cascade consists of a series of interconnected units, or stages. Study of the assembly as a whole is best made by focusing attention on the streams passing between the individual stages. An individual unit in a cascade receives two streams, one a V phase and one an L phase, from the two units adjacent to it, brings them into close contact, and delivers L and V

phases, respectively, to the same adjacent units. The fact that the contact units may be arranged either one above the other, as in the sieve-plate column, or side by side, as in a stage leaching plant, is important mechanically and may affect some of the details of operation of individual stages. The same material-balance equations, however, can be used for either arrangement.

Terminology for stage-contact plants

The individual contact units in a cascade are numbered serially, starting from one end. In this book, the stages are numbered in the direction of flow of the L phase, and the last stage is that discharging the L phase. A general stage in the system is the nth stage, which is number n counting from the entrance of the L phase. The stage immediately ahead of stage n in the sequence is stage $n-1$, and that immediately following it is stage $n+1$. Using a plate column as an example, Fig. 7.10 shows how the units in a cascade are numbered. The total number of stages is N, and the last stage in the plant is therefore the Nth stage.

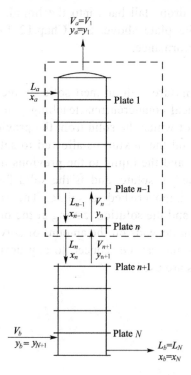

Fig. 7.10 Material-balance diagram for plate column.

To designate the streams and concentrations pertaining to any one stage, all streams originating in that stage carry the number of the unit as a subscript. Thus, for a two-component system, y_{n+1} is the mole fraction of component A in the V phase leaving stage $n+1$, and L_n is the molal flow rate of the L phase leaving the nth stage. The streams entering and leaving the cascade and those entering and leaving stage n in a plate tower are shown in Fig. 7.10. The figure could represent an absorber, a stripper, or the rectifying section of a distillation column. Quantities V_a, L_b, y_a, and x_b are equal to V_1, L_N, y_1, and x_N, respectively. This can be seen by reference to Fig. 7.10.

Material balances

Mole fractions are used in the following derivations, but similar equations based on mass fractions are often used for extraction problems. Consider the part of the cascade that includes stages 1 through n, as shown by the section enclosed by the dashed line in Fig. 7.10. The total input of material to this section is $L_a + V_{n+1}$ mol/h, and the total output is $L_n + V_a$ mol/h. Since, under steady flow, there is neither accumulation nor depletion, the input and the output are equal and

$$L_a + V_{n+1} = L_n + V_a \tag{7.2-1}$$

Equation (7.2-1) is a total material balance. Another balance can be written by equating input to output for component A. Since the number of moles of this component in a stream is the product of the flow rate and the mole fraction of A in the stream, the input of component A to the section under study, for a two-component system, is $L_a x_a + V_{n+1} y_{n+1}$ mol/h, the output is $L_n x_n + V_a y_a$ mol/h, and

$$L_a x_a + V_{n+1} y_{n+1} = L_n x_n + V_a y_a \qquad (7.2\text{-}2)$$

A material balance can also be written for component B, but such an equation is not independent of Eqs. (7.2-1) and (7.2-2), since if Eq. (7.2-2) is subtracted from Eq. (7.2-1), the result is the material-balance equation for component B. Equations (7.2-1) and (7.2-2) yield all the information that can be obtained from material balances alone written over the chosen section.

Overall balances covering the entire cascade are found in the same manner:

Total material balance:
$$L_a + V_b = L_b + V_a \qquad (7.2\text{-}3)$$

Component A balance:
$$L_a x_a + V_b y_b = L_b x_b + V_a y_a \qquad (7.2\text{-}4)$$

Enthalpy balances

In many equilibrium-stage processes the general energy balance can be simplified by neglecting mechanical potential energy and kinetic energy. If, in addition, the process is workless and adiabatic, a simple enthalpy balance applies. Then, for a two-component system, for the first n stages,

$$L_a H_{L,a} + V_{n+1} H_{V,n+1} = L_n H_{L,n} + V_a H_{V,a} \qquad (7.2\text{-}5)$$

where H_L and H_V are the enthalpies per mole of the L phase and V phase, respectively. For the entire cascade,

$$L_a H_{L,a} + V_b H_{V,b} = L_b H_{L,b} + V_a H_{V,a} \qquad (7.2\text{-}6)$$

Graphical methods for two-component systems

For systems containing only two components it is possible to solve many mass transfer problems graphically. The methods are based on material balances and equilibrium relationships; some more complex methods require enthalpy balances as well. These more complex methods are beyond scope of this book. The principles underlying the simple graphical methods are discussed in the following paragraphs. Their detailed applications to specific operations are covered in later chapters.

Operating-line diagram

For a binary system, the compositions of the two phases in a cascade can be shown on an arithmetic graph where x is the abscissa and y the ordinate. As shown by Eq. (7.2-2), the material balance at an intermediate point in the column involves x_n, the concentration of the L phase *leaving* stage n, and y_{n+1}, the concentration of the V phase *entering* that stage. Equation (7.2-2) can be written to show the relationship more clearly:

$$y_{n+1} = \frac{L_n}{V_{n+1}} x_n + \frac{V_a y_a - L_a x_a}{V_{n+1}} \qquad (7.2\text{-}7)$$

Equation (7.2-7) is the operating-line equation for the column; if the points x_n and y_{n+1} for all the stages are plotted, the line through these points is called the *operating line*. Note that if L_n and V_{n+1} are constant throughout the column, the equation is that of a straight line with slope L/V and intercept $y_a - (L/V)x_a$, and the line is easily located. For this case the operating line can also be drawn as a straight line connecting the terminal compositions (x_a, y_a) and (x_b, y_b). To understand why this is true, extend the dashed rectangle in Fig. 7.10 to include plate N, and consider that the stream V_b coming to the bottom stage is equivalent to a stream from a hypothetical stage $N + 1$, so that y_b corresponds to y_{N+1} and x_b to x_N. Similarly, the stream L_a at the top of the column can be considered to come from a hypothetical stage numbered 0 so that the point (x_0, y_1) or (x_a, y_a) locates the true upper end of the operating line.

In a packed column or other nonstagewise contacting device, such as the one described in Chap. 8 for gas absorption, x and y are continuous functions of height Z, whereas in staged towers x and y have only discrete values. The equilibrium line is of course continuous, but the operating line is drawn to connect a series of compositions, relating y_{n+1} to x_n. Usually, however, we are not sure of the values of y_n and x_n, so we draw the operating line as a solid line.

When the flow rates are not constant in the column, the operating line on a simple arithmetic plot is not straight. The terminal compositions may still be used to locate the ends of the line, and material-balance calculations over sections of the column are made to establish a few intermediate points. Often only one or two other points are needed because usually the operating line is only slightly curved.

The position of the operating line relative to the equilibrium line determines the direction of mass transfer and how many stages are required for a given separation. The equilibrium data are found by experiment, by thermodynamic calculations, or from published sources, and the equilibrium line is just a plot of the equilibrium values of x_e and y_e. For rectification in a distillation column, the operating line must lie below the equilibrium line, as shown in Fig. 7.11(a) Then the vapor coming to any plate contains less of the low boiler than the vapor in equilibrium with the liquid leaving the plate, so that vapor passing through the liquid will be enriched in the low-boiling component. The relative slopes of the lines are not important as long as the lines do not touch; the operating line could be less steep than the equilibrium line, and progressive enrichment of the vapor would still take place. The driving force for mass transfer is the difference $y_e - y_{n+1}$, as shown in Fig. 7.11(a).

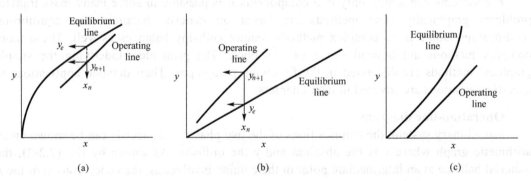

Fig. 7.11 Operating and equilibrium lines: (a) for rectification, (b) for gas absorption, (c) for desorption.

When one component is to be transferred from the V phase to the L phase, as in the absorption of soluble material from an inert gas, the operating line must lie above the equilibrium line, as in Fig. 7.11(b). The driving force for mass transfer is now $y_{n+1} - y_e$, or the difference between the actual vapor composition and the vapor composition in equilibrium with the liquid for that position in the column. In the design of gas absorbers the liquid rate is usually chosen to make the operating line somewhat steeper than the equilibrium line, which gives a moderately large driving force in the bottom part of the column and permits the desired separation to be made with relatively few stages.

In absorbing one component of the gas into a nonvolatile solvent, the total gas rate decreases and the total liquid rate increases as the two phases pass through the column. Therefore the operating line is usually curved, though the percentage change in slope, or L/V, is not as great as the change in either L or V, since both L and V are largest at the bottom of the column and smallest at the top. A method of calculating intermediate points on the operating line is shown later in Example7.4.

The reverse of gas absorption is called desorption or stripping, an operation carried out to recover valuable solute from the absorbing solution and regenerate the solution. The operating line must then lie below the equilibrium line, as in Fig. 7.11(c). Usually the temperature or pressure is changed to make the equilibrium curve much steeper than for the absorption process.

Ideal contact stages

The ideal stage is a standard to which an actual stage may be compared. In an ideal stage, the V phase leaving the stage is in equilibrium with the L phase leaving the same stage. For example, if plate n in Fig. 7.10 is an ideal stage, concentrations x_n and y_n are coordinates of a point on the curve of x_e versus y_e showing the equilibrium between the phases. In a plate column ideal stages are also called *perfect plates*.

To use ideal stages in design, it is necessary to apply a correction factor, called the *stage efficiency* or *plate efficiency*, which relates the ideal stage to an actual one. Plate efficiencies are discussed in Chap.9 and 12, and the present discussion is restricted to ideal stages.

Determining the number of ideal stages

A problem of general importance is that of finding the number of ideal stages required in an actual cascade to cover a desired range of concentration x_a to x_b or y_a to y_b. If this number can be determined, and if information on stage efficiencies is available, the number of actual stages can be calculated. This is the usual method of designing cascades.

A simple method of determining the number of ideal stages when there are only two components in each phase is a graphical construction using the operating-line diagram. Figure 7.12 shows the operating line and the equilibrium curve for a typical gas absorber. The ends of the operating line are point a, having coordinates (x_a, y_a), and point b, having coordinates (x_b, y_b). The problem of determining the number of ideal stages needed to accomplish the gas-phase concentration change y_b to y_a and the liquid-phase concentration change x_a to x_b is solved as follows.

The concentration of the gas leaving the top stage, which is stage 1, is y_a, or y_1. If the stage is ideal, the liquid leaving is equilibrium with the vapor leaving, so the point (x_1, y_1) must lie on the equilibrium curve. This fact fixes point m, found by moving horizontally from point a to the equilibrium curve. The abscissa of point m is x_1. The operating line is now used. It passes through all points having coordinates of the type (x_n, y_{n+1}), and since x_1 is known, y_2 is found by moving vertically from point m to the operating line at point n, the coordinates of which are (x_1, y_2). The step, or triangle, defined by points a, m, and n represents one ideal stage, the first one in this column. The second stage is located graphically on the diagram by repeating the same construction,

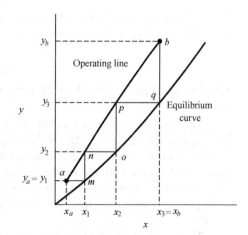

Fig. 7.12 Operating-line diagram for gas absorber.

passing horizontally to the equilibrium curve at point o, having coordinates (x_2, y_2), and vertically to the operating line again at point p, having coordinates (x_2, y_3). The third stage is found by again repeating the construction, giving triangle pqb. For the situation shown in Fig. 7.12, the third stage is the last, as the concentration of the gas entering that stage is y_b, and the liquid leaving it is x_b, which are the desired terminal concentrations. Three ideal stages are required for this separation.

The same construction can be used for determining the number of ideal stages needed in any cascade, whether it is used for gas absorption, stripping, distillation, leaching, or liquid extraction. The graphical step-by-step construction utilizing alternately the operating and equilibrium lines to find the number of ideal stages was first applied to the design of distillation columns, and is known as the *McCabe-Thiele method*. The construction can be started at either end of the column, and in general the last step will not exactly meet the terminal concentrations, as was the case in Fig. 7.12. A fractional step may be assigned, or the number of ideal stages may be rounded up to the nearest whole number.

EXAMPLE 7.4. By means of a plate column, acetone is absorbed from its mixture with air in a nonvolatile absorption oil. The entering gas contains 30 mole percent acetone, and the entering oil is acetone-free. Of the acetone in the air 97 percent is to be absorbed, and the concentrated liquor at the bottom of the tower is to contain 10 mole percent acetone. The equilibrium relationship is $y_e = 1.9 x_e$. Plot the operating line and determine the number of ideal stages.

Solution. Choose 100 mol of entering gas as a basis, and set this equal to V_b. The acetone entering is then $0.3 \times 100 = 30$ mol; the air entering is $100 - 30 = 70$ mol. With 97 percent absorbed, the acetone leaving is $0.03 \times 30 = 0.9$ mol and $y_a = 0.9/70.9 = 0.0127$; the acetone absorbed is $30 - 0.9 = 29.1$ mol. With 10 percent acetone in the leaving solution and no acetone in the entering oil, $0.1 L_b = 29.1$, and $L_b = 291$ mol. Then $L_a = 291 - 29.1 = 261.9$ mol.

To find an intermediate point on the operating line, make an acetone balance around the top part of the tower, assuming a particular value of yV, the moles of acetone left in the gas. For 10 mol left in the gas,

$$y = \frac{10}{10 + 70} = 0.125$$

The moles of acetone lost by the gas in this section, $10 - 0.9$, or 9.1, must equal the moles gained by the liquid. Hence where $y = 0.125$,

$$x = \frac{9.1}{261.9 + 9.1} = 0.0336$$

Similar calculations for $yV = 20$ give $y = 20/90 = 0.222$ and $x = 19.1/(261.9 + 19.1) = 0.068$.

The operating line is plotted in Fig. 7.13. Although it appears only slightly curved, the slope at the top is 1.57 times the slope at the bottom. The local slope is not equal to the local L/V ratio, since when y is large the change in y is not proportional to the amount transferred because of the change in the total flow. The L/V ratio changes by a factor of only 1.26 from bottom to top. With rich gas a correct design cannot be obtained using an average slope or an average L/V and the absorption factor method presented in the next section.

The number of ideal stages is 4 and a fraction. Based on the required change in x relative to the change that would be made in a full step, the fraction is l_1/l_2, or 0.27. A similar construction based on changes in y gives the fraction 0.33; the values differ because the operating and equilibrium lines are not parallel. The answer would be given as 4.3 stages.

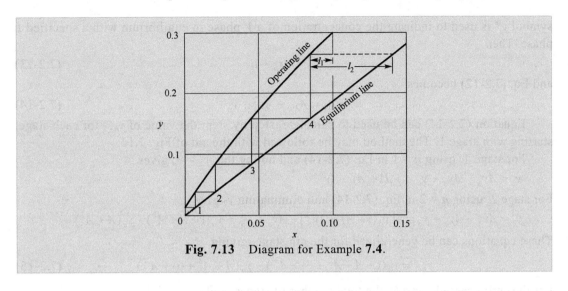

Fig. 7.13 Diagram for Example 7.4.

Absorption factor method for calculating the number of ideal stages

When the operating and equilibrium lines are both straight over a given concentration range x_a to x_b, the number of ideal stages can be calculated directly, and graphical construction is unnecessary. Formulas for this purpose are derived as follows.

Let the equation of the equilibrium line be

$$y_e = mx_e + B \qquad (7.2\text{-}8)$$

where, by definition, m and B are constant. If stage n is ideal,

$$y_n = mx_n + B \qquad (7.2\text{-}9)$$

Substitution for x_n into Eq. (7.2-7) gives, for ideal stages and constant L/V,

$$y_{n+1} = \frac{L(y_n - B)}{mV} + y_a - \frac{Lx_a}{V} \qquad (7.2\text{-}10)$$

It is convenient to define an absorption factor A by the equation

$$A = \frac{L}{mV} \qquad (7.2\text{-}11)$$

The absorption factor is the ratio of the slope of the operating line L/V to that of the equilibrium line m. It is a constant when both of these lines are straight. Equation (7.2-10) can be written

$$\begin{aligned} y_{n+1} &= A(y_n - B) + y_a - Amx_a \\ &= Ay_n - A(mx_a + B) + y_a \end{aligned} \qquad (7.2\text{-}12)$$

Normally A is made greater than 1.0 to permit nearly complete removal of solute from the V phase. The quantity $mx_a + B$ is, by Eq. (7.2-8), the concentration of the vapor that is in equilibrium with the inlet L phase, the concentration of which is x_a. This can be seen from Fig. 7.14. The

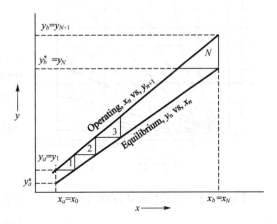

Fig. 7.14 Derivation of absorption factor equation.

symbol y^* is used to indicate the concentration of a V phase in equilibrium with a specified L phase. Then

$$y_a^* = mx_a + B \tag{7.2-13}$$

and Eq. (7.2-12) becomes

$$y_{n+1} = Ay_n - Ay_a^* + y_a \tag{7.2-14}$$

Equation (7.2-14) can be used to calculate, step by step, the value of y_{n+1} for each stage, starting with stage 1. The method may be followed with the aid of Fig. 7.14.

For stage 1, using $n = 1$ in Eq. (7.2-14) and noting that $y_1 = y_a$ gives

$$y_2 = Ay_a - Ay_a^* + y_a = y_a(1 + A) - Ay^*$$

For stage 2, using $n = 2$ in Eq. (7.2-14) and eliminating y_2 give

$$y_3 = Ay_2 - Ay_a^* + y_a = A[y_a(1 + A) - Ay^*] - Ay_a^* + y_a = y_a(1 + A + A^2) - y_a^*(A + A^2)$$

These equations can be generalized for the nth stage, giving

$$y_{n+1} = y_a(1 + A + A^2 + \cdots + A^N) - y_a^*(A + A^2 + \cdots + A^N) \tag{7.2-15}$$

For the entire cascade, $n = N$, the total number of stages, and

$$y_{n+1} = y_{N+1} = y_b$$

Then

$$y_b = y_a(1 + A + A^2 + \cdots + A^N) - y_a^*(A + A^2 + \cdots + A^N) \tag{7.2-16}$$

The sums in the parentheses of Eq. (7.2-16) are both sums of geometric series. The sum of such a series is

$$S_n = \frac{a_1(1 - r^n)}{1 - r}$$

where S_n = sum of first n terms of series
        a_1 = first term
        r = constant ratio of each term to preceding term

Equation (7.2-16) can then be written

$$y_b = y_a \frac{1 - A^{N+1}}{1 - A} - y_a^* A \frac{1 - A^N}{1 - A} \tag{7.2-17}$$

Equation (7.2-17) is a form of the *Kremser equation*. It can also shown as

$$N = \frac{1}{\ln A} \ln\left[\left(1 - \frac{1}{A}\right)\frac{y_b - mx_a}{y_a - mx_a} + \frac{1}{A}\right] \tag{7.2-17a}$$

It can be used as such or in the form of a chart relating N, A, and the terminal concentrations, as shown in Fig. 7.15.

It can also be put into a simpler form by the following method.

Equation (7.2-14) is, for stage N,

$$y_b = Ay_N - Ay_a^* + y_a \tag{7.2-18}$$

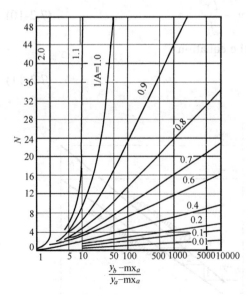

Fig. 7.15 Number of ideal stages

Figure 7.14 shows that $y_N = y_b^*$, and Eq. (7.2-18) can be written

$$y_a = y_b - A(y_b^* - y_a^*) \tag{7.2-19}$$

Collecting terms in Eq. (7.2-17) containing A^{N+1} gives

$$A^{N+1}(y_a - y_a^*) = A(y_b - y_a^*) + y_a - y_b \tag{7.2-20}$$

Substituting $y_a - y_b$ from Eq. (7.2-19) into Eq. (7.2-20) gives

$$A^N(y_a - y_a^*) = y_b - y_a^* - y_b^* + y_a^* = y_b - y_b^* \tag{7.2-21}$$

Taking logarithms of Eq. (7.2-21) and solving for N give

$$N = \frac{\ln\left[(y_b - y_b^*)/(y_a - y_a^*)\right]}{\ln A} \tag{7.2-22}$$

and from Eq. (7.2-19)

$$\frac{y_b - y_a}{y_b^* - y_a^*} = A \tag{7.2-23}$$

Equation (7.2-22) can be written

$$N = \frac{\ln\left[(y_b - y_b^*)/(y_a - y_a^*)\right]}{\ln\left[(y_b - y_a)/(y_b^* - y_a^*)\right]} \tag{7.2-24}$$

The various concentration differences in Eq. (7.2-24) are shown in Fig. 7.16.

When the operating line and the equilibrium line are parallel, A is unity and Eqs. (7.2-22) and (7.2-24) are indeterminate. In this case the number of steps is just the overall change in concentration divided by the driving force, which is constant. Thus,

$$N = \frac{y_b - y_a}{y_a - y_a^*} = \frac{y_b - y_a}{y_b - y_b^*} \tag{7.2-25}$$

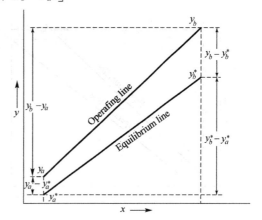

Fig. 7.16 Concentration differences in Eq. (7.2-24).

If the operating line has a lower slope than the equilibrium line, A is less than 1.0, but Eqs. (7.2-22) and (7.2-24) can still be used by inverting both terms to give

$$N = \frac{\ln\left[(y_a - y_a^*)/(y_b - y_b^*)\right]}{\ln A} \tag{7.2-26}$$

or

$$N = \frac{\ln\left[(y_a - y_a^*)/(y_b - y_b^*)\right]}{\ln\left[(y_b^* - y_a^*)/(y_b - y_a)\right]} \tag{7.2-27}$$

In the design of an absorber, as stated earlier, the liquid rate is usually chosen to make the operating line steeper than the equilibrium line or to make A greater than unity. Values of A less than 1.0 can arise when we are dealing with two or more absorbable components. If the value of A is slightly greater than 1.0 for the major solute, a second component with a much lower solubility (higher value of m) will have a value of A appreciably less than 1.0. If the gas stream and the solution are dilute, the preceding equations can be applied to each component independently.

L-phase form of Eq.(7.2-24)

The choice of y as the concentration coordinate rather than x is arbitrary. It is the conventional variable in gas absorption calculations. It may be used for stripping also, but in practice, equations in x are more common. They are

$$N = \frac{\ln\left[(x_a - x_a^*)/(x_b - x_b^*)\right]}{\ln\left[(x_a - x_b)/(x_a^* - x_b^*)\right]}$$

or

$$N = \frac{\ln\left[(x_a - x_a^*)/(x_b - x_b^*)\right]}{\ln S} \tag{7.2-28}$$

where x^* = equilibrium concentration corresponding to y
S = stripping factor

The stripping factor S is defined by

$$S \equiv \frac{1}{A} = \frac{mV}{L} \tag{7.2-29}$$

The stripping factor is the ratio of the slope of the equilibrium line to that of the operating line, and the conditions are usually chosen to make S greater than unity. The concentration differences in Eq. (7.2-28) are shown in Fig.7.17.

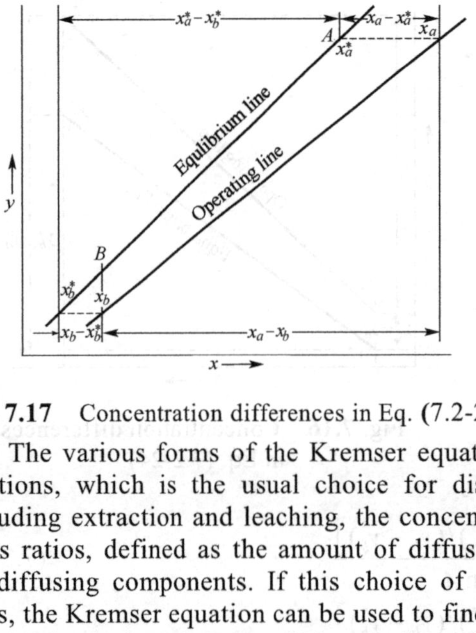

As shown in the derivations, it is not assumed that the linear extension of the equilibrium line passes through the origin. It is only necessary that the line be linear in the range where the steps representing the stages touch the line, as shown by line AB in Fig.7.17. Thus an equilibrium line that is almost linear near the origin but curves at higher concentrations is sometimes fitted by a straight line over part of its range to permit use of the Kremser equation.

Fig. 7.17 Concentration differences in Eq. (7.2-28).

The various forms of the Kremser equation were derived using concentrations in mole fractions, which is the usual choice for distillation or absorption. For some operations, including extraction and leaching, the concentrations may be expressed using mole ratios or mass ratios, defined as the amount of diffusing component divided by the amount of inert nondiffusing components. If this choice of units gives straight equilibrium and operating lines, the Kremser equation can be used to find the number of stages.

In the design of a plant, N is calculated from the proposed terminal concentrations and a selected value of A or S. Equation (7.2-22) or (7.2-24) is used for absorption and Eq. (7.2-28) for stripping. In estimating the effect of a change in operating conditions of an existing plant, Eq. (7.2-21) is used for absorption or its analog, Eq. (7.2-30), for stripping.

$$S^N = \frac{x_a - x_a^*}{x_b - x_b^*} \tag{7.2-30}$$

EXAMPLE7.5 Ammonia is stripped from a dilute aqueous solution by countercurrent contact with air in a column containing seven sieve trays. The equilibrium relationship is y

= $0.8x$, and when the molar flow of air is 1.5 times that of the solution, 90 percent of the ammonia is removed. (*a*) How many ideal stages does the column have, and what is the stage efficiency? (*b*) What percentage removal would be obtained if the air rate were increased to 2.0 times the solution rate?

Solution

(*a*) For a dilute solution and a dilute gas, L and V are assumed constant, and the stripping factor is

$$S = \frac{1}{A} = \frac{mV}{L} = 0.8 \times 1.5 = 1.2$$

All concentrations can be expressed in terms of x_a, the mole fraction of NH_3 in the entering solution:

$x_b = 0.1x_a$ $x_b^* = 0$ since $y_b = 0$

From an ammonia balance, $V\,\Delta y = V\,y_a = L\,\Delta x = L(0.9x_a)$. Hence

$$y_a = \frac{L}{V}(0.9x_a) = \frac{0.9}{1.5}x_a = 0.6x_a$$

also

$$x_a^* = \frac{y_a}{m} = \frac{0.6x_a}{0.8} = 0.75x_a$$

From Eq. (7.2-28),

$$N = \frac{\ln\left[(x_a - 0.75x_a)/(0.1x_a - 0)\right]}{\ln S} = \frac{\ln(0.25x_a/0.1x_a)}{\ln 1.2} = 5.02$$

The separation corresponds to 5.02 ideal stages, so the stage efficiency is $5.02/7 = 72$ percent.

(*b*) If V/L is increased to 2.0 and the number of ideal stages N does not change (same stage efficiency), $S = 0.8 \times 2.0 = 1.6$. Then from Eq. (7.2-30)

$$\ln \frac{x_a - x_a^*}{x_b} = 5.02 \times \ln 1.6 = 2.36$$

$$\frac{x_a - x_a^*}{x_b} = 10.59$$

Let η be the fraction of NH_3 removed. Then $x_b = (1 - \eta)\,x_a$. By a material balance,

$$y_a = \frac{L}{V}(x_a - x_b) = \frac{1}{2}[x_a - (1-\eta)x_a] = \frac{1}{2}\eta x_a$$

$$x_a^* = \frac{y_a}{m} = \frac{0.5\eta x_a}{0.8} = 0.625\eta x_a$$

Thus

$$x_a - x_a^* = (1 - 0.625\eta)x_a$$

also

$$x_a - x_a^* = 10.59 x_b = 10.59(1-\eta)x_a$$

From these, $\eta = 0.962$, or 96.2 percent is removed.

The conditions for the original case and the new case are sketched in Fig. 7.18.

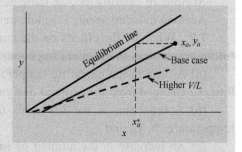

Fig. 7.18 Diagram for Example 7.5.

PROBLEMS

7.1 The data for the ammonia dissolved in the water at 101.3kPa and 20°C are listed in

table, do as follows:

(a) Plot curves of partial pressure p of NH_3 in the vapor in equilibrium with mole fraction x of NH_3 in the liquid, and y versus x, respectively.

(b) Calculate the Henry's law constant E in atm/mole fraction and the Henry's law constant m in mole frac/mole frac, respectively. And shows the range of which Henry's law applies to.

Partial pressure of ammonia p, kPa	0	0.40	0.80	1.20	1.60	2.00	2.43	3.32	4.23	6.67	9.28
Concentration of ammonia in water, g(NH_3)/100g(H_2O)	0	0.50	1.00	1.50	2.00	2.50	3.00	4.00	5.00	7.50	10.0

7.2 Using the data from table 7.1 to calculate the boiling point of the mixture of 50% benzene-toluene at a total pressure of 101.3 kPa.

7.3 Vapor-pressure data for a mixture of pentane (C_5H_{12}) and hexane (C_6H_{14}) are given by the table. Calculate the vapor and liquid composition in equilibrium for the pentane-hexane at 13.3kPa pressure on assuming that the vapor of mixture approaches ideal behavior and liquid follows Raoult's low.

t, K	260.6	265	270	275	280	285	289
p_A^0, kPa	13.3	17.3	21.9	26.5	34.5	42.5	48.9
p_B^0, kPa	2.83	3.5	4.26	5.0	8.53	11.2	13.3

7.4 Using the conditions in the problem 7.3 for the pentane-hexane mixture, do as follows:

(a) Calculate the relative volatility.

(b) Use the relative volatility to calculate the vapor and liquid composition in equilibrium and compare with the result given by the problem 7.3.

7.5 For the system benzene-toluene, do as follows, using the data from table 7.1:

(a) At 378.2 K, calculate y_A and x_A using Raoult's law.

(b) If a mixture has a composition of $x_A = 0.40$ and is at 358.2 K and 101.32 kPa pressure, will it boil? If not, at what temperature will it boil and what will be the composition of the vapor first coming off ?

7.6 Calculate the number of ideal stages for the system described in example 7.4, if the conditions are changed to the following:

Acetone in entering gas, 25 mole percent Acetone in bottoms liquor, 8 mole percent
Acetone in entering oil, 1.5 mole percent Acetone absorbed, 90 percent

7.7 What are the effects on the concentrations of the exit gas and liquid streams of the following changes in the operating conditions of the column of Example 7.5?

(a) A drop in the operating temperature that changes the equilibrium relationship to $y = 0.6x$. Unchanged from the original design: N, L/V, y_b, and x_a.

(b) A reduction in the L/V ratio from 1.5 to 1.25. Unchanged from original design: temperature, N, y_b, and x_a.

(c) An increase in the number of ideal stages from 5.02 to 8. Unchanged from original design: temperature, L/V, y_b, and x_a.

7.8 A column with eight plates and an estimated plate efficiency of 75 percent is used to remove component A from a dilute gas by absorption in water. By what factor must the L/V ratio exceed the slope of the equilibrium line to achieve 95 percent removal of component A?

7.9 If an ammonia absorber has 2 percent ammonia in the inlet gas and no ammonia in the incoming water, what fraction of the ammonia could be absorbed with an absorption factor of 0.9 and $N = 5$ or $N = 10$?

7.10 A toxic hydrocarbon is stripped from water with air in a column with eight ideal stages.

(a) What stripping factor is needed for 98 percent removal?

(b) What percentage removal could be achieved with a stripping factor of 2.0?

7.11 For the conditions of example 7.4, how many ideal stages would be needed for 97 percent absorption of the acetone of the incoming oil contained 0.005 mol fraction acetone?

7.12 A mixture of 5 percent butane and 95 percent air is fed to a sieve-plate absorber containing eight ideal plates. The absorbing liquid is a heavy, nonvolatile oil with a molecular weight of 250 and a specific gravity of 0.90. The absorption takes place at 1 atm and 15℃. The recovery of butane is to be 95 percent. The vapor pressure of butane at 15℃ is 1.92 atm, and the density of the liquid butane is 580 kg/m³ at 15℃.

(a) Calculate the cubic meters of fresh absorbing oil per cubic meter of butane recovered.

(b) Repeat, on the assumption that the total pressure is 3 atm and all other factors remain constant. Assume Raoult's and Dalton's laws apply.

7.13 Show how material balances and the equilibrium relationship can be used for a numerical solution of example 7.4. Calculate the vapor and liquid compositions for Stages 1 and 2 and compare the results with the graphical solution.

7.14 An aqueous solution with 25 ppm of a volatile organic compound (VOC) is stripped with nitrogen at $S = 0.80$. What fraction of the VOC is removed with 5 or with 10 ideal stages?

7.15 An air stripping column is used to remove 99 percent of A, a volatile hydrocarbon, from water.

(a) If the stripping factor is 1.8, how many ideal stages are needed?

(b) The water also contains traces of compound B, which is twice as volatile as A, and compound C, which is only half as volatile as A. What percent removal is expected for B and C?

名 词 术 语

英文 / 中文
absorber / 吸收设备
absorption factor / 吸收因子
azeotrope / 恒沸物
azeotropic composition / 恒沸组成
azeotropic distillation / 恒沸精馏
benzene-toluene / 苯-甲苯
boiling-point diagram / 沸点图
bottom product or bottoms / 塔底产品
bubble-point line / 泡点线
cascades / 多级操作
chlorinated hydrocarbon / 氯化碳氢化合物
countercurrent multistage / 多极逆流
degree of freedom / 自由度
dew-point line / 露点线

英文 / 中文
down-comer / 降液管
enthalpy balance / 焓衡算
equilibrium relate / 平衡关系
equilibrium stage / 平衡级
equilibrium-stage cascade / 多级平衡操作
geometric series / 几何级数
graphical method / 图解法
graphical step-by-step / 逐级图解
Henry's law / 亨利定律
hexane-heptane / 己烷-庚烷
horizontal tie line / 水平连接线
hydrocarbon / 碳氢化合物
ideal liquid / 理想液体
ideal stage / 理论级

英文 / 中文	英文 / 中文
ideal-solution law / 理想溶液定律	ordinary distillation / 普通精馏
inert gas / 惰性气体	overhead product / 塔顶产品
leaching / 浸提	partial condensation / 部分冷凝
low boiler / 低沸物	partial pressure / 分压
lower boiling constituent / 低沸物	perfect plate / 理论板
mass fraction / 质量分数	Raoult's law / 拉乌尔定律
mass transfer / 传质	reboiler / 再沸器
maximum boiling azeotrope / 最高恒沸点恒沸物	rectifying section / 精馏段
methyl alcohol-ethyl alcohol / 甲醇-乙醇	reflux / 回流
minimum boiling azeotrope / 最低恒沸点恒沸物	relative volatility / 相对挥发度
mole fraction / 摩尔分数	soluble material / 易溶物质
most volatile component / 易挥发组分	stage efficiency or plate efficiency / 板效率
nonstagewise contacting device / 非逐级接触设备	still / 釜
nonvolatile solvent / 不挥发性溶剂	stripping factor / 解吸因子
operating line / 操作线	stripping or enriching section / 提馏段
	total pressure / 总压
	two-phase region / 两相区
	weir / 堰

Chapter 8

Gas Absorption

Many chemical process materials and biological substances occur as mixtures of different components in the gas, liquid, or solid phase. In order to separate or remove one or more of the components from its original mixture, it must be contacted with another phase. The two phases are brought into more or less intimate contact with each other so that a solute or solutes can diffuse from one to the other. The two bulk phases are usually only somewhat miscible in each other. The two-phase pair can be gas-liquid, gas-solid, liquid-liquid, or liquid-solid. During the contact of the two phases the components of the original mixture redistribute themselves between the two phases. The phases are then separated by simple physical methods. By choosing the proper conditions and phases, one phase is enriched while the other is depleted in one or more components.

This chapter deals with the mass-transfer operations known as *gas absorption* and *stripping,* or *desorption*.

When the two contacting phases are a gas and a liquid, this operation is called *absorption*. A solute A or several solutes are absorbed from the gas phase into the liquid phase in absorption. This process involves molecular and turbulent diffusion or mass transfer of solute A through a stagnant, nondiffusing gas B into a stagnant liquid C. An example is absorption of ammonia A from air B by the liquid water C. Usually, the exit ammonia–water solution is distilled to recover relatively pure ammonia.

Another example is absorbing SO_2 from the flue gases by absorption in alkaline solutions. In the hydrogenation of edible oils in the food industry, hydrogen gas is bubbled into oil and absorbed. The hydrogen in solution then reacts with the oil in the presence of a catalyst. The reverse of absorption is called *stripping* or *desorption,* and the same theories and basic principles hold. An example is the steam stripping of nonvolatile oils, in which the steam contacts the oil and small amounts of volatile components of the oil pass out with the steam.

8.1 Principles of Absorption

Packed towers are used for continuous countercurrent contacting of gas and liquid in absorption. The tower in Fig. 8.1 consists of a cylindrical column containing a gas inlet and distributing space at the bottom, a liquid inlet and distributing device at the top, a gas outlet at the top, a liquid outlet at the bottom, and a packing or filling in the tower. The gas enters the distributing space below the packed section and rises upward through the openings or interstices in the packing and contacts the descending liquid flowing through the same openings. A large area of intimate contact between the liquid and gas is provided by the packing.

The diameter of a packed absorption tower depends on the quantities of gas and liquid handled, their properties, and the ratio of one stream to the other. The height of the tower, and hence the total volume of packing, depends on the magnitude of the desired concentration changes and on the rate of mass transfer per unit of packed volume. Calculations of the tower height, therefore, rest on material balances, enthalpy balances, and estimates of driving force and mass-transfer coefficients.

Equilibrium relations for gas-liquid systems in absorption were discussed in Chapter 7, and such data are needed for design of absorption towers. Some data are tabulated in Appendix19~21. Other, more extensive data are available in Perry and Green.

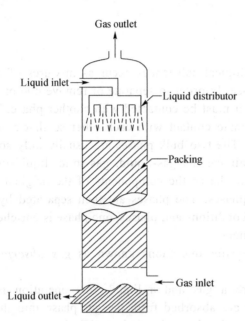
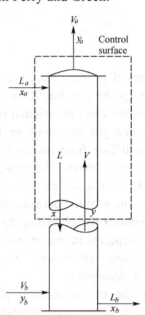

Fig. 8.1 Packed tower for absorption.   **Fig. 8.2** Material-balance diagram for packed column.

8.1.1 Material Balances

In a differential-contact plant such as the packed absorption tower illustrated in Fig.8.2, variations in composition are continuous from one end of the equipment to the other. For the case of solute A diffusing through a stagnant gas and then into a stagnant fluid, an overall material balance overall material-balance equations, based on the terminal streams, are:

Total material: $\quad\quad\quad\quad\quad\quad L_a + V_b = L_b + V_a \quad\quad\quad\quad\quad\quad$ (8.1-1)

Component A: $\quad\quad\quad\quad\quad L_a x_a + V_b y_b = L_b x_b + V_a y_a \quad\quad\quad\quad$ (8.1-2)

where V=molal flow rate of gas, mol/h; V_a, at outlet; V_b, at inlet

L=molal flow rate of liquid, mol/h; L_a, at liquid inlet; L_b, at liquid outlet

x=mole fraction of solute (component A) in liquid; x_a, at liquid inlet; x_b, at liquid outlet

y=mole fraction of solute (component A) in gas; y_a, at gas outlet; y_b, at gas inlet

Material balances for the portion of the column above an arbitrary section, as shown by the dashed line in Fig. 8.2, are as follows:

Total material: $\quad\quad\quad\quad\quad\quad L_a + V = L + V_a \quad\quad\quad\quad\quad\quad\quad$ (8.1-3)

Component A: $\quad\quad\quad\quad\quad L_a x_a + Vy = Lx + V_a y_a \quad\quad\quad\quad\quad$ (8.1-4)

where V is the molal flow rate of the gas phase (mol/h) and L that of the liquid phase(mol/h) at the same point in the tower.

The relationship between x and y at any point in the column, obtained by rearranging Eq. (8.1-4), is called the operating-line equation.

$$y = \frac{L}{V}x + \frac{V_a y_a - L_a x_a}{V} \tag{8.1-5}$$

The operating line can be plotted on an arithmetic graph along with the equilibrium curve, as shown in Fig. 8.3. The operating line must lie above the equilibrium line in order for absorption to take place, since this gives a positive driving force $y - y^*$ for absorption.

In Eq. (8.1-5), x and y represent the bulk compositions of the liquid and gas, respectively, in contact with each other at any given section through the column. It is assumed that the compositions at a given elevation are independent of position in the packing. The absorption of a soluble component from a gas mixture makes the total gas rate V decrease as the gas passes through the column, and the flow of liquid L increases. These changes make the operating line slightly curved, as shown in Fig. 8.3.

In the case of solute A diffusing through a stagnant gas (B) and then into a stagnant fluid, as in the absorption of acetone (A) from air (B) by water, the moles of inert or stagnant air and inert water remain constant throughout the entire tower. Using V' in mol inert air/h to describe the carrier gas flow rate and L' in mol inert solvent water/h to describe the flow rate of the nonvolatile solvent, the overall material balance on component A in Fig. 8.4 is:

$$L'(\frac{x_a}{1-x_a}) + V'(\frac{y_b}{1-y_b}) = L'(\frac{x_b}{1-x_b}) + V'(\frac{y_a}{1-y_a}) \tag{8.1-6}$$

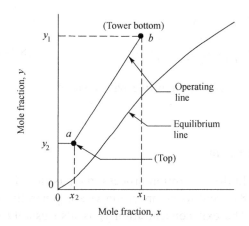
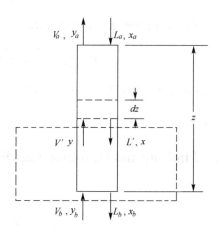

Fig. 8.3 Location of operating lines for absorption of A from V to L stream.

Fig. 8.4 Material balance for a countercurrent packed absorption tower.

where L' is mol inert liquid/h or mol inert liquid/h·m², V' is mol inert gas/h or mol inert gas/(h·m²), and y and x are mole fractions A in gas and liquid, respectively. The flows L' and V' are constant throughout the tower, but the total flows L and V are not constant.

In many instances more convenient expressions can be derived for evaluating the absorption process if a solute-free basis is used for compositions rather than mole fractions. The solute-free concentration in the liquid is:

$$X = \frac{x}{1-x} = \frac{\text{mole fraction of } A \text{ in the liquid}}{\text{mole fraction of non} - A \text{ compoents in the liquid}} \tag{8.1-7}$$

The solute-free basis for gas mixture is the same as that for liquid

$$Y = \frac{y}{1-y} = \frac{\text{mole fraction of } A \text{ in the gas}}{\text{mole fraction of non}-A \text{ compoents in the gas}} \tag{8.1-8}$$

If the carrier gas is considered to be completely insoluble in the solvent, and the solvent is considered to be completely nonvolatile, the carrier and solvent rates remain constant throughout the absorber. The overall material balance on solute A in Fig. 8.4 becomes:

$$L'X + V'Y_b = L'X_b + V'Y$$

or

$$Y = \frac{L'}{V'}X + Y_b - \frac{L'}{V'}X_b \tag{8.1-9}$$

Obviously, Eq.(8.1-9) can be used to relate compositions and flow rates between any two points in the countercurrent absorber. When plotted on X-Y coordinates, Eq.(8.1-9) gives a straight line with a slope of L'/V' and a Y intercept of $Y_b - \frac{L'}{V'}X_b$.

Overall material balance based on the mole fraction around the dashed-line box in Fig.8.4 gives:

$$L'(\frac{x}{1-x}) + V'(\frac{y_b}{1-y_b}) = L'(\frac{x_b}{1-x_b}) + V'(\frac{y}{1-y}) \tag{8.1-10}$$

For dilute mixtures, soluble component in gas less than 10 percent, the effects of changes in soluble components on the total flow are usually ignored; $(1-x)$ and $(1-y)$ can approximately be taken as 1.0 and Eq. (8.1-10) becomes:

$$L'x + V'y_b \cong L'x_b + V'y \tag{8.1-11}$$

Rearranging equation (8.1-11), and let $y_2 = y_b$ and $x_b = x_2$ give

$$y = \frac{L'}{V'}x + \frac{V'y_2 - Lx_2}{V} = \frac{L'}{V'}x + \left(y_2 - \frac{L'}{V'}x_2\right) \tag{8.1-12}$$

Equation (8.1-12), similar to the Eq.(8.1-9), is that of a straight line with a slope of L'/V' and intercept $y_2 - \frac{L'}{V'}x_2$.

8.1.2 Limiting and Optimum Gas-liquid Ratio

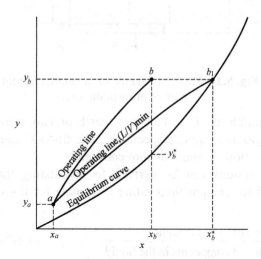

Fig. 8.5 Limiting gas-liquid ratio.

In the absorption process, the gas flow V (Fig. 8.4) and its composition y_b are generally set. The exit concentration y_a is also usually set by the designer, and the concentration x_a of the entering liquid is often fixed by process requirements. Hence, the amount of the entering liquid flow L_a or L is open to choose.

Equation (8.1-5) shows that the average slope of the operating line is L/V, the ratio of the molal flows of liquid and gas. Thus, for a given gas flow, a reduction in liquid flow decreases the slope of the operating line. Consider the operating line ab in Fig.8.5.

Assume that the gas rate and the

terminal concentrations x_a, y_a, and y_b are held constant and the liquid flow L decreased. The upper end of the operating line then shifts in the direction of the equilibrium line, and x_b, the concentration of the strong liquor, increases. The maximum possible liquor concentration and the minimum possible liquid rate are obtained when the operating line just touches the equilibrium line, as shown by line ab' in Fig. 8.5.

If the carrier gas flow rate V' and the flow rate of the nonvolatile solvent L' are used, and assuming that the dilute mixture, say soluble component less than 10 percent, is absorbed in the absorber, then, the operating line is approximately a straight line. The intersection of the operating and equilibrium lines gives a minimum slope of operating line and a minimum flow rate of nonvolatile solvent L'_{min}. As shown in Fig.8.6(a), the value of x_b is a maximum at x_{bmax} when L' is a minimum. At point b' the driving forces $y - y^*$, $y - y_i$, $x^* - x$, and $x_i - x$ are all zero. To solve for L'_{min}, the values y_b and x_{bmax} are substituted into the operating-line equation. At this condition, an infinitely deep packed section is necessary, as the concentration difference for mass transfer becomes zero at the bottom of the tower.

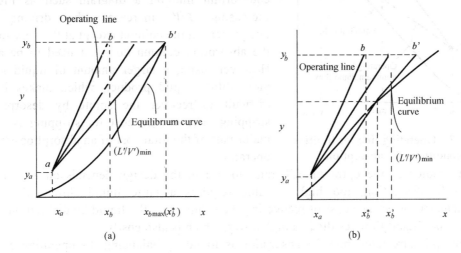

Fig. 8.6 Operating line for limiting conditions.

The minimum solvent-to-inert gas ratio L'/V' for case shown in Fig.8.6(a) can be calculated from the following expressions:

$$\left(\frac{L'}{V'}\right)_{min} = \frac{y_b - y_a}{x_{bmax} - x_a} \tag{8.1-13}$$

or

$$L'_{min} = V' \frac{y_b - y_a}{x_{bmax} - x_a} \tag{8.1-14}$$

Fig.8.6(b) shows a concave downward equilibrium line that could become tangent to the operating line at a sufficiently low solvent rate. In this case, the slope of the tangent must be determined from the graph. The minimum liquid-to-inert gas ratio for this case can be determined from Eq.(8.1-13) by replacing y_b and x_{bmax} with y_b^* and x_b^*.

$$\left(\frac{L'}{V'}\right)_{min} = \frac{y_b^* - y_a}{x_b^* - x_a} \qquad \left(\frac{L'}{V'}\right)_{min} = \frac{y_b - y_a}{x_b' - x_a} = \frac{y_b^* - y_a}{x_b^* - x_a} \tag{8.1-15}$$

or

$$L'_{min} = V' \frac{y_b^* - y_a}{x_b^* - x_a} \tag{8.1-16}$$

Assuming the equilibrium line is approximately straight over the range of concentrations used (in Fig.8.7), the equilibrium equation : $y = mx$, L'_{min} can be obtained from Eq. (8.1-17).

$$L'_{min} = V' \frac{y_b - y_a}{y_b/m - x_a} \tag{8.1-17}$$

In any actual tower the liquid rate must be greater than this minimum to achieve the specified change in gas composition. The L/V ratio is important in the economics of absorption in a countercurrent column. The driving force for mass transfer is $y - y^*$, which is proportional to the vertical distance between the operating line and the equilibrium line on a diagram such as Fig. 8.7. Increasing L/V increases the driving force everywhere in the column except at the very top, and the absorption column does not need to be as tall. However, using a larger amount of liquid gives a more dilute liquid product, which makes it more difficult to recover the solute by desorption or stripping. The energy cost for stripping is often a major part of the total cost of an absorption-stripping operation.

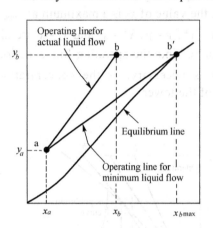

Fig. 8.7 Operating line for limiting conditions for absorption.

The choice of the optimum L/V ratio to use in the design depends on an economic balance. In absorption, too high a value requires a large liquid flow and hence a large-diameter tower. The cost of recovering the solute from the liquid by distillation will be high. A small liquid flow results in a high tower, which is also costly.

The optimum liquid rate for absorption is found by balancing the operating costs for both units against the fixed costs of the equipment. In general, the liquid rate for the absorber should be between 1.1 and 2.0 times the minimum rate, unless the liquid is to be discarded and not regenerated.

$$L_{opt} = 1.1 \sim 2.0 L_{min} \tag{8.1-18}$$

The conditions at the top of the absorber are often design variables that also have to be set, considering the balance between equipment and operating costs. For example, if tentative specifications call for 98 percent recovery of a product from a gas stream, the designer might calculate how much taller the column would have to be to get 99 percent recovery. If the value of the extra product recovered exceeds the extra costs, the optimum recovery is at least 99 percent, and the calculation should be repeated for even higher recovery. If the unremoved solute is a pollutant, its concentration in the vent gas may be set by emission standards, and the required percent recovery may exceed the optimum value based on product value and operating costs. The diagram in Fig. 8.4 shows a significant concentration of solute in the liquid fed to the column, and 99 percent removal from the gas would not be possible for this case. However, a lower value of x_a could be obtained by better stripping or more complete regeneration of the absorbing liquid. The value of x_a could be optimized, considering the extra equipment and operating costs for more complete regeneration and the savings from better operation of the absorber.

8.1.3 Rate of Absorption in Packed Towers

It is very difficult to measure experimentally the interfacial area A m^2 between phases L and V. Also, it is difficult to measure the film coefficients k_y and k_x and the overall coefficients K_y and K_x. Usually, experimental measurements in a packed tower yield a volumetric mass-transfer coefficient that combines the interfacial area and mass-transfer coefficient.

Defining a as interfacial area in m^2 per m^3 volume of packed section, the volume of packing in a height dz m (Fig.8.4) is Sdz, and the interfacial area is

$$dA = aSdZ \tag{8.1-19}$$

where S is m^2 cross-sectional area of tower. The volumetric film and overall mass-transfer coefficients are then defined as

$$k_x a = \frac{\text{mole}}{\text{s} \cdot \text{m}^3 \text{packing} \cdot \text{mole fraction}} \tag{8.1-20}$$

The rate of absorption can be expressed in four different ways using individual coefficients or overall coefficients based on the gas or liquid phases. Volumetric coefficients are used for most calculations, because it is more difficult to determine the coefficients per unit area and because the purpose of the design calculation is generally to determine the total absorber volume. In the following treatment the correction factors for one-way diffusion are omitted for simplicity, and the changes in gas and liquid flow rates are neglected. The equations are strictly valid only for lean gases but can be used with little error for mixtures with up to 10 percent solute. Absorption from rich gases is treated later as a special case.

The rate of absorption per unit volume of packed column is given by any of the following equations, where y and x refer to the mole fraction of the component A being absorbed:

$$G_A = k_y a(y - y_i) \tag{8.1-21}$$

$$G_A = k_x a(x_i - x) \tag{8.1-22}$$

$$G_A = K_y a(y - y^*) \tag{8.1-23}$$

$$G_A = K_x a(x^* - x) \tag{8.1-24}$$

The partial pressure difference $(p - p_i)$ can be used as the driving force for the gas phase, since it is proportional to $(y - y_i)$. Diagrams based on the mole ratios Y and X are sometimes used, since the operating line is then straight, but this approach is not recommended because Y and X are not valid measures of the driving force.

The individual coefficients $k_y a$ and $k_x a$ are based on a unit volume, as are the overall coefficients $K_y a$ and $K_x a$. The a in all these coefficients is the interfacial area per unit volume of the packed column or other device. It is hard to measure or to predict a, but in most cases it is not necessary to know its actual value since design calculations can be based on the volumetric coefficients.

Equation (6.1-19) holds for equimolar counter diffusion, or, when the solutions are dilute, Eqs. (8.1-21) and (8.1-22) are identical:

$$G_A = k_y a(y - y_i) = k_x a(x_i - x) \tag{8.1-25}$$

The interface composition (y_i, x_i) can be obtained from the operating-line diagram using Eq.

(8.1-25)

$$\frac{y - y_i}{x - x_i} = -\frac{k_x a}{k_y a} \qquad (8.1\text{-}26)$$

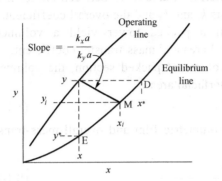

Fig. 8.8 Location of interface compositions.

Thus a line drawn from the operating line with a slope $-k_x a/k_y a$ will intersect the equilibrium line at (y_i, x_i), as shown in Fig.8.8. Usually it is not necessary to know the interface compositions, but these values are used for calculations involving rich gases or when the equilibrium line is strongly curved.

The overall driving forces are easily determined as vertical or horizontal lines on the y-x diagram. The overall coefficients are obtained from $k_y a$ and $k_x a$, using the local slope of the equilibrium curve m, as was shown in Chap.6 [Eq. (6.3-21)]:

$$\frac{1}{K_y a} = \frac{1}{k_y a} + \frac{m}{k_x a} \qquad (8.1\text{-}27)$$

The left-hand side of Eq. (8.1-27) is the total resistance based on the overall gas driving force and equals the gas film resistance $1/k_y a$ plus the liquid film resistance $m/k_x a$.

In a similar to equation (6.3-21) from Fig.8.8

$$x^* - x = x^* - x_i + x_i - x = \frac{y - y_i}{m} + x_i - x$$

Total resistance based on the overall liquid phase driving force is given by Eq.(8.1-28)

$$\frac{1}{K_x a} = \frac{1}{m k_y a} + \frac{1}{k_x a} \qquad (8.1\text{-}28)$$

In Eq. (8.1-28), the terms $1/(mk_y a)$ and $1/(k_x a)$ are the resistances to mass transfer in the gas film and the liquid film, respectively.

Several special cases of Eqs. (8.1-27) and (8.1-28) will now be discussed. When the coefficients $k_y a$ and $k_x a$ are of the same order of magnitude, the values of the slopes m is very important.

When the solubility of the gas is very high, such as with HCl in water, m is quite small, so that the equilibrium curve in Fig. 8.8 is almost horizontal, a small value of y in the gas will give a large value of x in equilibrium in the liquid. The gas solute A is then very soluble in the liquid phase, and hence the term $m/(k_x a)$ in Eq. (8.1-27) is very small. Then,

$$\frac{1}{K_y a} \cong \frac{1}{k_y a} \qquad (8.1\text{-}29)$$

and the gas-film resistance controls the rate of absorption or the "gas phase is controlling". With gases of intermediate solubility both resistances are important, but the term *controlling resistance* is sometimes used for the larger resistance. The absorption of NH_3 in water is often cited as an example of gas-film control, since the gas film has about 80 to 90 percent of the total resistance. The point M in Fig. 8.8 has moved down very close to E, so that:

$$y - y^* \cong y - y_i \tag{8.1-30}$$

Similarly, the solubility of the gas is very small, when the coefficients k_ya and k_xa are of the same order of magnitude, and m is very much greater than 1.0, the solute A is very insoluble in the liquid, $1/(mk_ya)$ becomes small, and

$$\frac{1}{K_xa} \cong \frac{1}{k_xa} \tag{8.1-31}$$

The liquid-film resistance is said to be controlling or "liquid phase is controlling" and $x_i \cong x_A^*$.

Systems for absorption of oxygen or CO_2 from air by water are similar to Eq. (8.1-31). This means that any change in k_xa has a nearly proportional effect on both K_ya and K_xa and on the rate of absorption, where as a change in k_ya has little effect. For example, Henry's law coefficient for CO_2 in water at 20℃ is 1430 atm/mol fraction, which corresponds to $m = 1430$ for absorption at 1 atm and $m = 143$ for absorption at 10 atm. Under these conditions the absorption of CO_2 in water is clearly liquid-film controlled. Increasing the gas velocity would increase k_ya but have a negligible effect on K_ya. Increasing the liquid velocity would increase the interfacial area a and probably also increase k_x both leading to an increase in K_xa and K_ya.

8.2 Calculation of Tower Height

8.2.1 Fundamental Calculation Equation of Packing Height

An absorber can be designed using any of the four basic rate equations, but the gas-film coefficients are often used, and the use of K_ya will be emphasized here. Choosing a gas-film coefficient does not require any assumption about the controlling resistance. Even if the liquid film controls, a design based on K_ya is as simple and accurate as one based on K_xa or k_xa.

Consider the packed column shown in Fig.8.9. The cross section is S, and the differential volume in height dZ is SdZ. For dilute gases the change in molar flow rate is neglected. For the differential height of tower dz in Fig.8.9, the moles of A leaving V equal the moles entering L:

$$d(Vy) = d(Lx) \tag{8.2-1}$$

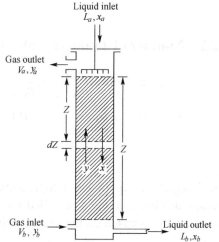

Fig. 8.9 Diagram of packed absorption tower.

where V = mol total gas/s, L = mol total liquid/s, and $d(Vy)=d(Lx)$=mol A transferred/s in height dZ m. The amount absorbed in section dZ is $-V\,dy$, which equals the absorption rate per unit volume times the differential volume. Eqs. (8.1-23) and (8.1-24) give the mol A transferred G_A using K_ya based on the overall gas phase driving force and K_xa based on the overall liquid phase driving force:

$$G_A = K_ya(y - y^*) = K_xa(x^* - x) \tag{8.2-2}$$

Multiplying the two sides of Eq.(8.2-2) by SdZ

$$G_A SdZ = K_ya(y - y^*)SdZ = K_xa(x^* - x)SdZ \tag{8.2-3}$$

where $G_A SdZ$ = mol A transferred/s in height dz m (mol/s).

Equating Eq. (8.2-1) to (8.2-3) and using y for the bulk gas phase and x for the bulk liquid phase

$$-Vdy = K_y a(y - y^*)SdZ \tag{8.2-4}$$

$$Ldx = K_x a(x^* - x)SdZ \tag{8.2-5}$$

The Eq.(8.2-4) is rearranged for integration, grouping the constant factors V, S, and $K_y a$ with dZ and reversing the limits of integration to eliminate the minus sign:

$$\int_0^Z dZ = \frac{V}{K_y aS} \int_{y_a}^{y_b} \frac{dx}{y - y^*} \tag{8.2-6}$$

In similar to Eq(8.2-6), the Eq(8.2-5) is rearranged for integration

$$\int_0^Z dZ = \frac{L}{K_x aS} \int_{x_a}^{x_b} \frac{dx}{x^* - x} \tag{8.2-7}$$

The right-hand side of Eq. (8.2-6) and (8.2-7) can be integrated directly for certain cases, or it can be determined numerically. We will examine some of these cases.

The equation for column height can be written as follows:

$$Z = \frac{V}{K_y aS} \int_{y_a}^{y_b} \frac{dy}{y - y^*} \tag{8.2-8}$$

or

$$Z = \frac{L}{K_x aS} \int_{x_a}^{x_b} \frac{dx}{x^* - x} \tag{8.2-9}$$

8.2.2 Number of Transfer Units and Height of A Transfer Unit

In Eq.(8.2-8), the unit of $\frac{V}{K_y aS}$ is $\frac{[kmol/s]}{[kmol/(m^3 \cdot s)][m^2]} = [m]$. It has the units of length and is called the *height of a transfer unit* (HTU) H_{Oy}.

$$H_{Oy} = \frac{V}{K_y aS} \tag{8.2-10}$$

The integral in Eq. (8.2-8) represents the change in vapor concentration divided by the average driving force and is called the *number of transfer units* (NTU) N_{Oy}. The subscripts show that N_{Oy} is based on the overall driving force for the gas phase.

$$N_{Oy} = \int_{y_a}^{y_b} \frac{dy}{y - y^*} \tag{8.2-11}$$

Thus a simple design method is to determine N_{Oy} from the y–x diagram and multiply it by H_{Oy} obtained from the literature or calculated from mass-transfer correlations:

$$Z = H_{Oy} \times N_{Oy} \tag{8.2-12}$$

The number of transfer units is somewhat like the number of ideal stages, discussed in Chap. 7, but the values are equal only if the operating line and equilibrium line are straight and parallel, as in Fig. 8.10(a). For this case,

$$N_{Oy} = \frac{y_b - y_a}{y - y^*} \tag{8.2-13}$$

When the operating line is straight but steeper than the equilibrium line, as in Fig.8.10(b), the number of transfer units is greater than the number of ideal stages. Note that for the example shown, the driving force at the bottom is $y_b - y_a$, the same as the change in vapor concentration across the tower, which has one ideal stage. However, the driving force at the top is y_a, which is several-fold smaller, so the average driving force is much less than $y_b - y_a$. When both the operating and equilibrium lines are straight, the proper average can be shown to be the logarithmic mean of the driving forces at the two ends of the column.

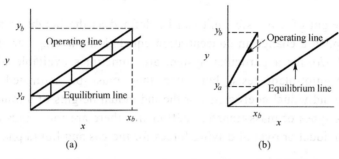

Fig. 8.10 Relationship between number of transfer units (NTU) and number of theoretical plates (NTP): (a) NTU = NTP; (b) NTU > NTP.

For straight operating and equilibrium lines, the number of transfer units is the change in concentration divided by the logarithmic mean driving force:

$$N_{Oy} = \frac{y_b - y_a}{\Delta y_m} \tag{8.2-14}$$

where Δy_m is the logarithmic mean of $y_b - y_b^*$ and $y_a - y_a^*$. Equation (8.2-14) is based on the gas phase.

$$\Delta y_m = \frac{(y_b - y_b^*) - (y_a - y_a^*)}{\ln\left[(y_b - y_b^*)/(y_a - y_a^*)\right]} \tag{8.2-15}$$

The corresponding equation based on the liquid phase is

$$H_{Ox} = \frac{L}{K_x a S} \tag{8.2-16}$$

$$N_{Ox} = \frac{x_b - x_a}{\Delta x_m} \tag{8.2-17}$$

where Δx_m is the logarithmic mean of $x_b^* - x_b$ and $x_a^* - x_a$. Eq. (8.2-16) is based on the liquid phase.

$$\Delta x_m = \frac{(x_b^* - x_b) - (x_a^* - x_a)}{\ln\left[(x_b^* - x_b)/(x_a^* - x_a)\right]} \tag{8.2-18}$$

The number of liquid-phase transfer units N_{Ox} is not the same as the number of gas-phase transfer units N_{Oy} unless the operating and equilibrium lines are straight and parallel. For absorption the operating line is usually steeper than the equilibrium line, which makes N_{Oy} greater than N_{Ox}, but this difference is offset by the difference between H_{Oy} and H_{Ox}, and the column height can be determined using either approach.

Combining the operating line with the integrals in Eq.(8.2-8)and (8.2-9), using the equilibrium-line equation $y = mx$, and letting $A = L/mV$, different forms of the equations for absorption with N_{Oy} and for stripping with N_{Ox} are obtained:

$$N_{Oy} = \frac{1}{1-1/A} \ln \left[(1-\frac{1}{A}) \left(\frac{y_b - mx_a}{y_a - mx_a} \right) + \frac{1}{A} \right] \tag{8.2-19}$$

or

$$N_{Ox} = \frac{1}{1-A} \ln \left[(1-A) \left(\frac{x_a - y_b/m}{x_b - y_b/m} \right) + A \right] \tag{8.2-20}$$

The overall height of a transfer unit can be defined as the height of a packed section required to accomplish a change in concentration equal to the average driving force in that section. Values of H_{Oy} for a particular system are sometimes available directly from the literature or from pilot-plant tests, but often they must be estimated from empirical correlations for the individual coefficients or the individual heights of a transfer unit. Just as there are four basic types of mass-transfer coefficients, there are four kinds of transfer units, those based on individual or overall driving forces for the gas and liquid phases. These are as follows:

Gas film: $\qquad H_y = \dfrac{V}{Sk_y a} \qquad N_y = \int \dfrac{dy}{y - y_i}$ (8.2-21)

Liquid film: $\qquad H_x = \dfrac{L}{Sk_x a} \qquad N_x = \int \dfrac{dx}{x_i - x}$ (8.2-22)

Overall gas: $\qquad H_{Oy} = \dfrac{V}{SK_y a} \qquad N_{Oy} = \int \dfrac{dy}{y - y^*}$ (8.2-23)

Overall liquid: $\qquad H_{Ox} = \dfrac{L}{SK_x a} \qquad N_{Ox} = \int \dfrac{dx}{x^* - x}$ (8.2-24)

The height of the packed tower is then:

$$Z = H_y N_y = H_x N_x = H_{Oy} N_{Oy} = H_{Ox} N_{Ox} \tag{8.2-25}$$

These equations are basically no different from those using mass-transfer coefficients. When the major resistance to mass transfer is in the gas phase, as in absorption of acetone from air by water, the overall number of transfer units based on the gas phase, N_{Oy} or the film N_y should be used. When the major resistance is in the liquid phase, as in absorption of O_2 by water or stripping of a slightly soluble solute from water, N_{Ox} or N_x should be employed. This was also discussed in detail in section 8.2.3. Often the film coefficients are not available, and then it is more convenient to use N_{Oy} or N_{Ox}.

8.2.3 Alternate Forms of Transfer Coefficients

The gas-film coefficients reported in the literature are often based on a partial pressure driving force instead of a mole-fraction difference and are written as $k_G a$ or $K_G a$. Their relationships to the coefficients used heretofore are simply:

$$k_G a = \frac{k_y a}{P} \tag{8.2-26}$$

and

$$K_G a = \frac{K_y a}{P} \tag{8.2-27}$$

where P is the total pressure. The units of $k_G a$ and $K_G a$ are commonly mol/(m^3·h·atm). Similarly liquid-film coefficients may be given as $k_L a$ or $K_L a$, where the driving force is a volumetric concentration difference; k_L is therefore the same as k_c defined by Eq. (6.3-1). Thus $k_L a$ and $K_L a$ are equal to $k_x a/\rho_M$ and $K_x a/\rho_M$, respectively, where ρ_M is the molar density of the liquid, mol/m^3. The units of $k_L a$ and $K_L a$ are usually mol/m^3·h·(mol/m^3) or h^{-1}. If V_y/M or V_M is substituted for V/S in Eqs. (8.2-21) and (8.2-23), and V_x/M for L/S in Eqs. (8.2-22) and (8.2-24), the equations for the height of a transfer unit may be written (since $M\rho_M = \rho_x$, the density of the liquid)

$$H_y = \frac{V_M}{k_G a P} \quad \text{and} \quad H_{Oy} = \frac{V_M}{K_G a P} \tag{8.2-28}$$

$$H_x = \frac{V_x/\rho_x}{k_L a} \quad \text{and} \quad H_{Ox} = \frac{V_x/\rho_x}{K_L a} \tag{8.2-29}$$

The terms H_G, H_L, N_G, and N_L often appear in the literature instead of H_y, H_x, N_y, and N_x, as well as the corresponding terms for overall values, but here the different subscripts do not signify any difference in either units or magnitude. If a design is based on N_{Oy}, the value of H_{Oy} can be calculated either from $K_y a$ or from values of H_y and H_x, as shown later. Starting with the equation for overall resistance, Eq.(8.1-27),

$$\frac{1}{K_y a} = \frac{1}{k_y a} + \frac{m}{k_x a} \tag{8.1-27}$$

Each term of Eq.(8.1-27) is multiplied by V_M, and the last term is multiplied by L_M/L_M, where $L_M = L/S = V_x/M$, the molar mass velocity of the liquid:

$$\frac{V_M}{K_y a} = \frac{V_M}{k_y a} + \frac{mV_M}{L_M} \times \frac{L_M}{k_x a} \tag{8.2-30}$$

Similarly, from Eq.(8.1-28)

$$\frac{1}{K_x a} = \frac{1}{k_x a} + \frac{1}{mk_y a} \tag{8.1-28}$$

Multiplying the two sides of Eq.(8.1-28) by L_M and both numerator and denominator of the last term on the right-hand side of the equation are multiplied by V_M, the result is

$$\frac{L_M}{K_x a} = \frac{L_M}{k_x a} + \frac{L_M}{mV_M} \times \frac{V_M}{k_y a} \tag{8.2-31}$$

From the definitions of HTU in Eqs. (8.2-21) to (8.2-23),

$$H_{Oy} = H_y + m\frac{V_M}{L_M}H_x \tag{8.2-32}$$

$$H_{Ox} = H_x + \frac{L_M}{mV_M}H_y \tag{8.2-33}$$

Note that often in the literature, the terms H_y, H_x, H_{Oy}, and H_{Ox} are used instead of H_G, H_L, H_{OG}, and H_{OL}; they are identical to each other.

EXAMPLE 8.1 A gas stream containing 3.0 percent A is passed through a packed column to remove 99 percent of the A by absorption in water. The absorber will operate at 25°C and 1 atm, and the gas and liquid rates are to be 20 mol/(h·m²) and 100 mol/(h·m²), respectively. Mass-transfer coefficients and equilibrium data are given below:

$$y^* = 3.1x \quad \text{at } 25°C$$
$$k_xa = 135 \text{ mol/(h·m}^3 \text{ · unit mol fraction)}$$
$$k_ya = 540 \text{ mol/(h·m}^3 \text{ · unit mol fraction)}$$

(a) Find N_{Oy}, H_{Oy}, and Z, assuming isothermal operation and neglecting changes in gas and liquid flow rates. What percent of the total resistance is in the gas phase?
(b) Calculate Z, using N_{Ox} and H_{Ox}.

Solution

(a) Assume $x_a = 0$. Since $V_M(x_b - x_a) = L_M(y_a - y_b)$

$$x_b = \frac{20 \times 0.03 \times 0.99}{100} = 0.00594$$

$$y_b^* = 3.1 \times 0.00594 = 0.01841$$

At the bottom of the column,

$$y_b - y_b^* = 0.03 - 0.01841 = 0.01159$$

At the top,

$$y_a - y_a^* = y_a = 0.0003$$

$$\Delta y_m = \frac{(y_b - y_b^*) - (y_a - y_a^*)}{\ln[(y_b - y_b^*)/(y_a - y_a^*)]} = \frac{0.01159 - 0.0003}{\ln 0.01159/0.0003} = 0.00309$$

$$N_{Oy} = \frac{0.03 \times 0.99}{0.00309} = 9.61$$

$$\frac{1}{K_ya} = \frac{1}{k_ya} + \frac{m}{k_xa} = \frac{1}{540} + \frac{3.1}{135} = 0.01315$$

$$K_ya = 76.056$$

$$H_{Oy} = \frac{20}{76.056} = 0.263 \text{m}$$

$$Z = 0.263 \times 9.61 = 2.53 \text{m}$$

$$x_b^* = \frac{0.03}{3.1} = 0.009677$$

The relative gas-film resistance is

$$\frac{1/135}{1/76.056} = 0.56 \text{ or } 56 \text{ percent}$$

(b) At the bottom of the column,

$$\Delta x = 0.009677 - 0.00594 = 0.003737$$

At the top, $x_a^* = \dfrac{0.0003}{3.1} = 9.677 \times 10^{-5}$ $x_a = 0$

$$\Delta x_m = \dfrac{(x_b^* - x_b) - (x_a^* - x_a)}{\ln\left[(x_b^* - x_b)/(x_a^* - x_a)\right]} = \dfrac{0.00374 - 0.000097}{\ln 0.00374/0.000097} = 9.96 \times 10^{-4}$$

$$N_{Ox} = \dfrac{0.00594}{9.96 \times 10^{-4}} = 5.96$$

$$\dfrac{1}{K_x a} = \dfrac{1}{k_x a} + \dfrac{1}{mk_y a} = \dfrac{1}{540} + \dfrac{1}{3.1 \times 135} = 4.24 \times 10^{-3} \quad K_x a = 235.85$$

$$H_{Ox} = \dfrac{100}{235.85} = 0.424\,\text{m}$$

$$Z = 5.96 \times 0.424 = 2.53\,\text{m}$$

Note that although N_{Oy} and N_{Ox} are quite different, the values of H_{Oy} and H_{Ox} are also different, so the values of Z are the same for the two methods of calculation.

8.2.4 Effect of Pressure

Absorption columns are often operated under pressure to give increased capacity and higher rates of mass transfer. The equilibrium partial pressure of the solute depends only on the liquid composition and the temperature, so the equilibrium mole fraction in the gas varies inversely with the total pressure

$$y_A = \dfrac{p_A}{P} \tag{8.2-34}$$

If the gas and liquid rates are kept constant so the operating line is not changed, going to a higher pressure increases the mole-fraction driving force, as shown in Fig. 8.11, and reduces the number of transfer units. At high pressure the minimum liquid rate is smaller, so the operating line could be changed to give a richer product, as indicated by the dashed line in Fig.8.11, and about the same number of transfer units as before.

The overall mass transfer coefficient $K_y a$ increases with pressure because the liquid-film resistance $m/k_x a$ decreases [see Eq. (8.1-27)]. The gas film coefficient $k_y a$ does not change much with pressure; the diffusivity varies inversely with pressure, but the concentration driving force for a given Δy_m is proportional to the pressure. Thus H_{Oy} will decrease if the liquid resistance is significant but will be unchanged if the gas film resistance controls.

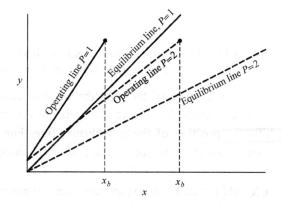

Fig. 8.11 Effect of pressure on absorption.

8.2.5 Temperature Variations in Packed Towers

When rich gas is fed to an absorption tower, the temperature in the tower varies appreciably from bottom to top. The heat of absorption of the solute raises the solution temperature, but evaporation of the solvent tends to lower the temperature. Usually the overall effect is an increase in the liquid temperature, but sometimes the temperature goes through a maximum near the bottom of the column. The shape of the temperature profile depends on the rates of solute absorption, evaporation or condensation of solvent, and heat transfer between the phases. Lengthy computations are needed to get the exact temperature profiles for liquid and gas, and in this text only simplified examples are presented. When the gas inlet temperature is close to the exit temperature of the liquid and the incoming gas is saturated, there is little solvent evaporation, and the rise in liquid temperature is roughly proportional to the amount of solute absorbed. The equilibrium line is then curved gradually upward, as shown in Fig.8.12(a), with increasing values of x corresponding to higher temperatures.

When the gas enters the columns 10 to 20°C below the exit liquid temperature and the solvent is volatile, evaporation will cool the liquid in the bottom part of the column, and the temperature profile may have a maximum, as shown in Fig. 8.12(b). When the feed gas is saturated, the temperature peak is not very pronounced, and for an approximate design, either the exit temperature or the estimated maximum temperature can be used to calculate equilibrium values for the lower half of the column.

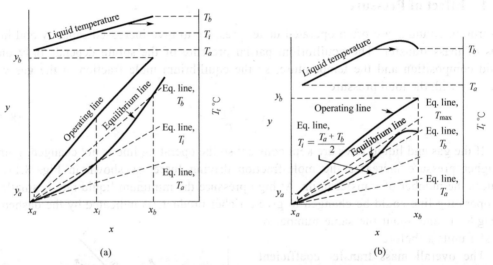

Fig. 8.12 Temperature profiles and equilibrium lines for adiabatic absorption:(a) no solvent evaporation:(b) significant solvent evaporation or cold gas feed.

The curvature of the equilibrium line complicates the determination of the minimum liquid rate, since decreasing the liquid rate increases the temperature rise of the liquid and shifts the position of the equilibrium line. For most cases, it is satisfactory to assume the pinch occurs at the bottom of the column to calculate L_{min}.

EXAMPLE 8. 2 A gas stream with 6.0 percent NH_3 (dry basis) and a flow rate of 4500 SCFM (ft³/min at 0°C, 1 atm) is to be scrubbed with water to lower the concentration to 0.02 percent. The absorber will operate at atmospheric pressure with inlet temperatures of 20 and 25°C for the gas and liquid, respectively. The gas is saturated with water vapor at the inlet temperature and can be assumed to leave as a saturated gas at 25°C. Calculate the value of N_{OG} if the liquid rate is 1.25 times the minimum.
Solution. The following solubility data are given by Perry.

x	$y_{20°C}$	$y_{30°C}$	$y_{40°C}$
0.0308	0.0239	0.0389	0.0592
0.0406	0.0328	0.0528	0.080
0.0503	0.0417	0.0671	0.1007
0.0735	0.0658	0.1049	0.1579

For $NH_3 \to NH_3(aq)$.
$\Delta H = -8.31\,kcal/g\,mol$

The temperature at the bottom of the column must be calculated to determine the minimum liquid rate.

Basis. 100 mol of dry gas in, containing 94 mol of air and 6 mol of NH_3. The outlet gas contains 94 mol of air.

The moles of ammonia in the outlet gas, since $y_a = 0.0002$, are

$$94\left(\frac{0.0002}{0.9998}\right) = 0.0188\,mol\,NH_3$$

The amount of ammonia absorbed is then
$$6 - 0.0188 = 5.98\,mol$$

Heat effects. The heat of absorption is $5.98 \times 8310 = 49690$ cal. Call this Q_a. Then

$$Q_a = Q_{sy} + Q_v + Q_{sx} \tag{8.2-35}$$

where Q_{sy} = sensible-heat change in gas
Q_v = heat of vaporization
Q_{sx} = sensible heat change in liquid

The sensible-heat changes in the gas are

$Q_{air} = 94\,mol \times 7.0\,cal/mol\cdot°C \times 5°C = 3290$ cal

$Q_{H_2O} = 2.4 \times 8.0 \times 5 = 96$ cal

$Q_{sy} = 3{,}290 + 96 = 3390$ cal

The amount of vaporization of water from the liquid is found as follows. At 20°C, $p_{H_2O} = 17.5$ mm Hg; at 25°C, $p_{H_2O} = 23.7$ mmHg. The amount of water in the inlet gas

$$100 \times \frac{17.5}{742.5} = 2.36\,mol$$

In the outlet gas it is

$$94.02 \times \frac{23.7}{736.3} = 3.03\,mol$$

The amount of water vaporized is therefore
$$(3.03 - 2.36) = 0.67\,mol.$$

Since the heat of vaporization $\Delta H_v = 583$ cal/g

$Q_v = 0.67 \times 583 \times 18.02 = 7040$ cal

Solving Eq. (8.2-34) for Q_{sx}, the sensible-heat change in the liquid, gives

$Q_{sx} = 49690 - 3390 - 7040 = 39260$ cal

The outlet temperature of the liquid T_b is found by trial. Assume that for the solution $C_p = 18$ cal/(g mol·°C); guess that $T_b = 40°C$ and $x_{max} = 0.031$, as estimated from the equilibrium solubility lines on Fig. 8.13. Then the total moles of liquid out L_b are

$$L_b = \frac{5.98}{0.031} = 192.9 \text{mol}$$

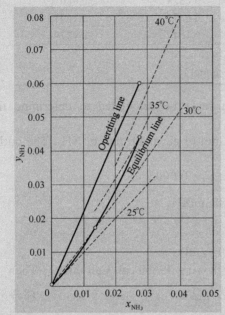

Since $T_a = 25℃$,

$$192.9 \times 18(T_b - 25) = 39260$$
$$T_b = 36.3℃$$

For a revised estimate of $T_b = 37℃$, $x_{max} = 0.033$

$$L_b = \frac{5.98}{0.033} = 181 \text{mol}$$
$$T_b - 25 = \frac{39260}{181 \times 18} = 12.1$$
$$T_b = 37℃$$

This procedure gives the minimum liquid rate; the minimum amount of water is
$$L_{min} = 181 - 6 = 175 \text{ mol H}_2\text{O}$$
For a water rate 1.25 times the minimum, $L_a = 1.25 \times 175 = 219$ mol, and $L_b = 219 + 6 = 225$ mol. Then, the temperature rise of the liquid is

$$T_b - 25 = \frac{39260}{225 \times 18} = 9.7℃$$

Fig.8.13 The y-x diagram for Example 8.2.

The liquid therefore leaves at 35℃, with $x_b = 5.98/225 = 0.0266$ and $y^* \approx 0.044$.

To simplify the analysis, the temperature is assumed to be a linear function of x, so that $T \approx 30℃$ at $x = 0.0137$. Using the data given for 30℃ and interpolating to get the initial slope for 25℃ and the final value of y^* for 35℃, the equilibrium line is drawn as shown in Fig.8.13. The operating line is drawn as a straight line, neglecting the slight change in liquid and gas rates. Because of the curvature of the equilibrium line, N_{OG} is evaluated by numerical integration or by applying Eq. (8.2-14) to sections of the column, which is the procedure used here.

y	Y^*	$y-y^*$	Δy_m	ΔN_{OG}
0.06	0.048	0.012	—	—
0.03	0.017	0.013	0.0125	2.4
0.01	0.0055	0.0045	0.0080	2.5
0.0002	0	0.0002	0.00138	7.1
				$N_{OG}=12.0$

8.3 Multicomponent Absorption

When more than one solute is absorbed from a gas mixture, separate equilibrium and operating lines are needed for each solute, but the slope of the operating line, which is L/V, is the same for all the solutes. A typical y–x diagram for absorption of two solutes is shown in

Fig. 8.14. In this example, B is a minor component of the gas, and the liquid rate was chosen to permit 95 percent removal of A with a reasonable packed height. The operating-line slope is about 1.5 times the slope of the equilibrium line for A, and $N_{Oy} \approx 5.5$. The operating line for B is parallel to that for A, and since the equilibrium line for B has a slope greater than L/V, there is a pinch at the bottom of the column, and only a small fraction of B can be absorbed. The operating line for B should be drawn to give the correct number of transfer units for B, which is generally about the same as N_{Oy} for A. However, in this example, x_{Bb} is practically the same as x_B^*, the equilibrium value for $y_{B,b}$,

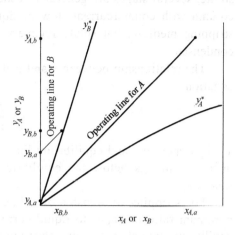

Fig. 8.14 Equilibrium and operating lines for muticomponent absorption.

and the fractional removal of B can be calculated directly from a material balance:

$$V(y_{B,b} - y_{B,a}) = L(x^*_{B,b} - x_{B,a}) \tag{8.3-1}$$

If nearly complete absorption of B is required, the operating line will have to be made steeper than the equilibrium line for B. Then the operating line will be much steeper than the equilibrium line for A, and the concentration of A in the gas will be reduced to a very low value. Examples of multicomponent absorption are the recovery of light hydrocarbon gases by absorption in heavy oil, the removal of CO_2 and H_2S from natural gas or coal gasifier products by absorption in methanol or alkaline solutions, and water scrubbing to recover organic products produced by partial oxidation. For some cases, the dilute solution approach presented here may have to be corrected for the change in molar flow rates or the effect of one solute on the equilibria for other gases, as shown in the analysis of a natural gasoline absorber.

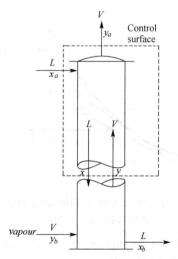

Fig. 8.15 Material-balance diagram for packed column.

8.4 Desorption or Stripping

In many cases, a solute that is absorbed from a gas mixture is desorbed from the liquid to recover the solute in more concentrated form and regenerate the absorbing solution. When the solute is being transferred from the L to the V stream, the process is called *stripping* in Fig. 8.15. To make conditions more favorable for desorption, the temperature may be increased or the total pressure reduced, or both these changes may be made. If the absorption is carried out under high pressure, a large fraction of the solute can sometimes be recovered simply by flashing to atmospheric pressure. However, for nearly complete removal of the solute, several stages are generally needed, and the desorption or stripping is carried out in a column with countercurrent flow of liquid and gas. Inert gas or steam can be used as the stripping medium, but solute recovery is easier if steam is used, since the steam can be condensed.

The relationship between x and y at any point in the column, is called the operating-line equation.

$$y = \frac{L}{V}x + \frac{V_a y_a - L_a x_a}{V} \quad (8.1\text{-}5)$$

Typical operating and equilibrium lines for stripping with steam are shown in Fig. 8.16. The operating line is below the equilibrium line, since this gives a driving force $y^* - y$ for *stripping*.

For stripping or transfer of solute from L to V, with x_a and x_b specified, there is a minimum ratio of vapor to liquid corresponding to the operating line that just touches the equilibrium line at some point, then the gas flow is at the minimum V_{min} The value of y_a is at a maximum at $y_{a,max}$. The pinch may occur in the middle of the operating line if the equilibrium line is curved upward, as in Fig. 8.17, or it may occur at the top of the column, at (y_a, x_a).

For simplicity the operating line is shown as a straight line, though it would generally be slightly curved because of the change in vapor and liquid rates. In an overall process of absorption and stripping, the cost of steam is often a major expense, and the process is designed to use as little steam as possible. The stripping column is operated at close to the minimum vapor rate, and some solute is left in the stripped solution, instead of trying for complete recovery. When the equilibrium line is curved upward, as in Fig. 8.17, the minimum steam rate becomes much higher as x_b approaches zero. As in absorption, the optimum gas flow rate is taken at about 1.5 times V_{min}, the same conditions for stripping hold for extracting solute from feed liquid L to solvent V.

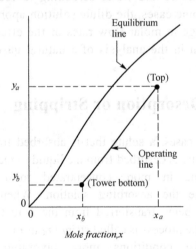
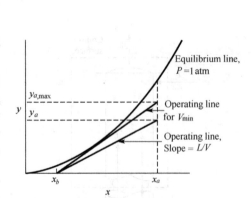

Fig. 8.16 Location of operating lines for stripping of A from L to V stream.

Fig. 8.17 Operating line for limiting conditions for a stripping column.

The height of a stripping column can be calculated from the number of transfer units and the height of a transfer unit, using the same equations as for absorption.

Often attention is focused on the liquid-phase concentration, and N_{Ox} and H_{Ox} are used:

$$Z = H_{Ox}N_{Ox} = H_{Ox}\int \frac{dx}{x^* - x} \tag{8.4-1}$$

Stripping with air is used in some cases to remove small amounts of gases such as ammonia or organic solvents from water. If there is no need to recover the solute in concentrated form, the optimum amount of air used may be much greater than the minimum, since it does not cost much to provide more air, and the column height is considerably reduced. The following example shows the effect of air rate in a stripping operation.

EXAMPLE 8.3 Water containing 6×10^{-6} trichloroethylene (TCE) is to be purified by stripping with air at 20°C. The product must contain less than 4.5×10^{-9} TCE to meet emission standards. Calculate the minimum air rate in standard cubic meters of air per cubic meter of water and the number of transfer units if the air rate is 1.5 to 5 times the minimum value.

Solution. Henry's law coefficient for TCE in water11 at 20°C is 0.0075 $m^3 \cdot atm/mol$. This can be converted to the slope of the equilibrium line in mole-fraction units as follows, since $P = 1$ atm and 1 m^3 of liquid weighs 10^6g:

$$m = 0.0075 \frac{atm \cdot m^3}{mol} \times \frac{1}{1 atm} \times \frac{10^6 \, molH_2O}{18 \, m^3} = 417$$

With this large value of m, the desorption is liquid-phase controlled. At the minimum air rate, the exit gas will be in equilibrium with the incoming solution. The molecular weight of TCE is 131.4, and

$$x_a = \frac{6 \times 10^{-6} \, molTCE}{131.4 \, gH_2O} \times \frac{18 g}{molH_2O} = 8.22 \times 10^{-7}$$

$$y_a = 417 \times 8.22 \times 10^{-7} = 3.43 \times 10^{-4}$$

Per cubic meter of solution fed, the TCE removed is

$$V_{TCE} = \frac{10^6 \times (6 \times 10^{-6} - 4.5 \times 10^{-9})}{131.4} = 4.56 \times 10^{-2} \, mol$$

The total amount of gas leaving is

$$V = \frac{4.56 \times 10^{-2}}{3.43 \times 10^{-4}} = 132.9 \, mol$$

Since 1 mol = 0.0224 std m^3 and since the change in gas flow rate is very small,

$$F_{min} = 132.9 \times 0.0224 = 2.98 \text{ std } m^3$$

The density of air at standard conditions is 1.295 kg/m^3, so the minimum rate on a mass basis is

$$\left(\frac{G_y}{G_x}\right)_{min} = \frac{2.98 \times 1.295}{1000} = 3.86 \times 10^{-3} \, kg \, air/kg \, water$$

If the air rate is 1.5 times the minimum value, then

$$y_a = \frac{3.43 \times 10^{-4}}{1.5} = 2.29 \times 10^{-4}$$

$$x_a^* = \frac{2.29 \times 10^{-4}}{417} = 5.49 \times 10^{-7}$$

$$C_a^* = 5.49 \times 10^{-7} \times \frac{131.4}{18} = 4.01 \times 10^{-6}$$

$$C_a - C_a^* = \Delta C_a = 6.0 - 4.01 = 1.99$$

At bottom,

$$C_b = 0.0045 \times 10^{-6} \qquad C_b^* = \Delta C_b = 0.0045 \times 10^{-6}$$

$$\Delta c_m = \frac{1.99 - 0.0045}{\ln(1.99/0.0045)} = 0.3259 \times 10^{-6}$$

Using concentrations in parts per million to calculate N_{OL}

$$N_{Ox} = \int \frac{dC}{C - C^*} = \frac{C_a - C_b}{\Delta C_m}$$

$$= \frac{6 - 0.0045}{0.3259}$$

$$= 18.4$$

Similar calculations for other multiples of the minimum flow rate give the following values. The packed height is based on an estimated value of $H_{Ox} = 0.91\text{m}$; this is somewhat greater than the values reported for 1-in. plastic Pall rings

Air rate	N_{Ox}	Z,m
$1.5V_{min}$	18.4	16.8
$2V_{min}$	13.0	11.9
$3V_{min}$	10.2	9.33
$5V_{min}$	8.7	7.96

Going from 1.5 to $2V_{min}$ or from 2 to $3V_{min}$ decreases the tower height considerably, and the reduction in pumping work for water is more than the additional energy needed to force air through the column. Further increase in V does not change Z very much, and the optimum air rate is probably in the range 3 to $5V_{min}$. Typical flow rates at $V = 3V_{min}$ might be $G_x = 49000$ kg/($m^2 \cdot$ h) and $G_y = 566$ kg/($m^2 \cdot$ h)).

8.5 Absorption from Rich Gases

When the solute being absorbed is present at moderate or high concentrations in the gas, there are several additional factors to consider in design calculations. The decrease in total gas flow and the increase in liquid flow must be accounted for in the material balance, and the correction factor for one-way diffusion should be included. Also, the mass-transfer coefficients will not be constant because of the changes in flow rate, and there may be an appreciable temperature gradient in the column, which will change the equilibrium line.

The amount of solute absorbed in a differential height dZ is $d(Vy)$, since both V and y decrease as the gas passes through the tower.

$$dN_A = d(Vy) = Vdy + ydV \tag{8.5-1}$$

If only A is being transferred, dN_A is the same as dV, so Eq. (8.5-1) becomes

$$dN_A = Vdy + ydN_A \tag{8.5-2}$$

or

$$dN_A = \frac{V}{1-y} \tag{8.5-3}$$

The effect of one-way diffusion in the gas film is to increase the mass-transfer rate for the gas film by the factor $1/\Delta y_m$, as shown by Eq. (6.3-8), so the effective overall coefficient $K_y a$ is somewhat larger than the normal value of $K_y a$:

$$\frac{1}{K'_y a} = \frac{\Delta y_m}{k_y a} + \frac{m}{k_x a} \tag{8.5-4}$$

In this treatment, the effect of one-way diffusion in the liquid film is neglected. The basic mass-transfer equation is then

$$dN_A = \frac{Vdy}{1-y} = K'_y a S dZ (y - y^*) \tag{8.5-5}$$

The column height can be found by a graphical integration, allowing for changes in V, $1 - y$, $y - y^*$, and $K'_y a$:

$$Z = \frac{1}{S} \int_{y_a}^{y_b} \frac{Vdy}{(1-y)(y-y^*)K'_y a} \tag{8.5-6}$$

If the process is controlled by the rate of mass transfer through the gas film, a simplified equation can be developed. The term Δy_m, which strictly applies only to the gas film, as shown in Eq. (8.5-4), is assumed to apply to the overall coefficient, since the gas film is controlling. The coefficient $K'_y a$ in Eq. (8.5-6) is replaced by $K_y a / \Delta y_m$, which leads to

$$Z = \frac{1}{S} \int_{y_a}^{y_b} \frac{V \Delta y_m dy}{K_y a (1-y)(y-y^*)} \tag{8.5-7}$$

Since $k_y a$ varies with about $V^{0.7}$, and $K_y a$ will show almost the same variation when the gas film controls, the ratio $V/K_y a$ does not change much. This term can be taken outside the integral and evaluated at the average flow rate, or the values at the top and bottom of the tower can be averaged. The term Δy_m is the logarithmic mean of $(1 - y)$ and $(1 - y_i)$, which is usually only slightly larger than $(1 - y)$. Therefore the terms Δy_m and $(1 - y)$ are assumed to cancel, and Eq. (8.5-7) becomes

$$Z = \frac{V}{SK_y a} \int_{y_a}^{y_b} \frac{dy}{y - y^*} \tag{8.5-8}$$

$$Z = H_{Oy} \times N_{Oy} \tag{8.5-9}$$

This is the same as Eq. (8.2-8) for dilute gases, except that the first term, which is H_{Oy}, is an average value for the column rather than a constant. Note that $K_y a$ from Eq. (8.1-27) is to be used here, and not $K'_y a$, since the Δy_m term was included in the derivation.

If the liquid film has the controlling resistance to mass transfer, gas-film coefficients could still be used for design calculations following Eq. (8.5-6). If liquid-film coefficients are used, and if the factor Δx_m is introduced to allow for one-way diffusion in the liquid, an equation similar to Eq. (18.41a) can be derived:

$$Z = \frac{L}{SK_x a} \int_{x_b}^{x_a} \frac{dx}{x^* - x} \tag{8.5-10}$$

or

$$Z = H_{Ox} N_{Ox} \qquad (8.5\text{-}11)$$

When the gas-film and liquid-film resistances are comparable in magnitude, there is no simple method of dealing with absorption of a rich gas. The recommended method is to base the design on the gas phase and use Eq. (8.5-6). Several values of y between y_a and y_b are chosen, and values of V, $K'_y a$, and $(y - y^*)$ are calculated. As a check, the value of Z from the integration should be compared with that based on the simple formula in Eq. (8.5-10), since the difference should not be great.

8.6 Absorption in Plate Columns

Gas absorption can be carried out in a column equipped with sieve trays or other types of plates normally used for distillation. A column with trays is sometimes chosen instead of a packed column to avoid the problem of liquid distribution in a large diameter tower and to decrease the uncertainty in scale up. Methods of determining the number of theoretical plates and estimating the average plate efficiency are discussed in Chap. 9.

8.7 Absorption with Chemical Reaction

Absorption followed by reaction in the liquid phase is often used to get more complete removal of a solute from a gas mixture. For example, a dilute acid solution can be used to scrub NH_3 from gas streams, and basic solutions are used to remove CO_2 and other acid gases. Reaction in the liquid phase reduces the equilibrium partial pressure of the solute over the solution, which greatly increases the driving force for mass transfer. If the reaction is essentially irreversible at absorption conditions, the equilibrium partial pressure is zero, and N_{Oy} can be calculated just from the change in gas composition. For $y^* = 0$,

$$N_{Oy} = \int_{y_a}^{y_b} \frac{dy}{y} = \ln \frac{y_b}{y_a} \qquad (8.7\text{-}1)$$

To illustrate the effect of a chemical reaction, consider the absorption of NH_3 in dilute HCl with a 300-fold reduction in gas concentration (6 to 0.02 percent). From Eq. (8.7-1), $N_{Oy} = \ln 300 = 5.7$, which can be compared with $N_{Oy} = 12$ for the same change in concentration using water at the conditions of Example 8.2.

A further advantage of absorption plus reaction is the increase in the mass transfer coefficient. Some of this increase comes from a greater effective interfacial area, since absorption can now take place in the nearly stagnant regions (static holdup) as well as in the dynamic liquid holdup. For NH_3 absorption in H_2SO_4 solutions, $K_G a$ was 1.5 to 2 times the value for absorption in water. Since the gas-film resistance is controlling, this effect must be due mainly to an increase in effective area. The values of $K_G a$ for NH_3 absorption in acid solutions were about the same as those for vaporization of water, where all the interfacial area is also expected to be effective. The factors $K_G a_{vap}/K_G a_{abs}$ and $K_G a_{react}/K_G a_{abs}$ decrease with increasing liquid rate and approach unity when the total holdup is much larger than +7the static holdup.

The factor $K_G a_{react}/K_G a_{abs}$ also depends on the concentration of reactant and is smaller when only a slight excess of reagent is present in the solution fed to the column. Data on liquid holdup and effective area have been published for Raschig rings and Berl saddles, but similar results for newer packings are not available.

When the liquid-film resistance is dominant, as in the absorption of CO_2 or H_2S in aqueous solutions, a rapid chemical reaction in the liquid can lead to a very large increase in the mass-transfer coefficient. The rapid reaction consumes much of the CO_2 very close to the gas-liquid interface, which makes the gradient for CO_2 steeper and enhances the process of mass transfer in the liquid. The ratio of the apparent value of k_L to that for physical absorption defines an enhancement factor φ, which ranges from 1.0 to 1000 or more. Methods of predicting φ from kinetic and mass-transfer data are given in other texts. When the value of φ is very large, the gas film may become the controlling resistance.

Although absorbing CO_2 in NaOH solution gives high rates of mass transfer, reagent costs and disposal problems make this approach impractical for large-scale use. Instead, CO_2 is removed by using aqueous solutions of amines or potassium carbonate, where the chemical reaction is reversible. Absorption in amine solutions can be carried out at 20 to 50℃ and the spent solutions regenerated with steam at 100 to 130℃. Complete regeneration is not necessary, since the equilibrium pressure of CO_2 is very low until about 20 percent of the amine has reacted.

Fig.8.18 shows the equilibrium and operating lines for one case from a recent study of CO_2 absorption in monoethanol amine solutions. The Rate Frac module of Aspen Plus was combined with thermodynamic and kinetic data to model the packed absorber and stripper, and help determine the optimum operating conditions. Only 85 percent removal is shown in Fig. 8.18, but over 95 percent removal can be obtained at the same L/V ratio with a taller column. When absorption is accompanied by a very slow reaction, the apparent values of K_Ga may be lower than with absorption alone. An example is the absorption of Cl_2 in water followed by hydrolysis of the dissolved chlorine. The slow hydrolysis reaction essentially controls the overall rate of absorption.

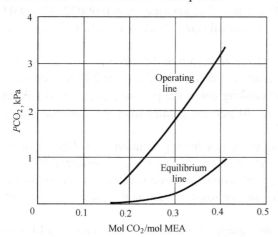 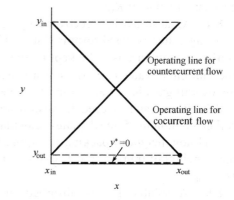

Fig. 8.18 CO_2 absorption in monomethylamine (MEA) solution. (*AfterFreguia and Rochelle.*)

Fig. 8.19 Operating lines for countercurrent flow and cocurrent flow for absorption plus an irreversible chemical reaction.

8.8 Cocurrent flow operation

When the chemical reaction is essentially irreversible and the equilibrium partial pressure of the solute is zero, the number of transfer units for a given separation is the same for countercurrent operation or for cocurrent flow of liquid and gas. Figure 8.19 shows typical operating lines for both cases. In this diagram x is the total solute absorbed and reacted, and not the amount of solute present in the original form. For cocurrent operation with the feeds at the top, the gas leaving at the bottom is exposed to rich liquid, which has absorbed a lot of solute; but if $y^*=0$, the driving force is just y, and N_{Oy} is calculated from Eq.(8.7-1), as for countercurrent flow.

The advantage of cocurrent operation is that there is no flooding limitation, and a gas flow rate much greater than normal can be used. This reduces the required column diameter, and the corresponding increase in liquid and gas mass velocity gives high mass-transfer coefficients. Liquid rates as high as 70 to 140 kg/(m^2 ·s) can be used. The values of K_Ga or H_{Oy} can be estimated approximately by extrapolating counterflow data for the same system to the higher mass velocities.

PROBLEMS

8.1 The gas stream from a chemical reactor contains 25mol % ammonia and the rest inert gases. The total flow is 181.4kmol/h to an absorption tower at 303K and 1.013×10^5 Pa pressure, where water containing 0.005mol frac ammonia is the scrubbing liquid. The outlet gas concentration is to be 2.0 mol % ammonia. What is the minimum flow L'_{min}? Using 1.5 times the minimum, plot the equilibrium and operating lines.

8.2 A gas stream contains 4.0 mol % NH_3 and its ammonia content is reduced to 0.5 mol % in a packed absorption tower at 293K and 1.013×10^5 Pa. The inlet pure water flow is 68.0 kg mol/h and the total inlet gas flow is 57.8 kmol/h. The tower diameter is 0.747 m. The film mass-transfer coefficients are $k'_y a = 0.0739$ kmol/s · m^{3}mol frac and $k'_x a = 0.169$ kmol/s · m^{3}mol frac. Using the design methods for dilute gas mixtures, calculate the tower height

8.3 In a tower 0.254 m in diameter absorbing acetone from air at 293 K and 101.32 kPa using pure water, the following experimental data were obtained. Height of 25.4mm Raschig rings = 4.88 m, V = 3.30 kmol air/h, y_b = 0.01053 mol frac acetone, y_a = 0.00072, L' = 9.03 kmol water/h, x_b = 0.00363 mol frac acetone, the equilibrium relate is $y_e=2x_e$. Calculate the experimental value of $K_y a$.

8.4 An absorber is to recover 99 percent of the ammonia in the air-ammonia stream fed to it, using water as the absorbing liquid. The ammonia content of the air is 20 mole percent. Absorber temperature is to be kept at 30°C by cooling coils; the pressure is 1 atm. (a) What is the minimum water rate? (b) For a water rate 40 percent greater than the minimum, how many overall gas-phase transfer units are needed?

8.5 Pure water is used to absorb ammonia in gas having ammonia of 2mol% in an absorber containing 5 ideal plates, calculate the recovery of ammonia if absorption factor $A=1$; and how height of packing is required for obtaining the same recovery if the empirical value of height of transfer unit $H_{Oy}=0.4$m in the packed tower? What is the ratio of actual L/V to $(L/V)_{min}$?

8.6 A soluble gas is absorbed in water using a packed tower. The equilibrium relationship may be taken as $y_e = 0.06x_e$. Terminal conditions are as follows:

	Top	Bottom
x	0	0.08
y	0.001	0.009

If H_x = 0.24 m and H_y = 0.36 m, what is the height of the packed section?

8.7 An absorption column is fed at the bottom with a gas containing 5 percent benzene and 95 percent air. At the top of the column a nonvolatile absorption oil is introduced, which contains 0.2 percent benzene by weight. Other data are as follows:
   Feed, 2,000 kg of absorption oil per hour
   Total pressure, 1 atm.

Temperature (constant), 26 ℃
Molecular weight of absorption oil, 230
Viscosity of absorbing oil, 4.0 cP.
Vapor pressure of benzene at 26 ℃, 100 mm Hg.
Volume of entering gas, 0.3 m^3/s
Tower packing, Intalox saddles, 1-in. nominal size
Fraction of entering benzene absorbed, 0.90
Mass velocity of entering gas, 1.1 kg/(m^2·s)

Calculate the height and diameter of the packed section of this tower. Assume Raoult's law applies.

8.8 An aqueous waste stream containing 1.0 weight percent NH$_3$ is to be stripped with air in a packed column to remove 99 percent of the NH$_3$. What is the minimum air rate, in kilograms of air per kilogram of water, if the column operates at 20 ℃? How many transfer units are required at twice the minimum air rate?

8.9 An absorber is to remove 99 percent of solute A from a gas stream containing 4 mol percent A. Solutions of A in the solvent follow Henry's law, and the temperature rise of the liquid can be neglected. (a) Calculate N_{Oy} for operation at 1 atm using solute-free liquid at a rate of 1.5 times the minimum value. (b) For the same liquid rate, calculate N_{Oy} for operation at 2 atm and at 4 atm. (c) Would the effect of pressure on N_{Oy} be partly offset by a change in H_{Oy}?

8.10 A packed-column gas absorber is designed to remove 98 percent of A from air at a liquid rate 1.2 times the minimum. The incoming solvent is free of A. (a) How many transfer units are required? (b) If the air contains a small amount of gas B, which is only half as soluble as gas A, what fraction of B is absorbed under the conditions of part (a)?

名 词 术 语

英文 / 中文

cocurrent flow / 并流
countercurrent / 逆流
desorption / 解吸
diffusivity / 扩散系数
equilibrium relations / 相平衡方程
equimolar counter diffusion / 等摩尔方向扩散
film coefficients / 膜系数
gas absorption / 气体吸收
height of a transfer unit（HTU）/传质单元高度
height of the packed section / 填料层高度
kinematic viscosity / 动力黏度
mass-transfer coefficients / 传质系数
molal heat capacity / 摩尔比热容
mole fraction of solute / 溶质的摩尔分数
mole ratio / 比摩尔
number of transfer units（NTU）/ 传质单元数
operating costs / 操作费
optimum gas-liquid ratio / 最适宜气液比

英文 / 中文

overall coefficients / 总系数
plate columns / 板式塔
pressure drop / 压降
rate of absorption / 吸收速率
regeneration / 再生
solute / 溶质
solvent / 溶剂
stripping / 脱吸
the average plate efficiency / 平均塔效率
the gas film resistance / 气膜阻力
the liquid film resistance / 液膜阻力
the logarithmic mean / 对数平均值
the packed column / 填料塔
the total resistance / 总阻力
the volumetric film mass-transfer coefficients / 膜体积传质系数
the volumetric overall mass-transfer coefficients / 总体积传质系数

Chapter 9

Distillation

In distillation operations, separation results from differences in vapor- and liquid-phase compositions arising from the partial vaporization of a liquid mixture or the partial condensation of a vapor mixture. The vapor phase becomes enriched in the more volatile components while the liquid phase is depleted of those same components.

In practice, distillation may be carried out by either of two principal methods. The first method is based on the production of a vapor by boiling the liquid mixture to be separated and condensing the vapors without allowing any liquid to return to the still. There is then no reflux. The second method is based on the return of part of the condensate to the still under such conditions that this returning liquid is brought into intimate contact with the vapors on their way to the condenser. Either of these methods may be conducted as a continuous process or as a batch process. The first sections of this chapter deal with continuous steady-state distillation processes, including single-stage partial vaporization without reflux (flash distillation) and continuous distillation with reflux (rectification) for systems containing only two components. Later section is concerned with batch distillation.

9.1 Flash Distillation

Flash distillation consists of vaporizing a definite fraction of the liquid in such a way that the evolved vapor is in equilibrium with the residual liquid, separating the vapor from the liquid, and condensing the vapor. Figure 9.1 shows the elements of a flash distillation plant. Feed is pumped by pump *a* through heater *b*, and the pressure is reduced through valve *c*. An intimate mixture of vapor and liquid enters the vapor separator *d*, in which sufficient time is allowed for the vapor and liquid portions to separate. Because of the intimacy of contact of liquid and vapor before separation, the separated streams are in equilibrium. Vapor leaves through line *e* and liquid through line *g*.

Flash distillation of binary mixtures

Flash distillation is used extensively in petroleum refining, in which petroleum fractions are heated in pipe stills and the heated fluid is flashed into vapor and residual liquid streams, each containing many components. Liquid from an absorber is often flashed to recover some of the solute; liquid from a high-pressure reactor may be flashed to a lower pressure, causing some vapor to be evolved.

Flash distillation of a two-component mixture shown in Fig. 9.1. Let the concentration of the feed be x_F, in mole fraction of the more volatile component. Let F be the molal rate of the feed, V be the molal rate of vapor withdrawn continuously, and L be the molal rate of the liquid. Let y_D and x_B be the concentrations of the vapor and liquid, respectively. By a material balance for the more volatile component, all that component in the feed must leave in the two exit streams, or

$$Fx_F = Vy_D + Lx_B \tag{9.1-1}$$

Let $f = V/F$ be the molal fraction of the feed that is vaporized and withdrawn continuously as vapor. Then $1-f$ is the molal fraction of the feed that leaves continuously as liquid.

$$x_F = \frac{V}{F} y_D + \frac{F-V}{F} x_B = f y_D + (1-f) x_B \tag{9.1-2}$$

There are two unknowns in Eq. (9.1-2): x_B and y_D. To use the equation, a second relationship between the unknowns must be available. Such a relationship is provided by the equilibrium curve or by an equation based on the *relative volatility*, α.

For binary mixtures the subscripts are usually omitted, since $x_B = 1 - x_A$ and $y_B = 1 - y_A$. Equation (7.1-9) can be converted to a more useful form directly relating y and x, understood to be y_{Ae} and x_{Ae}.

$$y = \frac{\alpha x}{1 + (\alpha - 1)x} \tag{9.1-3}$$

The molal fraction of the feed that is vaporized and withdrawn continuously as vapor in Eq. (9.1-2) is not fixed directly but depends on the enthalpy of the hot incoming liquid and the enthalpies of the vapor and liquid leaving the flash chamber.

$$FH_F = VH_y + LH_x \tag{9.1-4}$$

where H_F, H_y, and H_x are the enthalpies of the feed liquid, the vapor, and the liquid product, respectively.

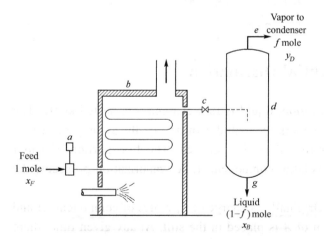

Fig. 9.1 Plant for flash distillation.

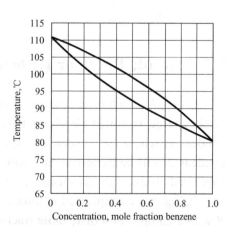

Fig. 9.2 Boiling-point diagram (system benzene-toluene at 1 atm).

EXAMPLE 9.1 A mixture of 50 mole percent benzene and 50 percent toluene is subjected to flash distillation at a separator pressure of 1 atm. The incoming liquid is heated to a temperature that will cause 40 percent of the feed to flash. (*a*) What are the compositions of the vapor and liquid leaving the flash chamber? (*b*) What is the required feed temperature?

Solution. The boiling point diagram for benzene-toluene is shown in Fig. 9.2. The boiling point of the feed is 92 ℃ and the temperature in the separator is assumed to be about 95 ℃. At 95 ℃, $\alpha = 2.45$, based on the vapor pressures of benzene and toluene.

(*a*) From Eqs. (9.1-2) and (9.1-3) with $V/F = 0.4$ and $x_F = 0.5$,

$$0.5 = 0.4\left(\frac{2.45x}{1+1.45x}\right) + 0.6x$$

$$(0.5 - 0.6x)(1 + 1.45x) = 0.98x$$

$$0.87x^2 + 0.855x - 0.5 = 0$$

$$x = 0.412$$

$$y = \frac{2.45 \times 0.412}{1 + 1.45 \times 0.412} = 0.632$$

From Fig. 9.2 for $x = 0.412$, $T = 95\,℃$.

(*b*) The heats of vaporization and specific heats of the liquid are

For benzene: $r = 7360$ cal/g mol, $C_p = 33$ cal/(mol · ℃)
For toluene: $r = 7960$ cal/g mol, $C_p = 40$ cal/(mol · ℃)

Choose liquid at 95 ℃ as the basis for enthalpy.

The average C_p for liquid is $(0.5 \times 33) + (0.5 \times 40) = 36.5$ cal/(mol · ℃).

The average r is $(0.632 \times 7360) + (0.368 \times 7960) = 7581$ cal/g mol.

From Eq. (9.1-4),

$$H_F = (T_F - 95) \times 36.5 = 0.4 \times 7581$$

$$T_F = 178\,℃$$

9.2 Simple Batch or Differential Distillation

In *simple batch* or *differential distillation*, liquid is first charged to a heated kettle. The liquid charge is boiled slowly and the vapors are withdrawn as rapidly as they form to a condenser, where the condensed vapor (distillate) is collected. The first portion of vapor condensed will be richest in the more volatile component A. As vaporization proceeds, the vaporized product becomes leaner in A.

In Fig. 9.3 a simple still is shown. Originally, a charge of L_1 moles of components A and B with a composition of x_1 mole fraction of A is placed in the still. At any given time, there are L moles of liquid left in the still with composition x, and the composition of the vapor leaving in equilibrium is y. A differential amount dL is vaporized.

The composition in the still pot changes with time. In deriving the equation for this process, we assume that a small amount of dL is vaporized. The composition of the liquid changes from x to $x - dx$ and the amount of liquid from L to $L - dL$. A material balance on A can be made, where the original amount = the amount left in the liquid + the amount of vapor:

$$Lx = (L - dL)(x - dx) + y\,dL \qquad (9.2\text{-}1)$$

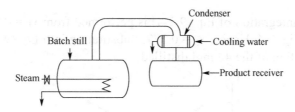

Fig. 9.3 Simple distillation in a batch still.

Multiplying out the right side,

$$Lx = Lx - xdL - Ldx + dxdL + ydL \tag{9.2-2}$$

Neglecting the term $dxdL$ and rearranging,

$$\frac{dL}{L} = \frac{dx}{y-x} \tag{9.2-3}$$

Integrating,

$$\int_{L_2}^{L_1} \frac{dL}{L} = \ln \frac{L_1}{L_2} = \int_{x_2}^{x_1} \frac{dx}{y-x} \tag{9.2-4}$$

where L_1 is the original moles charged, L_2 the moles left in the still, x_1 the original composition, and x_2 the final composition of liquid.

Equation (9.2-4) is known as the *Rayleigh equation*. The function $dx/(y-x)$ can be integrated graphically or numerically using tabulated equilibrium data or an equilibrium curve. The equilibrium curve gives the relationship between y and x. The integration of Eq. (9.2-4) can be done by calculating values of $f(x) = 1/(y-x)$ and numerically or graphically integrating Eq.(9.2-4) between x_1 and x_2. The average composition of total material distilled, y_{av}, can be obtained by material balance:

$$L_1 x_1 = L_2 x_2 + (L_1 - L_2) y_{av} \tag{9.2-5}$$

EXAMPLE 9.2 A mixture of 100 mol containing 50 mol % n-pentane and 50 mol % n-heptane is distilled under differential conditions at 101.3 kPa until 40 mol is distilled. What is the average composition of the total vapor distilled and the composition of the liquid left? The equilibrium data are as follows, where x and y are mole fractions of n-pentane:

x	1.000	0.867	0.594	0.398	0.254	0.145	0.059	0
y	1.000	0.984	0.925	0.836	0.701	0.521	0.271	0

Solution. The given values to be used in Eq. (9.2-4) are $L_1 = 100$ mol, $x_1 = 0.50$, $L_2 = 60$ mol, and V (moles distilled) = 40 mol. Substituting into Eq. (9.2-4),

$$\ln \frac{100}{60} = 0.510 = \int_{x_2}^{0.5} \frac{dx}{y-x}$$

The unknown is x_2, the composition of the liquid L_2 at the end of the differential distillation. To solve this by numerical integration, equilibrium values of y versus x are plotted so values of y can be obtained from this curve at small intervals of x. Alternatively, instead of plotting, the equilibrium data can be fitted to a polynomial function. For $x = 0.594$, the equilibrium value of $y = 0.925$. Then $f(x) = 1/(y-x) = 1/(0.925-0.594) = 3.02$. Other values of $f(x)$ are similarly calculated.

The numerical integration of Eq. (9.2-4) is performed from $x_1 = 0.5$ to x_2 such that the integral = 0.510 in Fig. 9.4. Hence, $x_2 = 0.277$. Substituting into Eq. (9.2-5) and solving for the average composition of the 40 mol distilled,

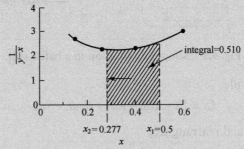

Fig. 9.4 Numerical integration for Example 9.2.

$$100 \times 0.50 = 60 \times 0.277 + 40 y_{av}$$

$$y_{av} = 0.835$$

A simple alternative to the Rayleigh equation can be derived for an ideal mixture based on the relative volatility. Although the temperature in the still increases during a batch distillation, the relative volatility, which is the ratio of vapor pressures, does not change much, and an average value can be used. From Eq. (7.1-6)

$$\alpha = \frac{y_A / x_A}{y_B / x_B}$$

If the mixture has L_A mol of A and L_B mol of B, the ratio L_A/L_B is equal to x_A/x_B; when dL mol is vaporized, the change in A is $y_A\, dL$ or dL_A, and the change in B is $y_B\, dL$ or dL_B. Substituting these terms into

$$\frac{y_A}{y_B} = \alpha \frac{x_A}{x_B} \tag{9.2-6}$$

gives

$$\frac{dL_A / dL}{dL_B / dL} = \frac{dL_A}{dL_B} = \alpha \frac{L_A}{L_B}$$

or

$$\frac{dL_A}{L_A} = \alpha \frac{dL_B}{L_B} \tag{9.2-7}$$

After integration between limits

$$\ln \frac{L_A}{L_{A1}} = \alpha \ln \frac{L_B}{L_{B1}} \tag{9.2-8}$$

or

$$\frac{L_B}{L_{B1}} = \left(\frac{L_A}{L_{A1}} \right)^{\frac{1}{\alpha}} \tag{9.2-9}$$

Equation (9.2-9) can be plotted as a straight line on logarithmic coordinates to help follow the course of a batch distillation, or it can be used directly if the recovery of one of the components is specified.

> **EXAMPLE 9.3** A batch of crude pentane contains 15 mole percent n-butane and 85 percent n-pentane. If a simple batch distillation at atmospheric pressure is used to remove 90 percent of the butane, how much pentane will be removed? What will be the composition of the remaining liquid?
>
> **Solution.** The final liquid is nearly pure pentane, and its boiling point is 36°C. The vapor pressure of butane at this temperature is 3.4 atm, giving a relative volatility of 3.4. For the initial conditions, the boiling point is about 27°C, and the relative volatility is 3.6. Therefore, an average value of 3.5 is used for α.
>
> Basis: 1 mol feed
>
> $$L_{A0} = 0.15 \text{ (butane)} \quad L_A = L_{A0}(1-0.9)=0.015 \quad L_{B0} = 0.85 \text{ (pentane)}$$
>
> From Eq. (9.2-9)
>
> $$\frac{L_B}{0.85} = 0.1^{\frac{1}{3.5}} = 0.518 \quad L_B = 0.518 \times 0.85 = 0.440$$
>
> $$L = 0.44 + 0.015 = 0.455 \text{ mol} \quad x_A = \frac{0.015}{0.455} = 0.033$$

9.3 Simple Steam Distillation

At atmospheric pressure high-boiling liquids cannot be purified by distillation, since the components of the liquid may decompose at the high temperatures required. Often the high-boiling substances are essentially insoluble in water, so a separation at lower temperatures can be obtained by *simple steam distillation*. This method is often used to separate a high-boiling component from small amounts of nonvolatile impurities.

If a layer of liquid water (A) and an immiscible high-boiling component (B) such as a hydrocarbon are boiled at 101.3 kPa abs pressure, then, by the phase rule, for three phases and two components,

$$F = 2-3+2 = 1 \, degree \, of \, freedom$$

Hence, if the total pressure is fixed, the system is fixed. Since there are two liquid phases, each will exert its own vapor pressure at the prevailing temperature and cannot be influenced by the presence of the other. When the sum of the separate vapor pressures equals the total pressure, the mixture boils and

$$p_A + p_B = P \tag{9.3-1}$$

where p_A is vapor pressure of pure water A and p_B is vapor pressure of pure B. Then the vapor composition is

$$y_B = \frac{p_B}{P} \tag{9.3-2}$$

As long as the two liquid phases are present, the mixture will boil at the same temperature, giving a vapor of constant composition y_A. The temperature is found by using the vapor-pressure curves for pure A and pure B.

Note that by steam distillation, as long as liquid water is present, the high-boiling component B vaporizes at a temperature well below its normal boiling point without using a vacuum. The vapors of water (A) and high-boiling component (B) are usually condensed in a condenser and the resulting two immiscible liquid phases separated. This method has the disadvantage that large amounts of heat must be used to evaporate the water simultaneously with the high-boiling compound.

The ratio moles of B distilled to moles of A distilled is

$$\frac{n_B}{n_A} = \frac{p_B}{p_A} \tag{9.3-3}$$

Steam distillation is sometimes used in the food industry for the removal of volatile taints and flavors from edible fats and oils. In many cases vacuum distillation is used instead of steam distillation to purify high-boiling materials. The total pressure is quite low so that the vapor pressure of the system reaches the total pressure at relatively low temperatures.

Van Winkle derives equations for steam distillation where an appreciable amount of a nonvolatile component is present with the high-boiling component. This involves a three-component system. He also considers other cases for binary batch, continuous, and multicomponent batch steam distillation.

9.4 Continuous Distillation with Reflux

Flash distillation is used most for separating components that boil at widely different temperatures. It is not effective in separating components of comparable volatility, which requires the use of distillation with reflux. For large-scale production, continuous distillation, as described in this section, is far more common than flash distillation and differential distillation.

9.4.1 Action on an Ideal Plate

On an ideal plate, by definition, the liquid and vapor leaving the plate are brought into equilibrium. Consider a single plate in an ideal cascade, such as plate n in Fig. 9.5. Assume that the plates are numbered serially from the top down and that the plate under consideration is the nth plate from the top. Then the plate immediately above plate n is plate $n - 1$, and that immediately below it is plate $n+1$. Subscripts are used on all quantities to show the point of origin of the quantity.

Two fluid streams enter plate n, and two leave it. A stream of liquid L_{n-1} mol/h from plate $n-1$ and a stream of vapor V_{n+1} mol/h from plate $n+1$ are brought into intimate contact. A stream of vapor V_n mol/h rises to plate $n - 1$, and a stream of liquid L_n mol/h descends to plate $n+1$. Since the vapor streams are the V phase, their concentrations are denoted by y. The liquid streams are the L phase, and their concentrations are denoted by x. Then the concentrations of the streams entering and leaving the nth plate are as follows:

Vapor leaving plate y_n Vapor entering plate y_{n+1}
Liquid leaving plate x_n Liquid entering plate x_{n-1}

Figure 9.6 shows the boiling-point diagram for the mixture being treated. The four concentrations given above are shown in this figure. By definition of an ideal plate, the vapor and liquid leaving plate n are in equilibrium, so x_n and y_n represent equilibrium concentrations. This is shown in Fig. 9.6. The vapor is enriched in more volatile component A as it travels up the column, and the liquid is depleted of A as it flows downward. Thus the concentration of A in both phases increases with the height of the column; x_{n-1} is greater than x_n, and y_n is greater than y_{n+1}. Although the streams leaving the plate are in equilibrium, those entering it are not. This can be seen from Fig. 9.6. When the vapor from plate $n+1$ and the liquid from plate $n-1$ are brought into intimate contact, their concentrations tend to move toward an equilibrium state, as shown by the arrows in Fig. 9.6. Some of the more volatile component A is vaporized from the liquid, decreasing the liquid concentration from x_{n-1} to x_n; and some of the less volatile component B is condensed from the vapor, increasing the vapor concentration from y_{n+1} to y_n. Since the liquids on the plates are at their boiling points, and there is only a slight temperature change from plate to plate, the heat needed to vaporize component A comes mainly from the heat released in the condensation of component B. Each plate in the cascade acts as an interchange apparatus in which component A is transferred to the vapor stream and component B to the liquid stream. Also, since the concentration of A in both liquid and vapor increases with column height, the temperature decreases, and the temperature of plate n is greater than that of plate $n-1$ and less than that of plate $n+1$.

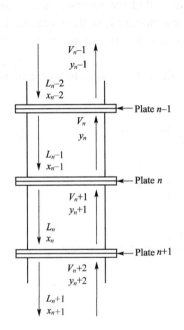

Fig. 9.5 Material-balance diagram for plate n.

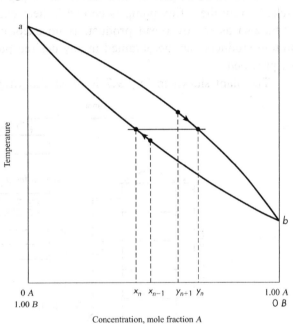

Fig. 9.6 Boiling-point diagram showing rectification on ideal plate.

9.4.2 Combination Rectification and Stripping

To produce nearly pure products at both the top and bottom of a continuous distillation column, the feed is admitted to a plate in the central portion of the column. If the feed is liquid, it flows down the column to the reboiler and is stripped of component A by the vapor rising from the reboiler. By this means a bottom product can be produced which is nearly pure B.

A typical continuous distillation column equipped with the necessary auxiliaries and containing rectifying and stripping sections is shown in Fig. 9.7. Column A is fed near its center with a steady flow of feed of definite concentration. Assume that the feed is a liquid at its boiling point. The action in the column is not dependent on this assumption, and other conditions of the feed will be discussed later. The plate on which the feed enters is called the *feed plate*. All plates above the feed plate constitute the rectifying section, and all plates below the feed, *including the feed plate itself*, constitute the stripping section. The feed flows down the stripping section to the bottom of the column, in which a definite level of liquid is maintained. Liquid flows by gravity to reboiler *B*. This is a steam-heated vaporizer that generates vapor and returns it to the bottom of the column. The vapor passes up the entire column. At one end of the reboiler is a weir. The bottom product is withdrawn from the pool of liquid on the downstream side of the weir and flows through the cooler *G*. This cooler also preheats the feed by heat exchange with the hot bottoms.

The vapors rising through the rectifying section are completely condensed in condenser *C*, and the condensate is collected in accumulator *D*, in which a definite liquid level is maintained. Reflux pump *F* takes liquid from the accumulator and delivers it to the top plate of the tower. This liquid stream is called *reflux*. It provides the downflowing liquid in the rectifying section that is needed to act on the upflowing vapor. Without the reflux, no rectification would occur in the rectifying section, and the concentration of the overhead product would be no greater than that of the vapor rising from the feed plate. Condensate not picked up by the reflux pump is cooled in heat exchanger *E*, called the *product cooler*, and withdrawn as the overhead product. If no azeotropes are encountered, both overhead and bottom products may be obtained in any desired purity if enough plates and adequate reflux are provided.

The plant shown in Fig. 9.7 is often simplified for small installations. In place of the

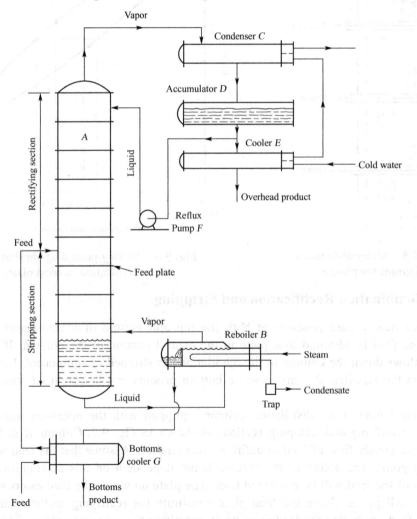

Fig. 9.7 Continuous fractionating column with rectifying and stripping sections.

reboiler, a heating coil may be placed in the bottom of the column to generate vapor from the pool of liquid there. The condenser is sometimes placed above the top of the column, and the reflux pump and accumulator are omitted. Reflux then returns to the top plate by gravity. A special valve, called a *reflux splitter*, may be used to control the rate of reflux return. The remainder of the condensate forms the overhead product.

9.4.3 Material Balances in Plate Columns

Overall material balances for two-component systems

Figure 9.8 is a material-balance diagram for a typical continuous distillation plant. The column is fed with F mol/h of concentration x_F and delivers D mol/h of overhead product of concentration x_D and B mol/h of bottom product of concentration x_B. Two independent overall material balances can be written.
Total-material balance

$$F = D + B \tag{9.4-1}$$

Component A balance

$$Fx_F = Dx_D + Bx_B \tag{9.4-2}$$

Eliminating B from these equations gives

$$\frac{D}{F} = \frac{x_F - x_B}{x_D - x_B} \tag{9.4-3}$$

Eliminating D gives

$$\frac{B}{F} = \frac{x_D - x_F}{x_D - x_B} \tag{9.4-4}$$

Equations (9.4-3) and (9.4-4) are true for all values of the flows of vapor and liquid within the column.

Net flow rates

Quantity D is the difference between the flow rates of the streams entering and leaving the top of the column. A material balance around the condenser and accumulator in Fig. 9.8 gives

$$D = V_a - L_a \tag{9.4-5}$$

The difference between the flow rates of vapor and liquid anywhere in the upper section of the column is also equal to D, as shown by considering the part of the plant enclosed by control surface I in Fig. 9.8. This part includes the condenser and all plates above $n+1$. A total material balance around this control surface gives

$$D = V_{n+1} - L_n \tag{9.4-6}$$

Thus quantity D is the *net flow rate* of material upward in the upper section of the column. Regardless of changes in V and L, their difference is constant and equal to D.
Similar material balances for component A give the equations

$$Dx_D = V_a y_a - L_a x_a = V_{n+1} y_{n+1} - L_n x_n \tag{9.4-7}$$

Quantity Dx_D is the net flow rate of component A upward in the upper section of the column. It, too, is constant throughout this part of the equipment.

In the lower section of the column, the net flow rates are also constant but are in a downward direction. The net flow rate of total material equals B; that of component A is Bx_B.

The following equations apply:

$$B = L_b - V_b = L_m - V_{m+1} \tag{9.4-8}$$

$$Bx_B = L_b x_b - V_b y_b = L_m x_m - V_{m+1} y_{m+1} \tag{9.4-9}$$

Subscript m is used in place of n to designate a general plate in the stripping section.

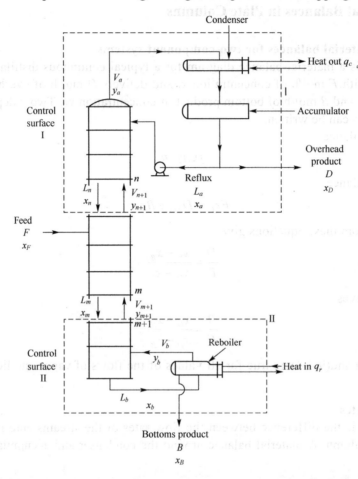

Fig. 9.8 Material-balance diagram for continuous fractionating column.

Operating lines

Because there are two sections in the column, there are also two operating lines, one for the rectifying section and the other for the stripping section. Consider first the rectifying section. As shown in Chap. 7 [Eq. (7.2-7)], the operating line for this section is

$$y_{n+1} = \frac{L_n}{V_{n+1}} x_n + \frac{V_a y_a - L_a x_a}{V_{n+1}} \tag{9.4-10}$$

Substitution for $V_a y_a - L_a x_a$ from Eq.(9.4-7) gives

$$y_{n+1} = \frac{L_n}{V_{n+1}} x_n + \frac{D x_D}{V_{n+1}} \tag{9.4-11}$$

The slope of the line defined by Eq.(9.4-11) is, as usual, the ratio of the flow of the liquid stream to that of the vapor stream. For further analysis it is convenient to eliminate V_{n+1} from

Eq.(9.4-11) by Eq.(9.4-6), giving

$$y_{n+1} = \frac{L_n}{L_n + D} x_n + \frac{Dx_D}{L_n + D} \qquad (9.4\text{-}12)$$

For the section of the column below the feed plate, a material balance over control surface II in Fig. 9.8 gives

$$V_{m+1} y_{m+1} = L_m x_m - B x_B \qquad (9.4\text{-}13)$$

In a different form, this becomes

$$y_{m+1} = \frac{L_m}{V_{m+1}} x_m - \frac{B x_B}{V_{m+1}} \qquad (9.4\text{-}14)$$

This is the equation for the operating line in the stripping section. Again the slope is the ratio of the liquid flow to the vapor flow. Eliminating V_{m+1} from Eq.(9.4-14) by Eq.(9.4-8) gives

$$y_{m+1} = \frac{L_m}{L_m - B} x_m - \frac{B x_B}{L_m - B} \qquad (9.4\text{-}15)$$

Equation(9.4-12) shows that the slope of the operating line in the rectifying section is always less than 1.0; in the stripping section, as shown by Eq.(9.4-15), the slope is always greater than 1.0.

9.4.4 Number of Ideal Plates; McCabe-Thiele Method

The number of plates required for a particular distillation problem can be found by computer design programs such as ASPEN, which usually use plate-to-plate calculations involving material and enthalpy balances. In such programs the number of plates is specified first; then, for a given overhead composition and reflux ratio, the composition of the bottoms is calculated. If this is not satisfactory, the reflux ratio or the number of plates is changed until the desired composition is found.

A simplified graphical procedure for calculating the number of plates is the McCabe-Thiele method. This method can also be adapted for computer calculations.

When the operating lines represented by Eqs.(9.4-12) and (9.4-15) are plotted with the equilibrium curve on the $x\ y$ diagram, the McCabe-Thiele step-by-step construction can be used to compute the number of *ideal* plates needed to accomplish a definite concentration difference in either the rectifying or the stripping. Equations(9.4-12) and(9.4-15), however, show that unless L_n and L_m are constant, the operating lines are curved and can be plotted only if the change in these internal streams with concentration is known. Enthalpy balances are required in the general case to determine the position of a curved operating line.

Constant molal overflow

For most distillations, the molar flow rates of vapor and liquid are nearly constant in each section of the column, and the operating lines are almost straight. This results from nearly equal molar heats of vaporization, so that each mole of high boiler that condenses as the vapor moves up the column provides energy to vaporize about 1 mol of low boiler. For example, the molar heats of vaporization of toluene and benzene are 7960 and 7360 cal/mol, respectively, so that 0.92 mol of toluene corresponds to 1.0 mol of benzene. The changes in enthalpy of the liquid and vapor streams and heat losses from the column often require slightly more vapor to be formed at the bottom, so the molar ratio of vapor flow at the bottom of a column section to that at the top is even closer to 1.0. In designing columns or interpreting plant performance the concept of *constant molal overflow* is often used, which means simply that in Eqs.(9.4-6)to(9.4-15), subscripts n, $n+1$, m, and $m+1$ on L and V may be dropped, and L and V now refer to flows in the upper part of the column, and L' and V' denote flows in the lower section. In this simplified model the material-balance equations are linear and the operating lines straight. An operating line can be plotted if the coordinates of two points on it are known. Then the McCabe-Thiele method is used without requiring enthalpy balances. The method can be modified, however, to include enthalpy balances.

Reflux ratio

The analysis of fractionating columns is facilitated by the use of a quantity called the *reflux ratio*. Two such quantities are used. One is the ratio of the reflux to the overhead product, and the other is the ratio of the reflux to the vapor. Both ratios refer to quantities in the rectifying section. The equations for these ratios are

$$R = \frac{L}{D} = \frac{V-D}{D} \quad \text{and} \quad R_V = \frac{L}{V} = \frac{L}{L+V} \tag{9.4-16}$$

In this text only R will be used.

If both numerator and denominator of the terms on the right-hand side of Eq.(9.4-12) are divided by D, the result is, for constant molal overflow,

$$y_{n+1} = \frac{R}{R+1} x_n + \frac{x_D}{R+1} \tag{9.4-17}$$

Equation(9.4-17) is an equation for the operating line of the rectifying section. Its slope is $R/(R+1)$; by substitution of $L = V - D$ from Eq.(9.4-16), it can be shown to be equal to L/V. The y intercept of this line is $x_D/(R+1)$. The value of x_D is set by conditions of the design, and R, the reflux ratio, is an operating variable that can be controlled at will by adjusting the split between reflux and overhead product or by changing the amount of vapor formed in the reboiler for a given flow rate of the overhead product. A point at the upper end of the operating line can be obtained by setting x_n equal to x_D in Eq.(9.4-17):

$$y_{n+1} = \frac{R}{R+1} x_D + \frac{x_D}{R+1} = x_D \tag{9.4-18}$$

The operating line for the rectifying section then intersects the diagonal at point (x_D, x_D). This is true for either a partial or a total condenser. (Partial condensers are discussed in the next section.)

Condenser and top plate

The McCabe-Thiele construction for the top plate does depend on the action of the condenser. Figure 9.9 shows material-balance diagrams for the top plate and the condenser. The concentration of the vapor from the top plate is y_1, and that for the reflux to the top plate is x_c. In accordance with the general properties of operating lines, the upper terminus of the line is at the point (x_c, y_1).

The simplest arrangement for obtaining reflux and liquid product, and one that is frequently used, is the single total condenser shown in Fig. 9.9(b), which condenses all vapor from the column and supplies both reflux and product. When such a single total condenser is used, the concentrations of the vapor from the top plate, of the reflux to the top plate, and of the overhead product are equal and can all be denoted by x_D. The operating terminus of the operating line becomes point (x_D, x_D), which is the intersection of the operating line with the diagonal. Triangle *abc* in Fig. 9.4-6(a) then represents the top plate in the column.

When a partial condenser is used, the liquid reflux does not have the same composition as the overhead product; that is, $x_c \neq x_D$. Sometimes two condensers are used in series, first a partial condenser to provide reflux, then a final condenser to provide liquid product. Such an arrangement is shown in Fig. 9.9(c). Vapor leaving the partial condenser has a composition y', which is the same as x_D. Under these conditions the diagram in Fig. 9.10(b) applies. The operating line passes through the point (x_D, x_D) on the diagonal, but as far as the column is concerned, the operating line ends at point a', which of course has the coordinates (x_c, y_1). Triangle $a'b'c'$ in Fig. 9.10(b) represents the top plate in the column. Since the vapor leaving a partial condenser is normally in equilibrium with the liquid condensate, the vapor composition y' is the ordinate value of the equilibrium curve where the abscissa is x_c, as shown in Fig. 9.10(b). The partial condenser, represented by the dotted triangle aba' in Fig. 9.10(b), is therefore equivalent to an additional theoretical stage in the distillation apparatus.

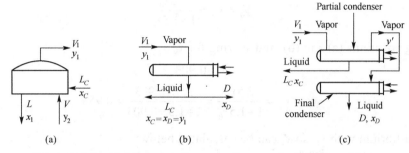

Fig. 9.9 Material-balance diagrams for top plate and condenser: (a) top plate; (b) total condenser; (c) partial and final condensers.

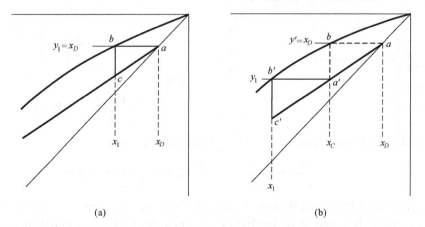

Fig. 9.10 Graphical construction for top plate: (a) using total condenser; (b) using partial and final condensers.

EXAMPLE 9.4 A continuous distillation column with reflux is used for the separation of a mixture of benzene and toluene. At the top of the column the vapor composition is y_1=0.94 (mole fraction of benzene). In the partial condenser 2/3 of the vapor (moles) is condensed and used as the reflux, the remainder of the vapor is condensed in the final condenser as overhead product. The relative volatility between benzene and toluene is 2.5 for the given operating conditions. Calculate:

(a) The compositions of overhead product (x_D) and reflux liquid (x_0).
(b) The vapor composition from the second theoretical plate (y_2).

Solution.

$$y_1 = 0.94 \quad \alpha = 2.5 \quad \frac{L}{V} = \frac{2}{3}$$

Making overall balance for the component benzene for the partial condenser

$$Vy_1 = Dx_D + Lx_0$$

$$y_1 = \frac{D}{V}x_D + \frac{L}{V}x_0 = 0.667x_D + 0.333x_0 = 0.94 \tag{a}$$

where $y_0 = x_D$.

The relationship between the concentration of liquid condensed x_0 and the concentration of the vapor remained y_0 can be depicted by the phase equilibrium

$$y_0 = x_D = \frac{2.5x_0}{1+1.5x_0} \tag{b}$$

Combining equations (a) and (b), and solving for x_0

$$x_0 = 0.903$$

$$y_0 = x_D = \frac{2.5x_0}{1+1.5x_0} = \frac{2.5 \times 0.903}{1+1.5 \times 0.903} = 0.958$$

y_1 is in equilibrium with x_1, so x_1 can be calculated below

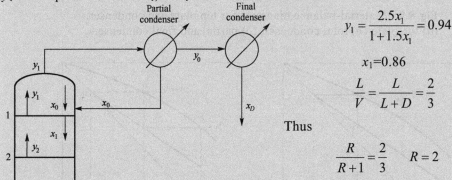

$$y_1 = \frac{2.5x_1}{1+1.5x_1} = 0.94$$

$$x_1 = 0.86$$

$$\frac{L}{V} = \frac{L}{L+D} = \frac{2}{3}$$

Thus

$$\frac{R}{R+1} = \frac{2}{3} \quad R = 2$$

According to the operating equation of the enriching section

$$y_2 = \frac{R}{R+1}x_1 + \frac{1}{R+1}x_D = \frac{2}{3} \times 0.86 + \frac{1}{3} \times 0.958 = 0.89$$

In the preceding treatment it is assumed that the condenser removes latent heat only and that the condensate is liquid at its bubble point. Then the reflux L is equal to L_c, the reflux from the condenser, and $V = V_1$. If the reflux is cooled below the bubble point, a portion of the vapor coming to plate 1 must condense to heat the reflux; so $V_1 < V$ and $L > L_c$. The additional amount ΔL that is condensed inside the column is found from the equation

$$\Delta L = \frac{L_c c_{pc}(t_1 - t_c)}{r_c} \tag{9.4-19}$$

where c_{pc} = specific heat of condensate
t_1 = temperature of liquid on top plate
t_c = temperature of returned condensate
r_c = heat of vaporization of condensate

The actual reflux ratio in the column is then

$$\frac{L}{D} = \frac{L_c + \Delta L}{D} = \frac{L_c[1 + c_{pc}(t_1 - t_c)/r_c]}{D} \qquad (9.4\text{-}20)$$

Temperature t_1 is not usually known, but it normally almost equals t_{bc}, the bubble-point temperature of the condensate. Thus t_{bc} is commonly used in place of t_1 in Eqs.(9.4-19)and(9.4-20). Subcooling the reflux makes the overhead vapor stream V_1 less than V_n. If the reboiler heat duty is unchanged and D is kept constant, less liquid is returned to the column than before, and the apparent reflux ratio R is decreased. The extra vapor that condenses to heat the reflux to its bubble point, however, increases the amount of liquid going down the column, so that the slope of the operating line $(L/V)_n$ is unchanged. A serious disadvantage may occur when an air-cooled condenser is used, for changes in air temperature may lead to fluctuations in t_c, and the operation of the column may be difficult to control.

Bottom plate and reboiler

The action at the bottom of the column is analogous to that at the top. Thus, Eq.(9.4-15), written for constant molal overflow, becomes, with L' and later V' used to denote flow rates in this section,

$$y_{m+1} = \frac{L'}{L' - B} x_m - \frac{Bx_B}{L' - B} \qquad (9.4\text{-}21)$$

If x_m is set equal to x_B in Eq.(9.4-21), y_{m+1} is also equal to x_B, so the operating line for the stripping section crosses the diagonal at point (x_B, x_B). This is true no matter what type of reboiler is used, as long as there is only one bottom product. The lower operating line could then be constructed using the point (x_B, x_B) and the slope $L'/(L'- B)$, which equals L'/V', but a more convenient method is described in the discussion on feed plates in the next section.

The material-balance diagram for the bottom plate and the reboiler is shown in Fig. 9.11 The lowest point on the operating line for the column itself is the point for the bottom plate (x_b, y_r), where x_b and y_r are the concentrations in the liquid leaving the bottom plate and the vapor coming from the reboiler. However, as shown earlier, the operating line can be extended to cross the diagonal at point (x_B, x_B).

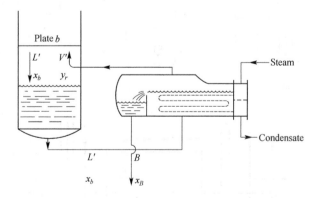

Fig. 9.11 Material-balance diagram for bottom plate and reboiler.

In the common type of reboiler shown in Figs. 9.7 and 9.11 the vapor leaving the reboiler is in equilibrium with the liquid leaving as bottom product. Then x_B and y_r are coordinates of a point on the equilibrium curve, and the reboiler acts as an ideal plate. Figure 9.12 shows the graphical construction for the reboiler (triangle *cde*) and the bottom plate (triangle *abc*). Such a reboiler is called a *partial reboiler*. Its construction is shown in detail in chapter 4.

Condition of feed

At the plate where the feed is admitted, the liquid rate or the vapor rate or both may change, depending on the thermal condition of the feed. Figure 9.13 shows diagrammatically the liquid and vapor streams into and out of the feed plate for various feed conditions. In Fig. 9.13(a), cold feed is assumed, and the entire feed stream adds to the liquid flowing down the column. In addition, some vapor condenses to heat the feed to the bubble point; this makes the liquid flow even greater in the stripping section and decreases the flow of vapor to the rectifying section.

In Fig. 9.13(b) the feed is assumed to be at its bubble point. No condensation is required to heat the feed, so $V = V'$ and $L' = F+L$. If the feed is partly vapor, as shown in Fig. 9.13(c), the liquid portion of the feed becomes part of L' and the vapor portion becomes part of V. If the feed is saturated vapor, as shown in Fig. 9.13(d), the entire feed becomes part of V, so $L = L'$ and $V = F+V'$. Finally, if the feed is superheated vapor, as shown in Fig. 9.13(e), part of the liquid from the rectifying column is vaporized to cool the feed to a state of saturated vapor. Then the vapor in the rectifying section consists of (1) the vapor from the stripping section, (2) the feed, and (3) the extra moles vaporized in cooling the feed. The liquid flow to the stripping section is less than that in the rectifying section by the amount of additional vapor formed.

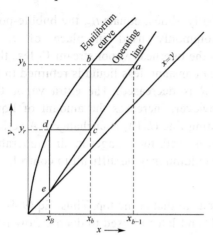

Fig. 9.12 Graphical construction for bottom plate and reboiler: triangle cde, reboiler; triangle abc, bottom plate.

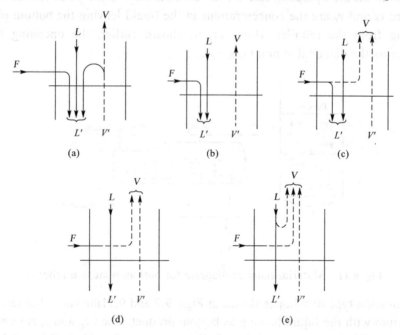

Fig. 9.13 Flow through feed plate for various feed conditions: (a) feed cold liquid; (b) feed saturated liquid; (c) feed partially vaporized; (d) feed saturated vapor; (e) feed superheated vapor.

All five of the feed types can be characterized by the use of a single factor, denoted by q and defined as the moles of liquid flow in the stripping section that result from the introduction of each mole of feed.

Then q has the following numerical limits for the various conditions:
Cold feed, $q > 1$
Feed at bubble point (saturated liquid), $q = 1$
Feed partially vapor, $0 < q < 1$
Feed at dew point (saturated vapor), $q = 0$
Feed superheated vapor $q < 0$

If the feed is a mixture of liquid and vapor, q is the fraction that is liquid. Such a feed may be produced by an equilibrium flash operation, so $q = 1 - f$, where f is the fraction of the original stream vaporized in the flash.

The value of q for cold-liquid feed is found from the equation

$$q = 1 + \frac{c_{pL}(t_b - t_F)}{r} \tag{9.4-22}$$

For superheated vapor the equation is

$$q = -\frac{c_{pV}(t_F - t_d)}{r} \tag{9.4-23}$$

where c_{pL}, c_{pV} = specific heats of liquid and vapor, respectively
t_F = temperature of feed
t_b, t_d = bubble point and dew point of feed, respectively
r = heat of vaporization

Feed line

Most columns operate with the feed as liquid at or near its boiling point. The feed is often preheated by heat exchange with the bottoms product or other hot liquids, or by the use of steam. This gives nearly the same vapor rate in both sections of the column.

The value of q obtained from Eq.(9.4-22) or(9.4-23) can be used with the material balances to find the locus for all points of intersection of the operating lines. The equation for this line of intersections can be found as follows.

The contribution of the feed stream to the internal flow of liquid is qF, so the total flow rate of reflux in the stripping section is

$$L' = L + qF \quad \text{and} \quad L' - L = qF \tag{9.4-24}$$

Likewise, the contribution of the feed stream to the internal flow of vapor is $F(1 - q)$, and so the total flow rate of vapor in the rectifying section is

$$V = V' + (1 - q)F \quad \text{and} \quad V - V' = (1 - q)F \tag{9.4-25}$$

For constant molal overflow, the material-balance equations for two sections are

$$Vy_n = Lx_{n+1} + Dx_D \tag{9.4-26}$$

$$V'y_m = L'x_{m+1} - Bx_B \tag{9.4-27}$$

To locate the point where the operating lines intersect, let $y_n = y_m$ and $x_{n+1} = x_{m+1}$ and subtract Eq.(9.4-27) from Eq.(9.4-26):

$$y(V - V') = (L - L')x + Dx_D + Bx_B \tag{9.4-28}$$

From Eq.(9.4-2), the last two terms in Eq.(9.4-28) can be replaced by Fx_F. Also, substituting for $L - L'$ from Eq.(9.4-24) and for $V - V'$ from Eq.(9.4-25) and simplifying lead to the result

$$y = \frac{q}{q-1}x - \frac{x_F}{q-1} \qquad (9.4\text{-}29)$$

Equation(9.4-29) represents a straight line, called the *feed line*, on which all intersections of the operating lines must fall. The position of the line depends only on x_F and q. The slope of the feed line is $q/(q - 1)$, and as can be demonstrated by substituting x for y in Eq.(9.4-29) and simplifying, the line crosses the diagonal at $x = x_F$.

Construction of operating lines

The simplest method of plotting the operating lines is to (1) locate the feed line; (2) calculate the y-axis intercept $x_D/(R+1)$ of the rectifying line and plot that line through the intercept and the point (x_D, x_D); (3) draw the stripping line through point (x_B, x_B) and the intersection of the rectifying line with the feed line. The operating lines in Fig. 9.14 show the result of this procedure.

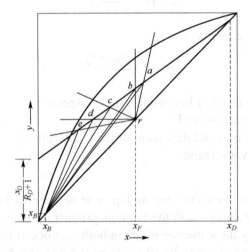

Fig. 9.14 Effect of feed condition on feed line: ra, feed cold liquid; rb, feed saturated liquid; rc, feed partially vaporized; rd, feed saturated vapor; re, feed superheated vapor.

In Fig. 9.14 are plotted operating lines for various types of feed, on the assumption that x_F, x_B, x_D, L, and D are all constant. The corresponding feed lines are shown. If the feed is a cold liquid, the feed line slopes upward and to the right; if the feed is a saturated liquid, the line is vertical; if the feed is a mixture of liquid and vapor, the line slopes upward and to the left, and the slope is the negative of the ratio of the liquid to the vapor; if the feed is a saturated vapor, the line is horizontal; and finally, if the feed is superheated vapor, the line slopes downward and to the left.

Feed plate location

After the operating lines have been plotted, the number of ideal plates is found by the usual step-by-step construction, as shown in Fig. 9.15. The construction can start either at the bottom of the stripping line or at the top of the rectifying line. In the following it is assumed that the construction begins at the top and that a total condenser is used. As the intersection of the operating lines is approached, it must be decided when the steps should transfer from the rectifying line to the stripping line. The change should be made in such a manner that the maximum enrichment per plate is obtained, so that the number of plates is as small as possible. Figure 9.15 shows that this criterion is met if the transfer is made immediately after a value of x is reached that is less than the x coordinate of the intersection of the two operating lines. The feed plate is always represented by the triangle that has one corner on the rectifying line and one corner on the stripping line. At the optimum position, the triangle representing the feed plate straddles the intersection of the operating lines.

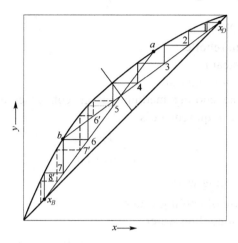

Fig. 9.15 Optimum feed plate location: ———, with feed on plate 5 (optimum location); ------, with feed on plate 7.

The transfer from one operating line to the other, and hence the feed plate location, can be made at any location between points a and b in Fig. 9.15; but if the feed plate is placed anywhere but at the optimum point, an unnecessarily large number of plates are called for. For example, if the feed plate in Fig. 9.15 is number 7, the smaller steps shown by the dashed lines make the number of ideal plates needed about 8 plus a reboiler, instead of 7 plus a reboiler when the feed is on plate number 5. Note that the liquid on the feed plate does not have the same composition as the feed except by coincidence, even when the feed plate location is optimum.

When we are analyzing the performance of a real column, the switch from one operating line to another must be made at a real feed plate. Because of changes in feed composition and uncertainties in plate efficiency, large columns are often operated with the feed entering a few plates above or below the optimum location. If large changes in feed composition are anticipated, alternate feed locations can be provided.

In an actual column with a fixed number of plates, feeding on the wrong plate may seriously affect the column performance. Feeding on too low a plate, for example, close to point b in Fig. 9.15, increases the number of plates in the rectifying section; but many of them are now operating in a pinched region where the driving force is small. These plates do very little separation. The diagram, therefore, must change, lowering the quality of both the top and bottom products to reflect the poorer performance of the plates. Feeding too high in the column leads to similar consequences.

Heating and cooling requirements

Heat loss from a large insulated column is relatively small, and the column itself is essentially adiabatic. The heat effects of the entire unit are confined to the condenser and the reboiler. If the average molal latent heat is r and the total sensible heat change in the liquid streams is small, the heat added in the reboiler q_r is $V'r$, in watts. When the feed is liquid at the bubble point ($q=1$), the heat supplied in the reboiler is approximately equal to that removed in the condenser; but for other values of q this is not true.

If saturated steam is used as the heating medium, the steam required at the reboiler is

$$\dot{m}_s = \frac{V'r}{r_s} \qquad (9.4\text{-}30)$$

where $\dot{m}_s$ = steam consumption
  V' = vapor rate from reboiler
  r_s = latent heat of steam
  r = molal latent heat of mixture

If water is used as the cooling medium in the condenser and the condensate is not subcooled, the cooling water requirement is

$$\dot{m}_w = \frac{Vr}{c_{pw}(t_2 - t_1)} \qquad (9.4\text{-}31)$$

where $\dot{m}_w$ = flow rate of cooling water
  $(t_2 - t_1)$ = temperature rise of cooling water
  c_{pw} = specific heat of cooling water

EXAMPLE 9.5 A continuous fractionating column is to be designed to separate 30000 kg/h of a mixture of 40 percent benzene and 60 percent toluene into an overhead product containing 97 percent benzene and a bottom product containing 98 percent toluene. These percentages are by weight. A reflux ratio of 3.5 mol to 1 mol of product is to be used. The molal latent heats of benzene and toluene are 7360 and 7960 cal/g mol, respectively. Benzene and toluene form a nearly ideal system with a relative volatility of about 2.5; the equilibrium curve is shown in Fig. 9.16. The feed has a boiling point of 95 ℃ at a pressure of 1 atm. (*a*) Calculate the moles of overhead product and bottom product per hour. (*b*) Determine the number of ideal plates and the position of the feed plate (*i*) if the feed is liquid and at its boiling point; (*ii*) if the feed is liquid and at 20 ℃ [specific heat 0.44 cal/(g · ℃)]; (*iii*) if the feed is a mixture of two-thirds vapor and one-third liquid. (*c*) If steam at 1.36 atm gauge is used for heating, how much steam is required per hour for each

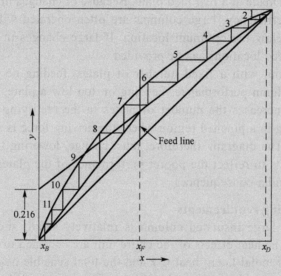

Fig. 9.16 Example **9.5**, part (*b*)(*i*).

of the above three cases, neglecting heat losses and assuming the reflux is a saturated liquid? (*d*) If cooling water enters the condenser at 25℃ and leaves at 40℃, how much cooling water is required, in cubic meters per hour?
Solution
(*a*) The molecular weight of benzene is 78, and that of toluene is 92. The concentrations of feed, overhead, and bottoms in mole fraction of benzene are

$$x_F = \frac{40/78}{40/78 + 60/92} = 0.440$$

$$x_D = \frac{97/78}{97/78 + 3/92} = 0.974$$

$$x_B = \frac{2/78}{2/78 + 98/92} = 0.0235$$

The average molecular weight of the feed is

$$\frac{100}{40/78 + 60/92} = 85.8$$

The average heat of vaporization of the feed is

$$r = 0.44 \times 7360 + 0.56 \times 7960 = 7696 \text{ cal/g mol}$$

The feed rate F is $30000/85.8 = 350$ kmol/h. By an overall benzene balance, using Eq.(9.4-3),

$$D = 350\left(\frac{0.440 - 0.0235}{0.974 - 0.0235}\right) = 153.4 \text{ kmol/h}$$

$$B = 350 - 153.4 = 196.6 \text{ kmol/h}$$

(*b*) Next we determine the number of ideal plates and position of the feed plate.

(*i*) The first step is to plot the equilibrium diagram and on it erect verticals at x_D, x_F, and x_B. These should be extended to the diagonal of the diagram. Refer to Fig. 9.16. The second step is to draw the feed line. Here $f = 0$, and the feed line is vertical and is a continuation of line $x = x_F$. The third step is to plot the operating lines. The intercept of the rectifying line on the *y* axis is, from Eq.(9.4-17), $0.974/(3.5+1) = 0.216$. This point is connected with the point x_D on the *y x* reference line. From the intersection of the rectifying operating line and the feed line, the stripping line is drawn.

The fourth step is to draw the rectangular steps between the two operating lines and the equilibrium curve. In drawing the steps, the transfer from the rectifying line to the stripping line is at the seventh step. By counting steps it is found that, besides the reboiler, 11 ideal plates are needed and feed should be introduced on the seventh plate from the top.

(*ii*) The latent heat of vaporization of the feed r is $7696/85.8 = 89.7$ cal/g. Substitution in Eq.(9.4-22) gives

$$q = 1 + \frac{0.44(95 - 20)}{89.7} = 1.37$$

From Eq.(9.4-29) the slope of the feed line is $-1.37/(1-1.37) = 3.70$. When steps are drawn for this case, as shown in Fig. 9.17, it is found that a reboiler and 10 ideal plates are needed and that the feed should be introduced on the sixth plate.

(*iii*) From the definition of *q* it follows that for this case $q = 1/3$ and the slope of the feed line is −0.5. The solution is shown in Fig. 9.18. It calls for a reboiler and 12 plates, with the feed entering on the seventh plate.

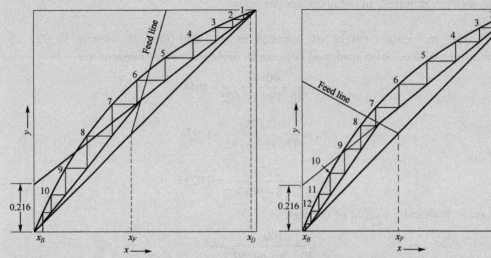

Fig. 9.17   Example 21.2, part (b)(ii).   Fig. 9.18   Example 21.2, part (b)(iii).

More plates are needed when the feed is largely vapor, partly because the feed line slants to the left and a few more plates are required in the rectifying section. The major reason, however, is that partially vaporized feed contributes less liquid to the stripping section than does a totally liquid feed, and the reflux ratio in the stripping section is reduced.

Table 9.1   Solution to Example 9.5, part (c)

Case	q	Reboiler steami $\dot{m}_s$ kg/h	Number of ideal plates	
			Graphical solution	Computed
(i)	1.0	10520	11	10.59
(ii)	1.37	12500	10	10.17
(iii)	0.333	6960	12	11.85

(c) The vapor flow V in the rectifying section, which must be condensed in the condenser, is 4.5 mol per mole of overhead product, or 4.5×153.4 = 690 kg mol/h. From Eq.(9.4-25)

$$V' = 690 - 350(1-q)$$

Using the heat of vaporization of toluene rather than that of benzene to be slightly conservative in design, r = 7960 cal/g mol. The heat from 0.4536kg of steam at 1.36 atm gauge, from App.6, is r_S = 522 cal/g. The steam required, from Eq.(9.4-30), in kg/h, is

$$\dot{m}_s = \frac{V'r}{r_s} = \frac{7960}{522}V' = 15.25[690 - 350(1-q)]$$

The results are given in Table 9.1.

Note: Using the equation $\alpha = 2.34+0.27x$ for the relative volatility of benzene-toluene mixtures, a computer program based on constant molal overflow indicated 10.59 plates required for part (i), 10.17 plates for part (ii), and 11.85 for part (iii).

(d) The cooling water needed, which is the same in all cases, is, from Eq.(9.4-31),

$$\dot{m}_w = \frac{7960 \times 690}{40-25} = 366160 \, \text{kg/h}$$

The density of water at 25°C, from App.5, is 996.3 kg/m³. The water requirement is 366160/996.3 = 367.5 m³/h.

The use of cold feed, case (ii), requires the smallest number of plates but the greatest amount of reboiler steam. The total energy requirement for the reboiler and the preheater is about the same for all three cases. The reasons for preheating the feed, in most cases, are to keep the vapor flow rate about the same in both sections of the column and to make use of the energy in a hot-liquid stream such as the bottom product.

In this example the total energy requirement was the same for each case because R was fixed at 3.5, but R/R_m ranged from 1.6 to 2.8. If R/R_m had been fixed, the energy requirements and the number of plates would have been different. For example, consider liquid or vapor feed in a benzene-toluene distillation with $x_F = 0.50$, $x_D = 0.98$, $x_B = 0.02$, $D/F = 0.5$, and $R/R_m = 1.20$. Using liquid feed ($q = 1.0$) would require $R = 1.52$, $V/F = 0.5 \times (1+1.52) = 1.26$, and 18 ideal plates. With vapor feed ($q = 0$), $R = 2.8$, and $V/F = 1.90$. The total energy used and the heat removed in the condenser are 50 percent greater than for liquid feed, though the reboiler heat duty is reduced, since $V_m/F = 0.90$, and only 15 ideal plates are needed.

Minimum number of plates

Since the slope of the rectifying line is $R/(R+1)$, the slope increases as the reflux ratio increases until, when R is infinite, $V = L$ and the slope is 1. The operating lines then both coincide with the diagonal. This condition is called *total reflux*. At total reflux the number of plates is a minimum, but the rates of feed and of both the overhead and bottom products are zero. Total reflux represents one limiting case in the operation of fractionating columns. The minimum number of plates required for a given separation may be found by constructing steps on an xy diagram between compositions x_D and x_B, using the 45° line as the operating line for both sections of the column. Since there is no feed in a column operating under total reflux, there is no discontinuity between the upper and lower sections.

For the special case of ideal mixtures, a simple method is available for calculating the value of $N_{\min}$ from the terminal concentrations x_B and x_D. This is based on the relative volatility of the two components α, Eq.(7.1-6), which is defined in terms of the equilibrium concentrations

$$\alpha = \frac{y_A/x_A}{y_B/x_B} \tag{7.1-6}$$

An ideal mixture follows Raoult's law, and the relative volatility is the ratio of the vapor pressures. Thus

$$\alpha = \frac{y_A/x_A}{y_B/x_B} = \frac{P_A/P}{P_B/P} = \frac{P_A}{P_B}$$

The ratio P_A/P_B does not change much over the range of temperatures encountered in a typical column, so the relative volatility is taken as constant in the following derivation.

For a binary system y_A/y_B and x_A/x_B may be replaced by $y_A/(1-y_A)$ and $x_A/(1-x_A)$, so Eq. (7.1-6) can be written for plate $n+1$ as

$$\frac{y_{n+1}}{1-y_{n+1}} = \alpha \frac{x_{n+1}}{1-x_{n+1}} \tag{9.4-32}$$

Since at total reflux $D = 0$ and $L/V = 1$, $y_{n+1} = x_n$. See Eq.(9.4-11), and note that the operating line is the 45° line; this leads to

$$\frac{x_n}{1-x_n} = \alpha \frac{x_{n+1}}{1-x_{n+1}} \tag{9.4-33}$$

At the top of the column, if a total condenser is used, $y_1 = x_D$, so Eq.(9.4-32) becomes

$$\frac{x_D}{1-x_D} = \alpha \frac{x_1}{1-x_1} \tag{9.4-34}$$

Writing Eq.(9.4-33) for a succession of n plates gives

$$\frac{x_1}{1-x_1} = \alpha \frac{x_2}{1-x_2} \tag{9.4-35}$$

$$\cdots$$

$$\frac{x_{n-1}}{1-x_{n-1}} = \alpha \frac{x_n}{1-x_n}$$

If Eq.(9.4-34) and all the equations in the set of Eqs.(9.4-35) are multiplied together and all the intermediate terms canceled, then

$$\frac{x_D}{1-x_D} = \alpha^n \frac{x_n}{1-x_n} \tag{9.4-36}$$

To reach the bottom discharge from the column, N_{min} plates and a reboiler are needed, and Eq(9.4-36) gives

$$\frac{x_D}{1-x_D} = \alpha^{N_{min}+1} \frac{x_B}{1-x_B}$$

Solving the equation for N_{min} by logarithms gives

$$N_{min} = \frac{\ln\frac{x_D(1-x_B)}{x_B(1-x_D)}}{\ln \alpha} - 1 \tag{9.4-37}$$

Equation(9.4-37) is the *Fenske equation*, which applies when α is constant. If the change in the value of α from the bottom of the column to the top is moderate, a geometric mean of the extreme values is recommended for α.

Minimum reflux

At any reflux less than total, the number of plates needed for a given separation is larger than at total reflux and increases continuously as the reflux ratio is decreased. As the ratio becomes smaller, the number of plates becomes very large, and at a definite minimum, called the *minimum reflux ratio*, the number of plates becomes infinite. All actual columns producing a finite amount of desired top and bottom products must operate at a reflux ratio between the minimum, at which the number of plates is infinity, and infinity, at which the number of plates is a minimum. If L_a/D is the operating reflux ratio and $(L_a/D)_{min}$ is the minimum reflux ratio, then

$$\left(\frac{L_a}{D}\right)_{min} < \frac{L_a}{D} < \infty \tag{9.4-38}$$

The minimum reflux ratio can be found by following the movement of the operating lines as the reflux is reduced. In Fig. 9.19 both operating lines coincide with the diagonal *afb* at total reflux. For an actual operation lines *ae* and *eb* are typical operating lines. As the reflux is further reduced, the intersection of the operating lines moves along the feed line toward the equilibrium curve, the area on the diagram available for steps shrinks, and the number of steps increases. When either one or both of the operating lines touch the equilibrium curve, the number of steps necessary to cross the point of contact becomes infinite. The reflux ratio corresponding to this situation is, by definition, the minimum reflux ratio.

For the normal type of equilibrium curve, which is concave downward throughout its length, the point of contact, at minimum reflux, of the operating and equilibrium lines is at the intersection of the feed line with the equilibrium curve, as shown by lines *ad* and *db* in Fig. 9.19. A further decrease in reflux brings the intersection of the operating lines outside of the equilibrium curve, as shown by lines *agc* and *cb*. Then even an infinite number of plates cannot pass point *g*, and the reflux ratio for this condition is less than the minimum.

The slope of operating line *ad* in Fig. 9.19 is such that the line passes through the points (x', y') and (x_D, x_D), where x' and y' are the coordinates of the intersection of the feed line and the equilibrium curve. Let the minimum reflux ratio be R_m. Then

$$\frac{R_m}{R_m+1} = \frac{x_D - y'}{x_D - x'} \qquad (9.4\text{-}39)$$

or

$$R_m = \frac{x_D - y'}{y' - x'}$$

Equation (9.4-39) cannot be applied to all systems. Thus, if the equilibrium curve has a concavity upward, as for example, the curve for ethanol and water shown in Fig. 9.20, it is clear that the rectifying line first touches the equilibrium curve between abscissas x_F and x_D and line *ac* corresponds to minimum reflux. Operating line *ab* is drawn for a reflux less than the minimum, even though it does intersect the feed line below point (x', y'). In such a situation the minimum reflux ratio must be computed from the slope of the operating line *ac* that is tangent to the equilibrium curve.

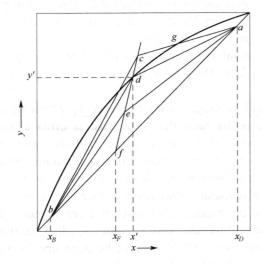

Fig. 9.19 Minimum reflux ratio.

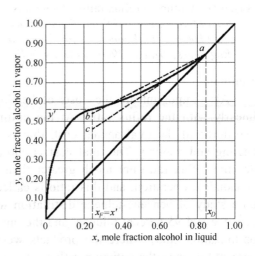

Fig. 9.20 Equilibrium diagram (system ethanol-water).

Invariant zone

At minimum reflux ratio an acute angle is formed at the intersection of an operating line and the equilibrium curve, as shown at point d in Fig. 9.19 or at the point of tangency in Fig. 9.20. In each angle an infinite number of steps are called for, representing an infinite number of ideal plates, in all of which there is no change in either liquid or vapor concentrations from plate to plate, so $x_{n-1} = x_n$ and $y_{n+1} = y_n$. The term *invariant zone* is used to describe these infinite sets of plates. The more descriptive term *pinch point* also is used.

With a normal equilibrium curve it is seen from Fig. 9.19 that at minimum reflux ratio the intersection of the q line and the equilibrium curve gives the concentrations of liquid and vapor at the feed plate (and at an infinite number of plates on either side of that plate). So an invariant zone forms at the bottom of the rectifying section and a second one at the top of the stripping section. The two zones differ only in that the liquid-vapor ratio is L/V in one and L'/V' in the other.

Optimum reflux ratio

As the reflux ratio is increased from the minimum, the number of plates decreases, rapidly at first and then more and more slowly until, at total reflux, the number of plates is a minimum. It will be shown later that the cross-sectional area of a column usually is approximately proportional to the flow rate of vapor. As the reflux ratio increases, both V and L increase for a given production, and a point is reached where the increase in column diameter is more rapid than the decrease in the number of plates. The cost of the unit is roughly proportional to the total plate area, that is, the number of plates times the cross-sectional area of the column; so the fixed charges for the column first decrease and then increase with reflux ratio. Fixed charges on the heat-exchange equipment—the reboiler and condenser—increase steadily with the reflux ratio. Curve 2 in Fig. 9.21 shows the total fixed charges, which drop sharply at first, then pass through a very shallow minimum.

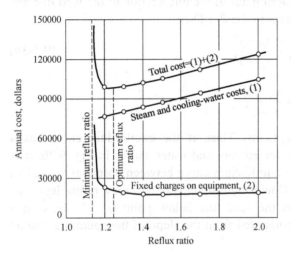

Fig. 9.21 Optimum reflux ratio. (*By permission from M. S. Peters and K. D. Timmerhaus, Plant Design and Economics for Chemical Engineers, 3rd ed., 1980, McGraw-Hill.*)

Also important are the costs of heating and cooling, shown by curve 1, which rises linearly with the reflux ratio. At the optimum reflux ratio the steam cost is often about two-thirds of the total annual cost, that is, the sum of the fixed charges and the costs of heating and cooling. The total cost is shown by the topmost curve. It is a minimum at a definite reflux ratio not much greater than the minimum reflux. This is the point of most economical operation, and this ratio is called the *optimum reflux ratio*. Figure 9.21 is based on a study of benzene-toluene distillation where the product compositions were 92 percent benzene and 95 percent toluene, and the optimum reflux ratio was 1.1 times R_m. In a similar optimization study where the products were very pure—99.97 percent benzene and 99.83 percent toluene—the optimum reflux ratio was $1.25R_m$. The optimum reflux ratio depends on the cost of energy; it will be closer to R_m when energy costs are relatively high, and farther from R_m when distillation equipment is made of expensive alloys. Actually, most plants are operated at reflux ratios somewhat above the optimum, because the total cost is not very sensitive to reflux ratio in this range, and better operating flexibility is obtained if a reflux ratio greater than the optimum is used.

> **EXAMPLE 9.6** What are (a) the minimum reflux ratio and (b) the minimum number of plates for cases (b)(i), (b)(ii), and (b)(iii) of Example 9.5?
> **Solution**
> (a) For minimum reflux ratio use Eq.(9.4-39). Here $x_D = 0.974$. The results are given in Table 9.2.
> (b) For minimum number of plates, the reflux ratio is infinite, the operating lines coincide with the diagonal, and there are no differences among the three cases. Use the Fenske equation [Eq.(9.4-37)]. Since $\alpha = 2.34 + 0.27x$, at $x = 0.024$, $\alpha = 2.35$; at $x = 0.974$, $\alpha = 2.60$
>
> $$\bar{\alpha} = \sqrt{2.35 \times 2.60} = 2.47$$
>
> From Eq.(9.4-34),
>
> **Table 9.2**
>
> | Case | x' | y' | R_m |
> |------|------|------|-------|
> | (b)(i) | 0.440 | 0.658 | 1.45 |
> | (b)(ii) | 0.521 | 0.730 | 1.17 |
> | (b)(iii) | 0.300 | 0.513 | 2.16 |
>
> $$N_{min} = \ln\left(\frac{0.974 \times 0.976}{0.024 \times 0.026}\right) - 1 = 8.105 - 1 = 7.1$$
>
> The minimum number of ideal plates is 7 plus a reboiler.

Number of ideal plates at operating reflux

A simple empirical method due to Gilliland is much used for preliminary estimates. The correlation requires knowledge only of the minimum number of plates at total reflux and the minimum reflux ratio. The correlation is given in Fig. 9.22 and is self-explanatory. The Gilliland correlation, however, is based mainly on calculations for systems with nearly constant relative volatility and may be considerably in error for nonideal systems. An alternate correlation in Fig. 9.23 shows that the ratio of N/N_{min} (for ideal binary systems) depends mainly on R/R_m for a wide range of relative volatilities. On the other hand, for the system methanol-water, where α changes from 7.5 for dilute solutions to 2.7 for nearly pure methanol, the values of N/N_{min} are much greater and change more rapidly than with ideal systems.

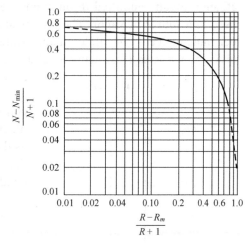

Fig. 9.22 Gilliland correlation.

Nearly pure products

When either the bottom or overhead product is nearly pure, a single diagram covering the entire range of concentrations is impractical as the steps near $x = 0$ and $x = 1$ become small. Auxiliary diagrams for the ends of the construction range may be prepared, on a large scale, so that the individual steps are large enough to be drawn. In practice, however, the calculations are usually done by computer, and the scale may easily be expanded to cover the desired range. (With computer calculations, of course, it is not necessary to display the

McCabe-Thiele diagram to find the number of plates needed, but it is often helpful to do so in order to visualize the solution to the problem.)

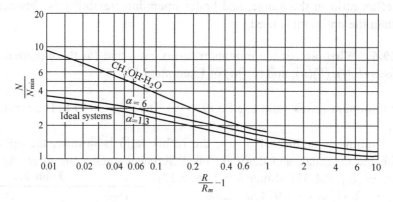

Fig. 9.23 Alternate correlation for N/N_{min}.

Another method of treating nearly pure products is based on the principle that Raoult's law applies to the major component and Henry's law to the minor component at each end of the equilibrium curve. In these regions, therefore, both the equilibrium and the operating lines are straight, so Eq. (7.2-27) can be used, and no graphical construction is required. The same equation may be used anywhere in the concentration range where both the operating and equilibrium lines are straight or nearly so.

EXAMPLE 9.7 A mixture of 2 mol percent ethanol and 98 mol percent water is to be stripped in a plate column to a bottom product containing not more than 0.01 mol percent ethanol. Steam, admitted through an open coil in the liquid on the bottom plate, is to be used as a source of vapor. The feed is at its boiling point. The steam flow is to be 0.2 mol per mole of feed. For dilute ethanol-water solutions, the equilibrium line is straight and is given by $y = 9.0x$. How many ideal plates are needed?

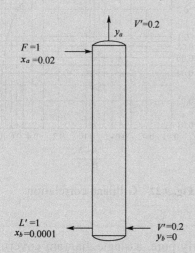

Fig. 9.24 Material-balance diagram for Example 9.7

Solution. Since both equilibrium and operating lines are straight, Eq. (7.2-27) rather than a graphical construction may be used. The material-balance diagram is shown in Fig. 9.24. No reboiler is needed, as the steam enters as a vapor. Also, the liquid flow in the tower equals the feed entering the column. By conditions of the problem

$$F = L' = 1 \qquad V' = 0.2 \qquad y_b = 0 \qquad x_a = 0.02$$

$$x_b = 0.0001 \qquad m = 9.0$$

$$y_a^* = 9.0 \times 0.02 = 0.18$$

$$y_b^* = 9.0 \times 0.0001 = 0.0009$$

To use Eq. (7.2-27), y_a, the concentration of the vapor leaving the column, is needed. This is found by an overall ethanol balance

$$V'(y_a - y_b) = L'(x_a - x_b)$$

$$0.2(y_a - 0) = 1(0.02 - 0.0001)$$

from which $y_a = 0.0995$. Substituting into Eq. (7.2-27) gives

$$N = \frac{\ln\left[\dfrac{(0.0995-0.18)}{(0-0.000995)}\right]}{\ln\left[\dfrac{(0.0009-0.18)}{(0-0.0995)}\right]} = \frac{\ln 89.4}{\ln 1.8} = 7.6 \text{ ideal plates}$$

9.4.5 Special Cases for Rectification Using McCabe-Thiele Method

Stripping-column distillation

In some cases the feed to be distilled is not supplied to an intermediate point in a column but is added to the top of the stripping column, as shown in Fig. 9.25(a). The feed is usually a saturated liquid at the boiling point, and the overhead product V_D is the vapor rising from the top plate, which goes to a condenser with no reflux or liquid returned to the tower.

The bottoms product B usually has a high concentration of the less volatile component B. Hence, the column operates as a stripping tower, with the vapor removing the more volatile A from the liquid as it flows downward. Assuming constant molar flow rates, a material balance of the more volatile component A around the dashed line in Fig. 9.25(a) gives, on rearrangement,

$$y_{m+1} = \frac{L_m}{V_{m+1}} x_m - \frac{Bx_B}{V_{m+1}} \qquad (9.4\text{-}40)$$

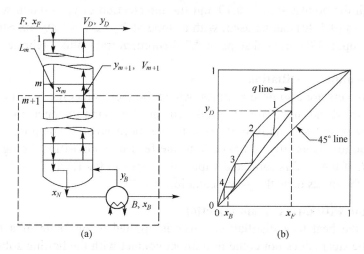

Fig. 9.25 Material balance and operating line for stripping tower:
(a) flows in tower, (b) operating and equilibrium line.

This stripping-line equation is the same as the stripping-line equation for a complete tower given as Eq.(9.4-14). It intersects the $y = x$ line at $x = x_B$, and the slope is constant at L_m/V_{m+1}.

If the feed is saturated liquid, then $L_m = F$. If the feed is cold liquid below the boiling point, the q line should be used and $q > 1$:

$$L_m = qF$$

In Fig. 9.25 the stripping operating-line Eq.(9.4-40) is plotted and the q line, Eq.(9.4-29), is also shown for $q = 1.0$. Starting at x_F, the steps are drawn down the tower.

EXAMPLE 9.8 A liquid feed at the boiling point of 400 kmol/h containing 70 mol % benzene (A) and 30 mol % toluene (B) is fed to a stripping tower at 101.3 kPa pressure. The bottoms product flow is to be 60 kmol/h containing only 10 mol % A and the rest B. Calculate the kmol/h overhead vapor, its composition, and the number of theoretical steps required.

Solution. Referring to Fig. 9.25(a), the known values are $F = 400$ kmol/h, $x_F = 0.70$, $B = 60$ kmol/h, and $x_B = 0.10$. The equilibrium data from Table 7.1 are plotted in Fig. 9.26. Making an overall material balance,

$$F = B + V_D$$

$$400 = 60 + V_D$$

Solving, $V_D = 340$ kmol/h. Making a component A balance and solving,

$$Fx_f = Bx_B + V_D y_D$$

$$400 \times 0.70 = 60 \times 0.10 + 340 y_D$$

$$y_D = 0.806$$

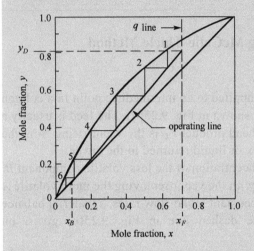

Fig. 9.26 Stripping tower for Example 9.8

For a saturated liquid, the q line is vertical and is plotted in Fig. 9.26. The operating line is plotted through the point $y = x_B = 0.10$ and the intersection of $y_D = 0.806$ with the q line. Alternatively, Eq.(9.4-40) can be used, with a slope of $L_m/V_{m+1} = 400/340$. Stepping off the trays from the top, 5.3 theoretical steps or 4.3 theoretical trays plus a reboiler are needed.

Enriching-column distillation

Enriching towers are also used at times, where the feed enters the bottom of the tower as a vapor. The overhead distillate is produced in the same manner as in a complete fractionating tower and is usually quite rich in the more volatile component A. The liquid product of the bottom is usually comparable to the feed in composition, being slightly leaner in component A. If the feed is saturated vapor, the vapor in the tower $V_n = F$. Enriching-line equation(9.4-12)holds, as does the q-line equation(9.4-29).

Rectification with direct steam injection

Generally, the heat to a distillation tower is applied to one side of a heat exchanger (reboiler) and the steam does not come into direct contact with the boiling solution, as shown in Fig. 9.11. However, when an aqueous solution of more volatile A and water B is being distilled, the heat required may be provided by the use of open steam injected directly at the bottom of the tower. The reboiler exchanger is then not needed.

The steam is injected as small bubbles into the liquid in the tower bottom, as shown in Fig. 9.27(a). The vapor leaving the liquid is then in equilibrium with the liquid if sufficient contact is obtained. Making an overall balance on the tower and a balance on A,

$$F + S = D + B \tag{9.4-41}$$

$$Fx_F + Sy_s = Dx_D + Bx_B \tag{9.4-42}$$

where $S =$ mol/h of steam and $y_S = 0 =$ mole fraction of A in steam. The enriching operating-line equation is the same as for indirect steam.

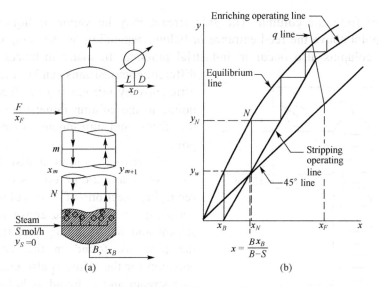

Fig. 9.27 Use of direct steam injection: (*a*) schematic of tower, (*b*) operating and equilibrium lines.

For the stripping-line equation, an overall balance and a balance on component A are as follows:

$$L_m + S = V_{m+1} + B \tag{9.4-43}$$

$$L_m x_m + S(0) = V_{m+1} y_{m+1} + B x_B \tag{9.4-44}$$

Solving for y_{m+1} in Eq.(9.4-44),

$$y_{m+1} = \frac{L_m}{V_{m+1}} x_m - \frac{B x_B}{V_{m+1}} \tag{9.4-45}$$

For saturated steam entering, $S = V_{m+1}$ and hence, by Eq.(9.4-43), $L_m = B$. Substituting into Eq. (9.4-45), the stripping operating line is

$$y_{m+1} = \frac{B}{S} x_m - \frac{B x_B}{S} \tag{9.4-46}$$

When $y = 0$, $x = x_B$. Hence, the stripping line passes through the point $y = 0$, $x = x_B$, as shown in Fig. 9.27(*b*), and is continued to the x axis. Also, for the intersection of the stripping line with the 45° line, when $y = x$ in Eq.(9.4-46),

$$x = \frac{B x_B}{B - S}$$

For a given reflux ratio and overhead distillate composition, the use of open steam rather than closed requires an extra fraction of a stage, since the bottom step starts below the $y = x$ line [Fig. 9.27(*b*)]. The advantage of open steam lies in simpler construction of the heater, which is a sparger.

Multiple feeds and sidestreams

Not all distillation columns operate with a single feed and two products. Multiple product towers are used in many situations. They are particularly useful for cases in which a small amount of a volatile contaminant is present in the feed. In this case the overhead product will be the contaminant, and side product somewhere in the rectifying section will be desired product from the column. The side stream may be vapor or liquid and may be removed at a point above the feed entrance or below, depending on the composition desired. Multiple feed columns also occur in industrial practice. In many instances two feeds of different compositions can be fractionated more efficiently by introducing the feeds at different points in the column rather than by combining the feeds and introducing them at a single point.

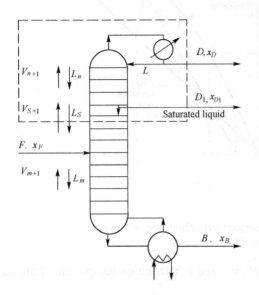

Figure 9.28 shows a distillation column from which a liquid product is withdrawn in the rectifying section of the column. The top enriching operating line above the liquid side stream and the stripping operating line below the feed are found in the usual way. The equation for the q line is also unaffected by the side stream and is found as before. The liquid side stream alters the liquid rate below it, and hence the material balance or operating line in the middle portion between the feed and liquid side stream plates as well.

A liquid side-draw product is shown in the dashed-line box in Fig. 9.28, making a total

Fig. 9.28 Process flow for a rectification tower with a liquid side stream.

material balance on the top portion of the tower gives

$$V_{s+1} = L_s + D_1 + D \tag{9.4-47}$$

where a stream of vapor V_{s+1} mol/h from next plate is brought into the dashed-line box and L_s out of it, D_1 is mol/h saturated liquid removed as a side stream. Since the liquid side stream is saturated,

$$L_n = L_s + D_1 \tag{9.4-48}$$

$$V_{s+1} = V_{n+1} \tag{9.4-49}$$

The side product has the composition of ether the vapor or the liquid leaving the tray. Also, the withdrawal of a side product causes an upset of either the vapor or liquid flow from the tray. A material balance around the rectifying section of the column shown in Fig.(9.4-24) for one component is

$$V_{s+1}y_{s+1} = L_s x_s + D_1 x_{D1} + D x_D \tag{9.4-50}$$

Solving for y_{s+1}, the operating line for the region between the side stream and the feed is

$$y_{s+1} = \frac{L_s}{V_{s+1}} x_s + \frac{D_1 x_{D1} + D x_D}{V_{s+1}} \tag{9.4-51}$$

This represents the operating-line equation for the section of the column between the side draw plate and the feed. This line will intersect the q line for the side product and operating line for the section of the column above the side draw at a common point. It will extend from this intersection to its y intercept, which is $(D_1 x_{D1} + D x_D)/V_{s+1}$. Inspection of the operating line equation for the section of the column above the feed reveals that the operating line for the section between the side draw and the feed always has a slope that is less than the slope of the operating line for the section above the side draw. Construction for stepping off plates with the side-draw column is given in Fig. 9.29.

9.4.6 Batch Distillation with Reflux

Batch distillation with only a simple still does not give a good separation unless the relative volatility is very high. In many cases, a rectifying column with reflux is used to improve the performance of the batch still. If the column is not too large, it may be mounted on top of the reboiler, as shown in Fig. 7.8, or it may be supported independently, with connecting pipes for the vapor and liquid streams.

The operation of a batch still and column can be analyzed using a McCabe-Thiele diagram, with the same operating-line equation that was used for the rectifying section of a continuous distillation [Eq.(9.4-17)]

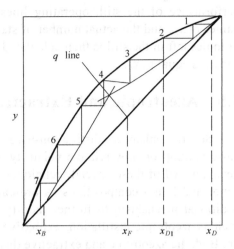

Fig. 9.29 McCabe-Thiele plot for a tower with a liquid side stream above the feed.

$$y_{n+1} = \frac{R}{R+1}x_n + \frac{x_D}{R+1}$$

The system may be operated to keep the top composition constant by increasing the reflux ratio as the composition of the liquid in the reboiler changes. The McCabe-Thiele diagram for this case would have operating lines of different slope positioned such that the same number of ideal stages was used to go from x_D to x_B at any time. Atypical diagram is shown in Fig. 9.30 for a still with five ideal stages including the reboiler. The upper operating line is for the initial conditions, when the concentration of low boiler in the still is about the same as the charge composition. (The concentration x_B is slightly lower than x_F because of the holdup of liquid on the plates.) The lower operating line and the dashed line steps show conditions when about one-third of the charge has been removed as overhead product.

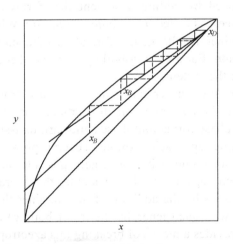

Fig. 9.30 McCabe-Thiele diagrams for a batch distillation. Upper operating line and solid lines: initial conditions; lower operating line and dashed lines: after one-third of the charge has been removed.

To determine the reflux ratio needed for a constant x_D and given x_B requires a trial-and-error calculation, since the last step on the assumed operating line must end exactly at x_B. However, once the initial reflux ratio is chosen by this method, the value of x_B for a later stage in the distillation can be obtained by assuming a value of R, constructing the operating line, and making the correct number of steps ending at x_B. By a material balance, Eqs.(9.4-3) and (9.4-4), the amount of product and remaining charge, can be calculated.

An alternative method of running a batch distillation is to fix the reflux ratio and let the overhead product purity vary with time, stopping the distillation when the amount of product or the average concentration in the total product reaches a certain value. To calculate the performance of the still, operating lines of constant slope are drawn starting at different values of x_D, and the actual number of stages is stepped off to determine x_B. The total number of moles left in the still is then calculated by integration of Eq. (9.2-4), where x_D is equal to y and x_1 is equal to x_B.

9.5 Azeotropic and Extractive Distillation

Several enhanced distillation-based separation techniques have been developed for close-boiling or low-relative-volatility systems, and for systems exhibiting azeotropic behavior. All of these special techniques are ultimately based on the same differences in the vapor and liquid compositions as ordinary distillation, but, in addition, they rely on some additional mechanism to further modify the vapor-liquid behavior of the key components, such as azeotropic distillation, extraxtive distillation, and pressure-swing distillation, and so on. Both the azeotropic and extractive distillations are briefly introduced in this section.

Azeotropic Distillation

The term azeotropic distillation has been applied to a broad class of fractional-distillation-based separation techniques in that specific azeotropic behavior is exploited to effect a separation. The agent that causes the specific azeotropic behavior, often called the *entrainer*, may already be present in the feed mixture (a self-entraining mixture) or may be an added mass-separation agent. Azeotropic distillation techniques are used throughout the petrochemical and chemical processing industries for the separation of close-boiling, pinched, or azeotropic systems for which simple distillation is ether too expensive or impossible. Separation of the original mixture may be enhanced by adding a solvent that forms an azeotrope with one of the key components. This process is called *azeotropic distillation*. With an azeotropic feed mixture, presence of the entrainer results in the formation of a more favorable azeotropic pattern for the desired separation. For a closing-boiling or pinched feed mixture, the entrainer changes the dimensionality of the system.

The approach of azeotropic distillation, choosing an entrainer to cause azeotrope formation in combination with liquid-liquid immiscibility, relies on distillation, but also exploits another physical phenomena, liquid-liquid phase formation(phase splitting), to assist in entrainer recovery. One powerful and versatile approach exploits several physical phenomena simultaneously including enhanced vapor-liquid behavior, where possible, and liquid-liquid behavior to bypass difficult distillation separations. For example, the overall separation of close-boiling mixtures can be made easier by the addition of the entrainer that forms a heterogeneous minimum-boiling azeotrope with one (generally the lower-boiling) of the key components. Two-liquid-phase formation provides a means of breaking this azeotrope, thus simplifying the entrainer recovery and recycle process.

The simplest case of combining vapor-liquid equilibrium and liquid-liquid equilibrium is the separation of a binary heterogeneous azeotropic mixture. One example is the dehydration of 1-butanol, a self-entraining system, in which butanol(117.7 ℃) and water form a minimum-boiling heterogeneous azeotrope(93.0 ℃). As shown in Fig. 9.31, the fresh feed may be added to either column C1 or C2, depending on whether the feed is on the organic-rich side or water-rich side of the azeotrope. The feed may also be added into the decanter directly if it doesn't move the overall composition of the decanter outside of the two-liquid-phase region. The column C1 produces anhydrous butanol as a bottoms product and a composition close to the butanol-water azeotrope as the distillate. After condensation, the azeotrope rapidly phase separates in the decanter. The upper layer, consisting of 78wt% butanol, is refluxed totally to C1 for further butanol recovery. The water layer, consisting of 92wt% water, is fed to C2. This column produces pure water as a bottoms product and, again, a composition close to the azeotrope as distillate for recycle to the decanter. Sparged steam may be used in C2, saving the cost of a reboiler.

A second example of the use of liquid-liquid nonmiscibilities in an azeotropic-distillation sequence is the separation of the ethanol-water minimum-boiling azeotrope. For this separation, a number of entrainers have been proposed, which are usually chosen to be

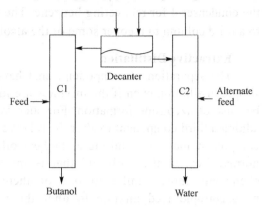

Fig. 9.31 Separation of butanol-water with heterogeneous azeotropic distillation.

immiscible with water, form a ternary minimum-boiling (preferably heterogeneous) azeotrope with ethanol and water (and, therefore, usually also binary minimum-boiling azeotropes with both ethanol and water). One three-column distillation sequence is shown in Fig. 9.32.

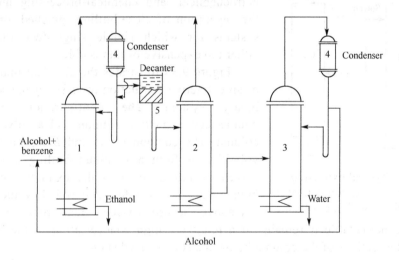

Fig. 9.32 Three-column sequence for dehydration with benzene.

An example of azeotropic distillation is the use of benzene, heptane, or cyclohexane to permit the separation of ethanol and water, which form a minimum boiling azeotrope with 95.6 weight percent alcohol (mole fraction of 89.4%). Adding benzene to the ethanol-water minimum-boiling azeotrope can form a ternary azeotrope with boiling point of 64.85℃, which contains benzene of 74.1 % mass fraction, ethanol of 18.5% mass fraction and water of 7.4% mass fraction). The alcohol-water mixture with about 95 weight percent alcohol, together with the benzene, is fed near the middle of the dehydrating column 1 in Fig. 9.32, and nearly pure alcohol is removed as the bottoms product. The overhead vapor is a ternary azeotrope, which is condensed in the condenser 4 and separated into two phases in the decanter 5. The upper benzene-rich layer(mass fraction: benzene 0.845, ethanol 0.145, water 0.01, is) is returned to the top of the dehydrating column 1 as reflux, and the water layer(mass fraction: benzene 0.11, ethanol 0.53, water 0.36) is sent to a stripping column 2, where the ternary azeotrope containing benzene entrainer is taken overhead and returned to the condenser 4 for recovering benzene. The stripper bottom is an aqueous stream that is sent to a third column to recover some of the alcohol.

Extractive Distillation

The separation of components that have nearly the same boiling points is difficult by simple distillation even if the mixtures are ideal, and complete separation may be impossible because of azeotrope formation. For such systems the separation can often be improved by adding a third component to alter the relative volatility of the original components. The added component may be a miscible, higher-boiling liquid, non-volatile mass-separation agent, normally called the "solvent" that is miscible with both of the key components but is chemically more similar to one of them. The solvent is added to an azeotropic or nonazeotropic feed mixture to alter the volatilities of the key components without the formation of any additional azeotropes. The key component that is more like the solvent will have a lower activity coefficient in the solution than the other component, so the separation is enhanced. This process is called *extractive distillation* and is like liquid-liquid extraction with an added vapor phase.

Extractive distillation is used throughout the petrochemical- and chemical-processing industries for the separation of close-boiling, pinched, or azeotropic systems for which simple single-feed distillation is either too expensive or impossible.

Figure 9.33 illustrates the classical implementation of an extractive distillation process for the separation of binary system. The configuration consists of double-feed extractive column and a solvent recovery column. The components A and B may a low relative volatility or form a minimum-boiling azeotrope. The solvent is introduced into the extractive column at a high concentration a few stages below the condenser, but above the primary-feed stage. Since the solvent is chosen to be nonvolatile it remains at a relatively high concentration in the liquid phase throughout the sections of the column below the solvent-feed stage.

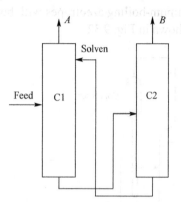

Fig. 9.33 Typical extracting distillation sequence, component A is less associated with the solvent.

One of the components, A (not necessarily the most volatile species of the original mixture), is withdrawn as an essentially pure distillate stream. Because the solvent is nonvolatile, at most a few stages above the solvent-feed stage are sufficient to rectify the solvent from the distillate. The bottom product, consisting of B and the solvent, is sent to the recovery column. The distillate from the recovery column is pure B, and the solvent-bottoms product is recycled back to the extractive column.

The solvent selectively alters the activity coefficients of the components being separated. To do this, a high concentration of solvent is necessary. Several features are essential:

(1) The solvent must be chosen to affect the liquid-phase behavior of the key components differently, otherwise no enhancement in separability will occur.

(2) The solvent must be higher boiling than the key components of the separation and must be relatively nonvolatile in the extractive column, in order to remain largely in the liquid phase.

(3) The solvent should not form additional azeotropes with the components in the

mixture to be separated.

(4) The extractive column must be a double-feed column, with the solvent feed above the primary feed; the column must have an extractive section.

One example of extractive distillation is the use of furfural to permit the separation of hexahydrobenzene from a mixture containing benzene and hexahydrobenzene, as shown in Fig. 9.34. The boiling points both the benzene and the hexahydrobenzene are 80.1℃ and 80.73℃ at the pressure of 1 atm, respectively. The boiling points are so close that it is very difficult to separate them from the mixture by an ordinary distillation. Furfural, which is a highly polar solvent, is added to the mixture of the benzene and the hexahydrobenzene to alter the volatility of components, as shown in Table 9.3.

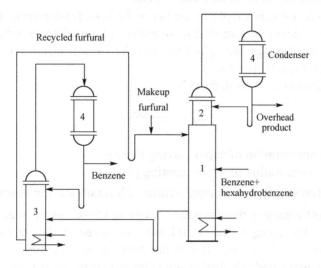

Fig. 9.34 Extractive distillation for the separation of benzene-hexahydrobenzene system by using furfural.

Table 9.3 Effect of furfural on the relative volatility of benzene-hexahydrobenzene system

Mole fraction of furfural in solution	0	0.2	0.4	0.5	0.6	0.7
Relative volatility of benzene-hexahydrobenzene	0.98	1.38	1.86	2.07	2.36	2.70

The benzene-hexahydrobenzene mixture is supplied to an intermediate point in the extractive column 1. Furfural is introduced into the top of the extractive column. The overhead vapor is the hexahydrobenzene, which is completely condensed in the condenser 4. The section 2 above the extractive column1 is used as recovering the furfural. The bottoms product containing the benzene and furfural is withdrawn from the reboiler and introduced into near center of the column 3 for separating the benzene from the benzene-furfural mixture. The benzene vapor, the boiling point of which is much lower than that of furfural, is completely condensed in the condenser 4 above the column 3; the furfural is withdrawn from the kettle of column 3 and recycled to the column 1 as a solvent.

9.6 Plate Efficiencies

In all the previous discussions of theoretical trays or stages in distillation, we assumed that the vapor leaving a tray was in equilibrium with the liquid leaving. However, if the time of contact and the degree of mixing on the tray are insufficient, the streams will not be in equilibrium. As a result the efficiency of the stage or tray will not be 100%. This means that we must use more actual trays for a given separation than the theoretical number of trays determined by calculation. To translate ideal plates to actual plates, the plate efficiency must be known. The following discussion applies to columns for gas absorption as well as to those for distillation.

Types of plate efficiency

Three kinds of plate efficiency are used: (1) overall efficiency, which concerns the entire column; (2) Murphree efficiency, which has to do with a single plate; and (3) local efficiency, which pertains to a specific location on a single plate.

The *overall efficiency* η_o is simple to use but is the least fundamental. It is defined as the ratio of the number of ideal plates needed in an entire column to the number of actual plates. For example, if six ideal plates are called for and the overall efficiency is 60 percent, the number of actual plates is $6/0.60 = 10$.

The *Murphree efficiency* η_M is defined by

$$\eta_M = \frac{y_n - y_{n+1}}{y_n^* - y_{n+1}} \tag{9.6-1}$$

where y_n = actual concentration of vapor leaving plate n

y_{n+1} = actual concentration of vapor entering plate n

y_n^* = concentration of vapor in equilibrium with liquid leaving downpipe from plate n

The Murphree efficiency is therefore the change in vapor composition from one plate to the next divided by the change that would have occurred if the vapor leaving were in equilibrium with the *liquid leaving*. The liquid leaving is generally not the same as the average liquid on the plate, and this distinction is important in comparing local and Murphree efficiencies.

The Murphree efficiency is defined using vapor concentrations as a matter of custom, but the measured efficiencies are rarely based on analysis of the vapor phase because of the difficulty in getting reliable samples. Instead, samples are taken of the liquid on the plates, and the vapor compositions are determined from a McCabe-Thiele diagram. A plate efficiency can be defined using liquid concentrations, but this is used only occasionally for desorption or stripping calculations.

Columns operated at high velocity will have significant entrainment, and this reduces the plate efficiency, because the drops of entrained liquid are less rich in the more volatile component than is the vapor. Although methods of allowing for entrainment have been published, most empirical correlations for the Murphree efficiency are based on liquid samples from the plates, and this includes the effect of entrainment.

The *local efficiency* η' is defined by

$$\eta' = \frac{y_n' - y_{n+1}'}{y_{en}' - y_{n+1}'} \tag{9.6-2}$$

where y_n' = concentration of vapor leaving specific location on plate n

y_{n+1}' = concentration of vapor entering plate n at same location

y_{en}' = concentration of vapor in equilibrium with liquid at same location

Since y_n' cannot be greater than y_{en}' a local efficiency cannot be greater than 1.00, or

100 percent.

Relation between Murphree and local efficiencies

In small columns, the liquid on a plate is sufficiently agitated by vapor flow through the perforations for there to be no measurable concentration gradients in the liquid as it flows across the plate. The concentration of the liquid in the downpipe x_n is that of the liquid on the entire plate. The change from concentration x_n to x_{n+1} occurs right at the exit from the downcomer, as liquid leaving the downcomer is vigorously mixed with the liquid on plate $n+1$. Since the concentration of the liquid on the plate is everywhere the same, so is that of the vapor from the plate and no gradients exist in the vapor streams. A comparison of the quantities in Eqs.(9.6-1) and (9.6-2) shows that $y_n = y'_n$, $y_{n+1} = y'_{n+1}$, and $y^*_n = y'_{en}$. Then $\eta_M = \eta'$ and the local and Murphree efficiencies are equal.

In larger columns, liquid mixing in the direction of flow is not complete, and a concentration gradient does exist in the liquid on the plate. The maximum possible variation is from a concentration of x_{n-1} at the liquid inlet to a concentration of x_n at the liquid outlet. To show the effect of such a concentration gradient, consider a portion of the McCabe-Thiele diagram, as shown in Fig. 9.35. This diagram corresponds to a Murphree efficiency of about 0.9, with y_n almost equal to y^*_n. However, if there is no horizontal mixing of the liquid, the vapor near the liquid inlet would contact liquid of composition x_{n-1} and be considerably richer than vapor contacting liquid of composition x_n near the exit. To be consistent with an average vapor composition y_n, the local vapor composition must range from y_a near the liquid exit to y_b near the liquid inlet. The local efficiency is therefore considerably lower than the Murphree efficiency, and η' would be about 0.6 for this example.

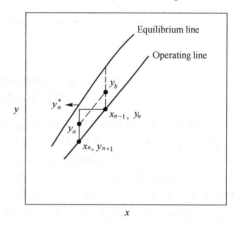

Fig. 9.35 Local and average vapor compositions for an unmixed plate.

When the local efficiency is high, say, 0.8 or 0.9, the presence of concentration gradients in the liquid sometimes gives an average vapor concentration greater than y^*_n, and the Murphree efficiency is then greater than 100 percent.

The relation between η_M and η' depends on the degree of liquid mixing and whether or not the vapor is mixed before going to the next plate.

Relation between Murphree and overall efficiencies

The overall efficiency of a column is not the same as the average Murphree efficiency of the individual plates. The relation between these efficiencies depends on the relative slopes of the equilibrium line and the operating line. When the equilibrium line is steeper than the operating line, which is typical for stripping columns, the overall efficiency is greater than the Murphree efficiency if η_M is less than 1.0. Consider a portion of the column where the liquid composition changes from x_{12} to x_{10}, a change requiring 1.0 ideal stage, as shown in Fig. 9.36(a). If two plates are actually required for this change, the overall efficiency for that portion of the column is 50 percent. However, if two partial steps are drawn assuming $\eta_M = 0.50$, as shown by the dashed line, then the predicted value of x_{10} is too high, since the first step would go halfway from x_{12} to x_{10} and the second step would be a much larger change.

The correct value of η_M is about 0.40, as shown in the steps drawn with a solid line.

When the equilibrium line is less steep than the operating line, as usually occurs near the top of the rectifying section, the overall efficiency is less than the Murphree efficiency, as illustrated in Fig. 9.36(b). For this case, two ideal plates are needed to go from x_1 to x_5, and if four actual plates are required, the overall efficiency is 0.50. By trial and error, a Murphree efficiency of 0.6 is found to give the correct value of x_5 after four partial steps.

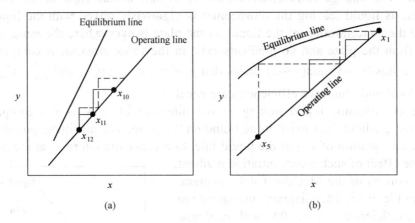

Fig. 9.36 Relationship between Murphree and overall efficiencies: (a) $\eta_M < \eta_o$; (b) $\eta_M > \eta_o$.

For columns with both a stripping and a rectifying section, the overall value of η_o may be fairly close to the average value of η_M, since the higher value of η_o in the stripping section, where $mV/L > 1$, tends to offset the lower value of η_o in the rectifying section, where $mV/L < 1$. For this reason, the difference between η_o and η_M is sometimes ignored in designing a column. However, in analyzing the performance of a real column or a section of the column, where the composition change over several plates is measured, the correct value of η_M should be determined by trial rather than just determining η_o and assuming $\eta_o = \eta_M$.

For the special case where the equilibrium and operating lines are straight, the following equation can be applied:

$$\eta_0 = \frac{\ln\left[1 + \eta_M\left(\frac{mV}{L} - 1\right)\right]}{\ln\frac{mV}{L}} \tag{9.6-3}$$

where m is the slope of the equilibrium line. Note that when $mV/L = 1.0$ or when $\eta_M \approx 1.0$, $\eta_M = \eta_o$.

PROBLEMS

9.1 A mixture of 100 mol containing 60 mol % n-pentane and 40 mol % n-heptane is vaporized at 101.32 kPa abs pressure until 40 mol of vapor and 60 mol of liquid in equilibrium with each other are produced. This occurs in a single-stage system, and the vapor and liquid are kept in contact with each other until vaporization is complete. The equilibrium data are given in Example 9.2. Calculate the composition of the vapor and the liquid.

9.2 A binary mixture which contains 60 mole percent more volatile component is subjected to flash distillation and differential distillation at a pressure of 1 atm, respectively. The mole fraction of the feed f that is vaporized is 1/3. what are the compositions of the overhead product and the bottoms product. Assume that the equilibrium relation is given by equation $y=0.46x+0.549$.

9.3 A mixture of 100 kmol which contains 60 mol % n-pentane (A) and 40 mol % n-heptane (B) is vaporized at 101.32 kPa pressure under differential conditions until 40 kg mol are distilled. Use equilibrium data from Example 9.2.

(a) What is the average composition of the total vapor distilled and the composition of the remaining liquid?

(b) If this same vaporization is done in an equilibrium or flash distillation and 40 kg mol are distilled, what is the composition of the vapor distilled and of the remaining liquid?

9.4 A rectification column is fed 100 kmol/h of a mixture of 50 mol % benzene and 50 mol % toluene at 101.32 kPa abs pressure. The feed is liquid at the boiling point. The distillate is to contain 90 mol % benzene and the bottoms 10 mol % benzene. The reflux ratio is 4.52:1. Calculate the kg mol/h distillate, kg mol/h bottoms, and the number of theoretical trays needed using the McCabe–Thiele method.

9.5 A binary mixture is fed to a continuous fractionating column at dew point, and the operating relations are given by equations: $y=0.723x+0.263$ (rectifying section); $y=1.25x - 0.0187$ (stripping section). Calculate: (a) the compositions of feed, distillate and bottom product, respectively; (b) the reflux ratio.

9.6 A liquid mixture containing 40% mole percent methanol and 60% mole percent water is to be separated in a continuous fractionating column at a pressure of 1atm. Calculate the value of q for the various conditions of feed: (a) the feeding is at 40°C; (b) the saturated liquid; (c) the saturated vapor.

9.7 A rectifying column containing the equivalent of three ideal plates is to be supplied continuously with a feed consisting of 0.4 mol % ammonia and 99.6 mol % water. Before entering the column, the feed is converted wholly to saturated vapor, and it enters between the second and third plates from the top of the column. The vapors from the top plate are totally condensed but not cooled. Per mole of feed, 1.35 mol of condensate is returned to the top plate as reflux, and the remainder of the distillate is removed as overhead product. The liquid from the bottom plate overflows to a reboiler, which is heated by closed steam coils. The vapor generated in the reboiler enters the column below the bottom plate, and bottom product is continuously removed from the reboiler. The vaporization in the reboiler is 0.7 mol per mole of feed. Over the concentration range involved in this problem, the equilibrium relation is given by the equation $y = 12.6x$. Calculate the mole fraction of ammonia in (a) the bottom product from the reboiler, (b) the overhead product, (c) the liquid reflux leaving the feed plate.

9.8 For the rectification given in Problem 9.4, where an equimolar liquid feed of benzene and toluene is being distilled to give a distillate of composition $x_D = 0.90$ and a bottoms of composition $x_B = 0.10$, calculate the following using graphical methods: (a) Minimum reflux ratio R_m, (b) Minimum number of theoretical plates at total reflux.

9.9 A saturated liquid feed of 200 mol/h at the boiling point containing 42 mol % heptane and 58% ethyl benzene is to be fractionated at 101.32 kPa abs to give a distillate containing 97 mol % heptane and a bottoms containing 1.1 mol % heptane. The reflux ratio used is 2.5:1. Calculate the mol/h distillate, mol/h bottoms, theoretical number of trays, and the feed tray number. Equilibrium data are given below at 101.32 kPa abs pressure for the mole fraction n-heptane x and y.

Temperature, °C	136.1	129.4	119.4	110.6	102.8	98.3
Concentration of heptane in liquid x	0.000	0.080	0.250	0.485	0.790	1.000
Concentration of heptane in vapor y	0.000	0.230	0.514	0.730	0.904	1.000

9.10 A total feed of 200 mol/h having an overall composition of 42 mol % heptane and 58 mol % ethyl benzene is to be fractionated at 101.3 kPa pressure to give a distillate containing 97 mol % heptane and a bottoms containing 1.1 mol % heptane. The feed enters the tower partially vaporized so that 40 mol % is liquid and 60 mol % vapor. Equilibrium data are given in Problem 9.9. Calculate the following: (*a*) Moles per hour distillate and bottoms; (*b*) Minimum reflux ratio R_m; (*c*) Minimum steps and theoretical trays at total reflux;

Theoretical number of trays required for an operating reflux ratio of 2.5:1. Compare with the results of Problem 9.9, which uses a saturated liquid feed.

9.11 An enriching tower is fed 100 kmol/h of a saturated vapor feed containing 40 mol % benzene (*A*) and 60 mol % toluene (*B*) at 101.32 kPa abs. The distillate is to contain 90 mol % benzene. The reflux ratio is set at 4.0:1. Calculate the kmol/h distillate *D* and bottoms *B* and their compositions. Also, calculate the number of theoretical plates required.

9.12 A liquid mixture containing 10 mol % *n*-heptane and 90 mol % *n*-octane is fed at its boiling point to the top of a stripping tower at 101.32 kPa abs. The bottoms are to contain 98 mol % *n*-octane. For every 3 mol of feed, 2 mol of vapor is withdrawn as product. Calculate the composition of the vapor and the number of theoretical plates required. The equilibrium data below are given as mole fraction *n*-heptane.

n-heptane *x*	0.284	0.097	0.067	0.039	0.012
n-heptane *y*	0.459	0.184	0.131	0.078	0.025

9.13 An equimolal mixture of *A* and *B* with a relative volatility of 2.3 is to be separated into a distillate product with 98.5 percent *A*, a bottoms product with 2 percent *A*, and an intermediate liquid product that is 80 percent *A* and has 40 percent of the *A* fed.

(*a*) Derive the equation for the operating line in the middle section of the column, and sketch the three operating lines on a McCabe-Thiele diagram.

(*b*) Calculate the amounts of each product per 100 moles of feed, and determine the minimum reflux rate if the feed is liquid at the boiling point.

(*c*) How much greater is the minimum reflux rate because of the withdrawal of the side-stream product?

9.14 A plant has two streams containing benzene and toluene, one with 37 percent benzene and one with 68 percent benzene. About equal amounts of the two streams are available, and a distillation tower with two feed points is proposed to produce 98 percent benzene and 99 percent toluene in the most efficient manner. However, combining the two streams and feeding at one point would be a simpler operation. For the same reflux rate, calculate the number of ideal stages required for the two cases.

9.15 It is desired to produce an overhead product containing 80 mol % benzene from a feed mixture of 68 mol % benzene and 32 mol % toluene. The following methods are considered for this operation. All are to be conducted at atmospheric pressure. For each method calculate the moles of product per 100 moles of feed and the number of moles vaporized per 100 mol feed. (*a*) Continuous flash distillation. (*b*) Continuous distillation in a still fitted with a partial condenser, in which 55 mol % of the entering vapors is condensed and returned to the still. The partial condenser is so constructed that vapor and liquid leaving it are in equilibrium and holdup in it is negligible.

9.16 A plant must distill a mixture containing 75 mol % methanol and 25 percent water. The overhead product is to contain 99.99 mol % methanol and the bottom product 0.002 mol %. The feed is cold, and for each mole of feed 0.15 mol of vapor is condensed at the feed plate. The reflux ratio at the top of the column is 1.4, and the reflux is at its bubble point. Calculate (*a*) the minimum number of plates; (*b*) the minimum reflux ratio; (*c*) the number of plates using a total condenser and a reboiler, assuming an average Murphree plate efficiency of 72 percent; (*d*) the number of plates using a reboiler and a partial condenser operating with the reflux in equilibrium with the vapor going to a final condenser. Equilibrium data are given below.

x	0.1	0.2	0.3	0.4	0.5	0.6	0.7	0.8	0.9	1.0
y	0.417	0.579	0.669	0.729	0.780	0.825	0.871	0.915	0.959	1.0

9.17 A liquid mixture of benzene and toluene is continuously fed to a plate column. Under the total reflux ratio condition, the compositions of liquid on three adjacent plates are 0.28, 0.41, and 0.57, respectively. Calculate the Murphree plate efficiency of the two lower plates. The equilibrium data for benzene-toluene are given as:

x	0.26	0.38	0.51
y	0.45	0.60	0.72

9.18 A two-component ideal solution contains 20 mol% more volatile component is to be separated in a stripping column. The mixture is added to the top of the column at the boiling point, as shown in figure. The stripping column is composed of a kettle and only one actual plate, and the kettle is considered as an ideal plate. The recovery of more volatile component in the overhead product is 80%. The composition of overhead product is 0.28, and the average relative volatility of the binary mixture is 2.5. Calculate the composition of bottom product and the Murphree plate efficiency of the actual plate, respectively.

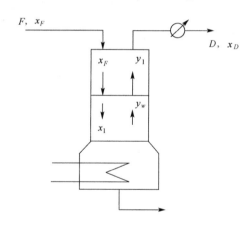

名 词 术 语

英文 / 中文
azeotrope / 恒沸物
azeotropic distillation / 恒沸精馏
boiling-point diagram / 沸点图
bottom product / 塔底产品
condensate / 冷凝液
condenser / 冷凝器
constant molal overflow / 恒摩尔流
continuous distillation / 连续蒸馏
differential distillation / 微分蒸馏
direct steam injection / 直接蒸汽加热
distillation / 蒸馏
enthalpy balance / 热衡算

英文 / 中文
entrainer / 夹带剂
extractive distillation / 萃取精馏
feed line / 进料线
feed plate location / 进料板位置
feed plate / 进料板
flash distillation / 闪蒸
graphical procedure / 图解法
heated kettle / 加热釜
high-boiling component / 高沸点组分
ideal plate / 理论板
less volatile component / 难挥发组分
liquid side-draw / 液相侧线抽料

英文 / 中文
local efficiency / 点效率
minimum reflux ratio / 最少回流比
more volatile component / 易挥发组分
multicomponent distillation / 多组分精馏
multiple feed column / 多点进料
Murphree efficiency / 默佛理效率
optimum reflux ratio / 最佳回流比
overall efficiency / 总板效率
overhead product / 塔顶产品
partial condensation / 部分冷凝
partial condenser / 分凝器
partial vaporization / 部分汽化
pinch point / 夹点
plate efficiency / 板效率
plate-to-plate calculation / 逐板计算法
Rayleigh equation / 雷利方程
reboiler / 再沸器
rectifying line / 精馏段操作线

英文 / 中文
rectifying section / 精馏段
reflux / 回流
reflux ratio / 回流比
reflux splitter / 回流分配器
separator / 分离器
sidestreams / 侧线
simple batch distillation / 简单间歇蒸馏
simple steam distillation / 简单水蒸气蒸馏
still / 釜
stripping line / 提馏段操作线
stripping section / 提馏段
stripping-column distillation / 提馏塔
subcooling / 过冷
superheated vapor / 过热蒸汽
thermal condition of the feed / 进料状态
total condenser / 全凝器
vacuum distillation / 真空蒸馏

Chapter 10

Leaching and Extraction

This chapter discusses the methods of removing one constituent from a solid or liquid by means of a liquid solvent. These techniques fall into two categories. The first, called *leaching* or *solid extraction,* is used to dissolve soluble matter from its mixture with an insoluble solid. The second, called *liquid extraction*, is used to recover a valuable product from a multicomponent solution by contact with an immiscible solvent that has a high affinity for the product. Liquid extraction can also be used to separate close-boiling liquids that would be difficult to separate by distillation.

10.1   Leaching

Leaching differs very little from the washing of filtered solids, as discussed in Chap.3, and leaching equipment strongly resembles the washing section of various filters. In leaching, the amount of soluble material removed is often greater than in ordinary filtration washing, and the properties of the solids may change considerably during the leaching operation. Coarse, hard, or granular feed solids may disintegrate into pulp or mush when their content of soluble material is removed.

10.1.1   Leaching Equipment

When the solids form an open, permeable mass throughout the leaching operation, solvent may be percolated through an unagitated bed of solids. With impermeable solids or materials that disintegrate during leaching, the solids are dispersed into the solvent and are later separated from it. Both methods may be either batch or continuous.

Leaching by percolation through stationary solid beds

Stationary solid-bed leaching is done in a tank with a perforated false bottom to support the solids and permit drainage of the solvent. Solids are loaded into the tank, sprayed with solvent until their solute content is reduced to the economical minimum, and excavated. In some cases the rate of solution is so rapid that one passage of solvent through the material is sufficient, but countercurrent flow of solvent through a battery of tanks is more common. In this method, fresh solvent is fed to the tank containing the solid that is most nearly extracted; it flows through the several tanks in series and is finally withdrawn from the tank that has been freshly charged. Such a series of tanks is called an *extraction battery*. The solid in any one tank is stationary until it is completely extracted. The piping is arranged so that fresh solvent can be introduced to any tank and strong solution withdrawn from any tank, making it possible to charge and discharge one tank at a time. The other tanks in the battery are kept in countercurrent operation by advancing the inlet and drawoff tanks one at a time as the material is charged and removed. Such a process is sometimes called a *Shanks process*.

In some solid-bed leaching the solvent is volatile, necessitating the use of closed vessels operated under pressure. Pressure is also needed to force solvent through beds of some less permeable solids. A series of such pressure tanks operated with countercurrent solvent flow is known as a *diffusion battery*.

Moving-bed leaching

In the machines illustrated in Fig. 10.1 the solids are moved through the solvent with little or no agitation. The Bollman extractor (Fig. 10.1(*a*)) contains a bucket elevator in a closed casing. There are perforations in the bottom of each bucket. At the top right-hand corner of the machine, as shown in the drawing, the buckets are loaded with flaky solids such as soybeans and are sprayed with appropriate amounts of *half miscella* as they travel downward. Half miscella is the intermediate solvent containing some extracted oil and some small solid particles. As solids and solvent flow concurrently down the right-hand side of the machine, the solvent extracts more oil from the beans. Simultaneously the fine solids are filtered out of the solvent, so that clean *full miscella* can be pumped from the right-hand sump at the bottom of the casing. As the partially extracted beans rise through the left side of the machine, a stream of pure solvent percolates countercurrently through them. It collects in the left-hand sump and is pumped to the half-miscella storage tank. Fully extracted beans are dumped from the buckets at the top of the elevator into a hopper from which they are removed by paddle conveyors. The capacity of typical units is 50 to 500 tons of beans per 24-h day.

In the Rotocel extractor, as illustrated in Fig. 10.1(*b*), a horizontal basket is divided into walled compartments with a floor that is permeable to the liquid. The basket rotates slowly about a vertical axis. Solids are admitted to each compartment at the feed point; the compartments then

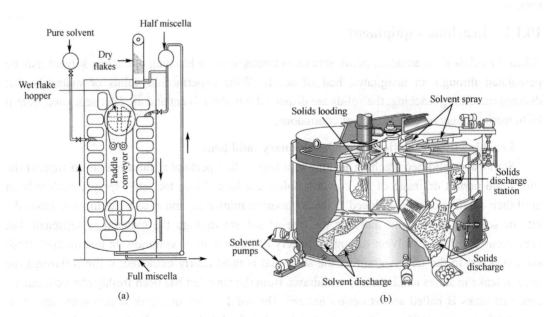

Fig. 10.1 Moving-bed leaching equipment: (*a*) Bollman extractor; (*b*) Rotocel extractor.

successively pass a number of solvent sprays, a drainage section, and a discharge point at which the floor of the compartment opens to discharge the extracted solids. The empty compartment moves to the feed point to receive its next load of solids. To give countercurrent extraction, fresh solvent is fed only to the last compartment before the discharge point, and the solids in each preceding compartment are washed with the effluent from the succeeding one.

Dispersed-solid leaching

Solids that form impermeable beds, either before or during leaching, are treated by dispersing them in the solvent by mechanical agitation in a tank or flow mixer. The leached residue is then separated from the strong solution by settling or filtration.

Small quantities can be leached batchwise in this way in an agitated vessel with a bottom drawoff for settled residue. Continuous countercurrent leaching is obtained with several gravity thickeners connected in series, as shown in Fig. 10.2, or when the contact in a thickener is inadequate by placing an agitator tank in the equipment train between each pair of thickeners. A still further refinement, used when the solids are too fine to settle out by gravity, is to separate the residue from the miscella in continuous solid-bowl helical-conveyor centrifuges. Many other leaching devices have been developed for special purposes, such as the solvent extraction of various oilseeds, with their specific design details governed by the properties of the solvent and of the solid to be leached. The dissolved material, or solute, is often recovered by crystallization or evaporation.

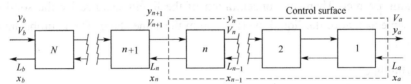

Fig. 10.2 Countercurrent leaching cascade.

10.1.2 Principles of Continuous Countercurrent Leaching

The most important method of leaching is the continuous countercurrent method using stages. Even in an extraction battery, where the solid is not moved physically from stage to stage, the charge in any one cell is treated by a succession of liquids of constantly decreasing concentration as if it were being moved from stage to stage in a countercurrent system.

Because of its importance, only the continuous countercurrent method is discussed here. Also, since the stage method is normally used, the differential-contact method is not considered. In common with other stage cascade operations, leaching may be considered, first, from the standpoint of ideal stages and, second, from that of stage efficiencies.

Ideal stages in countercurrent leaching

Figure 10.2 shows a material-balance diagram for a continuous countercurrent cascade. The stages are numbered in the direction of flow of the solid. The V phase is the liquid that overflows from stage to stage in a direction counter to that of the flow of the solid, dissolving solute as it moves from stage N to stage 1. The L phase is the liquid carried with the solid as it moves from stage 1 to stage N. Exhausted solids leave stage N, and concentrated solution overflows from stage 1.

It is assumed that the solute-free solid is insoluble in the solvent and that the flow rate of this solid is constant throughout the cascade. The solid is porous and inert (no adsorption) and carries with it an amount of solution that may or may not be constant. Let L refer to the flow of this retained liquid and V to the flow of the overflow solution. The flows V and L may be expressed in mass per unit time or may be based on a definite flow of dry solute-free solid. Also, in accordance with standard nomenclature, the terminal concentrations are as follows:

Solution retained by entering solid x_a
Solution retained by leaving solid x_b
Fresh solvent entering the system y_b
Concentrated solution leaving the system y_a

As in absorption and distillation, the quantitative performance of a countercurrent system can be analyzed by utilizing an equilibrium line and an operating line, and as before, the method to be used depends on whether these lines are straight or curved.

Equilibrium

In leaching, provided sufficient solvent is present to dissolve all the solute in the entering solid and there is no adsorption of solute by the solid, equilibrium is attained when the solute is completely dissolved and the concentration of the solution so formed is uniform. Such a condition may be obtained simply or with difficulty, depending on the structure of the solid. These factors are considered when stage efficiency is discussed. At present, it is assumed that the requirements for equilibrium are met. Then the concentration of the liquid retained by the solid leaving any stage is the same as that of the liquid overflow from the same stage. The equilibrium relationship is simply $x_e = y$.

Operating line

The equation for the operating line is obtained by writing material balances for that portion of the cascade consisting of the first n units, as shown by the control surface indicated by the dashed lines in Fig. 10.2. These balances are

Total solution:
$$V_{n+1} + L_a = V_a + L_n \tag{10.1-1}$$

Solute:
$$V_{n+1} y_{n+1} + L_a x_a = V_a y_a + L_n x_n \tag{10.1-2}$$

Solving for y_{n+1} gives the operating-line equation, which is the same as that derived earlier for the general case of an equilibrium-stage cascade [Eq. (7.2-27)]:

$$y_{n+1} = \left(\frac{L_n}{V_{n+1}}\right) x_n + \frac{V_a y_a - L_a x_a}{V_{n+1}} \tag{10.1-3}$$

As usual, the operating line passes through the points (x_a, y_a) and (x_b, y_b), and if the flow rates are constant, the slope is L/V.

Constant and variable underflow

Two cases are to be considered. If the density and viscosity of the solution change considerably with solute concentration, the solids from the lower-numbered stages may retain more liquid than those from the higher-numbered stages. Then, as shown by Eq. (10.1-3), the slope of the operating line varies from unit to unit. If, however, the mass of solution retained by the solid is independent of concentration, then L_n is constant and the operating line is straight. This condition is called *constant solution underflow*. If the underflow is constant, so is the overflow. Constant underflow and variable underflow are given separate consideration.

Number of ideal stages for constant underflow

When the operating line is straight, a McCabe-Thiele construction can be used to determine the number of ideal stages; but since in leaching the equilibrium line is always straight, Eq. (7.2-24) can be used directly for constant underflow. The use of this equation is especially simple here because $y_a^* = x_a$ and $y_b^* = x_b$.

Equation (7.2-24) cannot be used for the entire cascade if L_a, the solution entering with the unextracted solids, differs from L, the underflows within the system. Equations have been derived for this situation, but it is easy to calculate, by material balances, the performance of the first stage separately and then to apply Eq. (7.2-24) to the remaining stages.

Number of ideal stages for variable underflow

When the underflow and overflow vary from stage to stage, a modification of the McCabe-Thiele graphical method may be used for calculations. The terminal points on the operating line are determined using material balances. Assuming the amount of underflow L is known as a function of underflow composition, an intermediate value of x_n is chosen to fix L_n, and V_{n+1} is calculated from Eq. (10.1-1). The composition of the overflow y_{n+1} is then calculated from Eq. (10.1-2), and the point (x_n, y_{n+1}) is plotted along with the terminal compositions to give the curved operating line. Unless there is a large change in L and V or the operating line is very close to the equilibrium line, only one intermediate point need be calculated.

EXAMPLE 10.1 Oil is to be extracted from meal by means of benzene using a continuous countercurrent extractor. The unit is to treat 1,000 kg of meal (based on completely exhausted solid) per hour. The untreated meal contains 400 kg of oil and is contaminated with 25 kg of benzene. The fresh solvent mixture contains 10 kg of oil and 655 kg of benzene. The exhausted solids are to contain 60 kg of unextracted oil. Experiments carried out under conditions identical with those of the projected battery show that the solution retained depends on the concentration of the solution, as shown in Table 10.1. Find (*a*) the concentration of the strong solution, or extract; (*b*) the concentration of the solution adhering to the extracted solids; (*c*) the mass of solution leaving with the extracted meal; (*d*) the mass of extract; (*e*) the number of stages required. All quantities are given on an hourly basis.

Table 10.1 Data for Example 10.1

Concentration, kg oil /kg solution	Solution retained, kg/kg solid	Concentration, kg oil /kg solution	Solution retained, kg/kg solid
0.0	0.500	0.4	0.550
0.1	0.505	0.5	0.571
0.2	0.515	0.6	0.595
0.3	0.530	0.7	0.620

Solution. Let x and y be the mass fractions of oil in the underflow and overflow solutions.
  At the solvent inlet,
$$V_b = 10 + 655 = 665 \text{ kg solution/h}$$

$$y_b = \frac{10}{665} = 0.015$$

Determine the amount and composition of the solution in the spent solids by trial.

If $x_b = 0.1$, the solution retained, from Table 10.1, is 0.505 kg/kg. Then
$$L_b = 0.505 \times 1000 = 505 \text{ kg/h}$$
$$x_b = \frac{60}{505} = 0.119$$

From Table 10.1, the solution retained is 0.507 kg/kg:
$$L_b = 0.507(1000) = 507 \text{ kg/h}$$
$$x_b = \frac{60}{507} = 0.118 \text{ (close enough)}$$

Benzene in the underflow at L_b is 507−60 = 447 kg/h.

At the solid inlet,
$$L_a = 400 + 25 = 425 \text{ kg solution/h}$$
$$x_a = \frac{400}{425} = 0.941$$

Oil in extract = oil in −60 = 10 + 400 − 60 = 350 kg/h.
Benzene in extract = 655 +25 − 447 = 233 kg/h.
$$L_a = 350 + 233 = 583 \text{ kg/h}$$
$$y_a = \frac{350}{583} = 0.600$$

The answers to parts (*a*) to (*d*) are

(*a*) $y_a = 0.60$ (*b*) $x_b = 0.118$ (*c*) $L_b = 507$ kg/h (*d*) $V_a = 583$ kg/h

(*e*) Determine the inlet and exit concentrations for the first stage and locate the operating line for the remaining stages. Since $x_1 = y_a = 0.60$, solution retained is 0.595 kg/kg solid.
$$L_1 = 0.595 \times 1000 = 595$$

Overall material balance:
$$V_2 = L_1 + V_a - V_2 = 595 + 583 - 425 = 753 \text{kg/h}$$

Oil balance:
$$L_a x_a + V_2 y_2 = L_1 x_1 + V_a y_a$$
$$V_2 y_2 = 595 \times 0.60 + 583 \times 0.60 - 425 \times 0.941 = 307$$
$$y_2 = \frac{307}{753} = 0.408$$

The point $x_1 = 0.60$, $y_2 = 0.408$ is at one end of the operating line for the remaining stages. To determine an intermediate point on the operating line, choose $x_n = 0.30$.
$$L_n = \text{solution retained} = 0.53 \times 1000 = 530 \text{ kg/h}$$

By an overall balance,
$$V_{n+1} = 530 + 583 - 425 = 688 \text{ kg/h}$$

An oil balance gives
$$V_{n+1} y_{n+1} = L_n x_n + V_a y_a - L_a x_a$$
$$= 530 \times 0.30 + 583 \times 0.60 - 425 \times 0.941 = 108.8$$
$$y_{n+1} = \frac{108.8}{688} = 0.158$$

The points x_n, y_{n+1}, x_a, y_a, and x_b, y_b define a slightly curved operating line, as shown in

Fig. 10.3. Four ideal stages are required.

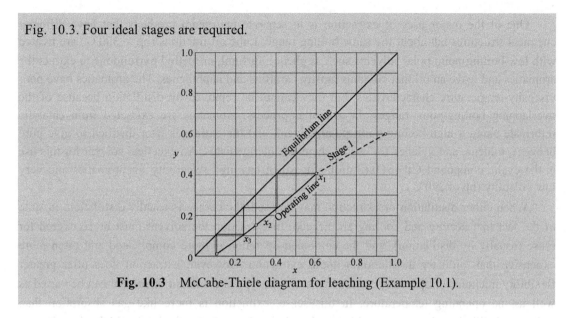

Fig. 10.3 McCabe-Thiele diagram for leaching (Example 10.1).

Saturated concentrated solution

A special case of leaching is encountered when the solute is of limited solubility and the concentrated solution reaches saturation. This situation can be treated by the above methods. The solvent input to stage N should be the maximum that is consistent with a saturated overflow from stage 1, and all liquids except that adhering to the underflow from stage 1 should be unsaturated. If too little solvent is used and saturation is attained in stages other than the first, all but one of the "saturated" stages are ineffective, and the solute concentration in the underflow from stage N is higher than it would be if more fresh solvent were used.

Stage efficiencies

In most leaching operations the solute is distributed through a more or less permeable solid. The rate of leaching is largely governed by the rate of diffusion through the solid, as discussed in Chap. 6, and the actual number of stages required may be considerably greater than the number of ideal stages. In washing impermeable solids, where the solute is confined to a film of strong solution on the solid surfaces, the approach to equilibrium is rapid, and the stage efficiency may be taken as unity.

10.2 Liquid Extraction

When separation by distillation is ineffective or very difficult, liquid extraction is one of the main alternatives to consider. Close-boiling mixtures or substances that cannot withstand the temperature of distillation, even under a vacuum, may often be separated from impurities by extraction, which utilizes chemical differences instead of vapor pressure differences. For example, penicillin is recovered from the fermentation broth by extraction with a solvent such as butyl acetate, after lowering the pH to get a favorable partition coefficient. The solvent is then treated with a buffered phosphate solution to extract the penicillin from the solvent and give a purified aqueous solution, from which penicillin is eventually produced by drying. Extraction is also used to recover acetic acid from dilute aqueous solutions; distillation would be possible in this case, but the extraction step considerably reduces the amount of water to be distilled.

One of the major uses of extraction is to separate petroleum products that have different chemical structures but about the same boiling range. Lube oil fractions (bp > 300℃) are treated with low-boiling-point polar solvents such as phenol, furfural, or methyl pyrrolidone to extract the aromatics and leave an oil that contains mostly paraffins and naphthenes. The aromatics have poor viscosity-temperature characteristics, but they cannot be removed by distillation because of the overlapping boiling-point ranges. In a similar process, aromatics are extracted from catalytic reformate using a high-boiling-point polar solvent, and the extract is later distilled to give pure benzene, toluene, and xylenes for use as chemical intermediates. An excellent solvent for this use is the cyclic compound $C_4H_8SO_2$ (Sulfolane), which has high selectivity for aromatics and very low volatility (bp of 290℃).

When either distillation or extraction may be used, the choice is usually distillation, in spite of the fact that heating and cooling are needed. In extraction the solvent must be recovered for reuse (usually by distillation), and the combined operation is more complicated and often more expensive than ordinary distillation without extraction. However, extraction does offer greater flexibility in choice of operating conditions, since the type and amount of solvent can be varied as well as the operating temperature. In this sense, extraction is more like gas absorption than ordinary distillation. In many problems, the choice between methods should be based on a comparative study of both extraction and distillation.

Extraction may be used to separate more than two components; and mixtures of solvents, instead of a single solvent, are needed in some applications. These more complicated methods are not treated in this text.

10.2.1 Extraction Equipment

In liquid-liquid extraction, as in gas absorption and distillation, two phases must be brought into good contact to permit transfer of material and then separated. In absorption and distillation, the mixing and separation are easy and rapid. In extraction, however, the two phases have comparable densities, so that the energy available for mixing and separation—if gravity flow is used—is small, much smaller than when one phase is a liquid and the other is a gas. The two phases are often hard to mix and harder to separate. The viscosities of both phases also are relatively high, and linear velocities through most extraction equipment are low. In some types of extractors, therefore, energy for mixing and separation is supplied mechanically.

Extraction equipment may be operated batchwise or continuously. A quantity of feed liquid may be mixed with a quantity of solvent in an agitated vessel, after which the layers are settled and separated. The extract is the layer of solvent plus extracted solute, and the raffinate is the layer from which solute has been removed. The extract may be lighter or heavier than the raffinate, and so the extract may be shown coming from the top of the equipment in some cases and from the bottom in others. The operation may, of course, be repeated if more than one contact is required; but when the quantities involved are large and several contacts are needed, continuous flow becomes economical. Most extraction equipment is continuous with either successive stage contacts or differential contacts. Representative types are mixer-settlers, vertical towers of various kinds that operate by gravity flow, agitated tower extractors, and centrifugal extractors. The characteristics of various types of extraction equipment are listed in Table 10.2. Liquid-liquid extraction can also be carried out using porous membranes. This method has promise for difficult separations.

Table 10.2 Performance of commercial extraction equipment

Type	Liquid capacity of combined streams ft³/ft²·h①	HTU, ft	Plate or stage efficiency, %	Spacing between plates or stages, in	Typical applications
Mixer-settler			75–100		Duo-Sol lube-oil process
Packed column	20–150	5–20			Phenol recovery
Perforated-plate column	10–200	1–20	6–24	30–70	Furfural lube-oil process
Baffle column	60–105	4–6	5–10	4–6	Acetic acid recovery
Agitated tower	50–100	1–2	80–100	12–24	Pharmaceuticals and organic chemicals

① ft² is the total cross-sectional area.

Mixer-settlers

For batchwise extraction the mixer and settler may be the same unit. A tank containing a turbine or propeller agitator is most common. At the end of the mixing cycle the agitator is shut off, the layers are allowed to separate by gravity, and extract and raffinate are drawn off to separate receivers through a bottom drain line carrying a sight glass. The mixing and settling times required for a given extraction can be determined only by experiment; 5 min for mixing and 10 min for settling are typical, but both shorter and much longer times are common.

For continuous flow the mixer and settler are usually separate pieces of equipment. The mixer may be a small agitated tank provided with inlets and a drawoff line and baffles to prevent short-circuiting, or it may be a motionless mixer or other flow mixer. The settler is often a simple continuous gravity decanter. With liquids that emulsify easily and have nearly the same density it may be necessary to pass the mixer discharge through a screen or pad of glass fiber to coalesce the droplets of the dispersed phase before gravity settling is feasible. For even more difficult separations, tubular or disk-type centrifuges are employed.

If, as is usual, several contact stages are required, a train of mixer-settlers is operated with countercurrent flow, as shown in Fig. 10.4(*a*). The raffinate from each settler becomes the feed to the next mixer, where it meets intermediate extract or fresh solvent. The principle is identical with that of the continuous countercurrent stage leaching system shown in Fig. 7.9. In Fig. 10.4(*b*) a combined mixer–settler is shown, which is sometimes used in the extraction of uranium salts or copper salts from aqueous solutions. Both types of mixer–settlers can be used in series for countercurrent or multiple-stage extraction. Typical stage efficiencies for a mixer–settler are 75% to 100%.

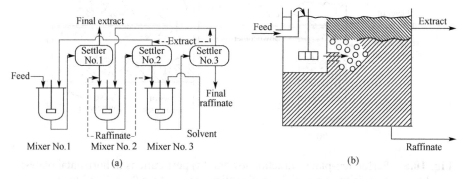

Fig. 10.4 Mixer-settler extraction system. (*a*) separate mixer–settler; (*b*) combined mixer–settler.

Packed extraction towers

Tower extractors give differential contacts, not stage contacts, and mixing and settling proceed continuously and simultaneously. Extraction can be carried out in an open tower, with drops of heavy liquid falling through the rising light liquid or vice versa; however, such towers, called spray towers, are rarely used because of pronounced axial mixing in the continuous phase. Instead, the tower is filled with packing such as rings or saddles (see chapter12), which causes the drops to coalesce and reform, and tends to limit axial dispersion.

In an extraction tower there is continuous transfer of material between phases, and the composition of each phase changes as it flows through the tower. At any given level, of course, equilibrium is not reached; indeed, it is the departure from equilibrium that provides the driving force for mass transfer. The design procedure for extraction towers is similar to that for packed absorption towers, but the height of a transfer unit is generally greater than for a typical absorber.

Flooding velocities in packed towers. If the flow rate of either the dispersed phase or the continuous phase is held constant and that of the other phase gradually increased, a point is reached where the dispersed phase coalesces, the holdup of that phase increases, and finally both phases leave together through the continuous-phase outlet. The effect, like the corresponding action in an absorption column, is called flooding. The larger the flow rate of one phase at flooding, the smaller is that of the other. A column obviously should be operated at flow rates below the flooding point. Flooding velocities in packed columns can be estimated from an empirical equation.

Perforated-plate towers

The axial mixing characteristic of an open tower can also be limited by using transverse perforated plate like those in the sieve-plate distillation towers described in Chaps.7 and 9. The perforations are typically $1\frac{1}{2}$ to $4\frac{1}{2}$ mm ($\frac{1}{16}$ to $\frac{3}{16}$) in diameter. Plate spacings range from 150 to 600 mm (6 to 24 in.). Usually the light liquid is the dispersed phase, and downcomers carry the heavy continuous phase from one plate to the next. As shown in Fig. 10.5(*a*), light liquid collects in a thin layer beneath each plate and jets into the thick layer of heavy liquid above. A modified design is shown in Fig. 10.5(*b*), in which the perforations are on one side of the plate only, alternating from left to right from one plate to the next. Nearly all the extraction takes place

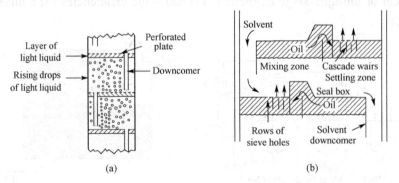

Fig. 10.5 Perforated-plate extraction towers: (*a*) perforations in horizontal plates; (*b*) cascade weir tray with mixing and settling zones. (*After Bushell* and *Fiocco*.)

in the mixing zone above the perforations, with the light liquid (oil) rising and collecting in a space below the next-higher plate, then flowing transversely over a weir to the next set of perforations. The continuous-phase heavy liquid (solvent) passes horizontally from the mixing zone to a settling zone in which any tiny drops of light liquid have a chance to separate and rise to the plate above. This design often greatly reduces the quantity of oil carried downward by the solvent and increases the effectiveness of the extractor.

Baffle towers

These extraction towers contain sets of horizontal baffle plates. Heavy liquid flows over the top of each baffle and cascades to the one beneath; light liquid flows under each baffle and sprays upward from the edge through the heavy phase. The most common arrangements are disk-and-doughnut baffles and segmental, or side-to-side, baffles. In both types the spacing between baffles is 100 to 150 mm (4 to 6 in.).

Baffle towers contain no small holes to clog or be enlarged by corrosion. They can handle dirty solutions containing suspended solids; one modification of the disk-and-doughnut towers even contains scrapers to remove deposited solids from the baffles. Because the flow of liquid is smooth and even, with no sharp changes in velocity or direction, baffle towers are valuable for liquids that emulsify easily. For the same reason, however, they are not effective mixers, and each baffle is equivalent to only 0.05 to 0.1 of an ideal stage.

Agitated tower extractors

Mixer-settlers supply mechanical energy for mixing the two liquid phases, but the tower extractors so far described do not. They depend on gravity flow both for mixing and for separation. In some tower extractors, however, mechanical energy is provided by internal turbines or other agitators, mounted on a central rotating shaft. In the rotating-disk contactor shown in Fig. 10.6(a), flat disks disperse the liquids and impel them outward toward the tower wall, where stator rings create quiet zones in which the two phases can separate. In other designs, sets of impellers are separated by calming sections to give, in effect, a stack of mixer-settlers one above the other. In the York-Scheibel extractor illustrated in Fig. 10.6(b), the regions surrounding the agitators are packed with wire mesh to encourage coalescence and separation of the phases. Most of the extraction takes place in the mixing sections, but some also occurs in the calming sections, so that the efficiency of each mixer-settler unit is sometimes greater than 100 percent. Typically each mixer-settler is 300 to 600 mm high, which means that several theoretical contacts can be provided in a reasonably short column. The problem of maintaining the internal moving parts, however, particularly where the liquids are corrosive, may be a serious disadvantage.

Pulse columns

Agitation may also be provided by external means, as in a pulse column. A reciprocating pump "pulses" the entire contents of the column at frequent intervals, so that a rapid reciprocating motion of relatively small amplitude is superimposed on the usual flow of the liquid phases. The tower may contain ordinary packing or special sieve plates. In a packed tower the pulsation disperses the liquids and eliminates channeling, and the contact between the phases is greatly improved. In sieve-plate pulse towers the holes are smaller than in nonpulsing towers, ranging from 1.5 to 3 mm in diameter, with a total open area in each plate of 6 to 23 percent of the cross-sectional area of the tower. Such towers are used almost entirely for processing highly corrosive radioactive liquids. No downcomers are used. Ideally the pulsation causes light liquid to be dispersed into the heavy phase on the upward stroke and the heavy phase to jet into the light phase on the downward stroke. Under these conditions the stage efficiency may reach 70 percent. This is possible, however, only when the volumes of the two phases are nearly the same and when there is almost no volume change during extraction. In the more usual case, the successive dispersions are less effective, and there is backmixing of one phase in one direction. The plate efficiency then drops to about 30 percent. Nevertheless, in both packed and sieve-plate pulse columns, the height required for a given number of theoretical contacts is often less than one-third that required in an unpulsed column.

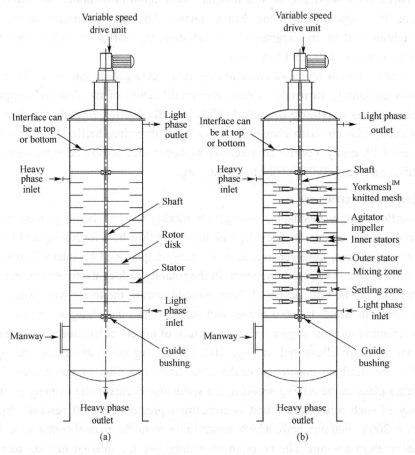

Fig. 10.6 Agitated extraction towers: (*a*) rotating-disk unit; (*b*) York-Scheibel extractor.

Centrifugal extractors

The dispersion and separation of the phases may be greatly accelerated by centrifugal force, and several commercial extractors make use of this. In the Podbielniak extractor a perforated spiral ribbon inside a heavy metal casing is wound about a hollow horizontal shaft through which the liquids enter and leave. Light liquid is pumped to the outside of the spiral at a pressure between 3 and 12 atm to overcome the centrifugal force; heavy liquid is fed to the center. The liquids flow countercurrently through the passage formed by the ribbons and the casing walls. Heavy liquid moves outward along the outer face of the spiral; light liquid is forced by displacement to flow inward along the inner face. The high shear at the liquid-liquid interface results in rapid mass transfer. In addition, some liquid sprays through the perforations in the ribbon and increases the turbulence. Up to 20 theoretical contacts may be obtained in a single machine, although 3 to 10 contacts are more common. Centrifugal extractors are expensive and find relatively limited use. They have the advantages of providing many theoretical contacts in a small space and of very short holdup times—about 4 s. Thus they are valuable in the extraction of sensitive products such as vitamins and antibiotics.

Auxiliary equipment

The dispersed phase in an extraction tower is allowed to coalesce at some point into a continuous layer from which one product stream is withdrawn. The interface between this layer and the predominant continuous phase is set in an open section at the top or bottom of a packed tower; in a sieve-plate tower it is set in an open section near the top of the tower when the light phase is dispersed. If the heavy phase is dispersed, the interface is kept near the bottom of the tower. The interface level may be automatically controlled by a vented overflow leg for the heavy phase, as in a continuous gravity decanter. In large columns the interface is often held at the desired point by a level controller actuating a valve in the heavy-liquid discharge line.

In liquid-liquid extraction the solvent must nearly always be removed from the extract or raffinate or both. Thus auxiliary stills, evaporators, heaters, and condensers form an essential part of most extraction systems and often cost much more than the extraction device itself. As mentioned at the beginning of this section, if a given separation can be done by either extraction or distillation, economic considerations usually favor distillation. Extraction provides a solution to problems that cannot be solved by distillation alone but does not usually eliminate the need for distillation or evaporation in some part of the separation system.

10.2.2 Principles of Extraction

Since most continuous extraction methods use countercurrent contacts between two phases, one a light liquid and the other a heavier one, many of the fundamentals of countercurrent gas absorption and of distillation carry over into the study of liquid extraction. Thus questions about ideal stages, stage efficiency, minimum ratio between the two streams, and size of equipment have the same importance in extraction as in distillation.

Triangular diagram

Phase rule. Generally in a liquid–liquid system we have three components, A, B, and S, and two phases in equilibrium. Substituting into the phase rule,

$$F = C - P + 2$$

Where P is the number of phases at equilibrium, C the number of total components in the two phases when no chemical reactions are occurring, and F is the number of variants or degrees of freedom of the system.

The number of degrees of freedom is 3. The variables are temperature, pressure, and four concentrations. Four concentrations occur because only two of the three mass-fraction concentrations in a phase can be specified. The third must make the total mass fractions equal to 1.0: $x_A + x_B + x_S = 1.0$. If pressure and temperature are set, which is the usual case, then, at equilibrium, setting one concentration in either phase fixes the system.

Material balance and lever-arm rule

This will be derived for use with the rectangular extraction-phase-diagram charts. In Fig. 10.7(a) two streams, R kg and E kg, containing components A, B, and S, are mixed (added) to give a resulting mixture stream M kg total mass. Writing an overall mass balance and a balance on A,

$$M = R + E \tag{10.2-1}$$

$$Mz_A = Rx_A + Ey_A \tag{10.2-2}$$

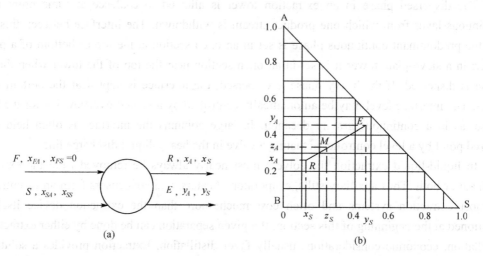

Fig. 10.7 Graphical addition and lever-arm rule: (a) process flow; (b) graphical addition.

where z_A is the mass fraction of A in the M stream. Writing a balance for component S,

$$Mz_S = Rx_S + Ey_S \tag{10.2-3}$$

Combining Eqs. (10.2-1), (10.2-2) and (10.2-3),

$$\frac{E}{R} = \frac{z_A - x_A}{y_A - z_A} = \frac{z_S - x_S}{y_S - z_S} \tag{10.2-4}$$

This shows that points E, M, and R must lie on a straight line. By using the properties of similar right triangles,

$$\frac{E}{R} = \frac{\overline{RM}}{\overline{EM}} \tag{10.2-5}$$

This is the *lever-arm rule*, which states that kg E divided by kg R is equal to the length of line $\overline{RM}$ divided by that of line $\overline{EM}$. Also,

$$\frac{E}{M} = \frac{\overline{RM}}{\overline{RE}} \tag{10.2-6}$$

The point M is the mixing point of point E and R; the point R is the difference between the points E and M; the point E is the difference between the points R and M. These same equations also hold for kmol and mol frac, lb$_m$, and so on.

Extraction of concentrated solutions; phase equilibria

In the extraction of concentrated solutions, the equilibrium relationships are more complicated than in other kinds of separation, because there are three or more components present and some of each component is present in each phase. The equilibrium data are often presented on a triangular diagram, such as those shown in Figs. 10.8 and 10.9.

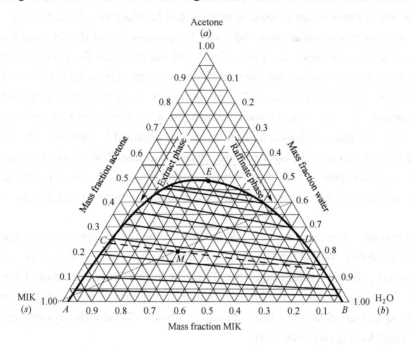

Fig. 10.8 System acetone–MIK–water at 25C. (*After Othmer, White, and Trueger.*)

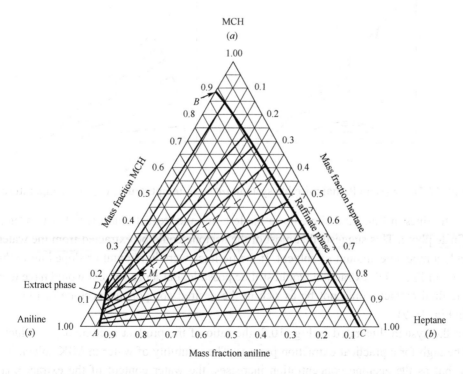

Fig. 10.9 System aniline–*n*-heptane–MCH at 25C: *a*, solute MCH; *b*, diluent, *n*-heptane; *s*, solvent, aniline.

Type I system. The system acetone–water–methyl isobutyl ketone (MIK), Fig. 10.8, is an example of a type I system, which shows partial miscibility of the solvent (MIK) and the diluent (water) but complete miscibility of the solvent and the component to be extracted (acetone).

Some of the features of an extraction process can be illustrated by using Fig. 10.8. When solvent is added to a mixture of acetone and water, the composition of the resulting mixture lies on a straight line between the point for pure solvent and the point for the original binary mixture. When enough solvent is added so that the overall composition falls under the dome-shaped curve, the mixture separates into two phases. The points representing the phase compositions can be joined by a straight tie line, which passes through the overall mixture composition. For clarity, only a few such tie lines are shown, and others can be obtained by interpolation. The line *ACE* shows compositions of the MIK layer (extract phase), and line *BDE* shows compositions of the water layer (raffinate phase). As the overall acetone content of the mixture increases, the compositions of the two phases approach each other, and they become equal at point *E*, the *plait point*.

Type II system. The solvent pairs B-S and A-S are partial miscible as shown in Fig. 10.10, and mutual solubility of which increase with increasing in temperature. For example, aniline–*n*-heptane– methyl cyclohexane (MCH) forms a type II system (Fig. 10.9), where the solvent (aniline) is only partially miscible with both the other components. Other examples are the system styrene (*A*)–ethylbenzene (*B*)–diethylene glycol (*S*) and the system chlorobenzene (*A*)–methyl ethyl ketone (*B*)–water (*S*).

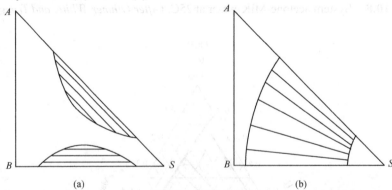

Fig. 10.10 System Partial miscible system. (*a*) high temperature; (*b*) low temperature.

The tie lines in Fig. 10.8 slope up to the left, and the extract phase is richer in acetone than the raffinate phase. This suggests that most of the acetone could be extracted from the water phase using only a moderate amount of solvent. If the tie lines were horizontal or sloped up to the right, as they are in Fig. 10.9, extraction would still be possible, but more solvent would have to be used, since the final extract would not be as rich in the desired component (acetone in Fig. 10.8 and MCH in Fig. 10.9).

For the system illustrated in Fig. 10.8, the ratio of the product acetone to the diluent water should be high for a practical extraction process. The solubility of water in MIK solvent is only 2 percent, but as the acetone concentration increases, the water content of the extract phase also increases. The data from Fig. 10.8 are replotted in Fig. 10.11 to show the gradual increase in water content y_{H_2O} with acetone content y_A. The ratio y_A/y_{H_2O} goes through a maximum at about

27 weight percent acetone in the extract phase. A higher concentration of acetone could be obtained, but the greater amount of water in the extract product would probably make operation at these conditions undesirable.

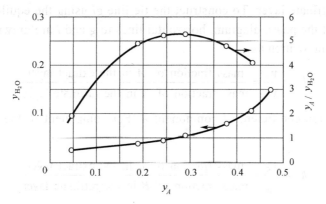

Fig. 10.11 Composition of extract phase for MIK-acetone-H$_2$O.

Equilibrium data on rectangular coordinates. Since triangular diagrams have some disadvantages due to the special coordinates, a more useful method of plotting the three-component data is to use rectangular coordinates. This is shown in Fig. 10.12 for the system acetic acid (A)–water (B)–isopropyl ether solvent (S). The solvent pair B and S are partially miscible. The concentration of component A is plotted on the vertical axis and that of S on the horizontal axis. The concentration of component B is obtained by difference from Eq. (10.2-7) or (10.2-8):

$$x_B = 1.0 - x_A - x_S \tag{10.2-7}$$

$$y_B = 1.0 - y_A - y_S \tag{10.2-8}$$

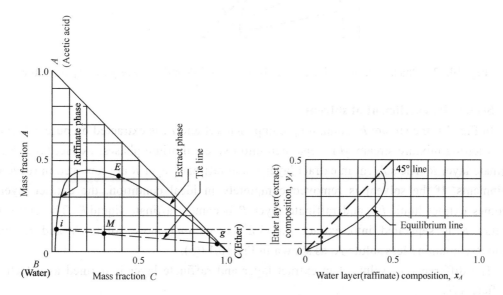

Fig. 10.12 Acetic acid (A)–water (B)–isopropyl ether (C) liquid–liquid phase diagram at 293 K (20 °C).

The two-phase region in Fig. 10.12 is inside the envelope and the one-phase region outside. A tie line *gi* is shown connecting the water-rich layer *i*, called the *raffinate layer*, and the ether-rich solvent layer *g*, called the *extract layer*. The raffinate composition is designated by *x* and the extract by *y*. Hence, the mass fraction of *S* is designated as y_S in the extract layer and as x_S in the raffinate layer. To construct the tie line *gi* using the equilibrium $y_A \sim x_A$ plot on the right side of the phase diagram, horizontal lines to *g* and *i* are drawn. The relationship of $y_A \sim x_A$ can also be written as:

$$k_A = \frac{y_A}{x_A} = \frac{\text{mass fraction of } A \text{ in the extract layer}}{\text{mass fraction of } A \text{ in the raffinate layer}} \qquad (10.2\text{-}9)$$

k_A is the distribution coefficient of component *A*. For component *B*, the similar expression is:

$$k_B = \frac{y_B}{x_B} = \frac{\text{mass fraction of } B \text{ in the extract layer}}{\text{mass fraction of } B \text{ in the raffinate layer}} \qquad (10.2\text{-}10)$$

On the both side of *plait point E*, the curve is *solubility curve*. In raffinate layer, the relationship of solubility curve is $x_S \sim x_A$, while in the extract layer, the relationship of solubility curve is $y_S \sim y_A$.

Another common type of phase diagram is shown in Fig. 10.13.

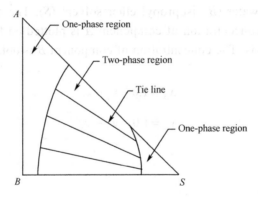

Fig. 10.13 Phase diagram where the solvent pairs *B-S* and *A-S* are partially miscible.

Selectivity coefficient of solvent

In Fig. 14, the stream *F*, containing components *A* and *B*, is extracted by pure solvent *S*. The resulting mixture stream *M* is separated into two equilibrium phases: extract layer *E* and raffinate layer *R*. The solvent in extract layer *E* and raffinate layer *R* are removed in recycling equipments. If the solvent is removed completely in ideal condition, the extract layer *E* becomes extract liquid E^0 and raffinate layer *R* becomes raffinate liquid R^0. Therefore, the mixture *F* is separated into extract liquid E^0 containing more solute A and raffinate liquid R^0 containing less solute A, as shown in Fig.10.14(*b*).

The difference of solute *A* in extract layer and raffinate layer is defined as selectivity coefficient β.

$$\beta = \frac{y_A/y_B}{x_A/x_B} = \frac{k_A}{k_B} \qquad (10.2\text{-}11)$$

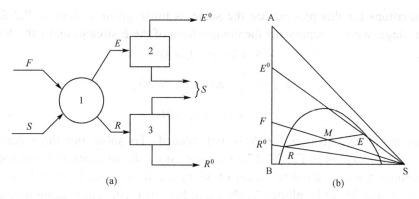

Fig. 10.14 Single-stage extraction process(1-extractor; 2, 3-solvent recovery equipment)

Because the ratio of mass fraction for component A and B in extract layer (y_A/y_B) is equal to that of in extract liquid (y_A^0/y_B^0), raffinate layer (x_A/x_B) is equal to that of in raffinate liquid (x_A^0/x_B^0). Thus

$$\beta = \frac{y_A/y_B}{x_A/x_B} = \frac{y_A^0/y_B^0}{x_A^0/x_B^0} \tag{10.2-12}$$

In extract liquid and raffinate liquid, $y_A^0 = 1 - y_B^0$, $x_A^0 = 1 - x_B^0$, the equation (10.2-12) leads to

$$y_A^0 = \frac{\beta x_A^0}{1 + (\beta - 1)x_A^0} \tag{10.2-13}$$

The selectivity coefficient β is similar to the relative volatility α in distillation, the value of β has relationship with the slope of equilibrium tie line. If tie line passes point S, the system can not be separated by extraction. If the value of β reaches infinite, the components B and S are immiscible. Only selectivity coefficient $\beta>1$, the extraction separation is possible.

Calculations for extraction processes

Single-stage equilibrium extraction. We now study the separation of A from a mixture of A and B by a solvent S in a single equilibrium stage. The process is shown in Fig. 10.15(*a*), where the solvent, as stream S, enters together with the stream F. The streams are mixed and equilibrated and the exit streams R and E leave in equilibrium with each other.

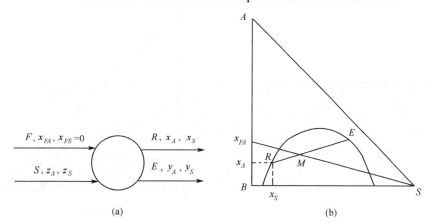

Fig. 10.15 Single-stage equilibrium liquid-liquid extraction:
(*a*) process flow diagram; (*b*) plot on phase diagram.

The equations for this process are the same as those given in Section 7.2 for a single equilibrium stage, where y represents the composition of the E streams and x the R streams:

$$F + S = R + E = M \quad (10.2\text{-}14)$$

$$Fx_{FA} + Sz_A = Rx_A + Ey_A = Mx_{AM} \quad (10.2\text{-}15)$$

$$Sz_S = Rx_S + Ey_S = Mx_{SM} \quad (10.2\text{-}16)$$

Since $x_A + x_B + x_S = 1$, an equation for B is not needed. To solve the three equations, the equilibrium-phase diagram in Fig. 10.15(b) is used. Since the amounts and compositions of F and S are known, we can calculate values of M, x_{AM}, and x_{SM} from Eqs.(10.2-14)- (10.2-16). The points F, S, and M can be plotted as shown in Fig. 10.15(b). Then, using trial and error, a tie line is drawn through point M, which locates the compositions of R and E. The amounts of R and E can be determined by substitution into Eqs.(10.2-14)-(10.2-16), or by using the *lever-arm rule*.

The phase compositions resulting from a single-stage extraction are easily obtained using the triangular diagram. For example, if a mixture with 40 percent acetone and 60 percent water is contacted with an equal mass of MIK solvent, the overall mixture M is obtained by the *lever-arm rule* in Fig. 10.8. A new tie line is drawn to show that the extract phase would be 0.232 acetone, 0.043 water, and 0.725 MIK. The raffinate phase would be 0.132 acetone, 0.845 water, and 0.023 MIK. Repeated contacting of the raffinate phase with fresh solvent, a process called crosscurrent extraction, would permit recovery of most of the acetone; but this would be less efficient than using a countercurrent cascade because of the large volume of solvent needed.

The ratio of acetone to water in the product is 5.4; that in the raffinate is 0.156. The corresponding numbers for point M in Fig. 10.9, the ratio of MCH to n-heptane, are 3.3 in the product and 1.6 in the raffinate. The separation is clearly more effective in the MIK-acetone-water system.

Continuous multistage crosscurrent extraction. The single-stage equilibrium contact was used to transfer solute A from one liquid to another liquid phase. To transfer more solute, single-stage contact can be repeated by bringing the exit R stream into contact with fresh solvent S, as shown in Fig. 10.16. In this way a greater percentage removal of solute A is obtained. However, this is wasteful of the solvent stream as well as giving a dilute product of A in the outlet solvent extract streams. In order to use less solvent and to obtain a more concentrated exit extract stream, countercurrent multistage contacting is often employed.

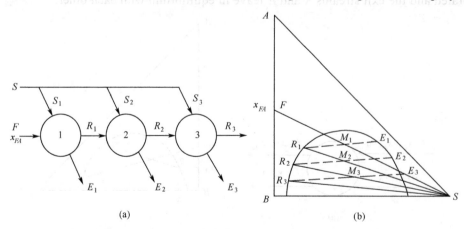

Fig. 10.16 Continuous multistage crosscurrent extraction:
(a) process flow diagram; (b) plot on phase diagram.

Continuous multistage countercurrent extraction and overall balance. The process flow for this extraction process is shown in Fig. 10.17. The feed stream containing the solute A to be extracted enters at one end of the process and the solvent stream enters at the other end. The extract and raffinate streams flow countercurrently from stage to stage, and the final products are the extract stream E_1 leaving stage 1 and the raffinate stream R_N leaving stage N.

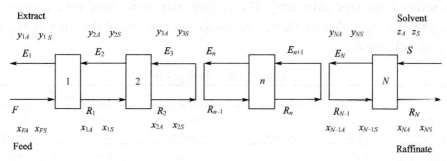

Fig. 10.17 Countercurrent-multistage-extraction-process flow diagram.

Making an overall balance on all N stages,

$$F + S = R_N + E_1 = M \tag{10.2-17}$$

where M represents total kg/h and is a constant, F the inlet feed flow rate in kg/h, S the inlet solvent flow rate in kg/h, E_1 the exit extract stream, and R_N the exit raffinate stream. Making an overall component balance on component S and component A,

$$Fx_{FS} + Sz_S = R_N x_{NS} + E_1 y_{1S} = Mx_{SM} \tag{10.2-18}$$

$$Fx_{FA} + Sz_A = R_N x_{NA} + E_1 y_{1A} = Mx_{AM}$$

Combining Eqs.(10.2-17) and (10.2-18), and rearranging,

$$x_{SM} = \frac{Fx_{FS} + Sz_S}{F + S} = \frac{R_N x_{NS} + E_1 y_{1S}}{R_N + E_1} \tag{10.2-19}$$

A similar balance on component A gives

$$x_{AM} = \frac{Fx_{FA} + Sz_A}{F + S} = \frac{R_N x_{NA} + E_1 y_{1A}}{R_N + E_1} \tag{10.2-20}$$

Eqs.(10.2-19) and (10.2-20) can be used to calculate the coordinates of point M on the phase diagram, which ties together the two entering streams F and S and the two exit streams E_1 and R_N. Usually, the flows and compositions of F and S are known and the desired exit composition x_{NA} is set. If we plot points F, S, and M as in Fig. 10.18, a straight line must connect these three points. Then R_N, M, and E_1 must lie on one line. Moreover, R_N and E_1 must also lie on the phase envelope, as shown. These balances also hold for lb_m and mass fraction, kg mol and mol fractions, and

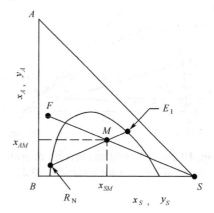

Fig.10.18 Use of the mixture point M for overall material balance in countercurrent solvent extraction.

so on.

EXAMPLE 10.2 Pure solvent isopropyl ether at the rate of $S = 600$ kg/h is being used to extract an aqueous solution of $F = 200$ kg/h containing 30 wt % acetic acid (A) by countercurrent multistage extraction. The desired exit acetic acid concentration in the aqueous phase is 4%. Calculate the compositions and amounts of the ether extract E_1 and the aqueous raffinate R_N.

Equilibrium relationship

Aqueous raffinate R			Ether extract E		
Acetic acid (A), %	Water (B), %	Isopropyl ether(S), %	Acetic acid (A), %	Water (B), %	Isopropyl ether(S), %
0.69	98.1	1.2	0.18	0.5	99.3
1.4	97.1	1.5	0.37	0.7	98.9
2.7	95.7	1.6	0.79	0.8	98.4
6.4	91.4	1.9	1.9	1.0	97.1
13.30	84.4	2.3	4.8	1.9	93.3
25.5	71.1	3.4	11.40	3.9	84.7
37.00	58.6	4.4	21.60	6.9	71.5
44.30	45.1	10.6	31.10	10.8	58.1
46.40	37.1	16.5	36.20	15.1	48.7

Solution. The given values are $S = 600$, $z_A = 0$, $z_S = 1.0$, $F = 200$, $x_{FA} = 0.30$, $x_{FB} = 0.70$, $x_{FS} = 0$, and $x_{NA} = 0.04$. In Fig. 10.19, S and F are plotted. Also, since R_N is on the phase boundary, it can be plotted at $x_{NA} = 0.04$. For the mixture point M, substituting into Eqs.(10.2-19) and (10.2-20),

$$x_{SM} = \frac{Fx_{FS} + Sz_S}{F+S} = \frac{200 \times 0 + 600 \times 1.0}{200+600} = 0.75$$

$$x_{AM} = \frac{Fx_{FA} + Sz_A}{F+S} = \frac{200 \times 0.30 + 600 \times 0}{200+600} = 0.075$$

Using these coordinates, the point M is plotted in Fig. 10.19. We locate E_1 by drawing a line from R_N through M and extending it until it intersects the phase boundary. This gives $y_{1A} = 0.08$ and $y_{1S} = 0.90$. For R_N a value of $x_{NS} = 0.017$ is obtained. By substituting into Eqs.(10.2-17) and (10.2-18) and solving, $R_N = 136$ kg/h and $E_1 = 664$ kg/h.

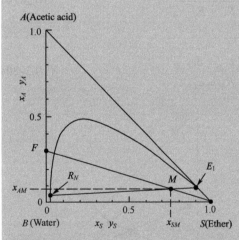

Fig. 10.19 Method to perform overall material balance for Example 10.2.

Use of McCabe-Thiele method

The separation achieved for a given number of ideal stages in a counterflow cascade can be determined by using a triangular diagram and special graphical techniques, but a modified McCabe-Thiele method, which is the approach used here, is simple to use and has satisfactory accuracy for most cases. The method focuses on the concentration of solute in the extract and raffinate phases, and the diagram does not show the concentration of the diluent in the extract or the concentration of solvent in the raffinate. However, these minor components of both phases are accounted for in determining the total flow of extract and raffinate, which affects the position of the operating line.

To apply the McCabe-Thiele method to extraction, the equilibrium data are shown on a rectangular graph, where the mass fraction of solute A in the extract or S phase is plotted as the ordinate and the mass fraction of solute A in the raffinate or B phase as the abscissa. For a type I system, the equilibrium line ends with equal compositions at the plait point. The use of only one concentration to characterize a ternary mixture may seem strange, but if the phases leaving a given stage are in equilibrium, only one concentration is needed to fix the compositions of both phases.

The operating line for the extraction diagram is based on Fig. 10.17 and Eq.(10.1-3), which gives the relationship between the solute concentration leaving stage n in the R phase and that coming from stage $n + 1$ in the E phase. The terminal points on the operating line (x_{FA}, y_{1A}) and (x_{NA}, z_A) are usually determined by an overall material balance, taking into account the ternary equilibrium data. Because of the decrease in the raffinate phase (R) and the increase in the extract phase (E) as they pass through the column, the operating line is curved. A material balance over a portion of the cascade is made to establish one or more intermediate points on the operating line. The number of ideal stages is then determined by drawing steps in the normal manner.

If the number of ideal stages is specified, the fraction of solute extracted and the final compositions are determined by trial and error. The fraction extracted or the final extract composition is assumed, and the curved operating line is constructed. If too many stages are required, a smaller fraction extracted is assumed and the calculations are repeated. Such calculations are generally done by computer.

EXAMPLE 10.3 A countercurrent extraction plant is used to extract acetone (A) from its mixture with water (B) by means of methyl isobutyl ketone (MIK) (S) at a temperature of 25 °C. The feed consists of 40 percent acetone and 60 percent water. Pure solvent equal in mass to the feed is used as the extracting liquid. How many ideal stages are required to extract 99 percent of the acetone fed? What is the extract composition after removal of the solvent?

Solution. Use the data in Fig. 10.8 to prepare a plot of the equilibrium relationship y_A versus x_A, which is the upper curve in Fig. 10.20. The terminal points for the operating line are determined by

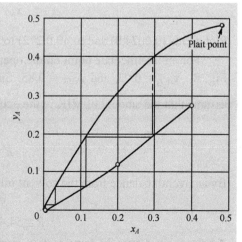

Fig. 10.20 McCabe-Thiele diagram for extraction (**Example 10.3**).

material balances with allowance for the amounts of water in the extract phase and MIK in the raffinate phase. Basis: $F = 100$ mass units per hour.

Let
$$n = \text{mass flow rate of } H_2O \ (B) \text{ in extract}$$
$$m = \text{mass flow rate of MIK } (S) \text{ in raffinate}$$

For 99 percent recovery of A, the extract has $0.99 \times 40 = 39.6A$, and the raffinate has $0.4A$.

The total flows are

At the top,
$$F = 100 = 40A + 60 B$$
$$E_1 = 39.6A + nB + (100 - m)S = 139.6 + n - m$$

At the bottom,
$$S = 100 \text{ MIK}$$
$$R_N = 0.4A + (60 - n)B + mS = 60.4 + m - n$$

Since n and m are small and tend to cancel in the summations for E_1 and F, the total extract flow E_1 is abou 140, which would make $y_A \cong 39.6/140 = 0.283$. The value of x_{NA} is about $0.4/60 = 0.0067$. These estimates are adjusted after calculating values of n and m.

From Fig. 10.8 for $y_A = 0.283$, $y_{H_2O} = 0.049$,

$$n = \frac{0.049}{1 - 0.049} \times (39.6 + 100 - m)$$

If m is very small, $n \cong (0.049/0.951) \times 139.6 = 7.2$.

From Fig. 10.8 for $x_A = 0.007$, $x_{MIK} = 0.02$,

$$m = \frac{0.02}{1 - 0.02} \times (0.4 + 60 - n)$$
$$\cong \frac{0.02}{0.98} \times (0.4 + 52.8) = 1.1$$

Revised $n = (0.049/0.951) \times (139.6 - 1.1) = 7.1$:
$$E_1 = 139.6 + 7.1 - 1.1 = 145.6$$
$$y_{1A} = \frac{39.6}{145.6} = 0.272$$
$$R_N = 60.4 + 1.1 - 7.1 = 54.4$$
$$x_{NA} = \frac{0.4}{54.4} = 0.0074$$

Plot points (0.0074,0) and (0.40,0.272) to establish the ends of the operating line.

For an intermediate point on the operating line, pick $y_A = 0.12$ and calculate E and R. From Fig. 8, $y_{H_2O} = 0.03$, and $y_{MIK} = 0.85$. Since the raffinate phase has only 2 to 3 percent MIK, assume that the amount of MIK in the extract is 100, the same as the solvent fed:

$$100 \cong E y_{MIK}$$
$$E \cong \frac{100}{0.85} = 117.6$$

By an overall balance from the solvent inlet (bottom) to the intermediate point,
$$S + R = R_N + E$$
$$R \cong 54.4 + 117.6 - 100 = 72.0$$

A balance on A over the same section gives x_A:
$$R x_A + S z_A = R_N x_{N_A} + E y_A$$

$$Rx_A \cong 0.4 + 117.6 \times 0.12 - 0$$
$$x_A \cong \frac{14.5}{72} = 0.201$$

This value is probably accurate enough, but corrected values of E, R, and x_A can be determined. For $x_A = 0.201$, $x_{MIK} \cong 0.03$ (Fig. 10.8). A balance on MIK from the solvent inlet to the intermediate point gives

$$S + Rx_{MIK} = R_N x_{MIK,b} + E y_{MIK}$$
$$E y_{MIK} = 100 + 72 \times 0.03 - 1.1$$

Revised
$$E = \frac{101.1}{0.85} = 118.9$$

Revised
$$R = 54.4 + 118.9 - 100 = 73.3$$

Revised
$$x_A = \frac{0.4 + 118.9 \times 0.12}{73.3} = 0.200$$

Plot $x_A = 0.20$, $y_A = 0.12$, which gives a slightly curved operating line. From Fig. 10.20, $N = 3.4$ stages.

Extraction of dilute solutions

If the solvent stream S contains components A and S and the feed stream F contains A and B, and if components B and S are relatively immiscible in each other, the stage calculations may be made more easily. The solute A is relatively dilute and is being transferred from F to S.

For batch or multistage extraction of dilute solutions, where changes in flow rate can be neglected and the distribution coefficient K is constant,

$$Y = KX \tag{10.2-21}$$

The distribution coefficient K is the ratio of the equilibrium concentration of the solute A in the extract Y (kg A /kg S) to that in the raffinate X (kg A /kg B).

Single-stage equilibrium extraction. For single stage extraction as showed in Fig. 10.21, the overall material balance for A is

$$S(Y - Z) = B(X_F - X) \tag{10.2-22}$$

Thus, the slope of operating line is $-\dfrac{B}{S} = \dfrac{Z - Y}{X_F - X}$ and the pass the points of H (X_F, Z).

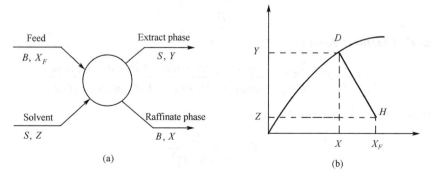

Fig. 10.21 Single-stage equilibrium extraction for immiscible system: (*a*) process flow diagram; (*b*) plot on *Y-X* diagram.

Continuous multistage crosscurrent extraction. It is convenient to use an extraction factor E which is equivalent to the stripping factor S defined by

$$S = \frac{1}{A} = \frac{mV}{L} \tag{7.2-29}$$

$$E \equiv \frac{1}{A} = \frac{KS}{B} \tag{10.2-23}$$

For a single-stage extraction with pure solvent the fraction of solute remaining is $1/(1+E)$ and the fraction recovered is $E/(1+E)$. While for multistage crosscurrent extraction in Fig. 10.22, the material balance for stage m is

$$B(X_{m-1} - X_m) = S_m(Y_m - Z) \tag{10.2-24}$$

And the operating line is $\dfrac{Y_m - Z}{X_m - X_{m-1}} = -\dfrac{B}{S_m}$.

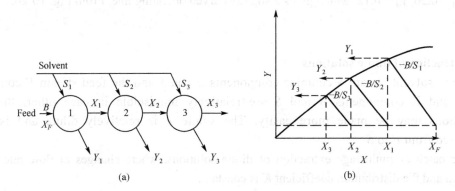

Fig. 10.22 Crosscurrent extraction for immiscible system:
(a) process flow diagram; (b) plot on Y-X diagram.

If pure solvent is used, then $Z = 0$, Eq. (10.2-24) becomes

$$B(X_{m-1} - X_m) = S_m Y_m \tag{10.2-25}$$

Phase equilibrium is

$$Y_m = K X_m \quad (K = \text{const.}) \tag{10.2-26}$$

Thus

$$X_m = \frac{X_{m-1}}{1 + \dfrac{S_m}{B} K} = \frac{X_{m-1}}{1 + \dfrac{1}{A_m}} \tag{10.2-27}$$

Extraction factor. Assume

$$\frac{1}{A_m} = \frac{S_m}{B} K = \frac{S_m Y_m}{B X_m} = \frac{\text{quantity of solute } A \text{ in extract phase}}{\text{quantity of solute } A \text{ in raffinate phase}} \tag{10.2-28}$$

When $S_1 = S_2 = \cdots = S_m = \cdots = S_N$, $1/A_m = 1/A$, then,

$$X_N = \frac{X_F}{(1 + 1/A)^N} \tag{10.2-29}$$

The number of stages is

$$N = \frac{\log(X_F/X_N)}{\log(1+1/A)} \qquad (10.2\text{-}30)$$

The fraction of solute remaining in raffinate is

$$\varphi = \frac{BX_N}{BX_F} = \frac{1}{(1+1/A)^N} = \frac{1}{(1+E)^N} \qquad (10.2\text{-}31)$$

Continuous multistage countercurrent extraction and overall balance. For multistage countercurrent extraction of dilute solutions in Fig. 10.23(a), making an overall balance for A over the first m stages, the operating line is

$$Y_{m+1} = \frac{B}{S}X_m + (Y_1 - \frac{B}{S}X_F) \qquad (10.2\text{-}32)$$

where B= kg inert B/h, S = kg inert S/h, Y = mass fraction A in S stream, and X = mass fraction A in B stream. This Eq.(10.2-32) is an operating-line equation whose slope B/S. If equilibrium line is plotted on an Y-X diagram. The numbers of stages are stepped off as shown previously for cases of distillation and absorption, as shown in Fig. 10.23(b).

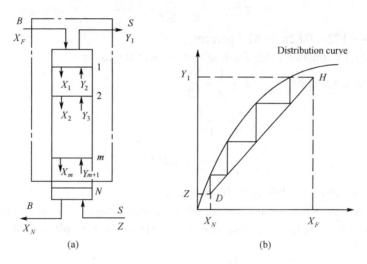

Fig. 10.23 Countercurrent extraction for immiscible system:
(a) process flow diagram, (b) plot on Y-X diagram.

The terminal points on the operating line H ($X=X_F$, $Y=Y_1$) and D ($X=X_N$, $Y=Z$) are usually determined by an overall material balance.

If the equilibrium line is $Y = KX$ and K is constant, the number of stages in countercurrent extraction process is

$$N = \frac{1}{\ln(1/A)} \ln\left[(1-A)\left(\frac{X_F - Z/K}{X_N - Z/K}\right) + A\right] \qquad (10.2\text{-}33)$$

or

$$N = \frac{1}{\ln E} \ln\left[(1-\frac{1}{E})\left(\frac{X_F - Z/K}{X_N - Z/K}\right) + \frac{1}{E}\right] \qquad (10.2\text{-}34)$$

The various forms of the Kremser equation [Eqs. (7.2-24), (7.2-25), (7.2-26), and (7.2-28)] can be used for a countercurrent extraction cascade.

EXAMPLE 10.4 Penicillin F is recovered from a dilute aqueous fermentation broth by extraction with amyl acetate, using 6 volumes of solvent per 100 volumes of the aqueous phase. At pH = 3.2 the distribution coefficient K is 80. (*a*) What fraction of the penicillin would be recovered in a single ideal stage? (*b*) What would be the recovery with two-stage extraction using fresh solvent in both stages? (*c*) How many ideal stages would be needed to give the same recovery as in part (*b*) if a counterflow cascade were used with $S/B = 0.06$?

Solution.

(*a*) single ideal stage extraction

By a material balance, since $Z = 0$,
$$B(X_F - X_1) = S(Y_1 - Z), \qquad B(X_F - X_1) = SY_1$$
$$Y_1 = KX_1$$
$$X_1\left(\frac{SK}{B} + 1\right) = X_F$$

The extraction factor is
$$E \equiv \frac{1}{A} = \frac{KS}{B} = \frac{80 \times 6}{100} = 4.8$$

Thus
$$\frac{X_1}{X_F} = \frac{1}{1+E} = \frac{1}{5.8} = 0.172$$

Recovery is $1 - 0.172 = 0.828$, or 82.8 percent.

(*b*) Two-stage extraction using fresh solvent in both stages

With the same value of
$$E = \frac{1}{A} = \frac{SK}{B}$$
$$\frac{X_2}{X_1} = \frac{1}{1+E}$$
$$\frac{X_2}{X_F} = \frac{1}{(1+E)^2} = 0.0297$$

Recovery is $1 - 0.0297 = 0.9703$, or 97.0 percent.

(*c*) With K and S/B constant, the number of ideal stages can be calculated from the stripping form of the Kremser equation [Eq. (7.2-28)], using E in place of its equivalent, the stripping factor S.
$$N = \frac{\ln\left[(X_F - X_F^*)/(X_N - X_N^*)\right]}{\ln E}$$

Let $X_F = X_0 = 100$, then $X_N = 3x_b = 3.0$, and $Y_1 = 97 \times 100/6 = 1617$.
$$X_F^* = \frac{Y_1}{K} = \frac{1617}{80} = 20.2, \qquad X_N^* = 0$$
$$N = \frac{\ln\left[(100 - 20.2)/3\right]}{\ln 4.8} = 2.09$$

Using a counterflow process requires only a slightly larger number of ideal stages than in part (*b*), but uses one-half as much solvent and increases the concentration of the extract.

Countercurrent extraction of type II systems using reflux

Just as in distillation, reflux can be used in countercurrent extraction to improve the separation of the components in the feed. This method is especially effective in treating type II systems, because with a center-feed cascade and the use of reflux, the two feed components can be separated into nearly pure products.

A flow diagram for countercurrent extraction with reflux is shown in Fig. 10.24. To emphasize the analogy between this method and fractionation, it is assumed that the cascade is a plate column. Any other kind of cascade, however, may be used. The method requires that sufficient solvent be removed from the extract leaving the cascade to form a raffinate, part of which is returned to the cascade as reflux, the remainder being withdrawn from the plant as a product. Raffinate is withdrawn from the cascade as bottoms product, and fresh solvent is admitted directly to the bottom of the cascade. None of the bottom raffinate needs to be returned as reflux, for the number of stages required is the same whether or not any of the raffinate is recycled to the bottom of the cascade.16 The situation is not the same as in continuous distillation, in which part of the bottoms must be vaporized to supply heat to the column.

The solvent separator, which is ordinarily a still, is shown in Fig. 10.24. As also shown in Fig. 10.24, solvent may be removed from both products by stripping, or in some cases by water washing, to give solvent-free products.

The close analogy between distillation and extraction, both using reflux, is shown in Table 10.3. Note that the solvent plays the same part in extraction that heat does in distillation.

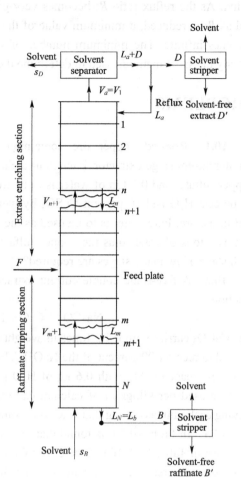

Fig. 10.24 Countercurrent extraction with reflux.

Table 10.3 Comparison of extraction with distillation, both using reflux

Distillation	Extraction
Vapor flow in cascade V	Extract flow in cascade V
Liquid flow in cascade L	Raffinate flow in cascade L
Overhead product D	Extract product D
Bottom product B	Raffinate product B
Condenser	Solvent separator
Bottom-product cooler	Raffinate solvent stripper
Overhead-product cooler	Extract solvent stripper
Heat to reboiler q_r	Solvent to cascade s_B
Heat removal in condenser q_c	Solvent removal in separator s_D
Reflux ratio $RD = L_a/D$	Reflux ratio $RD = L_a/D$
Rectifying section	Extract-enriching section
Stripping section	Raffinate-stripping section

Limiting reflux ratios

Just as in distillation, two limiting cases exist in operating a countercurrent extractor with reflux. As the reflux ratio R_D becomes very great, the number of stages approaches a minimum, and as R_D is reduced, a minimum value of the reflux ratio is reached where the number of stages becomes infinite. The minimum number of stages and the minimum reflux ratio are found by exactly the same methods used to determine the same quantities in distillation.

PROBLEMS

10.1 Roasted copper ore containing the copper as $CuSO_4$ is to be extracted in a countercurrent stage extractor. Each hour a charge consisting of 10 tons of inert solids, 1.2 tons of copper sulfate, and 0.5 ton of water is to be treated. The strong solution produced is to consist of 90 percent H_2O and 10 percent $CuSO_4$ by weight. The recovery of $CuSO_4$ is to be 98 percent of that in the ore. Pure water is to be used as the fresh solvent. After each stage, 1 ton of inert solids retains 2 tons of water plus the copper sulfate dissolved in that water. Equilibrium is attained in each stage. How many stages are required?

10.2 A five-stage countercurrent extraction battery is used to extract the sludge from the reaction

$$Na_2CO_3 + CaO + H_2O \rightarrow CaCO_3 + 2NaOH$$

The $CaCO_3$ carries with it 1.5 times its weight of solution in flowing from one unit to another. It is desired to recover 99 percent of the NaOH. The products from the reaction enter the first unit with no excess reactants but with 0.6 kg of H_2O per kilogram of $CaCO_3$. (*a*) How much wash water must be used per kilogram of calcium carbonate? (*b*) What is the concentration of the solution leaving each unit, assuming that $CaCO_3$ is completely insoluble?

10.3 In Prob. 10.2 it is found that the sludge retains solution varying with the concentration as shown in Table 10.4. If a 12 percent solution of the NaOH is to be produced, how many stages must be used to recover 97 percent of the NaOH?

Table 10.4

NaOH, wt %	0	5	10	15	20
Kg solution/kg $CaCO_3$	1.50	1.75	2.20	2.70	3.60

Table 10.5

Solution retained by 1 lb exhausted livers, gal	Solution concentration, gal oil/gal solution	Solution retained by 1 lb exhausted livers, gal	Solution concentration, gal oil/gal solution
0.035	0	0.068	0.4
0.042	0.1	0.081	0.5
0.050	0.2	0.099	0.6
0.058	0.3	0.120	0.68

10.4 Oil is to be extracted from halibut livers by means of ether in a countercurrent extraction battery. The entrainment of solution by the granulated liver mass was found by experiment to be as shown in Table 10.5. In the extraction battery, the charge per cell is to be 100 lb, based on completely exhausted livers. The unextracted livers contain 0.043 gal of oil per pound of exhausted material. A 95 percent recovery of oil is desired. The final extract is to contain 0.65

gal of oil per gallon of extract. The ether fed to the system is oil-free. (*a*) How many gallons of ether are needed per charge of livers? (*b*) How many extractors are needed?

10.5 In a continuous countercurrent train of mixer-settlers, 100 kg/h of a 40:60 acetone-water solution is to be reduced to 10 percent acetone by extraction with pure 1,1,2-trichloroethane at 25°C. (*a*) Find the minimum solvent rate. (*b*) At 1.8 times the minimum (solvent rate)/(feed rate), find the number of stages required. (*c*) For conditions of part (*b*) find the mass flow rates of all streams. Data are given in Table 10.6.

Table 10.6 Equilibrium data

Limiting solubility curve			Limiting solubility curve		
$C_2H_3Cl_3$, wt %	Water, wt %	Acetone, wt %	$C_2H_3Cl_3$, wt %	Water, wt %	Acetone, wt %
94.73	0.26	5.01	15.39	26.28	58.33
79.58	0.76	19.66	6.77	41.35	51.88
67.52	1.44	31.04	1.72	61.11	37.17
54.88	2.98	42.14	0.92	74.54	24.54
38.31	6.84	54.85	0.65	87.63	11.72
24.04	15.37	60.59	0.44	99.56	0.00

Tie lines					
Weight % in water layer			Weight % in trichloroethane layer		
$C_2H_3Cl_3$	Water	Acetone	$C_2H_3Cl_3$	Water	Acetone
0.52	93.52	5.96	90.93	0.32	8.75
0.73	82.23	17.04	73.76	1.10	25.14
1.02	72.06	26.92	59.21	2.27	38.52
1.17	67.95	30.88	53.92	3.11	42.97
1.60	62.67	35.73	47.53	4.26	48.21
2.10	57.00	40.90	40.00	6.05	53.95
3.75	50.20	46.05	33.70	8.90	57.40
6.52	41.70	51.78	26.26	13.40	60.34

10.6 A mixture containing 40 weight percent acetone and 60 weight percent water is contacted with an equal amount of MIK. (*a*) What fraction of the acetone can be extracted in a single-stage process? (*b*) What fraction of the acetone could be extracted if the fresh solvent were divided into two parts and two successive extractions used?

10.7 An antibiotic that has been extracted from a fermentation broth using amyl acetate at low pH is to be extracted back into clean water at pH = 6, where $K_D = 0.15$. If the water flow rate is set at 0.45 times the solvent rate, how many ideal stages would be needed for 98 percent recovery of the antibiotic in a countercurrent cascade?

10.8 Oil was extracted from small particles of rapeseed, averaging 0.58 mm in size, by contact with hexane. The particles originally contained 43.82 percent oil and 6.43 percent moisture. After drying, the oil content of the dried meal was reported for different extraction times. The results are given in the table below. Determine the effective diffusion coefficient for oil in the hexane-soaked particles using the equation for transient diffusion in spheres. Assume that a large excess of hexane was used and that the external mass transfer resistance was negligible.

Time, min	75	90	105	120
Oil in dry meal, kg/100 kg inert material	11.5	7.97	4.35	3.88

10.9 A solution containing 10 g/L of a valuable protein and 1 g/L of a protein impurity is extracted in a stirred vessel using an organic solvent. Distribution coefficient $K = 8$ for the valuable protein and 0.5 for the impurity. The initial volume is 500 L, and 400 L of solvent are used for the extraction. What are the final concentrations in the two phases, and what fraction of each protein is recovered in the solvent phase?

10.10 To show the effect of particle-size distribution on the rate of extraction, consider an oil-containing solid containing some particles half the average size and some 1.5 times the average size (say 25 percent 0.5 mm, 50 percent 1.0 mm, and 25 percent 1.5 mm). For an effective diffusion coefficient of 1×10^{-7} cm/s^2, determine the times for 50, 90, and 99 percent extraction and compare with the predicted times based on the average particle size.

10.11 An organic solute is to be extracted from a dilute aqueous solution using a solvent with a distribution coefficient of 6.8. For a continuous counterflow extractor, how many ideal stages are needed if the solvent flow is 0.35 times the solution flow and 99 percent recovery of the solute is required?

名 词 术 语

英文 / 中文

a battery of tank / 一组槽
agitated tower extractor / 搅拌塔萃取器
axial dispersion / 轴向扩散
baffle tower / 挡板塔
Bollman extractor / 立式吊篮（波耳曼）萃取机
buffered phosphate solution / 磷酸盐缓冲溶液
butyl acetate / 乙酸丁酯
centrifugal extractor / 离心萃取器
close-boiling liquid / 沸点相近的液体
close-boiling mixture / 沸点相近的混合物
complete miscibility / 完全互溶
continuous countercurrent leaching / 连续逆流浸提
continuous gravity decanter / 连续重力分层器
degrees of freedom / 自由度
differential-contact method / 微分接触法
diffusion battery / 扩散组
dispersed phase / 分散相
distribution coefficient / 分配系数
dome-shaped curve / 圆顶曲线
downcomer / 降液管
emulsify / 乳化
extract layer / 萃取层
extract / 萃取相
extraction battery / 萃取组
fermentation broth / 发酵液
flooding point / 泛点
heavy liquid / 重液体
high selectivity / 高选择性
high-boiling- point polar solvent / 高沸点极性溶剂

英文 / 中文

holdup / 持液
lever-arm rule / 杠杆定律
light liquid / 轻液体
liquid extraction / 液体萃取
liquid-liquid extraction / 液液萃取
low-boiling-point polar solvent / 低沸点极性溶剂
minimum reflux ratio / 最小回流比
mixer-settler / 混合澄清槽
mixing zone / 混合区
moving-bed continuous leaching 移动床式连续浸取
multiple-stage extraction / 多级萃取
multistage crosscurrent extraction / 多级错流萃取
overflow / 溢流
overlapping boiling-point range / 相互交叠的沸点范围
packed column / 填料柱
packed extraction tower / 填料萃取塔
paddle conveyor / 桨式输送机
partial miscibility / 部分互溶
perforated false bottom / 多孔假底
petroleum product / 石油产品
plait point / 临界混溶点（结点）
pulse column / 脉冲塔
raffinate / 萃余液
raffinate layer / 萃余层
rectangular coordinate / 直角坐标
rectangular extraction-phase-diagram / 直角形萃取相图
Rotocel extractor / 洛特赛萃取机

英文 / 中文	英文 / 中文
saddle packing / 鞍形填料	stationary solid-bed leaching / 静态床浸提
selectivity coefficient / 选择性	supercritical fluid extraction / 超临界萃取
settling zone / 沉降区	tie line / 联结线
short-circuiting / 短路	triangular diagram / 三角图
sieve-plate distillation tower / 筛板蒸馏塔	unagitated bed of solid / 无搅拌的固体床
sieve-plate pulse column / 筛板脉冲塔	underflow / 下溢
solid extraction / 固体萃取	vented overflow leg / 排放溢流腿
solid-bowl helical-conveyor centrifuge / 固体转筒螺旋式输送离心器	vertical tower / 垂直塔
solute-free solid / 无溶质固体	washing of filtered solid / 过滤后固体的洗涤

Chapter 11

Drying of Process Materials

11.1 Introduction and Methods of Drying

Purposes of drying

In general, drying means the removal of relatively small amounts of water or other liquid from the solid material to reduce the content of residual liquid to an acceptably low value. Drying is usually the final step in a series of operations, and the product from a dryer is often ready for final packaging.

The discussions of drying in this chapter are concerned with the removal of water from process materials and other substances. The term "drying" is also used to refer to removal of other organic liquid, such as benzene or organic solvents, from solids. Many of the types of equipment and calculation methods discussed for removal of water can also be used for removal of organic liquids.

Water or other liquids may be removed from solids mechanically by presses or centrifuges or thermally by vaporization. This chapter is restricted to drying by thermal vaporization. It is generally cheaper to remove liquid mechanically than thermally, and thus it is advisable to reduce the liquid content as much as practicable before feeding the material to a heated dryer.

Drying generally means removal of relatively small amounts of water from material. Evaporation refers to removal of relatively large amounts of water from material. In evaporation the water is removed as vapor at its boiling point. In drying the water is usually removed as a vapor by air.

In some cases water may be removed mechanically from solid materials by means of presses, centrifuging, and other methods. This is cheaper than drying by thermal means for removal of water, which will be discussed here. The moisture content of the final dried product varies depending upon the type of product. Dried salt contains about 0.5% water, coal about 4%, and many food products about 5%. Drying is usually the final processing step before packaging and makes many materials, such as soap powders and dyestuffs, more suitable for handling.

The liquid content of a dried substance varies from product to product. Occasionally the product contains no liquid and is called *bone-dry*. More commonly, the product does contain some liquid. Dried table salt, for example, contains about 0.5 percent water, dried coal about

4 percent, and dried casein about 8 percent. Dryness is a relative term, and drying means merely reducing the moisture content from an initial value to some acceptable final value.

Drying or dehydration of biological materials, especially foods, is used as a preservation technique. Microorganisms that cause food spoilage and decay cannot grow and multiply in the absence of water. Also, many enzymes that cause chemical changes in food and other biological materials cannot function without water. When the water content is reduced below about 10 wt %, the microorganisms are not active. However, it is usually necessary to lower the moisture content below 5 wt % in foods to preserve flavor and nutrition. Dried foods can be stored for extended periods of time.

General methods of drying

Drying methods and processes can be classified in several different ways. Drying processes can be classified as *batch*, where the material is inserted into the drying equipment and drying proceeds for a given period of time, or as *continuous*, where the material is continuously added to the dryer and dried material is continuously removed.

Drying processes can also be categorized according to the physical conditions used to add heat and remove water vapor: (1) in the first category, heat is added by direct contact with heated air at atmospheric pressure, and the water vapor formed is removed by the air; (2) in vacuum drying, the evaporation of water proceeds more rapidly at low pressures, and the heat is added indirectly by contact with a metal wall or by radiation (low temperatures can also be used under vacuum for certain materials that may discolor or decompose at higher temperatures); and (3) in freeze-drying, water is sublimed from the frozen material.

The solids to be dried may be in many different forms—flakes, granules, crystals, powders, slabs, or continuous sheets—and may have widely differing properties. The liquid to be vaporized may be on the surface of the solid, as in drying salt crystals; it may be entirely inside the solid, as in solvent removal from a sheet of polymer; or it may be partly outside and partly inside. The feed to some dryers is a liquid in which the solid is suspended as particles or is in solution. The dried product, such as salt and other inorganic solids, may be able to stand rough handling and high temperatures, or it may, like food or pharmaceuticals, require gentle treatment at low or moderate temperatures. Consequently, a multitude of types of dryers are on the market for commercial drying. They differ chiefly in the way the solids are moved through the drying zone and in the way heat is transferred.

Classification dryers

There is no simple way of classifying drying equipment. Some dryers are continuous, and others operate batchwise; some agitate the solids, and others are essentially unagitated. Operation under vacuum may be used to reduce the drying temperature. Some dryers can handle almost any kind of material, while others are severely limited in the type of feed they can accept.

Major divisions may be made among (1) dryers in which the solid is directly exposed to a hot gas (usually air), (2) dryers in which heat is transferred to the solid from an external medium such as condensing steam, usually through a metal surface with which the solid is in contact, and (3) dryers that are heated by dielectric, radiant, or microwave energy. Dryers that expose the solids to a hot gas are called *adiabatic* or *direct dryers;* those in which heat is transferred from an external medium are known as *nonadiabatic* or *indirect* dryers. Some units have more than one mode of heat transfer, such as hot gas plus a heated surface or hot gas plus radiation.

Solids handling in dryers

Most industrial dryers handle particulate solids during part of or all the drying cycle, although some, of course, dry large individual pieces such as ceramic ware or sheets of polymer. Here it is important only to describe the different patterns of motion of solid particles through dryers as a basis for understanding the principles of drying discussed in the next section.

In adiabatic dryers the solids are exposed to the gas in the following ways:

(1) Gas is blown across the surface of a bed or slab of solids or across one or both faces of a continuous sheet or film. This process is called *cross-circulation drying* [Fig.11.1(*a*)].

(2) Gas is blown through a bed of coarse granular solids that are supported on a screen. This is known as *through-circulation drying*. As in cross-circulation drying, the gas velocity is kept low to avoid any entrainment of solid particles [Fig.11.1(*b*)].

(3) Solids are showered downward through a slowly moving gas stream, often with some undesired entrainment of fine particles in the gas [Fig.11.1(*c*)]. Note that some rotary dryers are nonadiabatic.

(4) Gas passes through the solids at a velocity sufficient to fluidize the bed, as discussed in Chap. 3. Inevitably there is some entrainment of finer particles [Fig.11.1(*d*)].

(5) The solids are all entrained in a high-velocity gas stream and are pneumatically conveyed from a mixing device to a mechanical separator [Fig.11. 1(*e*)].

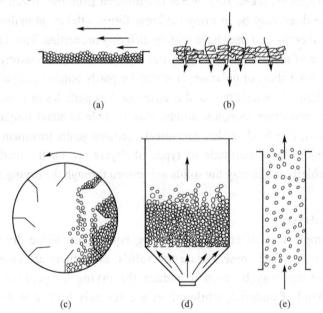

Fig. 11.1 Patterns of gas-solid interaction in dryers:
(*a*) gas flow across a static bed of solids; (*b*) gas passing through a bed of preformed solids;
(*c*) showering action in a rotary dryer; (*d*) fluidized solids bed;
(*e*) cocurrent gas-solid flow in a pneumatic-conveyor flash dryer.

(6) Drops of solution or slurry are evaporated by suspending them briefly in a hot gas stream, as in spray dryers (Fig. 11.23).

In nonadiabatic dryers the only gas to be removed is the vaporized water or solvent, although sometimes a small amount of "sweep gas" (often air or nitrogen) is passed through the unit. Nonadiabatic dryers differ chiefly in the ways in which the solids are exposed to the hot surface or other source of heat.

(1) Solids are spread over a stationary or slowly moving horizontal surface and "cooked" until dry. The surface may be heated electrically or by a heat-transfer fluid such as steam or hot water. Alternatively, heat may be supplied by a radiant heater above the solid.

(2) Solids are moved over a heated surface, usually cylindrical, by an agitator or a screw or paddle conveyor.

(3) Solids slide by gravity over an inclined heated surface or are carried upward with the surface for a time and then slide to a new location. (See "Rotary Dryers")

11.2 Properties of Moist Air and Humidity Chart

11.2.1 Humidity

In humidification operations, especially as applied to the system air-water, a number of rather special definitions are in common use. The usual basis for engineering calculations is a unit mass of vapor-free gas, where vapor means the gaseous form of the component that is also present as liquid and gas is the component present only in gaseous form. In this discussion a basis of a unit mass of vapor-free gas is used. In the gas phase the vapor will be referred to as component A and the fixed gas as component B. Because the properties of a gas-vapor mixture vary with total pressure, the pressure must be fixed. Unless otherwise specified, a total pressure of 1 atm is assumed. Also, it is assumed that mixtures of gas and vapor follow the ideal gas laws.

Definition of humidity

Humidity H is the mass of vapor carried by a unit mass of vapor-free gas. The humidity H of an air–water vapor mixture is defined as the kg of water vapor contained in 1 kg of dry air. The humidity so defined depends only on the partial pressure p_A of water vapor in the air and on the total pressure P (assumed throughout this chapter to be 101.325 kPa, 1.0 atm abs, or 760 mm Hg). If the partial pressure of the vapor is p_A atm, the molal ratio of vapor to gas at 1 atm is $p_A/(P - p_A)$. The humidity is therefore

$$H = \frac{M_A p_A}{M_B(P - p_A)} \quad (11.2\text{-}1)$$

where M_A and M_B are the molecular weights of components A and B, respectively.

Using the molecular weight of water (A) as 18.02 and of air as 28.97, the humidity H in kg H$_2$O/kg dry air, is as follows:

$$H \frac{\text{kg H}_2\text{O}}{\text{kg dry air}} = \frac{p_A}{P - p_A} \times \frac{\text{kmol H}_2\text{O}}{\text{kmol air}} \times \frac{18.02 \text{kg H}_2\text{O}}{\text{kmol H}_2\text{O}} \times \frac{1}{28.97 \text{kmol/kmol air}}$$

$$H = \frac{18.02}{28.97} \times \frac{p_A}{P - p_A} = 0.622 \frac{p_A}{P - p_A} \quad (11.2\text{-}2)$$

The humidity is related to the mole fraction in the gas phase by the equation:

$$y = \frac{H/M_A}{1/M_B + H/M_A} \quad (11.2\text{-}3)$$

Since H/M_A is usually small compared with $1/M_B$, often y may be considered to be directly proportional to H.

Saturated gas is gas in which the vapor is in equilibrium with the liquid at the gas temperature. The partial pressure of vapor in saturated gas equals the vapor pressure of the liquid at the gas temperature. If H_S is the saturation humidity and P_{AS} the vapor pressure of the liquid,

$$H_S = \frac{M_A}{M_B} \times \frac{p_{AS}}{P - p_{AS}} \quad (11.2\text{-}4)$$

$$H_S = \frac{18.02}{28.97} \times \frac{p_{AS}}{P - p_{AS}} = 0.622 \frac{p_{AS}}{P - p_{AS}} \quad (11.2\text{-}5)$$

Phase equilibria

In humidification and dehumidification operations the liquid phase is a single pure component. The equilibrium partial pressure of solute in the gas phase is therefore a unique function of temperature when the total pressure on the system is held constant. Also, at moderate pressures the equilibrium partial pressure is almost independent of total pressure and is virtually equal to the vapor pressure of the liquid. By Dalton's law the equilibrium partial pressure may be converted to the equilibrium mole fraction y_e in the gas phase. Since the liquid is pure, x_e is always unity. Equilibrium data are often presented as plots of y_e versus temperature at a given total pressure, as shown for the system air-water at 1 atm in Fig.11.2. The equilibrium mole fraction y_e is related to the saturation humidity by Eq. (11.2-3); thus

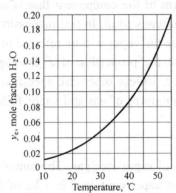

Fig. 11.2 Equilibria for the system air-water at 1 atm.

$$y_e = \frac{H_S/M_A}{1/M_B + H_S/M_A} \quad (11.2\text{-}6)$$

Relative Humidity, H_R

The amount of saturation of an air–water vapor mixture is also given as relative humidity H_R using partial pressures

Relative humidity H_R is defined as the ratio of the partial pressure of the vapor to the vapor pressure of the liquid at the gas temperature. It is usually expressed on a percentage basis, so 100 percent humidity means saturated gas and 0 percent humidity means vapor-free gas. By definition

$$H_R = \frac{p_A}{p_{AS}} \times 100\% \quad (11.2\text{-}7)$$

Percentage humidity, H_P

The percentage humidity H_P is the ratio of the actual humidity H to the saturation humidity H_S at the gas temperature, also on a percentage basis, or is defined as 100 times the actual humidity H of the air divided by the humidity H_S if the air were saturated at the same temperature and pressure:

$$H_P = \frac{H}{H_S} \times 100\% \tag{11.2-8}$$

Note that $H_R \neq H_P$, since H_P expressed in partial pressures by combining Eqs. (11.2-2), (11.2-5), and (11.2-8) is

$$H_P = \frac{H}{H_S} \times 100\% = \left(\frac{18.02}{28.97} \times \frac{p_A}{P-p_A}\right) \bigg/ \left(\frac{18.02}{28.97} \times \frac{p_{AS}}{P-p_{AS}}\right) = \frac{p_A}{p_{AS}} \times \frac{P-p_{AS}}{P-p_A} \tag{11.2-9}$$

This, of course, is not the same as Eq. (11.2-7). At all humidities other than 0 or 100 percent, the percentage humidity is less than the relative humidity.

EXAMPLE 11.1 The air in a room is at 26.7°C and a pressure of 101.325 kPa and contains water vapor with a partial pressure $p_A = 2.76$ kPa. Calculate the following:
(a) Humidity, H.
(b) Saturation humidity, H_S, and percentage humidity, H_P.
(c) Relative humidity, H_R.

Solution. From the steam tables at 26.7°C, the vapor pressure of water is $p_{AS} = 3.50$ kPa. Also, $p_A = 2.76$ kPa and $P = 101.3$ kPa. For part (a), using Eq. (11.2-2),

$$H = \frac{18.02}{28.97} \times \frac{p_A}{P-p_A} = \frac{18.02 \times 2.76}{28.97 \times (101.3-2.76)} = 0.01742 \text{ kg H}_2\text{O/kg air}$$

For part (b), using Eq. (11.2-5), the saturation humidity is

$$H_S = \frac{18.02}{28.97} \times \frac{p_{AS}}{P-p_{AS}} = \frac{18.02 \times 3.50}{28.97 \times (101.3-3.50)} = 0.02226 \text{ kg H}_2\text{O/kg air}$$

The percentage humidity, from Eq. (11.2-8), is

$$H_P = 100 \frac{H}{H_S} = \frac{100 \times 0.01742}{0.02226} = 78.3\%$$

For part (c), from Eq. (11.2-7), the relative humidity is

$$H_R = 100 \frac{p_A}{p_{AS}} = \frac{100 \times 2.76}{3.50} = 78.9\%$$

11.2.2 Humid Heat of an Air–Water Vapor Mixture

Humid heat c_s is the heat energy necessary to increase the temperature of 1kg of gas plus whatever vapor it may contain by 1°C. Thus

$$c_s = c_{pB} + Hc_{pA} \tag{11.2-10}$$

where c_{pB} and c_{pA} are the specific heats of dry gas and vapor, respectively.

The heat capacity of air and water vapor can be assumed constant over the temperature ranges usually encountered at 1.005 kJ/(kg dry air·°C) and 1.88 kJ/(kg water vapor·°C), respectively. Hence, for SI and English units,

$$c_s = 1.005 + 1.88H \qquad (11.2\text{-}11)$$

[In some cases c_s will be given as $(1.005 + 1.88H)10^3$ J/(kg · K).]

11.2.3 Humid Volume of an Air–Water Vapor Mixture

Humid volume v_H is the total volume of a unit mass of vapor-free gas plus what-ever vapor it may contain at 1 atm and the gas temperature. From the gas laws and the values of standard molar volume, v_H in SI units is related to humidity and temperature by the equation

$$v_H = \frac{0.0224T}{273}\left(\frac{1}{M_A} + \frac{H}{M_B}\right) \qquad (11.2\text{-}12)$$

where v_H is in cubic meters per gram and T is in Kelvins.

The humid volume v_H is the total volume in m³ of 1 kg of dry air plus the vapor it contains at 101.325 kPa (1.0 atm) abs pressure and the given gas temperature. Using the ideal gas law,

$$v_H \text{ m}^3/\text{kg dry air} = \frac{22.41}{273}\left(\frac{1}{28.97} + \frac{1}{18.02}H\right)T = (2.83\times10^{-3} + 4.56\times10^{-3}H)T \qquad (11.2\text{-}13)$$

For vapor-free gas $H=0$, and v_H is the specific volume of the fixed gas. For saturated gas $H=H_s$, and v_H becomes the *saturated volume*.

11.2.4 Total Enthalpy of an Air–Water Vapor Mixture

The total enthalpy of 1kg of air plus its water vapor is I J/kg or kJ/kg dry air. To calculate I, two reference states must be chosen, one for gas and one for vapor.

Let T_0 be the datum temperature chosen for both components, and base the enthalpy of component A on liquid A at T_0. (Temperature $T_0 = 0\,°\text{C}$ for most air-water problems.) Let the temperature of the gas be T and the humidity H. The total enthalpy is the sum of three items: the sensible heat of the vapor, the latent heat of the liquid at T_0, and the sensible heat of the vapor-free gas. Then

$$I = c_{pB}(T-T_0) + Hr_0 + c_{pA}H(T-T_0) \qquad (11.2\text{-}14)$$

where r_0 is the latent heat of the liquid at T_0. From Eq. (11.14) this becomes

$$I = c_s(T-T_0) + Hr_0 \qquad (11.2\text{-}15)$$

The total enthalpy is the sensible heat of the air–water vapor mixture plus the latent heat r_0 in kJ/kg water vapor of the water vapor at T_0. Note that $(T-T_0)\,°\text{C} = (T-T_0)$ K and that this enthalpy is referred to liquid water.

$$I \text{ kJ/kg dry air} = (1.005 + 1.88H)(T-T_0) + Hr_0 \qquad (11.2\text{-}16)$$

If the total enthalpy is referred to a base temperature T_0 of $0\,°\text{C}$, the equation for I becomes

$$I \text{ kJ/kg dry air} = (1.005 + 1.88H)T + 2501H \qquad (11.2\text{-}17)$$

11.2.5 Dew Point of an Air–Water Vapor Mixture

The temperature at which a given mixture of air and water vapor would be saturated is called the *dew-point temperature* T_d or simply the *dew point*. For example, at 26.7°C, the saturation vapor pressure of water is $p_{AS} = 3.50$ kPa. Hence, the dew point of a mixture containing water vapor having a partial pressure of 3.50 kPa is 26.7°C. If an air–water vapor mixture is at 37.8°C (often called the dry bulb temperature, since this is the actual temperature a dry thermometer bulb would indicate in this mixture) and contains water vapor of $p_A = 3.50$ kPa, the mixture would not be saturated. On cooling to 26.7°C, the air would be saturated, that is, at the dew point. On further cooling, some water vapor would condense, since the partial pressure cannot be greater than the saturation vapor pressure.

11.2.6 Adiabatic Saturation Temperatures

Consider the process shown in Fig.11.3, Water is sprayed into a stream of gas in a pipe or spray chamber to bring the gas to saturation. The pipe or chamber is insulated so that the gas leaves having a different humidity and temperature and the process is adiabatic. The gas, with an initial humidity H and temperature T, is cooled and humidified. If not all the water evaporates and there is sufficient time for the gas to come to equilibrium with the water, the temperature of the water being recirculated reaches a steady-state temperature called the *adiabatic saturation temperature*, T_S. The remaining liquid is also at T_S and can be recirculated to the spray nozzles. If the entering gas at temperature T having a humidity of H is not saturated, T_S will be lower than T.

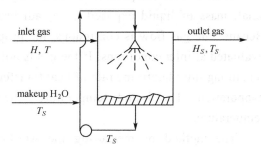

Fig. 11.3 Adiabatic air–water vapor saturator

The value of T_s depends on the temperature and initial humidity of the air and to a minor extent on the initial water temperature. To simplify the analysis, the water is often assumed to enter at T_s. An enthalpy balance can be written over this process. Pump work is neglected, and the enthalpy balance is based on temperature T_s as a datum.

Then the enthalpy of the makeup liquid is zero, and the total enthalpy of the entering gas equals that of the leaving gas. Since the latter is at datum temperature, its enthalpy is simply $H_s r_s$, where H_s is the saturation humidity and r_s is the latent heat, both at T_s. From Eq. (11.2-16) the total enthalpy of the entering gas is $c_s(T - T_s) + Hr_s$, and the enthalpy balance is

$$c_s(T - T_s) + Hr_s = c_s(T_s - T_s) + H_s r_s \qquad (11.2\text{-}18)$$

or, rearranging, and using Eq. (11.10) for c_s,

$$\frac{H_S - H}{T - T_s} = \frac{c_s}{r_s} = \frac{c_{pB} + c_{pA}H}{r_s}$$

or

$$\frac{H_S - H}{T - T_s} = \frac{c_s}{r_s} = \frac{1.005 + 1.88H}{r_s} \qquad (11.2\text{-}19)$$

434 | Unit Operations of Chemical Engineering

To find the adiabatic saturation temperature for gases other than air, a heat balance similar to Eq. (11.2-19) is used. Equation (11.2-19) cannot be solved directly for the adiabatic saturation temperature T_s, since H_s, c_s, and r_s are all functions of T_s. Thus T_s is obtained by a trial-and-error calculation, or, for the air-water system, by using *humidity charts*.

11.2.7 Dry-Bulb Temperature and Wet-Bulb Temperature

Dry-bulb temperature

Commonly, an uncovered thermometer is used along with the wet bulb to measure T, the actual gas temperature, and the gas temperature is usually called the *dry-bulb temperature*.

Wet-bulb temperature

The properties discussed above are static or equilibrium quantities. Equally important are the rates at which mass and heat are transferred between the gas and liquid phases that are not in equilibrium. The driving forces for mass and heat transfer are concentration and temperature differences, which can be predicted by using a quantity called the wet-bulb temperature.

The wet-bulb temperature is the steady-state, nonequilibrium temperature reached by a small mass of liquid exposed under adiabatic conditions to a continuous stream of gas. Because the gas flow is continuous, the properties of the gas are constant and are usually evaluated at inlet conditions. If the gas is not saturated, some liquid evaporates, cooling the remaining liquid until the rate of heat transfer to the liquid just balances the heat needed for evaporation. The liquid temperature when steady state is reached is the wet-bulb temperature.

The method of measuring the wet-bulb temperature is shown in Fig.11.4(*a*). A thermometer or other temperature-measuring device, such as a thermocouple, is covered by a wick, which is saturated with pure liquid and immersed in a stream of gas having a definite temperature T and humidity H. Assume that initially the temperature of the liquid is about that of the gas. Since the gas is not saturated, liquid evaporates; and because the process is adiabatic, the latent heat is supplied at first by cooling the liquid. As the temperature of the liquid decreases below that of the gas, sensible heat is transferred to the liquid. Ultimately a steady state is reached at such a liquid temperature that the heat needed to evaporate the liquid and heat the vapor to gas temperature is exactly balanced by the sensible heat flowing from the gas to the liquid. It is this steady-state temperature, denoted by T_w, that is called the *wet-bulb temperature*. It is a function of both T and H. The temperature and concentration gradients at steady state are shown in Fig. 11.4(*b*).

Theory of wet-bulb temperature

The adiabatic saturation temperature is the steady-state temperature attained when a large amount of water is contacted by the entering gas. The *wet bulb temperature* is the steady-state nonequilibrium temperature reached when a small amount of water is contacted under adiabatic conditions by a continuous stream of gas. Since the amount of liquid is small, the temperature and humidity of the gas are not changed, contrary to the case of adiabatic saturation, where the temperature and humidity of the gas are changed.

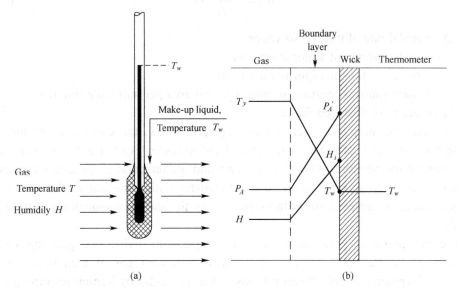

Fig. 11.4 (a) Wet-bulb thermometer; (b) Gradients in the gas boundary layer.

The method used to measure the wet bulb temperature is illustrated in Fig.11.4(a), where a thermometer is covered by a wick or cloth. The wick is kept wet by water and is immersed in a flowing stream of air-water vapor having a temperature of T (dry bulb temperature) and humidity H. At steady state, water is evaporating to the gas stream. The wick and water are cooled to T_w and stay at this constant temperature. The latent heat of evaporation is exactly balanced by the convective heat flowing from the gas stream at T to the wick at a lower temperature T_w.

At the wet-bulb temperature the rate of heat transfer from the gas to the liquid may be equated to the product of the rate of vaporization and the sum of the latent heat of evaporation at temperature T_w and the sensible heat of the vapor. Since radiation may be neglected, this balance may be written

$$q = M_A N_A [r_w + c_{pA}(T - T_w)] A \qquad (11.2\text{-}20)$$

where q = rate of sensible heat transfer to liquid
N_A = molal rate of vaporization
r_w = latent heat of liquid at wet-bulb temperature T_w
A = surface area of liquid

The rate of heat transfer may be expressed in terms of the area, the temperature drop, and an effective heat-transfer coefficient in the usual way, or

$$q = h_y (T - T_i) A \qquad (11.2\text{-}21)$$

where h_y = heat-transfer coefficient between gas and surface of liquid
T_i = temperature at interface
A = surface area of liquid

The rate of mass transfer may be expressed in terms of the mass-transfer coefficient, the area, and the driving force in mole fraction of vapor, or

$$N_A = \frac{k_y(y - y_i)}{\Delta y_m} \qquad (11.2\text{-}22)$$

where N_A = molal rate of transfer of vapor
y_i = mole fraction of vapor at interface
y = mole fraction of vapor in air stream
k_y = mass-transfer coefficient, mole per unit area per unit mole fraction
Δy_m = one-way diffusion factor

If the wick is completely wet and no dry spots show, the entire area of the wick is available for both heat and mass transfer and the areas in Eqs.(11.2-21)and (11.2-22) are equal. Since the temperature of the liquid is constant, no temperature gradients are necessary in the liquid to act as driving forces for heat transfer within the liquid, the surface of the liquid is at the same temperature as the interior, and the surface temperature of the liquid T_i equals T_w.

Since the liquid is pure, no concentration gradients exist, and granting interfacial equilibrium, y_i is the mole fraction of vapor in saturated gas at temperature T_w. It is convenient to replace the mole fraction terms in Eq. (11.2-22) by humidities through the use of Eq. (11.2-3), noting that y_i corresponds to H_w, the saturation humidity at the wet-bulb temperature. [See Eq. (11.2-6).] Following this by substituting q from Eq. (11.2-21) and N_A from Eq. (11.2-22) into Eq. (11.2-20) gives

$$h_y(T - T_w) = \frac{k_y}{\Delta y_m}\left(\frac{H_w}{1/M_B + H_w/M_A} - \frac{H}{1/M_B + H/M_A}\right) \times \left[r_w + c_{pA}(T - T_w)\right] \qquad (11.2\text{-}23)$$

Equation (11.2-23) may be simplified without serious error in the usual range of temperatures and humidities as follows: (1) the factor Δy_m is nearly unity and can be omitted; (2) the sensible-heat item $c_{pA}(T - T_w)$ is small in comparison with λ_w and can be neglected; (3) the terms H_w/M_A and H/M_A are small in comparison with $1/M_B$ and may be dropped from the denominators of the humidity terms. With these simplifications Eq. (11.2-23) becomes

$$h_y(T - T_w) = M_B k_y r_w (H_w - H)$$

or
$$\frac{H - H_w}{T - T_w} = -\frac{h_y}{M_B k_y r_w} \qquad (11.2\text{-}24)$$

For a given wet-bulb temperature, both r_w and H_w are fixed. The relation between H and T then depends on the ratio h_y/k_y. The close analogy between mass transfer and heat transfer provides considerable information on the magnitude of this ratio and the factors that affect it.

It has been shown in Chap. 4 that heat transfer by conduction and convection between a stream of fluid and a solid or liquid boundary depends on the Reynolds number DG/μ and the Prandtl number $c_p\mu/k$. Also, as shown in Chap. 4, the mass-transfer coefficient depends on the Reynolds number and the Schmidt number $\mu/(\rho D_v)$. As discussed in Chap. 4 and Chap.6, the rates of heat and mass transfer, when these processes are under the control of the same boundary layer, are given by equations that are identical in form. For turbulent flow of the gas stream these equations are

$$\frac{h_y}{c_p G} = b Re^n Pr^m \qquad (11.2\text{-}25)$$

and

$$\frac{\overline{M}k_y}{G} = bRe^n Sc^m \tag{11.2-26}$$

where b, n, m = constants
$\overline{M}$ = average molecular weight of gas stream

Substitution of h_y from Eq. (11.2-25) and k_y from Eq. (11.2-26) in Eq. (11.2-24), assuming $\overline{M} = M_B$, gives

$$\frac{H - H_w}{T - T_w} = -\frac{h_y}{M_B k_y r_w} = -\frac{c_p}{r_w}\left(\frac{Sc}{Pr}\right)^m \tag{11.2-27}$$

and
$$\frac{h_y}{M_B k_y} = c_p \left(\frac{Sc}{Pr}\right)^m \tag{11.2-28}$$

If m is taken as 2/3, the predicted value of $h_y/M_B k_y$ for air in water is $0.24(0.62/0.71)^{2/3}$, or 0.92 J/(g·℃). The experimental value is 1.09 J/(g·℃), somewhat larger than predicted, because of heat transfer by radiation. For organic liquids in air it is larger, in the range 1.6 to 2.0 J/(g·℃). The difference, as shown by Eq. (11.2-28), is the result of the differing ratios of Prandtl and Schmidt numbers for water and for organic vapors.

Measurement of wet-bulb temperature

To measure the wet-bulb temperature with precision, three precautions are necessary: (1) The wick must be completely wet, so no dry areas of the wick are in contact with the gas; (2) the velocity of the gas should be large enough (at least 5 m/s) to ensure that the rate of heat flow by radiation from warmer surroundings to the bulb is negligible in comparison with the rate of sensible heat flow by conduction and convection from the gas to the bulb; (3) if makeup liquid is supplied to the bulb, it should be at the wet-bulb temperature. When these precautions are taken, the wet-bulb temperature is independent of gas velocity over a wide range of flow rates.

The wet-bulb temperature superficially resembles the adiabatic saturation temperature T_s. Indeed, for air-water mixtures the two temperatures are nearly equal.

This is fortuitous, however, and is not true of mixtures other than air and water. The wet-bulb temperature differs fundamentally from the adiabatic saturation temperature. The temperature and humidity of the gas vary during adiabatic saturation, and the endpoint is a true equilibrium rather than a dynamic steady state.

11.2.8 Humidity Chart

A convenient diagram showing the properties of mixtures of a permanent gas and a condensable vapor is the humidity chart. A chart for mixtures of air and water at 1 atm is shown in Fig.11.5. Many forms of such charts have been proposed. Figure 11.5 is based on the Grosvenor chart. In this figure the humidity H is plotted versus the actual temperature of the air–water vapor mixture (dry bulb temperature).

Any point on the chart represents a definite mixture of air and water. The curved line marked 100 percent gives the humidity of saturated air as a function of air temperature. In Example 11.3, for 26.7℃, H_S was calculated as 0.02226 kg H_2O/kg air. Plotting this point for 26.7℃ and H_S = 0.02226 on Fig.11.5, it falls on the 100% saturated line.

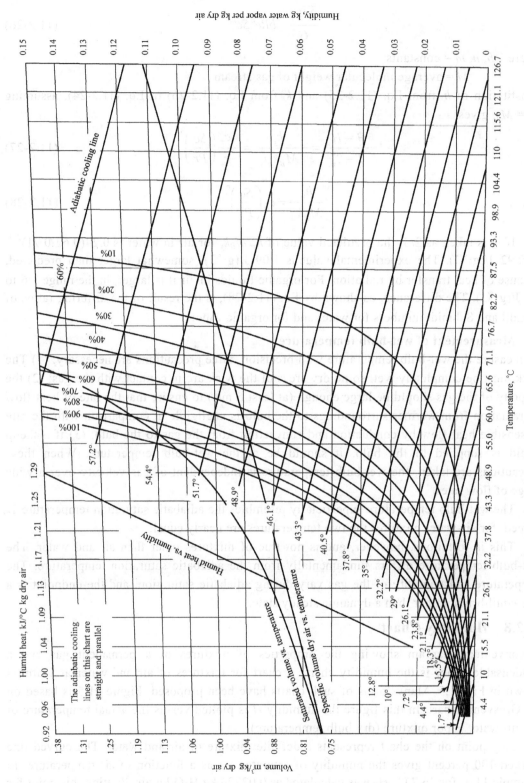

Fig. 11.5 Humidity chart. Air-water at 1 atm.

By using the vapor pressure of water, the coordinates of points on this line are found from Eq. (11.2-4). Any point above and to the left of the saturation line represents a mixture of saturated air and liquid water. This region is important only in checking fog formation. Any point below the saturation line represents undersaturated air, and a point on the temperature axis represents dry air. The curved lines between the saturation line and the temperature axis marked in even percents represent mixtures of air and water of definite *percentage humidities* H_p. As shown by Eq. (11.2-9), linear interpolation between the saturation line and the temperature axis can be used to locate the lines of constant percentage humidity.

The slanting lines running downward and to the right of the saturation line are called *adiabatic cooling lines*. They are plots of Eq. (11.2-19), each drawn for a given constant value of the adiabatic saturation temperature. For a given value of T_s, both H_s and r_s are fixed, and the line of H versus T can be plotted by assigning values to H and calculating corresponding values of T. Inspection of Eq. (11.2-19) shows that the slope of an adiabatic cooling line, if drawn on truly rectangular coordinates, is $-c_s/r_s$, and by Eq. (11.2-11), this slope depends on the humidity. On rectangular coordinates, then, the adiabatic cooling lines are neither straight nor parallel. In Fig. 11.5 the ordinates are sufficiently distorted to straighten the adiabatics and render them parallel, so interpolation between them is easy. The ends of the adiabatics are identified with the corresponding adiabatic saturation temperatures.

Lines are shown on Fig. 11.5 for the specific volume of dry air and the saturated volume. Both lines are plots of volume versus temperature. Volumes are read on the scale at the left. Coordinates of points on these lines are calculated by use of Eq. (11.2-12). Linear interpolation between the two lines, based on percentage humidity, gives the humid volume of unsaturated air. Also, the relation between the humid heat c_s and humidity is shown as a line on Fig. 11.5. This line is a plot of Eq. (11.2-10). The scale for c_s is at the top of the chart.

Use of humidity chart

The usefulness of the humidity chart as a source of data on a definite air-water mixture can be shown by reference to Fig. 11.6, which is a portion of the chart of Fig. 11.5. Assume, for example, that a given stream of undersaturated air is known to have a temperature T_1 and a percentage humidity H_{P1}. Point *a* represents this air on the chart. This point is the intersection of the constant-temperature line for T_1 and the constant-percentage-humidity line for H_{P1}. The humidity H_1 of the air is given by point *b*, the humidity coordinate of point *a*. The dew point is found by following the constant-humidity line through point *a* to the left to point *c* on the 100 percent line. The dew point temperature T_d is then read at point *d* on the temperature axis. The adiabatic saturation temperature is the temperature applying to the adiabatic cooling line through point *a*. The humidity at adiabatic saturation is found by following the adiabatic line through point *a* to its intersection *e* on the 100 percent line and reading humidity H_s at point *f* on the humidity scale. Interpolation between the adiabatic lines may be necessary. The adiabatic saturation temperature T_s is given by point *g*. If the original air is subsequently saturated at constant temperature, the humidity after saturation is found by following the constant-temperature line through point *a* to point *h* on the 100 percent line and reading the humidity at point *j*.

440 | Unit Operations of Chemical Engineering

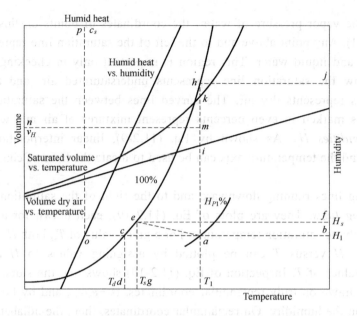

Fig. 11.6 Use of humidity chart.

The humid volume of the original air is found by locating points k and l on the curves for saturated and dry volumes, respectively, corresponding to temperature T_1. Point m is then found by moving along line lk a distance $(H_P/100)$ kl from point l, where kl is the line segment between points l and k. The humid volume v_H is given by point n on the volume scale. The humid heat of the air is found by locating point o, the intersection of the constant-humidity line through point a and the humid heat line, and reading the humid heat c_s at point p on the scale at the top.

Psychrometric line and Lewis relation

For a given wet-bulb temperature, Eq. (11.2-27) can be plotted on the humidity chart as a straight line having a slope of $-h_y/(M_B k_y r_w)$ and intersecting the 100 percent line at T_w. This line is called the *psychrometric line*. When both a psychrometric line from Eq. (11.2-27) and an adiabatic saturation line from Eq. (11.2-19), are plotted for the same point on the 100 percent curve, the relation between the lines depends on the relative magnitudes of c_s and $h_y/(M_B k_y)$.

For the system air-water at ordinary conditions the humid heat c_s is almost equal to the specific heat c_P, and the following equation is nearly correct:

$$\frac{h_y}{M_B k_y} \cong c_s \qquad (11.2\text{-}29)$$

Equation (11.2-29) is known as the *Lewis relation*. When this relation holds, the psychrometric line and the adiabatic saturation line become essentially the same. In Fig. 11.5 for air-water, therefore, the same line may be used for both. For other systems separate lines must be used for psychrometric lines. With nearly all mixtures of air and organic vapors, the psychrometric lines are steeper than the adiabatic saturation lines, and the wet-bulb temperature of any mixture other than a saturated one is higher than the adiabatic saturation temperature.

EXAMPLE 11.2 The temperature and dew point of the air entering a certain dryer are 65.6 and 15.6 ℃, respectively. What additional data for this air can be read from the humidity chart?

Solution. The dew point is the temperature coordinate on the saturation line corresponding to the humidity of the air. The saturation humidity for a temperature of 15.6 ℃ is 0.011g of water per g of dry air, and this is the humidity of the air. From the temperature and humidity of the air, the point on the chart for this air is located. At $H= 0.011$ and $T = 65.6℃$, the percentage humidity H_P is found by interpolation to be 5.2 percent.

The adiabatic cooling line through this point intersects the 100 percent line at 29.4℃, and this is the adiabatic saturation temperature. The humidity of saturated air at this temperature is 0.026 g water/g dry air. The humid heat of the air is 1.03 J/(g dry air · ℃). The saturated volume at 65.6 ℃ is 1.29 m³/kg dry air, and the specific volume of dry air at 65.6 ℃ is 0.958 m³/kg. The humid volume is, then, v_H is 0.978m³/kg.

EXAMPLE 11.3 Air entering a dryer has a temperature (dry bulb temperature) of 60℃ and a dew point of 26.7℃. Using the humidity chart, determine the actual humidity H, percentage humidity H_P, humid heat c_S, and humid volume v_H in SI units.

Solution. The dew point of 26.7℃ is the temperature when the given mixture is at 100% saturation. Starting at 26.7℃ (Fig. 11.5), and drawing a vertical line until it intersects the line for 100% humidity, a humidity of $H = 0.0225$ kg H₂O/kg dry air is read off the plot. This is the actual humidity of the air at 60℃. Stated in another way, if air at 60℃ and having a humidity $H = 0.0225$ kg H₂O/kg dry air is cooled, its dew point will be 26.7℃.

Locating this point where $H = 0.0225$ and $t = 60℃$ on the chart, the percentage humidity H_P is found to be 14%, by linear interpolation vertically between the 10 and 20% lines. The humid heat for $H = 0.0225$ is, from Eq. (9.3-6),

$$c_s = 1.005 + 1.88 \times 0.0225 = 1.047 \times 10^3 \text{J}/(\text{kg} \cdot ℃)$$

The humid volume at 60℃, from Eq. (9.3-7), is

$$v_H = (2.83 \times 10^{-3} + 4.56 \times 10^{-3} \times 0.0225) \times (60 + 273) = 0.977 \text{m}^3/\text{kg dry air}$$

Humidity charts for systems other than air-water

A humidity chart may be constructed for any system at any desired total pressure. The data required are the vapor pressure and latent heat of vaporization of the condensable component as a function of temperature, the specific heats of pure gas and vapor, and the molecular weights of both components. If a chart on a mole basis is desired, all equations can easily be modified to the use of molal units. If a chart at a pressure other than 1 atm is wanted, obvious modifications in the above equations may be made. Charts for several common systems besides air-water have been published.

11.2.9 Measurement of Humidity

The humidity of a stream or mass of gas may be found by measuring either the dew point or the wet-bulb temperature or by direct absorption methods.

Dew-point methods

If a cooled, polished disk is inserted into gas of unknown humidity and the temperature of the disk gradually lowered, the disk reaches a temperature at which mist condenses on the polished surface. The temperature at which this mist just forms is the temperature of equilibrium between the vapor in the gas and the liquid phase. It is therefore the dew point. A check on the reading is obtained by slowly increasing the disk temperature and noting the temperature at which the mist just disappears. From the average of the temperatures of mist formation and disappearance, the humidity can be read from a humidity chart.

Psychrometric methods

A very common method of measuring the humidity is to determine simultaneously the wet-bulb and dry-bulb temperatures. From these readings the humidity is found by locating the psychrometric line intersecting the saturation line at the observed wet-bulb temperature and following the psychrometric line to its intersection with the ordinate of the observed dry-bulb temperature.

Direct methods

The vapor content of a gas can be determined by direct analysis, in which a known volume of gas is drawn through an appropriate analytical device.

11.3 Principles of Drying

Because of the wide variety of materials that are dried in commercial equipment and the many types of equipment used, there is no single theory of drying that covers all materials and dryer types. Variations in shape and size of stock, in moisture equilibria, in the mechanism of flow of moisture through the solid, and in the method of providing the heat required for the vaporization—all prevent a unified treatment. General principles used in a semiquantitative way are relied upon. Dryers are seldom designed by the user but are bought from companies that specialize in the engineering and fabrication of drying equipment.

11.3.1 Temperature patterns in dryers

The way in which temperatures vary in a dryer depends on the nature and liquid content of the feedstock, the temperature of the heating medium, the drying time, and the allowable final temperature of the dry solids. The pattern of variation, however, is similar from one dryer to another. Typical patterns are shown in Fig. 11.7.

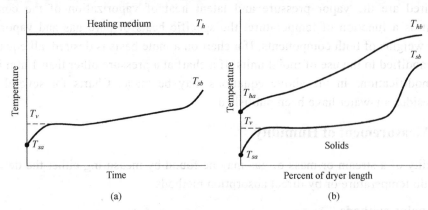

Fig. 11.7 Temperature patterns in dryers:
(a) batch dryer; (b) continuous countercurrent adiabatic dryer.

In a batch dryer with a heating medium at constant temperature (Fig. 11.7a), the temperature of the wet solids rises rather quickly from its initial value T_{sa} to the initial vaporization temperature T_v. In a nonadiabatic dryer with no sweep gas, T_v is essentially the boiling point of the liquid at the pressure prevailing in the dryer. If a sweep gas is used or if the dryer is adiabatic, T_v is at or near the wet-bulb temperature of the gas (which equals the adiabatic saturation temperature if the gas is air and water is the liquid being evaporated). Drying may occur at T_v for a considerable time, but often after a short time the temperature of the wet solids gradually rises as a zone of dry solids forms near the surface. The vaporization temperature then depends on the heat- and mass-transfer resistances in the dry zone, as well as on those in the external boundary layer. In the final stages of drying, the solids temperature rises rapidly to some higher value T_{sb}.

The drying time indicated in Fig. 11.7(a) may range from a few seconds to many hours. The solids may be at T_v for most of the drying cycle or for only a small fraction of it. The temperature of the heating medium is often constant, as shown, or it may be programmed to change as drying proceeds.

In an ideal continuous dryer, each particle or element of the solid passes through a cycle similar to that shown in Fig. 11.7(a) on its way from the inlet to the outlet of the dryer. In steady-state operation the temperature at any given point in a continuous dryer is constant, but it varies along the length of the dryer. Figure 11.7(b) shows a temperature pattern for an adiabatic countercurrent dryer. The solids inlet and gas outlet are on the left; the gas inlet and solids outlet are on the right. Again the solids are quickly heated from T_{sa} to T_v. The vaporization temperature may change as drying proceeds, even though the wet-bulb temperature remains the same. Near the gas inlet the solids may be heated to a temperature well above T_v in a relatively short length of the dryer, since the energy needed to heat the dry solids is small compared to that needed for vaporization. For heat-sensitive materials the dryer would be designed to keep T_{sb} close to T_v. Hot gas enters the dryer at T_{hb}, usually with low humidity. The temperature profile of the gas may have a complex shape because of the variation of the temperature driving force and the change in overall heat-transfer coefficient during the drying process.

11.3.2 Phase Equilibria

As in other transfer processes, such as mass transfer, the process of drying of materials must be approached from the viewpoint of the equilibrium relationships together with the rate relationships. In most of the drying apparatus discussed in 11.8, material is dried in contact with an air–water vapor mixture. The equilibrium relationships between the air–water vapor and the solid material will be discussed in this section.

An important variable in the drying of materials is the humidity of the air in contact with a solid of given moisture content. Suppose that a wet solid containing moisture is brought into contact with a stream of air having a constant humidity H and temperature. A large excess of air is used, so its conditions remain constant. Eventually, after exposure of the solid sufficiently long for equilibrium to be reached, the solid will attain a definite moisture content. This is known as the *equilibrium moisture content* of the material under the specified humidity and temperature of the air.

Equilibrium data for moist solids are commonly given as relationships between the relative humidity of the gas and the liquid content of the solid, in mass of liquid per unit mass of bone-dry solid. Examples of equilibrium relationships are shown in Fig. 11.8. Curves of this type are nearly independent of temperature. The abscissas of such curves are readily converted to absolute humidities, in mass of vapor per unit mass of dry gas. The moisture content is usually expressed on a dry basis as kg of water per kg of moisture-free (bone-dry) solid or kg H_2O/100 kg dry solid.

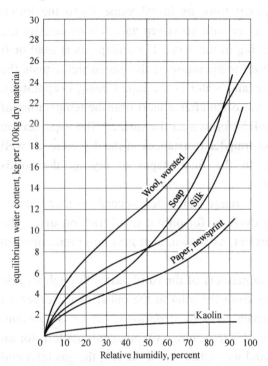

Fig. 11.8 Equilibrium-moisture curves at 25 ℃.

The remainder of the discussion in this section is based on the air-water system, but it should be remembered that the underlying principles apply equally well to other gases and liquids.

When a wet solid is brought into contact with air of lower humidity than that corresponding to the moisture content of the solid, as shown by the humidity equilibrium curve, the solid tends to lose moisture and dry to equilibrium with the air. When the air is more humid than the solid in equilibrium with it, the solid absorbs moisture from the air until equilibrium is attained.

For some solids the value of the equilibrium moisture content depends on the direction from which equilibrium is approached. The different value for the equilibrium moisture content is obtained according to whether a wet sample is allowed to dry by desorption or a dry sample adsorbs moisture by adsorption. For drying calculations it is the desorption equilibrium that is the larger value and is of particular interest

Porous solids such as catalysts or adsorbents often have an appreciable equilibrium moisture content at moderate relative humidity. Liquid water in fine capillaries exerts an abnormally low vapor pressure because of the highly concave surface of the meniscus. Adsorbents such as silica or alumina have monolayers of water strongly adsorbed on the surface, and this water has a much lower vapor pressure than liquid water. Beds of nonporous particles such as sand have negligible equilibrium moisture content in humid air unless the particles are so small that the fillets of liquid where particles touch have a very small radius of curvature.

In fluid phases diffusion is governed by concentration differences expressed in mole fractions. In a wet solid, however, the term *mole fraction* may have little meaning, and for ease in drying calculations the moisture content is nearly always expressed in mass of water per unit mass of bone-dry solid. This practice is followed throughout this chapter.

Free and equilibrium moisture of a substance

The air entering a dryer is seldom completely dry but contains some moisture and has a definite relative humidity. For air of definite humidity, the moisture content of the solid leaving the dryer cannot be less than the equilibrium moisture content corresponding to the humidity of the entering air. That portion of the water in the wet solid that cannot be removed by the inlet air, because of the humidity of the latter, is called the *equilibrium moisture*.

The free water is the difference between the total water content of the solid and the equilibrium water content. Free moisture content in a sample is the moisture above the equilibrium moisture content. This free moisture is the moisture that can be removed by drying under the given percent relative humidity.

Thus, if X_T is the total moisture content and if X^* is the equilibrium moisture content, the free moisture X is

$$X = X_T - X^* \tag{11.3-1}$$

It is X, rather than X_T, that is of interest in drying calculations.

For example, in Fig. 11.9 silk has an equilibrium moisture content of 8.5kg H_2O/100kg dry material in contact with air of 50% relative humidity and 25℃. If a sample contains 10kg H_2O/100kg dry material, only 10.0−8.5, or 1.5, kg H_2O/100kg material is removable by drying; this is the free moisture of the sample under these drying conditions. In many texts and references, the moisture content is given as percent moisture on a dry basis. This is exactly the same as the kg H_2O/100kg dry material multiplied by 100.

Bound and unbound water in solids

If an equilibrium curve like those in Fig. 11.8 or 11.9 is continued to its intersection with the axis for 100 percent humidity, the moisture content so defined is the minimum moisture this material can carry and still exert a vapor pressure as great as that exerted by liquid water at the same temperature. If such a material contains

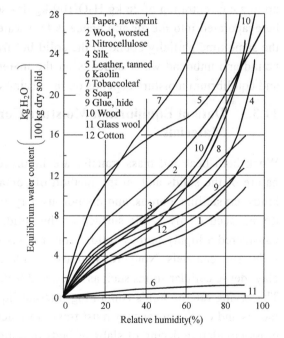

Fig. 11.9 Typical equilibrium moisture contents of some solids at approximately 298 K (25℃).

more water than that indicated by this intersection, it can still exert only the vapor pressure of water at the solids temperature. This makes possible a distinction between two types of water held by a given material. The water corresponding to concentrations lower than that indicated by the intersection of the curves in Fig.11.8 with the line for 100 percent humidity is called bound water, because it exerts a vapor pressure less than that of liquid water at the same temperature. Substances containing *bound water* are often called *hygroscopic* substances.

Bound water may exist in several conditions. Liquid water in fine capillaries exerts an abnormally low vapor pressure because of the highly concave curvature of the surface; moisture in cell or fiber walls may suffer a vapor pressure lowering because of solids dissolved in it; water in natural organic substances is in physical and chemical combination, the nature and strength of which vary with the nature and moisture content of the solid.

Water corresponding to concentrations greater than that indicated by the intersections is called *unbound water*. Unbound water, on the other hand, exerts its full vapor pressure and is largely held in the voids of the solid. Large nonporous particles, such as coarse sand, contain only unbound water.

The terms employed in this discussion may be clarified by reference to Fig.11.9. If the sample contains 30 kg H_2O/100 kg dry solid, for example , 4 kg H_2O/100 kg dry solid is unbound and 26 kg H_2O/100 kg dry solid is bound. Assume, now, that this sample is to be dried with air of 30 percent relative humidity. Curve 2 shows that the lowest moisture content that can be reached under these conditions is 9 kg H_2O/100 kg dry solid. This, then, is the equilibrium moisture content for this particular set of conditions. If a sample containing 30 kg H_2O/100 kg dry solid is to be dried with air at 30 percent relative humidity, it contains 21 kg H_2O/100 kg dry solid and 9 kg H_2O/100 kg dry solid equilibrium moisture. Any amount up to a concentration of 26 kg H_2O/100 kg dry solid is still bound water, but most of this can be evaporated into the air and hence is free water. Thus water can be both bound and free at the same time partially bound to the solid but free to be evaporated. The distinction between bound and unbound water depends on the material itself, while the distinction between free and equilibrium moisture depends on the drying conditions.

11.3.3 Data of Equilibrium Moisture Content for Inorganic and Biological Materials

When both heat and mass transfer are involved, the mechanism of drying depends on the nature of the solids and on the method of contacting the solids and gas. Solids are of three kinds: crystalline, porous, and nonporous. Crystalline particles contain no interior liquid, and drying occurs only at the surface of the solid. A bed of such particles, of course, can be considered a highly porous solid. Truly porous solids such as catalyst pellets, contain liquid in interior channels. Nonporous solids include colloidal gels such as soap, glue, and plastic clay; dense cellular solids such as wood and leather; and many polymeric materials.

The drying rate of solids containing internal liquid, however, depends on the way the liquid moves and on the distance it must travel to reach the surface. This is especially important in cross-circulation drying of slabs or beds of solids. Drying by this method is slow, is usually done batchwise, and has been displaced by other faster methods in most large-scale drying operations; it remains important, however, in the production of pharmaceuticals and fine chemicals, especially when drying conditions must be carefully controlled.

Typical data for various materials

If the material contains more moisture than its equilibrium value in contact with a gas of a given humidity and temperature, it will dry until it reaches its equilibrium value. If the material contains less moisture than its equilibrium value, it will adsorb water until it reaches its equilibrium value. For air having 0% humidity, the equilibrium moisture value of all materials is zero.

The equilibrium moisture content varies greatly with the type of material for any given percent relative humidity, as shown in Fig.11.8 for some typical materials at room temperature. Nonporous insoluble solids tend to have equilibrium moisture contents which are quite low, as shown for glass wool and kaolin. Certain spongy, cellular materials of organic and biological origin generally show large equilibrium moisture contents. Examples of these in Fig. 11.8 are wool, leather, and wood.

Typical food materials

In Fig. 11.8 the equilibrium moisture contents for some typical food materials are plotted versus percent relative humidity. These biological materials also show large values for equilibrium moisture content. Data in this figure and in Fig. 11.9 for biological materials show that at high percent relative humidities of about 60% to 80%, the equilibrium moisture content increases very rapidly with increases in relative humidity.

In general, at low relative humidities the equilibrium moisture content is greatest for food materials high in protein, starch, or other high-molecular-weight polymers and lower for food materials high in soluble solids. Crystalline salts and sugars as well as fats generally adsorb small amounts of water.

Effect of temperature

The equilibrium moisture content of a solid decreases somewhat with an increase in temperature. For example, for raw cotton at a relative humidity of 50%, the equilibrium moisture content decreased from 7.3 kg H_2O/100 kg dry solid at 37.8°C (311 K) to about 5.3 at 93.3°C (366.5 K), a decrease of about 25%. Often, for moderate temperature ranges, the equilibrium moisture content will be assumed constant when experimental data are not available at different temperatures.

At present, theoretical understanding of the structure of solids and surface phenomena does not enable us to predict the variation of equilibrium moisture content of various materials from first principles. However, by using models such as those used for adsorption isotherms of multilayers of molecules and others, attempts have been made to correlate experimental data. Henderson gives an empirical relationship between equilibrium moisture content and percent relative humidity for some agricultural materials. In general, empirical relationships are not available for most materials, and equilibrium moisture contents must be determined experimentally. Also, equilibrium moisture relationships often vary from sample to sample of the same kind of material.

11.4 Rate-of-Drying Curves

11.4.1 Method Using Experimental Drying Curve

In the drying of various types of process materials from one moisture content to another, it is usually desired to estimate the size of dryer needed, the various operating conditions of humidity and temperature for the air used, and the time needed to perform the amount of drying required. As discussed in Section 11.3, equilibrium moisture contents of various materials cannot be predicted and must be determined experimentally. Similarly, since our knowledge of the basic mechanisms of rates of drying is quite incomplete, it is necessary in most cases to obtain some experimental measurements of drying rates.

Experimental determination of rate of drying

To experimentally determine the rate of drying for a given material, a sample is usually placed on a tray. If it is a solid material it should fill the tray so that only the top surface is exposed to the drying air stream. By suspending the tray from a balance in a cabinet or duct through which the air is flowing, the loss in weight of moisture during drying can be determined at different intervals without interrupting the operation.

In doing batch-drying experiments, certain precautions should be observed to obtain usable data under conditions that closely resemble those to be used in the large-scale operations. The sample should not be too small in weight and should be supported in a tray or frame similar to the large-scale one. The ratio of drying to nondrying surface (insulated surface) and the bed depth should be similar. The velocity, humidity, temperature, and direction of the air should be the same and constant to simulate drying under constant drying conditions.

11.4.2 Rate of Drying Curves for Constant-Drying Conditions

Constant drying conditions

Consider a bed of wet solids, perhaps 50 to 75 mm deep, over which air is circulated. Assume that the temperature, humidity, and velocity and direction of flow of the air across the drying surface are constant. This is called drying under *constant drying conditions*. Note that only the conditions in the air stream are constant, as the moisture content and other factors in the solid are changing with time and position in the bed.

Conversion of data to rate-of-drying curve

Data obtained from a batch-drying experiment are usually obtained as W total weight of the wet solid (dry solid plus moisture) at different times t hours in the drying period. These data can be converted to rate-of-drying data in the following ways. First, the data are recalculated. If W is the weight of the wet solid in kg total water plus dry solid and W_S is the weight of the dry solid in kg,

$$X_t = \frac{W - W_S}{W_S} \left(\frac{\text{kg total water}}{\text{kg dry solid}} \right) \tag{11.4-1}$$

For the given constant drying conditions, the equilibrium moisture content X^* kg equilibrium moisture/kg dry solid is determined. Then the free moisture content X in kg free water/kg dry solid is calculated for each value of X_t:

$$X = X_t - X^* \tag{11.4-2}$$

Using the data calculated from Eq. (11.4-2), a plot of free moisture content X versus time t in h is made, as in Fig. 11.10(a). To obtain the rate-of-drying curve from this plot, the slopes of the tangents drawn to the curve in Fig. 11.10(a) can be measured, which give values of dX/dt at given values of t. The rate R is calculated for each point by

$$R = -\frac{L_S}{A} \times \frac{dX}{dt} \tag{11.4-3}$$

where R is drying rate in kg H$_2$O/(h·m^2), L_S kg of dry solid used, and A exposed surface area for drying in m^2. For obtaining R from Fig. 11.10(a), a value of L_S/A of 21.5 kg/m^2 was used.

The drying-rate curve is then obtained by plotting R versus the moisture content, as in Fig. 11.10(b).

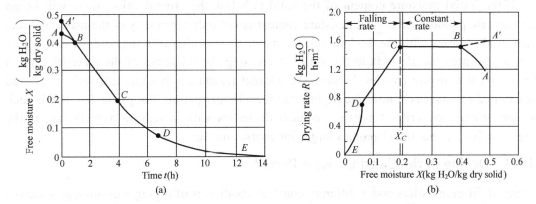

Fig. 11.10 Typical drying-rate curve for constant drying conditions: (a) plot of data as free moisture versus time; (b) rate of drying curve as rate versus free moisture content.

Another method for obtaining the rate-of-drying curve is to first calculate the weight loss ΔX for a Δt time. For example, if $X_1 = 0.350$ at a time $t_1 = 1.68$ h and $x_2 = 0.325$ at a time $t_2 = 2.04$ h, $\Delta X/\Delta t = (0.350-0.325)/(2.04-1.68)$. Then, using Eq. (11.4-4) and $L_S/A = 21.5$,

$$R = -\frac{L_S}{A} \times \frac{\Delta X}{\Delta t} = 21.5 \times \frac{0.350-0.325}{2.04-1.68} = 1.493 \quad (11.4\text{-}4)$$

This rate R is the average over the period 1.68 to 2.04 h and should be plotted at the average concentration $X = (0.350 + 0.325)/2 = 0.338$.

Plot of rate-of-drying curve

In Fig. 11.10(b) the rate-of-drying curve for constant-drying conditions is shown. At zero time the initial free moisture content is shown at point A. In the beginning the solid is usually at a colder temperature than its ultimate temperature, and the evaporation rate will increase. Eventually, at point B, the surface temperature rises to its equilibrium value. Alternatively, if the solid is quite hot to start with, the rate may start at point A'. This initial unsteady-state adjustment period is usually quite short and it is often ignored in the analysis of times of drying.

From point B to point C in Fig. 11.10(a) the line is straight, and hence the slope and rate are constant during this period. This *constant-rate-of-drying period* is shown as line BC in Fig. 11.10(b). At point C on both plots, the drying rate starts to decrease in the *falling-rate period* until it reaches point D. In this first falling-rate period, the rate shown as line CD in Fig. 11.10(b) is often linear. At point D the rate of drying falls even more rapidly, until it reaches point E, where the equilibrium moisture content is X^* and $X = X^*-X^* = 0$. In some materials being dried, the region CD may be missing completely, or it may constitute all of the falling-rate period.

11.4.3 Critical Moisture Content

The point at which the constant-rate period ends (whether or not the drying rate is truly constant) is called the *critical moisture content*. Sometimes it is clearly identifiable, as shown by point C in Fig.11.10; more often it is approximate. It represents the moisture content below which insufficient liquid can be transferred from the interior of the solid to maintain a continuous or nearly continuous liquid film on the surface.

If the initial moisture content of the solid is below the critical value, there will be no constant-rate period. The critical moisture content is not only a property of the material being dried. It also varies with the thickness of the material, the rate of drying, and the resistances to heat and mass transfer within the solid. Decreasing the thickness of the material gives a lower critical moisture content, because the internal resistances become relatively small and the external resistances control the drying rate for a longer period. For particulate solids such as sand or clay, the critical moisture content decreases with increasing particle size in the range 5 to 200μm, but then becomes larger for coarser particles.

11.4.4 Drying in the Constant-Rate Period

Drying of different solids under different constant conditions of drying will often give curves of different shapes in the falling-rate period, but in general the two major portions of the drying-rate curve—constant-rate period and falling-rate period—are present.

In the constant-rate drying period, the surface of the solid is initially very wet and a continuous film of water exists on the drying surface. This water is entirely unbound water and it acts as if the solid were not present. The rate of evaporation under the given air conditions is independent of the solid and essentially the same as the rate from a free liquid surface. Increased roughness of the solid surface, however, may lead to higher rates than from a flat surface.

If the solid is porous, most of the water evaporated in the constant-rate period is supplied from the interior of the solid. This period continues only as long as the water is supplied to the surface as fast as it is evaporated. Evaporation during this period is similar to that in determining the wet bulb temperature, and in the absence of heat transfer by radiation or conduction, the surface temperature is approximately the same as the wet bulb temperature.

11.4.5 Drying in the Falling-Rate Period

In Fig. 11.10(b) point C (*critical free moisture content* X_C), there is insufficient water on the surface to maintain a continuous film of water. The entire surface is no longer wetted, and the wetted area continually decreases in this first falling-rate period until the surface is completely dry, at point D in Fig. 11.10(b).

The second falling-rate period begins at point D when the surface is completely dry. The plane of evaporation slowly recedes from the surface. Heat for the evaporation is transferred through the solid to the zone of vaporization. Vaporized water moves through the solid into the air stream.

In some cases no sharp discontinuity occurs at point D, and the change from partially wetted to completely dry conditions at the surface is so gradual that no distinct change is detectable.

The amount of moisture removed in the falling-rate period may be relatively small, but the time required may be long. This can be seen in Fig. 11.10. The period BC for constant-rate drying lasts for about 3.0 h and reduces X from 0.40 to about 0.19, a reduction of 0.21 kg H_2O/kg dry solid. The falling-rate period CE lasts about 9.0 h and reduces X only from 0.19 to 0.

11.4.6 Moisture Movements in Solids During Drying in the Falling-Rate Period

When drying occurs by evaporation of moisture from the exposed surface of a solid, moisture must move from the depths of the solid to the surface. The mechanisms of this movement affect the drying during the constant-rate and falling-rate periods. Some of the theories advanced to explain the various types of falling-rate curves will be briefly reviewed.

Liquid diffusion theory

According to this theory, diffusion of liquid moisture occurs when there is a concentration difference between the depths of the solid and the surface. This method of transport of moisture is usually found in nonporous solids where single-phase solutions are formed with the moisture, such as in paste, soap, gelatin, and glue. It is also found in drying the last portions of moisture from clay, flour, wood, leather, paper, starches, and textiles. In drying many food materials, the movement of water in the falling-rate period also occurs by diffusion.

The shapes of the moisture-distribution curves in the solid at given times are qualitatively consistent The moisture diffusivity D_{AB} usually decreases with decreased moisture content, so that the diffusivities are usually average values over the range of concentrations used. Materials drying in this way are usually said to be drying by diffusion, although the actual mechanisms may be quite complicated. Since the rate of evaporation from the surface is quite fast, that is, the resistance is quite low, compared to the diffusion rate through the solid in the falling-rate period, the moisture content at the surface is at the equilibrium value.

The shape of a diffusion-controlled curve in the falling-rate period is similar to Fig. 11.11(a). If the initial constant-rate drying is quite high, the first falling-rate period of unsaturated surface evaporation may not appear. If the constant-rate drying is quite low, the period of unsaturated surface evaporation is usually present in region CD in Fig. 11.10(b) and the diffusion-controlled curve is in region DE.

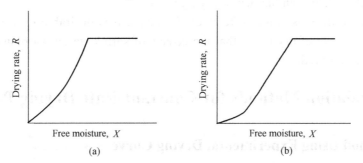

Fig. 11.11 Typical drying-rate curves: (a) diffusion-controlled falling-rate period; (b) capillary-controlled falling-rate period in a fine porous solid.

Capillary movement in porous solids

When granular and porous solids such as clays, sand, soil, paint pigments, and minerals are being dried, unbound or free moisture moves through the capillaries and voids of the solids by capillary action, not by diffusion. This mechanism, involving surface tension, is similar to the movement of oil in a lamp wick.

A porous solid contains interconnecting pores and channels of varying pore sizes. As water is evaporated, a meniscus of liquid water is formed across each pore in the depths of the solid. This sets up capillary forces by the interfacial tension between the water and the solid. These capillary forces provide the driving force for moving water through the pores to the surface. Small pores develop greater forces than do large pores.

At the beginning of the falling-rate period at point C in Fig. 11.10(b), the water is being brought to the surface by capillary action, but the surface layer of water starts to recede below the surface. Air rushes in to fill the voids. As the water is continuously removed, a point is reached where there is insufficient water left to maintain continuous films across the pores, and the rate of drying suddenly decreases at the start of the second falling-rate period at point D. Then the rate of diffusion of water vapor in the pores and rate of conduction of heat in the solid may become the main factors in drying.

In fine pores in solids, the rate-of-drying curve in the second falling-rate period may conform to the diffusion law; the curve is concave upward, as shown in Fig. 11.11(b). For very porous solids, such as a bed of sand, where the pores are large, the rate-of-drying curve in the second falling-rate period is often straight, and hence the diffusion equations do not apply.

11.4.7 Effect of Shrinkage

A factor often greatly affecting the drying rate is the shrinkage of the solid as moisture is removed. Rigid solids do not shrink appreciably, but colloidal and fibrous materials such as vegetables and other foodstuffs do undergo shrinkage. The most serious effect is that there may be developed a hard layer on the surface which is impervious to the flow of liquid or vapor moisture and slows the drying rate; examples are clay and soap. In many foodstuffs, if drying occurs at too high a temperature, a layer of closely packed, shrunken cells, which are sealed together, forms at the surface. This presents a barrier to moisture migration and is known as *case hardening*. Another effect of shrinkage is to cause the material to warp and change its structure. This can happen in drying wood.

Sometimes, to decrease these effects of shrinkage, it is desirable to dry with moist air. This decreases the rate of drying so that the effects of shrinkage on warping or hardening at the surface are greatly reduced.

11.5 Calculation Methods for Constant-Rate Drying Period

11.5.1 Method Using Experimental Drying Curve

Probably the most important factor in drying calculations is the length of time required to dry a material from a given initial free moisture content X_1 to a final moisture content X_2. For drying in the constant-rate period, we can estimate the time needed by using experimental batch-drying curves or by using predicted mass- and heat-transfer coefficients.

Method using drying curve

To estimate the time of drying for a given batch of material, the best method is based on actual experimental data obtained under conditions where the feed material, relative exposed surface area, gas velocity, temperature, and humidity are essentially the same as in the final drier. Then the time required for the constant-rate period can be determined directly from the drying curve of free moisture content versus time.

> **EXAMPLE 11.4 Time of Drying from Drying Curve.**
> A solid whose drying curve is represented by Fig. 11.10(a) is to be dried from a free moisture content $X_1 = 0.38$ kg H$_2$O/kg dry solid to $X_2 = 0.25$ kg H$_2$O/kg dry solid. Estimate the time required.
> **Solution.** From Fig. 11.10(a) for $X_1 = 0.38$, t_1 is read off as 1.28 h. For $X_2 = 0.25$, $t_2 = 3.08$ h. Hence, the time required is
> $$t = t_2 - t_1 = 3.08 - 1.28 = 1.80 \text{ h}$$

11.5.2 Method Using Rate-of-Drying Curve for Constant-Rate Period

Instead of using the drying curve, the rate-of-drying curve can be used. The drying rate R is defined as

$$R = -\frac{L_S}{A} \times \frac{dX}{dt} \tag{11.5-1}$$

This can be rearranged and integrated over the time interval for drying from X_1 at $t_1 = 0$ to X_2 at $t_2 = t$:

$$t = \int_{t_1=0}^{t_2=t} dt = \frac{L_S}{A} \int_{X_2}^{X_1} \frac{dX}{R} \tag{11.5-2}$$

If the drying takes place within the constant-rate period, so that both X_1 and X_2 are greater than the critical moisture content X_C, then $R = $ constant $= R_C$. Integrating Eq. (11.5-2) for the constant-rate period,

$$t = \frac{L_S}{AR_C}(X_1 - X_2) \tag{11.5-3}$$

> **EXAMPLE 11.5 Drying Time from Rate-of-Drying Curve.**
> Repeat Example 11.4 but use Eq. (11.5-3) and Fig. 11.10(b).
> **Solution.** As stated previously, a value of 21.5 for L_S/A was used to prepare Fig. 11.10(b) from Fig. 11.10(a). From Fig. 11.10(b), $R_C = 1.51$ kg H$_2$O/(h · m^2). Substituting into Eq. (11.5-3),
> $$t = \frac{L_S}{AR_C}(X_1 - X_2) = \frac{21.5}{1.51} \times (0.38 - 0.25) = 1.85 \text{ h}$$
> This is close to the value of 1.80 h in Example 11.4.

11.5.3 Method Using Predicted Transfer Coefficients for Constant-Rate Period

In the constant-rate period of drying, the surfaces of the grains of solid in contact with the drying air flow remain completely wetted. As stated previously, the rate of evaporation of moisture under a given set of air conditions is independent of the type of solid and is essentially the same as the rate of evaporation from a free liquid surface under the same conditions. However, surface roughness may increase the rate of evaporation.

During this constant-rate period, the solid is so wet that the water acts as if the solid were not there. The water evaporated from the surface is supplied from the interior of the solid. The rate of evaporation from a porous material occurs by the same mechanism as that occurring at a wet bulb thermometer, which is essentially constant-rate drying.

Equations for predicting constant-rate drying

Drying of a material occurs by mass transfer of water vapor from the saturated surface of the material through an air film to the bulk gas phase or environment. The rate of moisture movement within the solid is sufficient to keep the surface saturated. The rate of removal of the water vapor (drying) is controlled by the rate of heat transfer to the evaporating surface, which furnishes the latent heat of evaporation for the liquid. At steady state, the rate of mass transfer balances the rate of heat transfer.

To derive the equation for drying, we neglect heat transfer by radiation to the solid surface and also assume no heat transfer by conduction from metal pans or surfaces. Assuming only heat transfer to the solid surface by convection from the hot gas to the surface of the solid and mass transfer from the surface to the hot gas (Fig.11.12), we can write equations which are the same as those for deriving the wet bulb temperature T_W in Eq. (11.2-24).

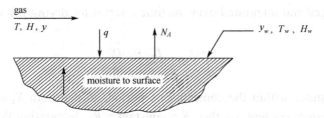

Fig.11.12 Heat and mass transfer in constant-rate drying.

The rate of convective heat transfer q in W (J/s) from the gas at T °C to the surface of the solid at T_W °C, where $(T - T_W)$ °C $= (T - T_W)$ K is

$$q = h(T - T_W) \tag{11.5-4}$$

where h is the heat-transfer coefficient in W/(m² · K) and A is the exposed drying area in m². The equation of the flux of water vapor from the surface is

$$N_A = k_y(y_W - y) \tag{11.5-5}$$

$$y \cong \frac{HM_B}{M_A} \tag{11.5-6}$$

Using the approximation from Eq. (11.5-6) and substituting into Eq. (11.5-5),

$$N_A = k_y \frac{M_B}{M_A}(H_W - H) \tag{11.5-7}$$

The amount of heat needed to vaporize N_A kg mol/(s · m²) water, neglecting the small sensible heat changes:

$$q = M_A N_A r_W A \tag{11.5-8}$$

where r_W is the latent heat at T_W in J/kg.

Equating Eqs. (11.5-4) and (11.5-8) and substituting Eq. (11.5-7) for N_A,

$$R_C = \frac{q}{Ar_W} = \frac{h(T - T_W)}{r_W} = k_y M_B (H_W - H) \tag{11.5-9}$$

Equation (11.5-9) is identical to Eq. (11.2-24) for the wet bulb temperature. Hence, in the absence of heat transfer by conduction and radiation, the temperature of the solid is at the wet bulb temperature of the air during the constant-rate drying period. Thus, the rate of drying R_C can be calculated using the heat-transfer equation $h(T - T_W)/r_W$ or the mass-transfer equation $k_y M_B(H_W - H)$. However, it has been found more reliable to use the heat-transfer equation (11.5-8), since an error in determining the interface temperature T_W at the surface affects the driving force $(T - T_W)$ much less than it affects $(H_W - H)$.

$$R_C \text{ kgH}_2\text{O}/(\text{s} \cdot \text{m}^2) = \frac{h}{r_W}(T - T_W) \tag{11.5-10}$$

To predict R_C in Eq. (11.5-10), the heat-transfer coefficient must be known. However, because the shape of the leading edge of the drying surface causes more turbulence, the following can be used for an air temperature of 45~150°C and a mass velocity G of 2450~29300 kg/(h·m²) or a velocity of 0.61~7.6 m/s:

$$h = 0.0204 G^{0.8} \tag{11.5-11}$$

where G is the product $u\rho$ of the velocity u of air flowing parallel to the surface and the density ρ of air in kg/(h · m²) and h is W/(m² · K). Air flows perpendicular to the surface for a G of 3900~19 500 kg/(h·m²) or a velocity of 0.9~4.6 m/s.

$$h = 1.17 G^{0.37} \tag{11.5-12}$$

Equations (11.5-10)~(11.5-12) can be used to estimate the rate of drying during the constant-rate period. However, if possible, experimental measurements of the drying rate are preferred.

To estimate the time of drying during the constant-rate period, substituting Eq. (11.5-9) into (11.5-3),

$$t = \frac{L_S r_W (X_1 - X_2)}{Ah(T - T_W)} = \frac{L_S (X_1 - X_2)}{Ak_y M_B (H_w - H)} \tag{11.5-13}$$

EXAMPLE 11.6 Prediction of Constant-Rate Drying.
An insoluble wet granular material is dried in a pan 0.457 m×0.457 m and 25.4 mm deep. The material is 25.4 mm deep in the pan, and the sides and bottom can be considered to be insulated. Heat transfer is by convection from an air stream flowing parallel to the surface at a velocity of 6.1 m/s. The air is at 65.6°C and has a humidity of 0.010 kg H₂O/kg dry air. Estimate the rate of drying for the constant-rate period using SI.
Solution. For a humidity $H = 0.010$ and dry bulb temperature of 65.6°C, using the humidity chart, Fig. 11.5, the wet bulb temperature T_W is found to be 28.9°C and $H_W = 0.026$ by following the adiabatic saturation line (the same as the wet bulb line) to the saturated humidity. Using Eq. (11.2-13) to calculate the humid volume,

$$v_H = (2.83 \times 10^{-3} + 4.56 \times 10^{-3} H)T$$
$$= (2.83 \times 10^{-3} + 4.56 \times 10^{-3} \times 0.01) \times (273 + 65.6) = 0.974 \text{ m}^3/\text{kg dry air}$$

The density for 1.0 kg dry air + 0.010 kg H_2O is
$$\rho = \frac{1.0 + 0.010}{0.974} = 1.037 \text{ kg/m}^3$$

The mass velocity G is
$$G = V\rho = 6.1 \times 1.037 \times 3600 = 22770 \text{ kg/(h} \cdot \text{m}^2)$$

Using Eq. (11.5-11),
$$h = 0.0204 G^{0.8} = 0.0204 \times (22770)^{0.8} = 62.45 \text{ W/(m}^2 \cdot \text{K)}$$

At $T_W = 28.9°C$, $r_W = 2433$ kJ/kg from steam tables.

Substituting into Eq. (11.5-10) and noting that $(65.6 - 28.9)°C = (65.6 - 28.9)$ K,
$$R_C = \frac{h}{r_W}(T - T_W) \times 3600 = \frac{62.45}{2433 \times 1000}(65.6 - 28.9) \times 3600 = 3.39 \text{ kg/(h} \cdot \text{m}^2)$$

The total evaporation rate for a surface area of 0.457×0.457 m^2 is
$$\text{total rate} = R_C A = 3.39 \times 0.457 \times 0.457 = 0.708 \text{ kg H}_2\text{O/h}$$

11.5.4 Effect of Process Variables on Constant-Rate Period

As stated previously, experimental measurements of the drying rate are usually preferred over using the equations for prediction. However, these equations are quite helpful in predicting the effect of changing the drying-process variables when limited experimental data are available.

Effect of air velocity. When conduction and radiation heat transfer are not present, the rate R_C of drying in the constant-rate region is proportional to h and hence to $G^{0.8}$ as given by Eq. (11.5-11) for air flow parallel to the surface. The effect of gas velocity is less important when radiation and conduction are present.

Effect of gas humidity. If the gas humidity H is decreased for a given T of the gas, then from the humidity chart the wet bulb temperature T_W will decrease. Then, using Eq. (11.5-9), R_C will increase. For example, if the original conditions are R_{C1}, T_1, T_{W1}, H_1, and H_{W1}, then if H_1 is changed to H_2 and H_{W1} is changed to H_{W2}, R_{C2} becomes

$$R_{C2} = R_{C1} \frac{T - T_{W2}}{T - T_{W1}} \frac{r_{W1}}{r_{W2}} = R_{C1} \frac{H_{W2} - H_2}{H_{W1} - H_1} \tag{11.5-14}$$

However, since $r_{W1} \approx r_{W2}$,

$$R_{C2} = R_{C1} \frac{T - T_{W2}}{T - T_{W1}} = R_{C1} \frac{H_{W2} - H_2}{H_{W1} - H_1} \tag{11.5-15}$$

Effect of gas temperature. If the gas temperature T is increased, T_W is also increased somewhat, but not as much as the increase in T. Hence, R_C increases as follows:

$$R_{C2} = R_{C1} \frac{T_2 - T_{W2}}{T_1 - T_{W1}} = R_{C1} \frac{H_{W2} - H_2}{H_{W1} - H_1} \tag{11.5-16}$$

Effect of thickness of solid being dried. For heat transfer by convection only, the rate

R_C is independent of the thickness x_1 of the solid. However, the time t for drying between fixed moisture contents X_1 and X_2 will be directly proportional to the thickness x_1. This is shown by Eq. (11.5-3), where increasing the thickness with a constant A will directly increase the amount of L_S kg dry solid.

Experimental effect of process variables. Experimental data tend to bear out the conclusions reached on the effects of material thickness, humidity, air velocity, and $T-T_W$.

11.6 Calculation Methods for Falling-rate Drying Period

11.6.1 Method Using Numerical Integration

In the falling-rate drying period, as shown in Fig. 11.10(b), the rate of drying R is not constant but decreases when drying proceeds past the critical free moisture content X_C. When the free moisture content X is zero, the rate drops to zero.

The time of drying for any region between X_1 and X_2 has been given by Eq. (11.5-2):

$$t = \frac{L_S}{A} \int_{X_2}^{X_1} \frac{dX}{R} \tag{11.5-2}$$

If the rate is constant, Eq. (11.5-2) can be integrated to give Eq. (11.5-3). However, in the falling-rate period, R varies. For any shape of falling-rate drying curve, Eq. (11.5-2) can be integrated by plotting $1/R$ versus X and determining the area under the curve using graphical integration or numerical integration with a spreadsheet.

EXAMPLE 11.7 A batch of wet solid whose drying-rate curve is represented by Fig. 11.10(b) is to be dried from a free moisture content of $X_1 = 0.38$ kg H_2O/kg dry solid to $X_2 = 0.04$ kg H_2O/kg dry solid. The weight of the dry solid is $L_S = 399$ kg dry solid and $A = 18.58$ m² of top drying surface. Calculate the time for drying. Note that $L_S/A = 399/18.58 = 21.5$ kg/m².

Solution. From Fig. 11.10(b), the critical free moisture content is $X_C = 0.195$ kg H_2O/kg dry solid. Hence, the drying occurs in the constant-rate and falling-rate periods.

For the constant-rate period, $X_1 = 0.38$ and $X_2 = X_C = 0.195$. From Fig. 11.10(b), $R_C = 1.51$ kg H_2O/(h·m²). Substituting into Eq. (11.5-2),

$$t = \frac{L_S}{AR_C}(X_1 - X_2) = \frac{399 \times (0.38 - 0.195)}{18.58 \times 1.51} = 2.63 \text{h}$$

For the falling-rate period, reading values of R for various values of X from Fig. 11.10(b), the following table is prepared:

X	R	1/R	X	R	1/R
0.195	1.51	0.663	0.065	0.71	1.41
0.150	1.21	0.826	0.050	0.37	2.70
0.100	0.90	1.11	0.040	0.27	3.70

To determine this area by numerical integration using a spreadsheet, the calculations are given in the following table:

X	R	$1/R$	ΔX	$(1/R)_{av}$	$(\Delta X)(1/R)_{av}$
0.195	0.51	0.663	0.045	0.745	0.0335
0.150	1.21	0.826	0.050	0.969	0.0485
0.100	0.90	1.111	0.035	1.260	0.0441
0.065	0.71	1.408	0.015	2.055	0.0308
0.050	0.37	2.702	0.010	3.203	0.0320
0.040	0.27	3.704			Total=0.1889

The area of the first rectangle is the average height $(0.663 + 0.826)/2$, or 0.745, times the width ΔX (0.195–0.150), or 0.045, giving 0.0335. Other values are similarly calculated, giving a total of 0.1889.

Substituting into Eq. (11.5-1),

$$t = \frac{L_S}{A} \int_{X_2}^{X_1} \frac{dX}{R} = \frac{399}{18.58} \times 0.1889 = 4.06\text{h}$$

The total time is $2.63 + 4.06 = 6.69$ h.

11.6.2 Calculation Methods for Special Cases in Falling-Rate Region

In certain special cases in the falling-rate region, the equation for the time of drying, Eq. (11.6-1), can be integrated analytically.

Rate is a linear function of X

If both X_1 and X_2 are less than X_C and the rate R is linear in X over this region,

$$R = aX + b \tag{11.6-1}$$

where a is the slope of the line and b is a constant. Differentiating Eq. (11.6-1) gives $dR = adX$. Substituting this into Eq. (11.5-1),

$$t = \frac{L_S}{aA} \int_{R_2}^{R_1} \frac{dR}{R} = \frac{L_S}{aA} \ln \frac{R_1}{R_2} \tag{11.6-2}$$

Since $R_1 = aX_1 + b$ and $R_2 = aX_2 + b$,

$$a = \frac{R_1 - R_2}{X_1 - X_2} \tag{11.6-3}$$

Substituting Eq. (11.6-3) into (11.6-2),

$$t = \frac{L_S(X_1 - X_2)}{A(R_1 - R_2)} \ln \frac{R_1}{R_2} \tag{11.6-4}$$

Rate is a linear function through origin

In some cases a straight line from the critical moisture content passing through the origin adequately represents the whole falling-rate period. In Fig. 11.10(*b*) this would be a straight line from C to E at the origin. Often, for lack of more detailed data, this assumption is made. Then, for a straight line through the origin, where the rate of drying is directly proportional to the free moisture content,

$$R = aX \tag{11.6-5}$$

Differentiating, $dX = dR/a$. Substituting into Eq. (11.5-2),

$$t = \frac{L_S}{aA}\int_{R_2}^{R_1}\frac{dR}{R} = \frac{L_S}{aA}\ln\frac{R_1}{R_2} \quad (11.6\text{-}6)$$

The slope a of the line is R_C/X_C, and for $X_1 = X_C$ at $R_1 = R_C$,

$$t = \frac{L_S X_C}{AR_C}\ln\frac{R_C}{R_2} \quad (11.6\text{-}7)$$

Noting also that $R_C/R_2 = X_C/X_2$,

$$t = \frac{L_S X_C}{AR_C}\ln\frac{X_C}{X_2} \quad (11.6\text{-}8)$$

or

$$R = R_C\frac{X}{X_C} \quad (11.6\text{-}9)$$

EXAMPLE 11.8 Repeat Example 11.7, but as an approximation assume a straight line for the rate R versus X through the origin from point X_C to $X = 0$ for the falling-rate period.
Solution. $R_C = 1.51$ kg $H_2O/(h \cdot m^2)$ and $X_C = 0.195$. Drying in the falling-rate region is from X_C to $X_2 = 0.040$. Substituting into Eq. (11.6-8),

$$t = \frac{L_S X_C}{AR_C}\ln\frac{X_C}{X_2} = \frac{399 \times 0.195}{18.58 \times 1.51}\ln\frac{0.195}{0.040} = 4.40\text{h}$$

This value of 4.40 h compares favorably with the value of 4.06 h obtained in Example 11.7 by numerical integration.

11.7 Equations for Various Types of Dryers

11.7.1 Through-Circulation Drying in Packed Beds

For through-circulation drying, where the drying gas passes upward or downward through a bed of wet granular solids, both a constant-rate period and a falling-rate period of drying may result. Often the granular solids are arranged on a screen so that the gas passes through the screen and through the open spaces or voids between the solid particles.

Derivation of equations

To derive the equations for this case, no heat losses will be assumed, so the system is adiabatic. The drying will be for unbound moisture in the wet granular solids. We shall consider a bed of uniform cross-sectional area A m^2, where a gas flow of G kg dry gas/(h $\cdot$ m^2 cross section) enters with a humidity of H_1. By a material balance on the gas at a given time, the gas leaves the bed with a humidity H_2. The amount of water removed from the bed by the gas is equal to the rate of drying at this time:

$$W = G(H_2 - H_1) \quad (11.7\text{-}1)$$

where $W =$ kg $H_2O/(h \cdot m^2$ cross section) and $G =$ kg dry air/(h $\cdot$ m^2 cross section).

In Fig. 11.13 the gas enters at T_1 and H_1 and leaves at T_2 and H_2. Hence, the temperature T and humidity H both vary through the bed. Making a heat balance over the short section dz m of the bed,

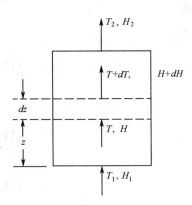

Fig. 11.13 Heat and material balances in a through-circulation dryer in a packed bed.

$$dq = -Gc_s A dT \quad (11.7\text{-}2)$$

where A is cross-sectional area in m², q is the heat-transfer rate in W (J/s), and c_s is the humid heat of the air–water vapor mixture in Eq. (11.2-11). Note that G in this equation is in kg/(s · m²). The heat-transfer equation gives

$$dq = ha A dZ(T - T_W) \quad (11.7\text{-}3)$$

where T_W is wet-bulb temperature of solid, h is the heat-transfer coefficient in W/(m²·K), and a is m² surface area of solids/m³ bed volume. Equating Eq. (11.7-2) to (11.7-3), rearranging, and integrating,

$$\frac{ha}{Gc_s}\int_0^z dz = -\int_{T_1}^{T_2}\frac{dT}{T - T_W} \quad (11.7\text{-}4)$$

$$\frac{haz}{Gc_s} = \ln\frac{T_1 - T_W}{T_2 - T_W} \quad (11.7\text{-}5)$$

where z = bed thickness = δ m.

For the constant-rate period of drying by air flowing parallel to a surface, Eq. (11.7-6) was derived:

$$t = \frac{L_s r_W (X_1 - X_2)}{Ah(T - T_W)} = \frac{L_s (X_1 - X_2)}{A k_y M_B (H_W - H)} \quad (11.7\text{-}6)$$

Using the definition of a, we obtain $L_s = x_1 A \rho_s$,

$$\frac{L_s}{A} = \frac{\rho_s}{a} \quad (11.7\text{-}7)$$

where ρ_s = solid density kg dry solid/m³.

Substituting Eq. (11.7-7) into (11.7-6) and setting $X_2 = X_C$ for drying to X_C, we obtain the equation for through-circulation drying in the constant-rate period:

$$t = \frac{\rho_s r_W (X_1 - X_C)}{ah(T - T_W)} = \frac{\rho_s (X_1 - X_C)}{a k_y M_B (H_W - H)} \quad (11.7\text{-}8)$$

In a similar manner, Eq. (11.7-8) for the falling-rate period, which assumes that R is proportional to X, becomes, for through-circulation drying,

$$t = \frac{\rho_s r_W X_C \ln(X_C/X)}{ah(T - T_W)} = \frac{\rho_s X_C \ln(X_C/X)}{a k_y M_B (H_W - H)} \quad (11.7\text{-}9)$$

Both Eqs. (11.7-8) and (11.7-9), however, hold for only one point in the bed in Fig. 11.18, since the temperature T of the gas varies throughout the bed. Hence, in a manner similar to the derivation in heat transfer, a log mean temperature difference can be used as an approximation for the whole bed in place of $T - T_W$ in Eqs. (11.7-8) and (11.7-9):

$$(T-T_W)_{LM} = \frac{(T_1-T_W)-(T_2-T_W)}{\ln\left[(T_1-T_W)/(T_2-T_W)\right]} \qquad (11.7\text{-}10)$$

Substituting Eq. (11.7-6) for the denominator of Eq. (11.7-10) and substituting the value of T_2 from Eq. (11.7-6) into Eq.(11.7-10),

$$(T-T_W)_{LM} = \frac{(T_1-T_W)\left(1-e^{\frac{-haz}{Gc_S}}\right)}{haz/Gc_S} \qquad (11.7\text{-}11)$$

Substituting Eq. (11.7-11) into Eq.(11.7-8) for the constant-rate period and setting $x_1 = z$,

$$t = \frac{\rho_S r_W x_1 (X_1 - X_C)}{Gc_S (T_1-T_W)\left(1-e^{\frac{-hax_1}{Gc_S}}\right)} \qquad (11.7\text{-}12)$$

Similarly, for the falling-rate period an approximate equation is obtained:

$$t = \frac{\rho_S r_W x_1 X_C \ln(X_C/X)}{Gc_S (T_1-T_W)\left(1-e^{\frac{-hax_1}{Gc_S}}\right)} \qquad (11.7\text{-}13)$$

A major difficulty with the use of Eq. (11.7-13) is that the critical moisture content is not easily estimated. Different forms of Eqs. (11.7-12) and (11.7-13) can also be derived, using humidity instead of temperature.

Heat-transfer coefficients

For through-circulation drying, where the gases pass through a bed of wet granular solids, the following equations for estimating h for adiabatic evaporation of water can be used:

$$\frac{D_p G_t}{\mu} > 350 \qquad h = 0.151 \frac{G_t^{0.59}}{D_p^{0.41}} \qquad (11.7\text{-}14)$$

$$\frac{D_p G_t}{\mu} < 350 \qquad h = 0.214 \frac{G_t^{0.49}}{D_p^{0.51}} \qquad (11.7\text{-}15)$$

where h is in W/(m²·K), D_p is diameter of a sphere having the same surface area as the particle in the bed in m, G_t is the total mass velocity entering the bed in kg/(h·m²), and μ is viscosity in kg/(m·h).

Geometry factors in a bed

To determine the value of a, m² surface area/m³ of bed, in a packed bed for spherical particles having a diameter D_p m,

$$a = \frac{6(1-\varepsilon)}{D_p} \qquad (11.7\text{-}16)$$

where ε is the void fraction in the bed. For cylindrical particles,

$$a = \frac{4(1-\varepsilon)(h+0.5D_C)}{D_C h} \qquad (11.7\text{-}17)$$

where D_c is diameter of cylinder in m, and h is length of cylinder in m. The value of D_p to use in Eqs. (11.7-14) and (11.7-15) for a cylinder is the diameter of a sphere having the same surface area as the cylinder, as follows:

$$D_P = \left(D_C h + 0.5 D_C^2\right)^{1/2} \tag{11.7-18}$$

Equations for very fine particles

The equations derived for the constant- and falling-rate periods in packed beds hold for particles of about 3~19 mm in diameter in shallow beds about 10~65 mm thick. For very fine particles of 10~200 mesh (1.66~0.079 mm) and bed depth greater than 11 mm, the interfacial area a varies with the moisture content. Empirical expressions are available for estimating a and the mass-transfer coefficient.

EXAMPLE 11.9 Through-Circulation Drying in a Bed.
A granular paste material is extruded into cylinders with a diameter of 6.35 mm and length of 25.4 mm. The initial total moisture content X_{t1} = 1.0 kg H_2O/kg dry solid and the equilibrium moisture is X^* = 0.01. The density of the dry solid is 1602 kg/m^3. The cylinders are packed on a screen to a depth of x_1 = 50.8 mm. The bulk density of the dry solid in the bed is ρ_S = 641 kg/m^3. The inlet air has a humidity H_1 = 0.04 kg H_2O/kg dry air and a temperature T_1 = 121.1°C. The gas superficial velocity is 0.811 m/s and the gas passes through the bed. The total critical moisture content is X_{tC} = 0.50. Calculate the total time to dry the solids to X_t = 0.10 kg H_2O/kg dry solid.

Solution.
For the solid,
$$X_1 = X_{t1} - X^* = 1.00 - 0.01 = 0.99 \text{ kg } H_2O/\text{kg dry solid}$$
$$X_C = X_{tC} - X^* = 0.50 - 0.01 = 0.49$$
$$X = X_t - X^* = 0.10 - 0.01 = 0.09$$

For the gas, T_1 = 121.1°C and H_1 = 0.04 kg H_2O/kg dry air. The wet-bulb temperature T_W = 47.2°C and H_W = 0.074. The solid temperature is at T_W if radiation and conduction are neglected. The density of the entering air at 121.1°C and 1 atm is as follows:

$$v_H = (2.83 \times 10^{-3} + 4.56 \times 10^{-3} \times 0.04) \times (273 + 121.1) = 1.187 \text{ m}^3/\text{kg dry air}$$

$$\rho = \frac{1.00 + 0.04}{1.187} = 0.876 \text{ kg dry air} + H_2O/\text{m}^3$$

The mass velocity of the dry air is

$$G = v\rho \left(\frac{1.0}{1.0 + 0.04}\right) = 0.811 \times 3600 \times 0.876 \times \frac{1}{1.04} = 2459 \text{ kg dry air}/(\text{h} \cdot \text{m}^3)$$

Since the inlet H_1 = 0.040 and the outlet will be less than 0.074, an approximate average H of 0.05 will be used to calculate the total average mass velocity. The approximate average G_t is

$$G_t = 2459 + 2459 \times 0.05 = 2582 \text{ kg air} + H_2O/(\text{h} \cdot \text{m}^2)$$

For the packed bed, the void fraction ε is calculated as follows for 1 m^3 of bed containing solids plus voids. A total of 641 kg dry solid is present. The density of the dry solid is 1602 kg dry solid/m^3 solid. The volume of the solids in 1 m^3 of bed is then 641/1602, or 0.40 m^3 solid. Hence, ε = 1 − 0.40 = 0.60. The solid cylinder length h = 0.0254 m. The diameter D_c = 0.00635 m. Substituting into Eq. (11.7-16),

$$a = \frac{4(1-\varepsilon)(h+0.5D_c)}{D_c h} = \frac{4(1-0.6)\times[0.0254+0.5(0.00635)]}{0.00635\times 0.0254}$$

$$= 283.5 \text{ m}^2 \text{ surface area}/\text{m}^3 \text{ bed volume}$$

To calculate the diameter D_p of a sphere with the same area as the cylinder using Eq. (11.7-17),

$$D_p = (D_c h + 0.5 D_c^2)^{1/2} = [0.00635\times 0.0254 + 0.5\times(0.00635)^2]^{1/2} = 0.0135 \text{ m}$$

The bed thickness $x_1 = 50.8$ mm $= 0.0508$ m.

To calculate the heat-transfer coefficient, the Reynolds number is first calculated. Assuming an approximate average air temperature of 93.3°C, the viscosity of air $\mu = 2.15 \times 10^{-5}$ kg/(m·s) $= 2.15 \times 10^{-5} \times 3600 = 7.74 \times 10^{-2}$ kg/(m·h). The Reynolds number is

$$N_{Re} = \frac{D_p G_t}{\mu} = \frac{0.0135 \times 2582}{7.74 \times 10^{-2}} = 450$$

Using Eq. (11.7-13),

$$h = 0.151 \frac{G_t^{0.59}}{D_p^{0.41}} = \frac{0.151 \times (2582)^{0.59}}{(0.0135)^{0.41}} = 90.9 \text{ W}/(\text{m}^2 \cdot \text{K})$$

For $T_W = 47.2$°C, $r_W = 2389$ kJ/kg, or 2.389×10^6 J/kg (1027 btu/lb$_m$), from steam tables. The average humid heat is

$$c_S = 1.005 + 1.88H = 1.005 + 1.88(0.05) = 1.099 \text{ KJ}/(\text{kg dry air}\cdot\text{K}) = 1.099 \times 10^3 \text{ J}/(\text{kg}\cdot\text{K})$$

To calculate the time of drying for the constant-rate period using Eq. (11.7-11) and $G = 2459/3600 = 0.6831$ kg/(s·m^2),

$$t = \frac{\rho_S r_W x_1 (X_1 - X_C)}{G c_S (T_1 - T_W)(1 - e^{\frac{-hax_2}{Gc_S}})}$$

$$= \frac{641 \times 2.389 \times 10^6 \times 0.0508 \times (0.99 - 0.49)}{0.683 \times 1.099 \times 10^3 \times 121.1 - 47.2 \times (1 - e^{\frac{-90.9\times 283.5\times 0.0508}{0.683\times 1.099\times 10^3}})}$$

$$= 850 \text{s} = 0.236 \text{h}$$

For the time of drying for the falling-rate period, using Eq. (11.7-12),

$$t = \frac{\rho_S r_W x_1 X_C \ln(X_C/X)}{G c_S (T_1 - T_W)(1 - e^{\frac{-hax_1}{Gc_S}})}$$

$$= \frac{641 \times 2.389 \times 10^6 \times 0.0508 \times 0.49 \times \ln(0.49/0.09)}{0.6831 \times 1.099 \times 10^3 \times 121.1 - 47.2 \times (1 - e^{\frac{-90.9\times 283.5\times 0.0508}{0.683\times 1.099\times 10^3}})}$$

$$= 1412 \text{s} = 0.392 \text{h}$$

$$\text{total time } t = 0.236 + 0.392 = 0.628 \text{h}$$

11.7.2 Tray Drying with Varying Air Conditions

For drying in a compartment or tray dryer where the air passes in parallel flow over the surface of the tray, the air conditions do not remain constant. Heat and material balances similar to those for through circulation must be made to determine the exit-gas temperature and humidity.

In Fig. 11.14 air is shown passing over a tray. It enters having a temperature of T_1 and humidity H_1 and leaves at T_2 and H_2. The spacing between the trays is b m and dry air flow is G kg dry air/(s · m^2 cross-sectional area). Writing a heat balance over a length dL_t of tray for a section 1 m wide,

$$dq = Gc_s(1 \times b)dT \qquad (11.7\text{-}19)$$

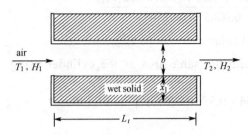

Fig. 11.14 Heat and material balances in a tray dryer.

The heat-transfer equation is

$$dq = h(1 \times dL_t b)(T - T_W) \qquad (11.7\text{-}20)$$

Rearranging and integrating,

$$\frac{hL_t}{Gc_s b} = \ln \frac{T_1 - T_W}{T_2 - T_W} \qquad (11.7\text{-}21)$$

Defining a log mean temperature difference similar to Eq. (11.7-10) and substituting into Eqs. (11.5-11) and (11.6-8), we obtain the following. For the constant-rate period,

$$t = \frac{x_1 \rho_S L_t r_W (X_1 - X_C)}{Gc_s b(T_1 - T_W)(1 - e^{\frac{-hL_t}{Gc_s b}})} \qquad (11.7\text{-}22)$$

For the falling-rate period, an approximate equation is obtained:

$$t = \frac{x_1 \rho_S L_t r_W X_C \ln(X_C / X)}{Gc_s b(T_1 - T_W)(1 - e^{\frac{-hL_t}{Gc_s b}})} \qquad (11.7\text{-}23)$$

11.7.3 Material and Heat Balances for Continuous Dryers

Simple heat and material balances

In Fig. 11.15 a flow diagram is given for a continuous-type dryer where the drying gas flows countercurrently to the solids flow. The solid enters at a rate of L_S kg dry solid/h, having a free moisture content X_1 and a temperature T_{S1}. It leaves at X_2 and T_{S2}. The gas enters at a rate G kg dry air/h, having a humidity H_1 kg H$_2$O/kg dry air and a temperature of T_{G1}. The gas leaves at T_{G2} and H_2.

For a material balance on the moisture,

$$GH_2 + L_S X_1 = GH_1 + L_S X_2 \qquad (11.7\text{-}24)$$

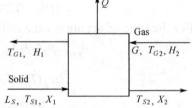

Fig. 11.15 Process flow for a counter-current continuous dryer.

For a heat balance a datum of T_0 °C is selected. A convenient temperature is 0 °C (32°F). The enthalpy of the wet solid is composed of the enthalpy of the dry solid plus that of the liquid as free moisture. The heat of wetting is usually neglected. The enthalpy of the gas I_G in kJ/kg dry air is

$$I_G = c_S(T_G - T_0) + Hr_0 \qquad (11.7\text{-}25)$$

where r_0 is the latent heat of water at T_0 °C, 2501 kJ/kg at 0 °C, and c_S is the humid heat, given as kJ/(kg dry air · K)

$$c_s = 1.005 + 1.88H \qquad (11.2\text{-}11)$$

The enthalpy of the wet solid I_S in kJ/kg dry solid, where $(T_S - T_0)°C = (T_S - T_0)$ K, is

$$I_S = c_{pS}(T_S - T_0) + Xc_{pA}(T_S - T_0) \tag{11.7-26}$$

where c_{pS} is the heat capacity of the dry solid in kJ/(kg dry solid · K) and c_{pA} is the heat capacity of liquid moisture in kJ/(kg H$_2$O · K). The heat of wetting or adsorption is neglected.

A heat balance on the dryer is

$$GI_{G2} + L_s I_{s1} = GI_{G1} + L_s I_{s2} + Q \tag{11.7-27}$$

where Q is the heat loss in the dryer in kJ/h. For an adiabatic process $Q = 0$, and if heat is added, Q is negative.

EXAMPLE 11.10 Heat Balance on a Dryer.
A continuous countercurrent dryer is being used to dry 453.6 kg dry solid/h containing 0.04 kg total moisture/kg dry solid to a value of 0.002 kg total moisture/kg dry solid. The granular solid enters at 26.7°C and is to be discharged at 62.8°C. The dry solid has a heat capacity of 1.465 kJ/(kg · K), which is assumed constant. Heating air enters at 93.3°C, having a humidity of 0.010 kg H$_2$O/kg dry air, and is to leave at 37.8°C. Calculate the air flow rate and the outlet humidity, assuming no heat losses in the dryer.

Solution. The flow diagram is given in Fig. 11.15. For the solid, $L_S = 453.6$ kg/h dry solid, $c_{pS} = 1.465$ kJ/(kg dry solid · K), $X_1 = 0.040$ kg H$_2$O/(kg dry solid), $c_{pA} = 4.187$ kJ/(kg H$_2$O · K), $T_{S1} = 26.7$°C, $T_{S2} = 62.8$°C, $X_2 = 0.002$. (Note that X values used are X_t values.) For the gas, $T_{G2} = 93.3$°C, $H_2 = 0.010$ kg H$_2$O/kg dry air, and $T_{G1} = 37.8$°C. Making a material balance on the moisture using Eq. (11.7-24),

$$GH_2 + L_S X_1 = GH_1 + L_S X_2$$
$$G \times 0.010 + 453.6 \times 0.040 = GH_1 + 453.6 \times 0.002 \tag{a}$$

For the heat balance, the enthalpy of the entering gas at 93.3°C, using 0°C as a datum, is, by Eq. (11.7-24), ΔT °C = ΔT K and $r_0 = 2501$ kJ/kg,

$$I_{G2} = c_S(T_{G2} - T_0) + H_2 r_0$$
$$= (1.005 + 1.88 \times 0.010) \times (93.3 - 0) + 0.010 \times 2501 = 120.5 \text{kJ/kg dry air}$$

For the exit gas,

$$I_{G1} = c_S(T_{G1} - T_0) + H_1 r_0$$
$$= (1.005 + 1.88 H_1) \times (37.8 - 0) + H_1 \times 2501 = 37.99 + 2572 H_1$$

For the entering solid, using Eq. (11.7-25),

$$I_{S1} = c_{pS}(T_{S1} - T_0) + X_1 c_{pA}(T_{S1} - T_0)$$
$$= 1.465 \times (26.7 - 0) + 0.040 \times 4.187 \times (26.7 - 0) = 43.59 \text{kJ/kg dry solid}$$

$$I_{S2} = c_{pS}(T_{S2} - T_0) + X_2 c_{pA}(T_{S2} - T_0)$$
$$= 1.465 \times (62.8 - 0) + 0.002 \times 4.187 \times (62.8 - 0) = 92.53 \text{kJ/kg dry solid}$$

Substituting into Eq. (11.7-27) for the heat balance with $Q = 0$ for no heat loss,

$$G \times 120.5 + 453.6 \times 43.59 = G(37.99 + 2572 H_1) + 453.6 \times 92.53 + 0 \tag{b}$$

Solving Eqs. (a) and (b) simultaneously,

$$G = 1166 \text{kg dry air/h}, \quad H_1 = 0.0248 \text{kg H}_2\text{O/kg dry air}$$

Air recirculation in dryer

In many dryers it is desired to control the wet bulb temperature at which the drying of the solid occurs. Also, since steam costs are often important in heating the drying air, recirculation of the drying air is sometimes used to reduce costs and control humidity. Part of the moist, hot air leaving the dryer is recirculated (recycled) and combined with the fresh air. This is shown in Fig.11.16. Fresh air having a temperature T_{G1} and humidity H_1 is mixed with recirculated air at T_{G2} and H_2 to give air at T_{G3} and H_3. This mixture is heated to T_{G4} with $H_4 = H_3$. After drying, the air leaves at a lower temperature T_{G2} and a higher humidity H_2.

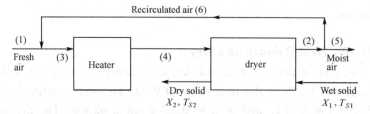

Fig.11.16 Process flow for air recirculation in drying.

The following material balances on the water can be made. For a water balance on the heater, noting that $H_6 = H_5 = H_2$,

$$G_1H_1 + G_6H_2 = (G_1+G_6)H_4 \tag{11.7-28}$$

Making a water balance on the dryer,

$$(G_1+G_6)H_4 + L_SX_1 = (G_1+G_6)H_2 + L_SX_2 \tag{11.7-29}$$

In a similar manner heat balances can be made on the heater and dryer and on the overall system.

Thermal efficiency

The thermal efficiency of a dryer can be defined as the percentage of the energy supplied that is used in evaporating the water or solvent. The efficiency can be calculated from the energy losses, which include the sensible heat of the warm moist air that is vented, the sensible heat of the discharged solids, and the heat loss to the surroundings.

11.7.4 Continuous Countercurrent Drying

Introduction and temperature profiles

Drying continuously offers a number of advantages over batch-drying. Smaller sizes of equipment can often be used, and the product has a more uniform moisture content. In a continuous dryer the solid is moved through the dryer while in contact with a moving gas stream that may flow parallel or countercurrent to the solid. In countercurrent adiabatic operation, the entering hot gas contacts the leaving solid, which has been dried. In parallel adiabatic operation, the entering hot gas contacts the entering wet solid.

In Fig.11.17 typical temperature profiles of the gas T_G and the solid T_S are shown for a continuous countercurrent dryer. In the preheat zone, the solid is heated up to the wet bulb or adiabatic saturation temperature. Little evaporation occurs here, and for low-temperature drying this zone is usually ignored. In the constant-rate zone, I, unbound and surface moisture are evaporated, and the temperature of the solid remains essentially constant at the adiabatic saturation temperature if heat is transferred by convection. The rate of drying would be constant here but the gas temperature is changing, as well as the humidity. The moisture content falls to the critical value X_C at the end of this period.

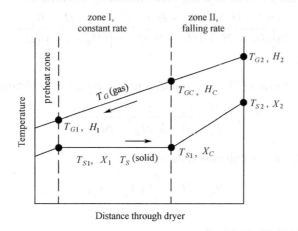

Fig.11.17 Temperature profiles for a continuous countercurrent dryer.

In zone II, unsaturated surface and bound moisture are evaporated and the solid is dried to its final value X_2. The humidity of the gas entering zone II is H_2 and it rises to H_C. The material-balance equation (11.7-24) may be used to calculate H_C as follows:

$$L_S(X_C - X_2) = G(H_C - H_2) \tag{11.7-30}$$

where L_S is kg dry solid/h and G is kg dry gas/h.

Equation for constant-rate period

The rate of drying in the constant-rate region in zone I would be constant if it were not for the varying gas conditions. The rate of drying in this section is given by an equation similar to Eq. (11.5-9):

$$R = k_y M_B (H_W - H) = \frac{h}{r_W}(T_G - T_W) \tag{11.7-31}$$

The time for drying is given by Eq. (11.5-2) using limits between X_1 and X_C:

$$t = \frac{L_S}{A} \int_{X_C}^{X_1} \frac{dX}{R} \tag{11.7-32}$$

where A/L_S is the exposed drying surface m²/kg dry solid. Substituting Eq. (11.7-30) into (11.7-31) and $(G/L_S) \, dH$ for dX,

$$t = \frac{G}{L_S} \times \frac{L_S}{A} \frac{1}{k_y M_B} \int_{H_C}^{H_1} \frac{dH}{H_W - H} \tag{11.7-33}$$

where G = kg dry air/h, L_S = kg dry solid/h, and A/L_S = m²/kg dry solid. This can be integrated graphically or numerically.

For the case where T_W or H_W is constant for adiabatic drying, Eq. (11.7-33) can be integrated:

$$t = \frac{G}{L_S} \times \frac{L_S}{A} \times \frac{1}{k_y M_B} \ln \frac{H_W - H_C}{H_W - H_1} \tag{11.7-34}$$

The above can be modified by use of a log mean humidity difference:

$$\Delta H_{LM} = \frac{(H_W - H_C) - (H_W - H_1)}{\ln[(H_W - H_C)/(H_W - H_1)]} = \frac{H_1 - H_C}{\ln[(H_W - H_C)/(H_W - H_1)]} \tag{11.7-35}$$

Substituting Eq. (11.7-35) into (11.7-34), an alternative equation is obtained:

$$t = \frac{G}{L_S} \times \frac{L_S}{A} \times \frac{1}{k_y M_B} \times \frac{H_1 - H_C}{\Delta H_{LM}} \tag{11.7-36}$$

From Eq. (11.7-30), H_C can be calculated as follows:

$$H_C = H_2 + \frac{L_S}{G}(X_C - X_2) \tag{11.7-37}$$

Equation for falling-rate period

For the situation where unsaturated surface drying occurs, H_W is constant for adiabatic drying, the rate of drying is directly dependent upon X as in Eq. (11.6-9), and Eq. (11.7-31) applies:

$$R = R_C \frac{X}{X_C} = k_y M_B (H_W - H) \frac{X}{X_C} \tag{11.7-38}$$

Substituting Eq. (11.7-38) into Eq. (11.5-2),

$$t = \frac{L_S}{A} \times \frac{X_C}{k_y M_B} \int_{X_2}^{X_C} \frac{dX}{(H_W - H)X} \tag{11.7-39}$$

Substituting GdH/L_S for dX and $(H - H_2)G/L_S + X_2$ for X,

$$t = \frac{G}{L_S} \times \frac{L_S}{A} \times \frac{X_C}{k_y M_B} \int_{H_2}^{H_C} \frac{dH}{(H_W - H)[(H - H_2)G/L_S + X_2]} \tag{11.7-40}$$

$$t = \frac{G}{L_S} \times \frac{L_S}{A} \times \frac{X_C}{k_y M_B} \times \frac{1}{(H_W - H_2)G/L_S + X_2} \ln \frac{X_C(H_W - H_2)}{X_2(H_W - H_C)} \tag{11.7-41}$$

Again, to calculate H_C, Eq. (11.7-37) can be used.

These equations for the two periods can also be derived using the last part of Eq. (11.7-30) and temperatures instead of humidities.

11.8 Drying Equipment

Of the many types of commercial dryers available, only a small number of important types are considered here. The first and larger group comprises dryers for rigid or granular solids and semisolid pastes; the second group consists of dryers that can accept slurry or liquid feeds.

11.8.1 Dryers for Solids and Pastes

Typical dryers for solids and pastes include tray and screen-conveyor dryers for materials that cannot be agitated and tower, rotary, screw-conveyor, fluid-bed, and flash dryers where agitation is permissible. In the following treatment these types are ordered, as far as possible, according to the degree of agitation and the method of exposing the solid to the gas or contacting it with a hot surface, as discussed at the beginning of this chapter. The ordering is complicated, however, by the fact that some types of dryers may be either adiabatic or nonadiabatic or a combination of both.

Tray dryers

A typical batch tray dryer is illustrated in Fig.11.18. It consists of a rectangular chamber of sheet metal containing two trucks that support racks H. Each rack carries a number of shallow trays, perhaps 750 mm square and 50 to 150 mm deep that are loaded with the material to be dried. Heated air is circulated at 2 to 5 m/s between the trays by fan C and motor D and passes over heaters E. Baffles G distribute the air uniformly over the stack of trays. Some moist air is continuously vented through exhaust duct B; makeup fresh air enters through inlet A. The racks are mounted on truck wheels I, so that at the end of the drying cycle the trucks can be pulled out of the chamber and taken to a tray-dumping station.

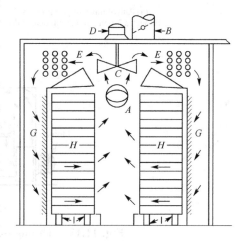

Fig. 11.18 Tray dryer.

Tray dryers are useful when the production rate is small. They can dry almost anything, but because of the labor required for loading and unloading, they are expensive to operate. They find most frequent application for valuable products like dyes and pharmaceuticals. Drying by circulation of air across stationary layers of solid is slow, and drying cycles are long: 3 to 48 h per batch. Occasionally through circulation drying is used, but this is usually neither economical nor necessary in batch dryers because shortening the drying cycle does not reduce the labor required for each batch. Energy savings may be significant, however.

Tray dryers may be operated under vacuum, often with indirect heating. The trays may rest on hollow metal plates supplied with steam or hot water or may themselves contain spaces for a heating fluid. Vapor from the solid is removed by an ejector or vacuum pump. A small flow of nitrogen is sometimes added to help carry away the vapors, but most of the heat of vaporization comes from conduction through the tray and the moist solid. Vacuum dryers are considerably more expensive than dryers operating at atmospheric pressure, but they are preferred for heat sensitive materials. Also, recovery of organic solvents is easier than if a large flow of air is used in an adiabatic dryer.

Screen-conveyor dryers

A typical through-circulation screen-conveyor dryer is shown in Fig. 11.19. A layer 25 to 150 mm thick of material to be dried is slowly carried on a traveling metal screen through a long drying chamber or tunnel. The chamber consists of a series of separate sections, each with its own fan and air heater. At the inlet end of the dryer, the air usually passes upward through the screen and the solids; near the discharge end, where the material is dry and may be dusty, air is passed downward through the screen. The air temperature and humidity may differ in the various sections, to give optimum conditions for drying at each point.

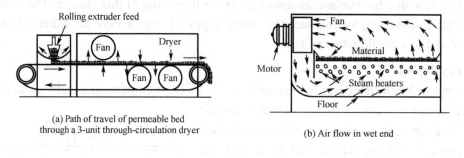

(a) Path of travel of permeable bed through a 3-unit through-circulation dryer

(b) Air flow in wet end

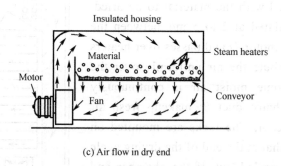

(c) Air flow in dry end

Fig. 11.19 Through-circulation screen-conveyor dryer.

Screen-conveyor dryers are typically 2 m wide and 4 to 50 m long, giving drying times of 5 to 120 min. The minimum screen size is about 30-mesh. Coarse granular, flaky, or fibrous materials can be dried by through circulation without any pretreatment and without loss of material through the screen. Pastes and filter cakes of fine particles, however, must be preformed before they can be handled on a screen-conveyor dryer. The aggregates usually retain their shape while being dried and do not dust through the screen except in small amounts. Provision is sometimes made for recovering any fines that do sift through the screen.

Screen-conveyor dryers handle a variety of solids continuously and with a very gentle action; their cost is reasonable, and their steam consumption is low, typically 2 kg of steam per kilogram of water evaporated. Air may be recirculated through, and vented from, each section separately or passed from one section to another countercurrently to the solid. These dryers are particularly applicable when the drying conditions must be appreciably changed as the moisture content of the solid is reduced.

Tower dryers

A tower dryer contains a series of circular trays mounted one above the other on a central rotating shaft. Solid feed dropped on the topmost tray is exposed to a stream of hot air or gas that passes across the tray. The solid is then scraped off and dropped to the tray below.

It travels in this way through the dryer, discharging as dry product from the bottom of the tower. The flow of solids and gas may be either parallel or countercurrent. Some mixing of the solids occurs as material is scraped off each tray, so the final solids are more uniform than with other tray dryers.

The *turbodryer* illustrated in Fig. 11.20 is a tower dryer with internal recirculation of the heating gas. Turbine fans circulate the air or gas outward between some of the trays, over heating elements, and inward between other trays. Gas velocities are commonly 0.6 to 2.4 m/s. The bottom two trays of the dryer shown in Fig. 11.20 constitute a cooling section for dry solids. Preheated air is usually drawn in the bottom of the tower and discharged from the top, giving countercurrent flow. A turbodryer functions partly by cross-circulation drying, as in a tray dryer, and partly by showering the particles through the hot gas as they tumble from one tray to another.

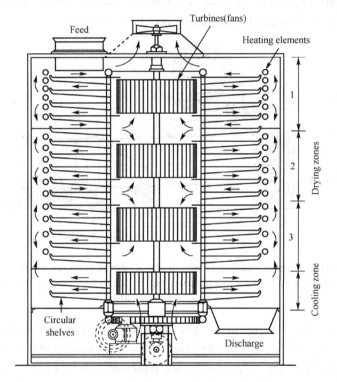

Fig. 11.20 Turbodrver.

Rotary dryers

A rotary dryer consists of a revolving cylindrical shell, horizontal or slightly inclined toward the outlet. Wet feed enters one end of the cylinder; dry material discharges from the other. As the shell rotates, internal flights lift the solids and shower them down through the interior of the shell. Rotary dryers are heated by direct contact of gas with the solids, by hot gas passing through an external jacket, or by steam condensing in a set of longitudinal tubes mounted on the inner surface of the shell. The last of these types is called a steam-tube rotary dryer. In a direct-indirect rotary dryer, hot gas first passes through the jacket and then the shell where it comes into contact with the solids.

A typical adiabatic countercurrent air-heated rotary dryer is shown in Fig. 11.21. A rotating shell A made of sheet steel is supported on two sets of rollers B and driven by a gear and pinion C. At the upper end is a hood D, which connects through fan E to a stack and a spout F, which brings in wet material from the feed hopper. Flights G, which lift the material being dried and shower it down through the current of hot air, are welded inside the shell. At the lower end the dried product discharges into a screw conveyor H. Just beyond the screw conveyor is a set of steam-heated extended-surface pipes that preheat the air. The air is moved through the dryer by a fan, which may, if desired, discharge into the air heater so that the whole system is under a positive pressure. Alternatively, the fan may be placed in the stack as shown, so that it draws air through the dryer and keeps the system under a slight vacuum. This is desirable when the material tends to dust. Rotary dryers of this kind are widely used for salt, sugar, and all kinds of granular and crystalline materials that must be kept clean and may not be directly exposed to very hot flue gases.

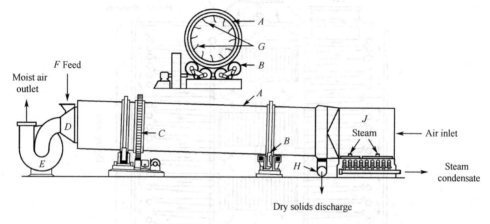

Fig. 11.21 Countercurrent air-heated rotary dryer: A, dryer shell; B, shell-supporting rolls; C, drive gear; D, air discharge hood; E, discharge fan; F, feed chute; G, lifting flights; H, product discharge; J, air heater.

The allowable mass velocity of the gas in a direct-contact rotary dryer depends on the dusting characteristics of the solid being dried and ranges from 2,000 to 25,000 kg/(m² · h) for coarse particles. Inlet gas temperatures are typically 120 to 175 °C for steam-heated air and 550 to 800 °C for flue gas from a furnace. Dryer diameters range from 1 to 3 m; the peripheral speed of the shell is commonly 20 to 25 m/min.

Screw-conveyor dryers

A screw conveyor is a continuous indirect-heat dryer, consisting essentially of a horizontal screw conveyor (or paddle conveyor) enclosed in a cylindrical jacketed shell. Solid fed in one end is conveyed slowly through the heated zone and discharges from the other end. The vapor evolved is withdrawn through pipes set in the roof of the shell. The shell is 75 to 600 mm in diameter and up to 6 m long; when greater length is required, several conveyors are set one above another in a bank. Often the bottom unit in such a bank is a cooler in which water or another coolant in the jacket lowers the temperature of the dried solids before they are discharged.

The rate of rotation of the conveyor is slow, from 2 to 30 rpm. Heat-transfer coefficients are based on the entire inner surface of the shell, even though the shell runs only 10 to 60 percent full. The coefficient depends on the loading in the shell and on the conveyor speed. It ranges, for many solids, between 15 and 60 W/(m$^2 \cdot$ °C).

Screw-conveyor dryers handle solids that are too fine and too sticky for rotary dryers. They are completely enclosed and permit recovery of solvent vapors with little or no dilution by air. When provided with appropriate feeders, they can be operated under moderate vacuum. Thus they are adaptable to the continuous removal and recovery of volatile solvents from solvent-wet solids, such as spent meal from leaching operations. For this reason they are sometimes known as *desolventizers*.

A related type of equipment is described later under "Thin-Film Dryers."

Fluid-bed dryers

Dryers in which the solids are fluidized by the drying gas find applications in a variety of drying problems. The particles are fluidized by air or gas in a boiling-bed unit, as shown in Fig. 11.22. Mixing and heat transfer are very rapid. Wet feed is admitted to the top of the bed; dry product is taken out from the side, near the bottom. In the dryer shown in Fig.11.22 there is a random distribution of residence times; the average time a particle stays in the dryer is typically 30 to 120 s when only surface liquid is vaporized and up to 15 to 30 min if there is also internal diffusion. Small particles are heated essentially to the exit dry-bulb temperature of the fluidizing gas; consequently, thermally sensitive materials must be dried in a relatively cool suspending medium. Even so, the inlet gas may be hot, for it mixes so rapidly that the temperature is virtually uniform, at the exit gas temperature, throughout the bed. If fine particles are present, either from the feed or from particle breakage in the fluidized bed, there may be considerable solids carryover with the exit gas, and cyclones and bag filters are needed for fines recovery.

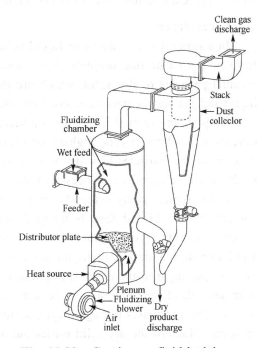

Fig. 11.22 Continuous fluid-bed dryer.

Some rectangular fluid-bed dryers have separately fluidized compartments through which the solids move in sequence from inlet to outlet. These are known as *plug flow dryers;* in them the residence time is almost the same for all particles. Drying conditions can be changed from one compartment to another, and often the last compartment is fluidized with cold gas to cool the solids before discharge. Because of the rather complex temperature patterns, the true average temperature difference for the dryer as a whole is not easy to define. Usually, in fact, the outlet temperatures of solids and gas are so nearly the same that the difference between them cannot be measured.

Flash dryers

In a flash dryer, a wet pulverized solid is transported for a few seconds in a hot gas stream. Drying takes place during transportation. The rate of heat transfer from the gas to the suspended solid particles is high, and drying is rapid, so that no more than 3 or 4 s is required to evaporate substantially all the moisture from the solid. The temperature of the gas is high—often about 650 ℃ at the inlet—but the time of contact is so short that the temperature of the solid rarely rises more than 50 ℃ during drying. Flash drying may therefore be applied to sensitive materials that in other dryers would have to be dried indirectly by a much cooler heating medium.

Sometimes a pulverizer is incorporated in the flash drying system to give simultaneous drying and size reduction.

11.8.2 Dryers for Solutions and Slurries

A few types of dryers evaporate solutions and slurries entirely to dryness by thermal means. Typical examples are spray dryers, thin-film dryers, and drum dryers.

Spray dryers

In a spray dryer, a slurry or liquid solution is dispersed into a stream of hot gas in the form of a mist of fine droplets. Moisture is rapidly vaporized from the droplets, leaving residual particles of dry solid, which are then separated from the gas stream. The flow of liquid and gas may be cocurrent, countercurrent, or a combination of both in the same unit.

Droplets are formed inside a cylindrical drying chamber by pressure nozzles, two-fluid nozzles, or, in large dryers, high-speed spray disks. In all cases it is essential to prevent the droplets or wet particles of solid from striking solid surfaces before drying has taken place, so that drying chambers are necessarily large. Diameters of 2.5 to 9 m are common.

In the typical spray dryer shown in Fig. 11.23, the chamber is a cylinder with a short conical bottom. Liquid feed is pumped into a spray-disk atomizer set in the roof of the chamber. In this dryer the spray disk is about 300 mm in diameter and rotates at 5,000 to 10,000 r/min. It atomizes the liquid into tiny drops, which are thrown radially into a stream of hot gas entering near the top of the chamber. Cooled gas is drawn by an exhaust fan through a horizontal discharge line set in the side of the chamber at the bottom of the cylindrical section. The gas passes through a cyclone separator where any entrained particles of solid are removed. Much of the dry solid settles out of the gas into the bottom of the drying chamber, from which it is removed by a rotary valve and screw conveyor and combined with any solid collected in the cyclone.

The chief advantages of spray dryers are the very short drying time, which permits drying of highly heat-sensitive materials, and the production of solid or hollow spherical particles. The desired consistency, bulk density, appearance, and flow properties of some products, such as foods or synthetic detergents, may be difficult or impossible to obtain in any other type of dryer. Spray dryers also have the advantage of yielding from a solution, slurry, or thin paste, in a single step, a dry product that is ready for the package. A spray dryer may combine the functions of an evaporator, a crystallizer, a dryer, a size reduction unit, and a classifier. Where one can be used, the resulting simplification of the overall manufacturing process may be considerable.

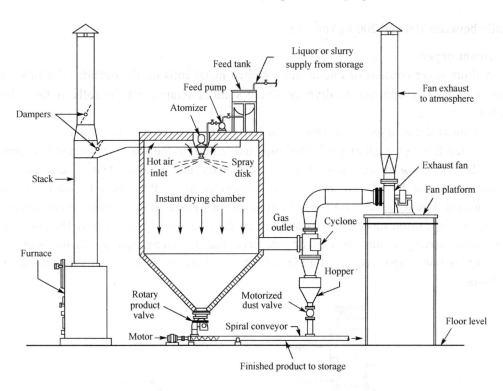

Fig. 11.23 Spray dryer with parallel flow.

Considered as dryers alone, spray dryers are not highly efficient. Much heat is ordinarily lost in the discharged gases. They are bulky and very large, often 25 m or more high, and are not always easy to operate. The bulk density of the dry solid—a property of especial importance in packaged products—is often difficult to keep constant, for it may be highly sensitive to changes in the solids content of the feed, to the inlet gas temperature, and to other variables.

In spray-drying solutions, the evaporation from the surface of the drops leads to initial deposition of solute at the surface before the interior of the drop reaches saturation. The rate of diffusion of solute back into the drop is slower than the flow of water from the interior to the surface, and the entire solute content accumulates at the surface. The final dry particles are often hollow, and the product from a spray dryer is quite porous.

Thin-film dryers

Competitive with spray dryers in some situations are thin-film dryers which can accept a liquid or a slurry feed and produce a dry, free-flowing solid product. They normally are built in two sections, the first of which is a vertical agitated evaporator-dryer Here most of the liquid is removed from the feed, and partially wet solid is discharged to the second section, as illustrated in Fig. 11.24, in which the residual liquid content of the material from the first section is reduced to the desired value.

The thermal efficiency of thin-film dryers is high, and there is little loss of solids, since little or no gas needs to be drawn through the unit. They are useful in removing and recovering solvents from solid products. They are relatively expensive and are somewhat limited in heat-transfer area. With both aqueous and solvent feeds the acceptable feed rate is usually between 100 and 200 kg/(m² · h).

Drum dryers

A drum dryer consists of one or more heated metal rolls on the outside of which a thin layer of liquid is evaporated to dryness. Dried solid is scraped off the rolls as they slowly revolve.

A typical drum dryer—a double-drum unit with center feed—is shown in Fig. 11.25. Liquid is fed from a trough or perforated pipe into a pool in the space above and between the two rolls. The pool is confined there by stationary end plates. Heat is transferred by conduction to the liquid, which is partly concentrated in the space between the rolls. Concentrated liquid issues from the bottom of the pond as a viscous layer cover the remainder of the drum surfaces. Substantially all the liquid is vaporized from the solid as the drums turn, leaving a thin layer of dry material that is scraped off by doctor blades into conveyors below. Vaporized moisture is collected and removed through a vapor hood above the drums.

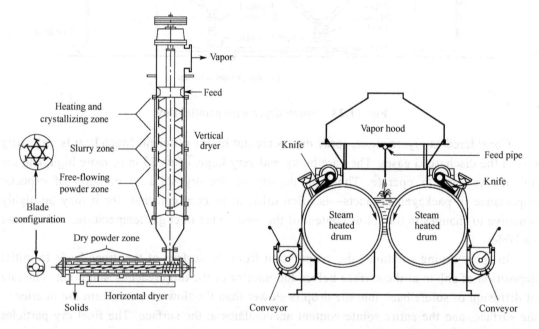

Fig. 11.24 "Combi" thin-film dryer.   **Fig. 11.25** Double-drum dryer with center feed.

Double-drum dryers are effective with dilute solutions, concentrated solutions of highly soluble materials, and moderately heavy slurries. They are not suitable for solutions of salts with limited solubility or for slurries of abrasive solids that settle out and create excessive pressure between the drums.

The rolls of a drum dryer are 0.6 to 3 m in diameter and 0.6 to 4 m long, revolving at 1 to 10 rpm. The time that the solid is in contact with hot metal is 6 to 15 s, which is short enough to result in little decomposition even of heat-sensitive products. The heat-transfer coefficient is high, from 1,200 to 2,000 W/(m² · ℃) under optimum conditions, although it may be only one-tenth of these values when conditions are adverse. The drying capacity is proportional to the active drum area; it is usually between 5 and 50 kg per square meter of drying surface per hour.

11.8.3 Selection of Drying Equipment

The first consideration in selecting a dryer is its operability; above all else, the equipment must produce the desired product in the desired form at the desired rate. Despite the variety of commercial dryers on the market, the various types are largely complementary, not competitive, and the nature of the drying problem dictates the type of dryer that must be used, or at least limits the choice to perhaps two or three possibilities. The final choice is then made on the basis of capital and operating costs. Attention must be paid, however, to the costs of the entire isolation system, not just the drying unit alone.

Complete isolation train; evaporator-dryers

Many industrial processes involve the isolation of solids from solution in water or other solvents to yield a dry, purified granular product suitable for packaging and sale. The process steps needed to accomplish this are typically evaporation, crystallization, filtration or centrifuging, drying, size reduction, and classification or screening. Five or six pieces of equipment are needed. Sometimes an evaporator dryer can eliminate the need for many of these. Spray dryers, drum dryers, and thin-film dryers can accept liquid feed and convert it directly into dry product ready for packaging. The added cost of removing moisture thermally instead of mechanically is more than made up by the economies of installing and operating one piece of equipment instead of many.

It may, of course, be necessary to treat the liquid feed prior to drying to remove impurities that cannot be permitted to appear in the dried product. It is essential, therefore, that the engineer pay attention to the whole isolation process and not focus narrowly on the drying step alone.

General considerations

There are some general guidelines for selecting a dryer, but it should be recognized that the rules are far from rigid and exceptions not uncommon. Batch dryers, for example, are most often used when the production rate of dried solid is less than 150 to 200kg/h; continuous dryers are nearly always chosen for production rates greater than 1 or 2 tons/h. At intermediate production rates other factors must be considered. Thermally sensitive materials must be dried at low temperature under vacuum, with a low-temperature heating medium, or very rapidly as in a flash or spray dryer. Fragile crystals must be handled gently as in a tray dryer, a screen-conveyor dryer, or tower dryer.

The dryer must also operate reliably, safely, and economically. Operating and maintenance costs must not be excessive; pollution must be controlled; energy consumption must be minimized. As with other equipment, these considerations may conflict with one another, and a compromise must be reached in finding the optimum dryer for a given service.

As far as the drying operation itself is concerned, adiabatic dryers are generally less expensive than nonadiabatic dryers, in spite of the lower thermal efficiency of adiabatic units. Unfortunately there is usually a lot of dust carryover from adiabatic dryers, and these entrained particles must be removed almost quantitatively from the drying gas. Elaborate particle removal equipment may be needed, equipment that may cost as much as the dryer itself. This often makes adiabatic dryers less economical than a "buttoned-up" nonadiabatic system in which little or no gas is used. Rotary dryers are an example; they were once the most common type of continuous dryer, but because of the inevitable entrainment, other types of dryers which avoid the problem of dust carryover would now, if possible, be selected in their place. Nonadiabatic dryers are always chosen for very fine particles or for solids that are too chemically reactive to be exposed to a stream of gas. They are also widely used for solvent removal and recovery.

PROBLEMS

11.1 The air in a room is at 37.8°C and a total pressure of 101.3 kPa abs containing water vapor with a partial pressure $p_A = 3.59$ kPa. Calculate: (a) Humidity. (b) Saturation humidity and percentage humidity. (c) Relative humidity.

11.2 The air entering a dryer has a temperature of 65.6°C and dew point of 15.6°C. Using the humidity chart, determine the actual humidity and percentage humidity. Calculate the humid volume of this mixture and also calculate c_s using SI units.
Ans: $H = 0.0113$ kg H_2O/kg dry air, $H_P = 5.3\%$, $c_S = 1.026$ kJ/(kg·K),
$v_H = 0.976$ m³ air + water vapor/kg dry air

11.3 Air at 82.2°C having a humidity $H = 0.0655$ kg H_2O/kg dry air is contacted in an adiabatic saturator with water. It leaves at 80% saturation. (a) What are the final values of H and T°C? (b) For 100% saturation, what would be the values of H and T?

11.4 The air in a room has a humidity H of 0.021 kg H_2O/kg dry air at 32.2°C and 101.3 kPa abs pressure. Calculate: (a) percentage humidity η; (b) percentage relative humidity φ.

11.5 The air entering a dryer has a temperature of 65.6°C and dew point of 15.6°C. Using the humidity chart, determine the actual humidity and percentage humidity. Calculate the humid volume of this mixture and also calculate c_S.

11.6 An air–water vapor mixture going to a drying process has a dry bulb temperature of 57.2°C and a humidity of 0.030 kg H_2O/kg dry air. Using the humidity chart and appropriate equations, determine the percentage humidity, saturation humidity at 57.2°C, dew point, humid heat, and humid volume.

11.7 Air enters an adiabatic saturator having a temperature of 76.7°C and a dew-point temperature of 40.6°C. It leaves the saturator 90% saturated. What are the final values of H and T°C?

11.8 An air-water vapor mixture has a dry bulb temperature of 65.6°C and a wet bulb temperature of 32.2°C. What is the humidity of the mixture?

11.9 The humidity of an air–water vapor mixture is $H = 0.030$ kg H_2O/kg dry air. The dry bulb temperature of the mixture is 60°C. What is the wet bulb temperature?

11.10 Air having a dry bulb temperature of 37.8°C and a wet bulb temperature of 26.7°C is to be dried by first cooling to 15.6°C to condense water vapor and then heating to 23.9°C. (a) Calculate the initial humidity and percentage humidity. (b) Calculate the final humidity and percentage humidity.
Hint: Locate the initial point on the humidity chart. Then go horizontally (cooling) to the 100% saturation line. Follow this line to 15.6°C. Then go horizontally to the right to 23.9°C.

11.11 Air entering an adiabatic cooling chamber has a temperature of 32.2°C and a percentage humidity of 65%. It is cooled by a cold water spray and saturated with water vapor in the chamber. After leaving, it is heated to 43°C. The final air has a percentage humidity of 40%. (a) What is the initial humidity of the air? (b) What is the final humidity after heating?

11.12 A batch of wet solid was dried on a tray dryer using constant drying conditions and a thickness of material on the tray of 25.4 mm. Only the top surface was exposed. The drying rate during the constant-rate period was $R = 2.05$ kg $H_2O/(h \cdot m^2)$. The ratio L_S/A used was 24.4 kg dry solid/m² exposed surface. The initial free moisture was $X_1 = 0.55$ and the critical moisture content $X_C = 0.22$ kg free moisture/kg dry solid.

Calculate the time to dry a batch of this material from $X_1 = 0.45$ to $X_2 = 0.30$ using the same drying conditions but a thickness of 50.8 mm, with drying from the top and bottom surfaces.

Hint: First calculate L_S/A for this new case.

11.13 Using the conditions in Example 11.6 for the constant-rate drying period, do as follows: (a) Predict the effect on R_C if the air velocity is only 3.05 m/s. (b) Predict the effect if the gas temperature is raised to 76.7°C and H remains the same. (c) Predict the effect on the time t for drying between moisture contents X_1 to X_2 if the thickness of material dried is 38.1 mm instead of 25.4 mm and the drying is still in the constant-rate period.

11.14 A granular insoluble solid material wet with water is being dried in the constant-rate period in a pan 0.61m×0.61m and the depth of material is 25.4mm. The sides and bottom are insulated. Air flows parallel to the top drying surface at a velocity of 3.05m/s and has a dry bulb temperature of 60°C and wet bulb temperature of 29.4°C. The pan contains 11.34 kg of dry solid having a free moisture content of 0.35 kg H_2O/kg dry solid, and the material is to be dried in the constant-rate period to 0.22 kg H_2O/kg dry solid. (a) Predict the drying rate and the time in hours needed. (b) Predict the time needed if the depth of material is increased to 44.5 mm.

11.15 A wet solid is to be dried in a tray dryer under steady-state conditions from a free moisture content of $X_1 = 0.40$ kg H_2O/kg dry solid to $X_2 = 0.02$ kg H_2O/kg dry solid. The dry solid weight is 99.8 kg dry solid and the top surface area for drying is 4.645 m². The drying-rate curve can be represented by Fig. 11.10(b). (a) Calculate the time for drying using numerical integration in the falling-rate period. (b) Repeat but use a straight line through the origin for the drying rate in the falling-rate period.

11.16 In order to test the feasibility of drying a certain foodstuff, drying data were obtained in a tray dryer with air flow over the top exposed surface having an area of 0.186 m². The bone-dry sample weight was 3.765 kg dry solid. At equilibrium after a long period, the wet sample weight was 3.955 kg H_2O+solid. Hence, 3.955–3.765, or 0.190, kg of equilibrium moisture was present. The following sample weights versus time were obtained in the drying test:

Time, h	Weight, kg	Time, h	Weight, kg	Time, h	Weight, kg
0	4.944	2.2	4.554	7.0	4.019
0.4	4.885	3.0	4.404	9.0	3.978
0.8	4.808	4.2	4.241	12.0	3.955
1.4	4.699	5.0	4.150		

(a) Calculate the free moisture content X kg H_2O/kg dry solid for each data point and plot X versus time. *Hint*: For 0 h, 4.944–0.190–3.765 = 0.989 kg free moisture in 3.765 kg dry solid. Hence, $X = 0.989/3.765$. (b) Measure the slopes, calculate the drying rates R in kg $H_2O/(h \cdot m^2)$, and plot R versus X. (c) Using this drying-rate curve, predict the total time to dry the sample from $X = 0.20$ to $X = 0.04$. Use numerical integration for the falling-rate period. What are the drying rate R_C in the constant-rate period and X_C?

11.17 A material was dried in a tray-type batch dryer using constant-drying conditions. When the initial free moisture content was 0.28 kg free moisture/kg dry solid, 6.0 h was required to dry the material to a free moisture content of 0.08 kg free moisture/kg dry solid. The critical free moisture content is 0.14. Assuming a drying rate in the falling-rate region, where the rate is a straight line from the critical point to the origin, predict the time to dry a sample from a free moisture content of 0.33 to 0.04 kg free moisture/kg dry solid.

Hint: First use the analytical equations for the constant-rate and the linear falling-rate periods with the known total time of 6.0 h. Then use the same equations for the new conditions.

11.18 A granular biological material wet with water is being dried in a pan 0.305m × 0.305 m and 38.1 mm deep. The material is 38.1 mm deep in the pan, which is insulated on the sides and bottom. Heat transfer is by convection from an air stream flowing parallel to the top surface at a velocity of 3.05 m/s, having a temperature of 65.6°C and humidity $H = 0.010$ kg H_2O/kg dry air. The top surface receives radiation from steam-heated pipes whose surface temperature $T_R = 93.3$°C. The emissivity of the solid is $\varepsilon = 0.95$. It is desired to keep the surface temperature of the solid below 32.2°C so that decomposition will be kept low. Calculate the surface temperature and the rate of drying for the constant-rate period.

11.19 Repeat Example 11.9, making heat and material balances, but with the following changes. The solid enters at 15.6°C and leaves at 60°C. The gas enters at 87.8°C and leaves at 32.2°C. Heat losses from the dryer are estimated as 2931 W.

11.20 The wet solid is to be dried from water content 20% to 2% (wet basis) in an adiabatic dryer under atmospheric pressure. The feed of wet solid into the dryer is 600kg/h. After being preheated to 100°C, the fresh air with a dry-bulb temperature of 20°C and a humidity of 0.013 kg water/kg dry air is sent to the dryer, and leaves the dryer at 60°C.

(a) What is the volume of fresh air required per unit time (in m^3/h)?

(b) How much heat is obtained by air when passing through the preheater (in kJ/h)?

[Hint: $v_H = (0.772 + 1.244 H_0) \times \dfrac{t_0 + 273}{273}$]

11.21 The wet feed material to a continuous dryer contains 50 wt % water on a wet basis and is dried to 27 wt % by countercurrent air flow. The dried product leaves at the rate of 907.2 kg/h. Fresh air to the system is at 25.6°C and has a humidity of $H = 0.007$ kg H_2O/kg dry air. The moist air leaves the dryer at 37.8°C and $H = 0.020$ kg H_2O/kg dry air and part of it is recirculated and mixed with the fresh air before entering a heater. The heated mixed air enters the dryer at 65.6°C and $H = 0.010$ kg H_2O/kg dry air. The solid enters at 26.7°C and leaves at 26.7°C. Calculate the fresh-air flow, the percent air leaving the dryer that is recycled, the heat added in the heater, and the heat loss from the dryer.

名 词 术 语

英文 / 中文

- absolute humidities / 绝对湿度
- adiabatic saturation temperature / 绝热饱和温度
- adiabatic / 绝热
- adsorbents / 吸附剂
- adsorption isotherms / 吸附等温线
- agitator / 搅拌器
- alumina / 氧化铝
- atomizer / 喷雾器
- batch dryer / 间歇干燥器
- batchwise / 间歇式
- biological / 生物的
- boiling point / 沸点
- bone-dry / 绝干
- bound water / 结合水
- boundary layer / 边界层
- bulk / 主体
- capillary forces / 毛细作用力
- capillary / 毛细管
- catalysts / 催化剂
- colloidal gels / 凝胶体
- compartment dryer / 厢式干燥器
- concave / 凹面的
- concentration differences / 浓度差
- consistency / 浓度
- constant-drying conditions / 恒定干燥条件
- constant-rate period / 恒速阶段
- countercurrently / 逆流
- counterflow / 逆流
- critical moisture content / 临界湿含量
- cross-circulation drying / 交叉循环干燥
- crystalline / 结晶的
- crystals / 结晶体
- cyclone / 旋风分离器
- dehumidification / 去湿
- dehydration / 脱水
- detergents / 洗涤剂
- dew point / 露点
- dew-point temperature / 露点温度
- diffusivity / 扩散系数
- dry bulb temperature / 干球温度
- drying / 干燥
- empirical / 经验的

英文 / 中文

- energy balance / 能量衡算
- enthalpy / 焓
- enzymes / 酶
- equilibrium curve / 平衡曲线
- equilibrium moisture content / 平衡湿含量
- equilibrium water / 平衡水分
- exhaust fan / 排气扇
- exit / 出口
- Fahrenheit / 华氏温度
- falling-rate period / 降速阶段
- feedstock / 进料
- fermentation / 发酵
- fine chemicals / 精细化学品
- flakes / 薄片
- flash dryers / 气流干燥器
- fluid-bed / 流化床
- fluidize / 液化
- free moisture content / 自由湿含量
- free water / 自由水分
- freeze-drying / 冷冻干燥
- geometry factors / 几何因子
- granular / 颗粒状的
- granules / 颗粒
- hardening / 硬化
- heat of vaporization / 汽化热
- heating medium / 加热介质
- heat-sensitive / 热敏性的
- humid air / 湿空气
- humid heat / 湿比热容
- humid volume / 湿比容
- humidification / 加湿
- humidity chart / 湿度图
- humidity / 湿度
- hygroscopic / 吸湿性的
- inlet / 入口
- inorganic / 无机的
- interfacial tension / 界面张力
- interior channels / 内部通道
- latent heat / 潜热
- log mean temperature difference / 对数平均温度差
- lyophilization / 冻干法
- material balances / 物料衡算

英文 / 中文	英文 / 中文
mechanical separator / 机械分离器	shell / 壳
mechanism of drying / 干燥机理	shrinkage / 收缩
meniscus / 半月形的	silica / 硅土
microorganisms / 微生物	slabs / 厚片
moist air / 湿空气	slurry / 浆体
moisture content / 湿含量	spray disk / 喷雾圆盘
moisture equilibria / 湿度平衡	spray dryers / 喷雾干燥机
moisture-free (bone-dry) solid / 绝干固体	spray nozzles / 喷头
monolayer / 单层	steady-state operation / 稳态操作
multilayers / 多层	sterilization / 杀菌
nonadiabatic / 非绝热	stock / 物料
nonporous / 无孔的	sublimation / 升华
numerical integration / 数值积分	surface tension / 表面张力
operating line / 操作线	the saturation vapor pressure / 饱和蒸汽压
pastes / 糊状	the specific heats / 比热容
percentage humidity / 湿度百分数	the total moisture content / 总湿含量
pharmaceuticals / 医药品	thermal efficiency / 热效率
polymeric material / 聚合材料	thermometer / 温度计
porous / 多孔的	thin-film dryers / 薄膜干燥器
powders / 粉体	through-circulation drying / 循环干燥
pulverized / 研磨成粉状的	tray drying / 盘式干燥
rate-of-drying curve / 干燥速率曲线	trial-and-error / 试差法
refrigerant / 制冷剂	turbodryer / 涡轮干燥器
relative humidity / 相对湿度	unbound water / 非结合水
required packed height / 所需的填料层高度	undersaturated air / 不饱和空气
rotary dryers / 旋转干燥器	vacuum drying / 真空干燥
rum dryers / 转筒干燥器	vapor pressure of water / 水的蒸汽压
saturated gas / 饱和空气	vapor-free gas / 不含蒸汽的空气、绝干空气
saturation humidity / 饱和湿度	vitamins / 维生素
screen / 筛	void fraction / 空隙率
screen-conveyor dryers / 带式干燥器	volumetric heat-transfer coefficient for gas / 气体体积传热系数
screw-conveyor / 螺旋杆	warp / 弯曲
semiquantitative / 半定量的	wet-bulb temperature / 湿球温度
sensible heat / 显热	
settles / 沉降	

Chapter 12

Tray and Packed Tower Design

12.1 Packed Tower Design

12.1.1 Packed Tower and Packings

A common apparatus used in gas absorption and certain other operations is the packed tower, an example of which is shown in Fig. 12.1. The device consists of a cylindrical column, or tower, equipped with a gas inlet and distributing space at the bottom; a liquid inlet and distributor at the top; gas and liquid outlets at the top and bottom, respectively; and a supported mass of inert solid shapes, called *tower packing.* The packing support is typically a screen, corrugated to give it strength, or an upper grid or riser plate, as shown in Fig. 12.2, which prevents packing movement and is with a large open area so that flooding does not occur at the support. The inlet liquid, which may be pure solvent or a dilute solution of solute in the solvent and which is called the *weak liquor*, is distributed over the top of the packing by the distributor and, in ideal operation, uniformly wets the surfaces of the packing. The distributor shown in Fig. 12.1 is a set of perforated pipes. In large towers, spray nozzles or distributor plates with overflow weirs (Fig. 12.3) are more common. For very large towers, up to 9 m in diameter, Nutter Engineering advertises a plate distributor with individual drip tubes.

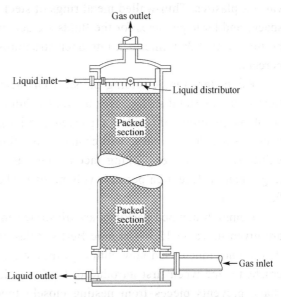

Fig. 12.1 Packed tower.

The solute-containing gas, or rich gas, enters the distributing space below the packing and flows upward through the interstices in the packing countercurrent to the flow of the liquid. The packing provides a large area of contact between the liquid and gas and encourages intimate contact between the phases. The solute in the rich gas is absorbed by the fresh liquid entering the tower, and dilute, or lean, gas leaves the top. The liquid is enriched in solute as it flows down the tower, and concentrated liquid, called the *strong liquor*, leaves the bottom of the tower through the liquid outlet.

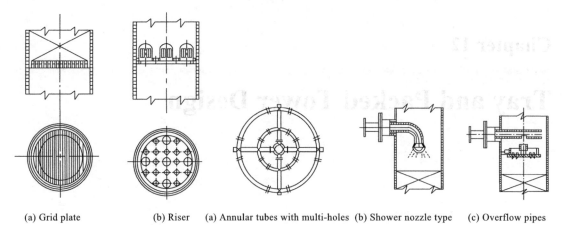

(a) Grid plate (b) Riser (a) Annular tubes with multi-holes (b) Shower nozzle type (c) Overflow pipes

Fig. 12.2 Packing supports. **Fig. 12.3** Liquid distributors.

Tower packings are divided into three principal types: those that are dumped at random into the tower, those that must be stacked by hand, and those known as structured or ordered packings. Dumped packings consist of units 6 to 75 mm(1/4 to 3 in.) in major dimension; packings smaller than 25 mm are used mainly in laboratory or pilot-plant columns. In stacked packings the units are 50 to 200 mm (2 to 8 in.) in size. They are much less commonly used than dumped packings and are not discussed here.

Dumped tower packings are made of cheap, inert materials such as clay, porcelain, or various plastics. Thin-walled metal rings of steel or aluminum are sometimes used. High void spaces and large passages for the fluids are achieved by making the packing units irregular or hollow, so that they interlock into open structures with a porosity or void fraction of 60 to 90 percent.

Parameters for describing physical characteristics of packings mainly include Specific Surface area σ (total surface area per unit volume of packed section), porosity or void fraction ε(volume fraction of voids in packed section), packing factor F_p (which is σ/ε^3 when packings are dry, σ and ε will change if packings are wetted by the spraying liquid and the value σ/ε^3 is the wet packing factor) and bulk density ρ_p. Usually good packings are with large total surface area per unit volume of packed section, and large void fraction, and low bulk density.

Common dumped packings are illustrated in Fig. 12.4, and their physical characteristics are given in Table 12.1. Ceramic Berl saddles and Raschig rings are older types of packing that are not much used now, although they were a big improvement over ceramic spheres or crushed stone when first introduced. Intalox saddles are somewhat like Berl saddles, but the shape prevents pieces from nesting closely together, and this increases the bed porosity. Super Intalox saddles are a slight variation with scalloped edges; they are available in plastic or ceramic form. Pall rings are made from thin metal with portions of the wall bent inward or from plastic with slots in the wall and stiffening ribs inside. Hy-pak metal packing and Flexirings (not shown) are similar in shape and performance to metal Pall rings. Beds of Pall rings have over 90 percent void fraction and lower pressure drop than most other packings of the same nominal size. Norton's new IMTP (Intalox Metal Tower Packing) has a very open structure and even lower pressure drop than Pall rings. Additional pressure-drop packing factors for many commercial packings are given by Robbins and, in SI units, by Perry.

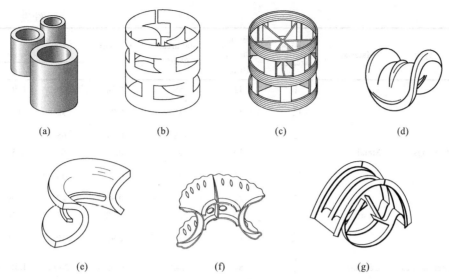

Fig. 12.4 Common tower packings: (*a*) Raschig rings; (*b*) metal Pall ring; (*c*) plastic Pall ring; (*d*) Berl saddle; (*e*) ceramic Intalox saddle; (*f*) plastic Super Intalox saddle; (*g*) metal Intalox saddle.

Structured packings with ordered geometry evolved from the Stedman packing of the late 1930s, but they found few industrial uses until the Sulzer packing was developed about 1965. Early structured packings were fabricated from wire gauze; most current ones are made of sheets of perforated corrugated metal, with adjacent sheets arranged so that liquid spreads over their surfaces while vapor flows through channels formed by the corrugations. The channels are set at an angle of 45° with the horizontal; the angle alternates in direction in successive layers, as shown schematically in Fig. 12.5. Each layer is a few inches thick. Various proprietary packings differ in the size and arrangement of the corrugations and the treatment of the packing surfaces. Typically the triangular corrugations are 25 to 40 mm across the base, 17 to 25 mm on the side, and 10 to 15 mm high. The porosity ranges from 0.93 to 0.97, and the specific surface area from 200 to 250 m^2/m^3. The Sulzer BX packing, fabricated from metal gauze, provides a specific surface area of 500 m^2/m^3 with a porosity of 0.90.

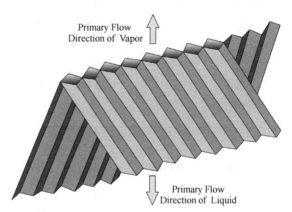

Fig. 12.5 Schematic drawing of structured packing.

Table 12.1 Characteristics of dumped tower packings

Type	Material	Nominal size, in.	Bulk density,[1] lb/ft³	Total area,[1] ft²/ft³	Porosity ε	Packing factors[2] F_p	f_p
Raschig rings	Ceramic	$\frac{1}{2}$	55	112	0.64	580	1.52[3]

Type	Material	Nominal size, in.	Bulk density,[1] lb/ft³	Total area,[1] ft²/ft³	Porosity ε	Packing factors[2] F_p	f_p
Pall rings	Metal	1	42	58	0.74	155	1.36[3]
		$1\frac{1}{2}$	43	37	0.73	95	1.0
		2	41	28	0.74	65	0.92[3]
		1	30	63	0.94	56	1.54
		$1\frac{1}{2}$	24	39	0.95	40	1.36
		2	22	31	0.96	27	1.09
	Plastic	1	5.5	63	0.90	55	1.36
		$1\frac{1}{2}$	4.8	39	0.91	40	1.18
Berl saddles	Ceramic	$\frac{1}{2}$	54	142	0.62	240	1.58[3]
		1	45	76	0.68	110	1.36[3]
		$1\frac{1}{2}$	40	46	0.71	65	1.07[3]
Intalox saddles	Ceramic	$\frac{1}{2}$	46	190	0.71	200	2.27
		1	42	78	0.73	92	1.54
		$1\frac{1}{2}$	39	59	0.76	52	1.18
		2	38	36	0.76	40	1.0
		3	36	28	0.79	22	0.64
Super Intatox saddles	Ceramic	1	—	—	—	60	1.54
		2	—	—	—	30	1.0
IMTP	Metal	1	—	—	0.97	41	1.74
		$1\frac{1}{2}$	—	—	0.98	24	1.37
		2	—	—	0.98	18	1.19
Hy-Pak	Metal	1	19	54	0.96	45	1.54
		$1\frac{1}{2}$	—	—	—	29	1.36
		2	14	29	0.97	26	1.09
Tri-Pac	Plastic	1	6.2	85	0.90	28	—
		2	4.2	48	0.93	16	—

[1] Bulk density and total area are given per unit volume of column.
[2] Factor F_p is a pressure drop factor and f_p a relative mass-transfer coefficient.
[3] Based on NH_3–H_2O data; other factors based on CO_2–NaOH data.

12.1.2 Fluid Mechanics of Packed Tower

Contact between liquid and gas

The requirement of good contact between liquid and gas is the hardest to meet, especially in large towers. Ideally the liquid, once distributed over the top of the packing, flows in thin films over all the packing surface all the way down the tower. Actually the films tend to grow thicker in some places and thinner in others, so that the liquid collects into small rivulets and flows along localized paths through the packing. Especially at low liquid rates much of the packing surface may be dry or, at best, covered by a stagnant film of liquid. This effect is known as *channeling;* it is the chief reason for the poor performance of large packed towers.

Channeling is severe in towers filled with stacked packing, which is the main reason they are not much used. It is less severe in dumped packings. In towers of moderate size, channeling can be minimized by having the diameter of the tower at least 8 times the packing diameter. If the ratio of tower diameter to packing diameter is less than 8 to 1, the liquid tends to flow out of the packing and down the walls of the column. Even in small towers filled with packings that meet this requirement, however, liquid distribution and channeling have a major effect on column performance. In large towers the initial distribution is especially important, but even with good initial distribution it is necessary to include redistributors for the liquid every 5 to 10 m in the tower, immediately above each packed section (as shown in Fig. 12.6). Improved liquid distribution has made possible the effective use of packed towers as large as 9 m in diameter.

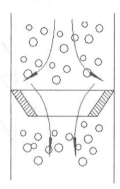

Wettability. Only wetted surfaces of packings are effective mass transfer area. If very low liquid rates have to be used, the packing wetting rates and spray densities should be checked to make sure it is above the minimum recommended by the packing manufacturer.

Liquid spray density is defined as liquid volume sprayed per unit time per unit cross sectional area of the column [m^3/(m$^2 \cdot$ s)] and wetting rate is volumetric liquid rate per unit length of packing periphery in the cross sectional area.

Fig. 12.6 Liquid redistributor.

Relation between wetting rate and spray density is as follows

$$U_{\min} = L_W \sigma \qquad (12.1\text{-}1)$$

where $U_{\min}$ = minimum spray density, m^3/(m$^2 \cdot$ s)
  L_w = minimum wetting rate, m^3/(m $\cdot$ s)
  σ = specific surface area of packings, m^2/m^3

When diameter of packings is less than 75 mm, $(L_W)_{\min} = 0.08$ m^3/(m $\cdot$ h); when diameter greater than 75 mm, $(L_W)_{\min} = 0.012$ m^3/(m $\cdot$ h).

In practice, liquid spray density must be greater than the minimum spray density. If the design liquid flow rate is too low, the diameter of the column should be reduced. For some processes liquid can be recycled to increase the flow over the packings.

Pressure drop and limiting flow rates

Figure 12.7 shows typical data for the pressure drop in a packed tower. The pressure drop per unit packing depth comes from fluid friction; it is plotted on logarithmic coordinates versus the gas flow rate G_y, expressed in mass of gas per hour per unit of cross-sectional area, based on the empty tower. Therefore, G_y is related to the superficial gas velocity by the equation $G_y = u_o \rho_y$, where ρ_y is the density of the gas. When the packing is dry, the line so obtained is straight and has a slope of about 1.8. The pressure drop therefore increases with the 1.8 power of the velocity. If the packing is irrigated with a constant flow of liquid, the relationship between pressure drop and gas flow rate initially follows a line parallel to that for dry packing. The pressure drop is greater than that in dry packing, because the liquid in the tower reduces the space available for gas flow. The void fraction, however, does not change with gas flow. At moderate gas velocities the line for irrigated packing gradually becomes steeper, because the gas now impedes the downflowing liquid and the liquid holdup increases with gas rate. The point at which the liquid holdup starts to increase, as judged by a change in the slope of the pressure drop line, is called the *loading point*. However, as is evident from Fig. 12.7, it is not easy to get an accurate value for the loading point.

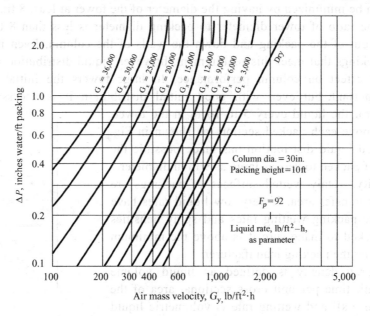

Fig. 12.7 Pressure drop in a packed tower for air-water system with 1-in. Intalox saddles.(1,000 lb/ft^2 · h=1.356 kg/m^2 · s; 1 in. H$_2$O/ft=817Pa/m)

With still further increase in gas velocity, the pressure drop rises even more rapidly, and the lines become almost vertical when the pressure drop is about 150 to 250 mm of water per meter of packing. In local regions of the column, the liquid becomes the continuous phase, and the flooding point has been reached. Higher gas flows can be used temporarily, but then liquid rapidly accumulates, and the entire column may fill with liquid.

The gas velocity in an operating packed column must obviously be lower than the flooding velocity. However, as flooding is approached, most of or all the packing surface is wetted, maximizing the contact area between gas and liquid. The designer must choose a velocity far enough from the flooding velocity to ensure safe operation but not so low as to require a much larger column. Lowering the design velocity increases the tower diameter without much change in the required height, since lower gas and liquid velocities lead to a nearly proportional reduction in mass-transfer rate. Decreased pressure drop is one benefit of low gas velocity, but the cost of power consumed is usually not a major factor in optimizing the design. The gas velocity is sometimes chosen as one-half the predicted flooding velocity obtained from a generalized correlation. This might seem too conservative, but there is considerable scatter in published data for flooding velocities, and the generalized correlations are not very accurate. A closer approach to flooding may be used if detailed performance data are available for the packing selected. Packed towers may also be designed on the basis of a definite pressure drop per unit height of packing.

The flooding velocity depends strongly on the type and size of packing and the liquid mass velocity. Figure 12.8 shows data for Intalox saddles taken from Fig. 12.7 and similar curves for other sizes. Flooding was assumed to occur at a pressure drop of 2.0 in. H$_2$O/ft of packing, since the pressure drop curves are vertical or nearly so at this point. For low liquid rates the flooding velocity varies with about the −0.2 to −0.3 power of the liquid rate and the 0.6 to 0.7 power of the packing size. The effects of liquid rate and packing size become more pronounced at high liquid mass velocities.

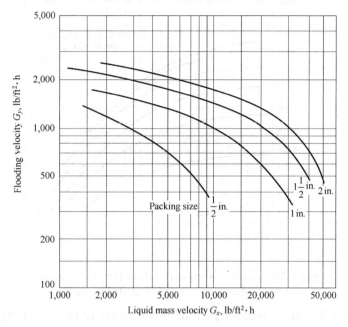

Fig. 12.8 Flooding velocities in ceramic Intalox saddles, air-water system. (1,000 lb/ft² · h=1.356 kg/m² · s)

Several generalized correlations have been proposed for the pressure drop and flooding velocity in packed columns. Most of these use a log-log plot with $(G_x/G_y)(\rho_y/\rho_x)^{0.5}$ on the abscissa and a function containing G_y^2 on the ordinate. Usually the flow ratio G_x/G_y is set from equilibrium and economic considerations as explained in chapter 8, and G_y can then be determined directly, whereas trial-and-error solution is needed if G_y and G_x are on separate axes, as in Fig. 12.8. The packing characteristics are accounted for by a packing factor F_p, which decreases with increasing packing size or increasing void fraction. Packing factors cannot be predicted from theory using the Ergun equation [Eq.(3.1-14)] because of the complex shapes, and they are determined empirically. Unfortunately, no single correlation for pressure drop gives a good fit to all packings, and values of F_p based on fitting the data for low pressure drops may differ significantly from values obtained by fitting the data for high pressure drops or by fitting the flooding data.

A widely used correlation for estimating pressure drops in dumped packings is given in Fig. 12.9 where G_x and G_y are in lb/(ft² · s), μ_x is in cP, ρ_x and ρ_y are in lb/ft³, and g_c is 32.174 lb · ft/(lbf · s²). Earlier versions of this correlation included a flooding line above the line for $\Delta P = 1.5$ in. H₂O/ft of packing, but recent studies show flooding at pressure drops of only 0.7 to 1.5 in. H₂O/ft of packing for 2-or 3-in. packings. An empirical equation for the limiting pressure drop is

$$\Delta P_{\text{flood}} = 0.115 F_p^{0.7} \tag{12.1-2}$$

where ΔP_{flood} = pressure drop at flooding, in. H₂O/ft of packing

490 | Unit Operations of Chemical Engineering

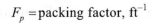
F_p = packing factor, ft^{-1}

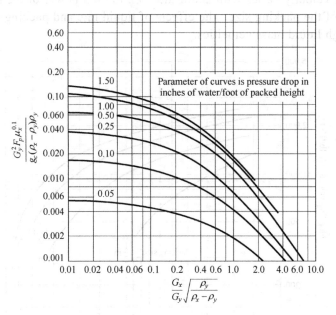

Fig. 12.9 Generalized correlation for pressure drop in packed columns. (1 in.H$_2$O/ft=817 Pa/m)

Equation (12.1-2) can be used for packing factors from 10 to 60. For higher values of F_p, the pressure drop at flooding can be taken as 2.0 in H$_2$O/ft.

An alternate correlation for the pressure drop in packed columns was proposed by Strigle and is shown in Fig. 12.10. The abscissa is essentially the same as for Fig. 12.9, but the ordinate includes the capacity factor $C_s = u_0\sqrt{\rho_y/(\rho_x - \rho_y)}$, where u_0 is the superficial velocity in feet per second. The kinematic viscosity of the liquid v is in centistokes. The semilog plot permits easier interpolation than the log-log plot, though both correlations are based on the same set of data.

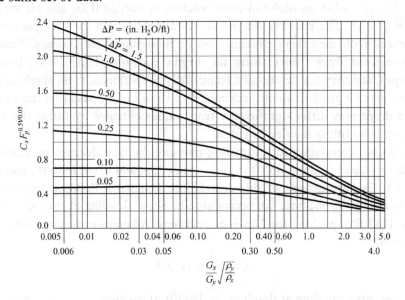

Fig. 12.10 Alternate generalized pressure drop correlation. (1 in. H$_2$O/ft=817Pa/m)

Pressure drop of structured packings

Flooding velocities in several commercial structured packings are shown in Fig. 12.11. The pressure drop in structured packings may be predicted from some rather complicated equations given by Fair and Bravo, but relatively little experimental information is available. Towers containing structured packing are best designed in collaboration with the packing manufacturer. Spiegel and Meier state that most structured packings reach their maximum capacity at a pressure drop of about 1000 $(N/m^2)/m$ (1.22 in. H_2O per ft), at a vapor velocity of 90 to 95 percent of the flooding velocity.

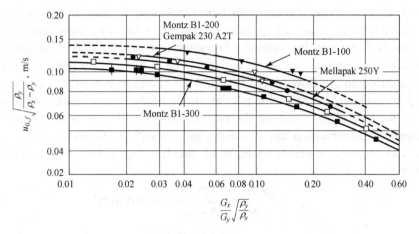

Fig. 12.11 Flooding velocities in structured packings.

12.1.3 Calculations of Packed Tower

Packed-bed height

For the design of packed distillation columns it is simpler to treat the separation as a staged process, and use the concept of the height of an equivalent equilibrium stage to convert the number of ideal stages required to a height of packing. The methods for estimating the number of ideal stages given in chapter 7 can then be applied to packed columns.

The column height Z is usually based on the number of theoretical plates and the height equivalent to a theoretical plate (HETP), that is

$$Z = \text{HETP} \times N_T \tag{12.1-3}$$

Though packed absorption and stripping columns can also be designed as staged process, it is usually more convenient to use the integrated form of the differential equations set up by considering the rates of mass transfer at a point in the column, that is, for example

$$Z = H_{OG} N_{OG} \tag{12.1-4}$$

The equation above is derived and discussed in chapter 8.

Diameter of packed column

The capacity of a packed column is determined by its cross-sectional area. Normally, the column will be designed to operate at the highest economical pressure drop, to ensure good liquid and gas distribution and avoid flooding. The diameter of packed column D is determined by the following equations

$$D = \sqrt{\frac{4V_S}{\pi u}} \quad (12.1\text{-}5)$$

$$u = (0.5 \sim 0.85)u_f \quad (12.1\text{-}6)$$

where V_S = volumetric gas rate, m³/s
u_f = gas velocity at flooding, m/s

At the same time, $D_{\text{tower}}/d_{\text{packing}}$ must be kept more than 8 to avoid channeling flow.

EXAMPLE 12.1 A tower packed with 1-in. (25.4-mm) ceramic Intalox saddles is to be built to treat 25000 ft³ (708 m³) of entering gas per hour. The ammonia content of the entering gas is 2 percent by volume. Ammonia-free water is used as absorbent. The temperature of the entering gas and the water is 68°F (20°C); the pressure is 1 atm. The ratio of liquid flow to gas flow is 1.25 lb liquid per pound of gas. (*a*) If the design pressure drop is 0.5 in. H₂O per foot of packing, what should be the mass velocity of the gas and the diameter of the tower? (*b*) Estimate the ratio of the gas velocity to the flooding velocity using a generalized correlation and also using specific data for Intalox saddles.

Solution. The average molecular weight of the entering gas is (29×0.98)+(17×0.02)= 28.76. Then

$$\rho_y = \frac{28.76}{359} \times \frac{492}{460+68} = 0.07465 \text{ lb/ft}^3$$

$$\rho_x = 62.3 \text{ lb/ft}^3$$

(*a*) Use the pressure-drop correlation in Fig. 12.10.

$$\frac{G_x}{G_y}\left(\frac{\rho_y}{\rho_x}\right)^{0.5} = 1.25 \times \left(\frac{0.07465}{62.3}\right)^{0.5} = 0.0433$$

For $\Delta P = 0.5$, $C_s F_p^{0.5} v^{0.05} = 1.38$

From Table 12.1, $F_p = 92$. At 20°C, $v = 1.0$ cSt

$$C_s = \frac{1.38}{92^{0.5} \times 1.0^{0.05}} = 0.144$$

$$u_0 = C_S \left(\frac{\rho_x - \rho_y}{\rho_y}\right)^{0.5} = 4.16 \text{ft/s}$$

$$G_y = 4.16 \times 0.07465 \times 3{,}600 = 1120 \text{ lb/(ft}^2 \cdot \text{h)}$$

$$G_x = 1.25 \ G_y = 1400 \text{ lb/(ft}^2 \cdot \text{h)}$$

In this case, the value of G_y can be checked using the data for 1-in. Intalox saddles in Fig. 12.7. Interpolation between the curves for dry packing and $G_x = 3{,}000$ lb/(ft² · h), gives $G_y \approx 1{,}000$ lb/(ft² · h). This value is used for the design.

Total gas flow: 25000 × 0.0746 = 1865 lb/h
Cross-sectional area of the tower:

$$S = 1865/1000 = 1.865 \text{ ft}^2$$

$$D = \left(\frac{4 \times 1.865}{\pi}\right)^{0.5} = 1.54 \text{ft (470mm)}$$

(b) The generalized correlation of Fig. 12.10 does not include a line for flooding, but the pressure drop at flooding is about 2.0 in. H$_2$O/ft, and extrapolation to this value indicates an ordinate of 1.95 compared to the design value of 1.38. Therefore

$$\frac{G_{y,\text{design}}}{G_{y,\text{flooding}}} = \frac{1.38}{1.95} = 0.71$$

This ratio applies when G_x/G_y is held constant. The ratio is lower with a lower G_x. If G_x is kept at 1400 lb/(ft^2 · h), Fig. 12.8 indicates flooding at G_y =1 650 lb/(ft^2 · h). Then

$$\frac{G_{y,\text{design}}}{G_{y,\text{flooding}}} = \frac{1000}{1650} = 0.61$$

EXAMPLE 12. 2 Substitute Gempak 230 A2T packing, manufactured by Glitsch, for the 1-in. Intalox saddles in the tower of Example 12.1. What increase in gas mass velocity is expected?

Solution. Use Fig. 12.11. From Example 12.1, G_x/G_y =1.25 and

$$\frac{G_x}{G_y}\left(\frac{\rho_y}{\rho_x}\right)^{0.5} = 1.25 \times \left(\frac{0.07465}{62.3}\right)^{0.5} = 0.0433$$

From Fig. 12.11, $u_{0,f}\sqrt{\rho_y/(\rho_x-\rho_y)} = 0.11$, The superficial vapor velocity at flooding is therefore

$$u_{0,f} = C_S\left(\frac{62.3-0.07}{0.07465}\right)^{0.5} = 3.175 \text{m/s}$$

The allowable vapor velocity, at 60 percent of flooding, is

$$u_0 = 3.175 \times 0.6 = 1.905 \text{ m/s or } 6.25 \text{ ft/s}$$

The corresponding mass velocity is

$$G = 6.25 \times 0.7465 \times 3,600 = 1,680 \text{ lb/(ft}^2 \cdot \text{h)}$$

The increase in gas mass velocity is (1 680/1 000) − 1 = 0.68 or 68 percent.

12.2 Plate Tower Design

12.2.1 Introduction

To translate ideal plates to actual plates, a correction for the efficiency of the plates must be applied. There are other important decisions, some at least as important as fixing the number of plates, that must be made before a design is complete. These include the type of trays, the size and pattern of the holes in the tray, the downcomer size, the tray spacing, the weir height, the allowable vapor and liquid rates and pressure drop per tray, and the column diameter. A mistake in these decisions results in poor fractionation, lower-than-desired capacity, poor operating flexibility, and with extreme errors, an inoperative column. Correcting such errors after a plant has been built can be costly. Since many variables that influence plate efficiency depend on the design of the individual plates, the fundamentals of plate design are discussed first.

The extent and variety of rectifying columns and their applications are enormous. The largest units are usually in the petroleum industry, but large and very complicated distillation plants are encountered in fractionating solvents, in treating liquefied air, and in general chemical processing. Tower diameters may range from 300 mm to more than 9 m and the number of plates from a few to about a hundred. Plate spacings may vary from 150 mm or less to 1 or 2 m. Formerly bubble-cap plates (Fig. 12.12) were most common; today most columns contain sieve trays or valve trays (Fig. 12.13). Columns may operate at high pressures or low, from temperatures of liquid gases up to 900℃ reached in the rectification of sodium and potassium vapors. The materials distilled can vary greatly in viscosity, diffusivity, corrosive nature, tendency to foam, and complexity of composition. Plate towers are as useful in absorption as in rectification, and the fundamentals of plate design apply to both operations.

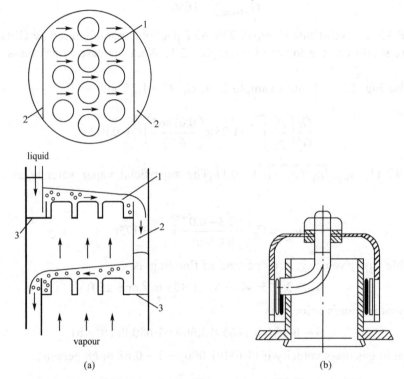

Fig. 12.12 Bubble-cap: (*a*) Bubble-cap tray; (*b*) Single bubble-cap
1—Bubble-cap; 2—downcomer; 3—plate

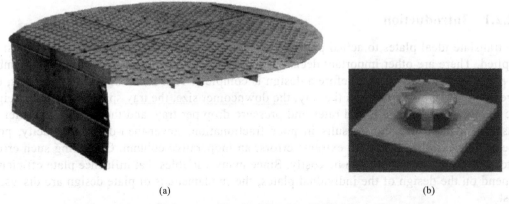

Fig 12.13 (*a*) Valve tray (Koch-Glitsch); (*b*) Single valve (open).

Designing fractionating columns, especially large units and those for unusual applications, is best done by experts. Although the number of ideal plates and the heat requirements can be computed quite accurately without much previous experience, other design factors are not precisely calculable, and a number of equally sound designs can be found for the same problem. In common with most engineering activities, sound design of fractionating columns relies on a few principles, on a number of empirical correlations (which are in a constant state of revision), and on much experience and judgment.

The following discussion is limited to the usual types of column, equipped with sieve plates or valve-trays, operating at pressures not far from atmospheric, and treating mixtures having ordinary properties.

12.2.2 Fluid Mechanics of Plate Columns

Normal operation of plate columns

A sieve plate is designed to bring a rising stream of vapor into intimate contact with a descending stream of liquid. The liquid flows across the plate and passes over a weir to a downcomer leading to the plate below. The flow pattern on each plate is therefore crossflow rather than countercurrent flow, but the column as a whole is still considered to have countercurrent flow of liquid and vapor. The fact that there is crossflow of liquid on the plate is important in analyzing the hydraulic behavior of the column and in predicting the plate efficiency.

There is another class of tray which has no separate downcomers and yet it still employs a tray type of construction giving a hydrodynamic performance between that of a packed and a plate column. But it is less used and not discussed here.

Figure 12.14 shows a plate in a sieve-tray column in normal operation. The downcomers are the segment-shaped regions between the curved wall of the column and the straight chord weir. Each downcomer usually occupies 10 to 15 percent of the column cross section, leaving 70 to 80 percent of the column area for bubbling or contacting. In small columns the downcomer may be a pipe welded to the plate and projecting up above the plate to form a circular weir. For very large columns, additional downcomers may be provided at the middle of the plate to decrease the length of the liquid flow path. In some cases an underflow weir or tray inlet weir is installed as shown in Fig. 12.14 to improve the liquid distribution and to prevent vapor bubbles from entering the downcomer.

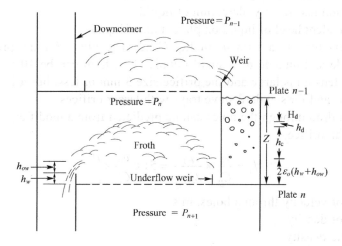

Fig 12.14 Normal operation of sieve plate.

The vapor passes through the perforated region of the plate, which occupies most of the space between downcomers. The holes are usually 5 to 12 mm in size and arranged in a triangular pattern. One or two rows of holes may be omitted near the overflow weir to permit some degassing of the liquid before it passes over the weir. Some holes may also be omitted near the liquid inlet to keep vapor bubbles out of the downcomer. Under normal conditions, the vapor velocity is high enough to create a frothy mixture of liquid and vapor that has a large surface area for mass transfer. The average density of the froth may be as low as 0.2 of the liquid density, and the froth height is then several times the value corresponding to the amount of liquid actually on the plate.

In a valve-tray column (Fig. 12.13) the openings in the plate are quite large, typically 38 mm in diameter. The openings are covered with lids or "valves" which rise and fall as the vapor rate varies, providing a variable area for vapor flow. Downcomers and crossflow of the liquid are used as with ordinary sieve trays. Valve trays are more expensive than conventional trays but have the advantage of a large *turndown ratio* (the ratio of the maximum allowable vapor velocity to the lowest velocity at which the column will operate satisfactorily), up to 10 or more, so the operating range of the column is large. Recent developments in valve trays have increased the available bubbling area and improved the distribution of vapor flowing through the plates.

Vapor pressure drop

The flow of vapor through the holes and the liquid on the plate requires a difference in pressure. The pressure drop across a single plate is usually 50 to 100 mm H_2O, and the pressure drop over a 40-plate column is then about 2 to 4 m H_2O. The pressure required is automatically developed by the reboiler, which generates vapor at a pressure sufficient to overcome the pressure drop in the column and condenser. The overall pressure drop is calculated to determine the pressure and temperature in the reboiler, and the pressure drop per plate must be checked to make sure the plate will operate properly, without weeping or flooding.

The pressure drop across the plate can be divided into two parts, the friction loss in the holes and the pressure drop due to the holdup of liquid on the plate. The pressure drop is usually given as an equivalent head in millimeters of liquid:

$$h_p = h_c + h_l \tag{12.2-1}$$

where h_p = total pressure drop per plate, mm of liquid
h_c = friction loss for dry plate, mm of liquid
h_l = equivalent head of liquid on plate, mm of liquid

Some workers include a term h_σ in Eq. (12.2-1) to allow for the pressure difference between the inside and outside of a small bubble. This term can be 10 to 20 mm of liquid when the surface tension is large and the orifice size 3 mm or less, but it is usually neglected for organic liquids and for standard sieve trays with larger orifices.

The pressure drop through the holes can be predicted from a modification of Eq. (1.6-12) for flow through an orifice:

$$h_c = \frac{u_0^2}{C_0^2} \times \frac{\rho_V}{2g\rho_L} = 51.0 \frac{u_0^2}{C_0^2} \times \frac{\rho_V}{\rho_L} \tag{12.2-2}$$

where u_0 = vapor velocity through holes, m/s
ρ_V = vapor density
ρ_L = liquid density
C_0 = orifice coefficient

Equation (12.2-2) gives h_c in millimeters of liquid. The coefficient comes from

$$\frac{1000\text{mm/m}}{2\times 9.81\text{m/s}^2} = 51.0$$

If u_0 is expressed in feet per second and h_c in inches, the coefficient becomes

$$\frac{12}{2\times 32.2} = 0.186$$

Orifice coefficient C_0 depends on the fraction open area (the ratio of the total cross-sectional area of the holes to the column cross section) and on the ratio of tray thickness to hole diameter, as shown in Fig. 12.15. The increase in C_0 with open area is similar to the change in C_0 for single orifices as the ratio of orifice diameter to pipe diameter increases. The coefficients vary with plate thickness, but for most sieve plates the thickness is only 0.1 to 0.3 times the hole size. For these thicknesses and the typical fraction open area of 0.08 to 0.10, the value of C_0 is 0.66 to 0.72.

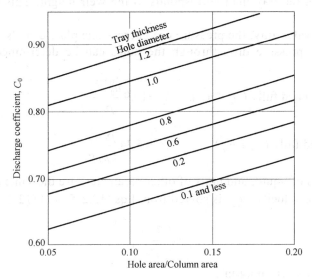

Fig. 12.15 Discharge coefficients for vapor flow, sieve trays. [*I. Liebson, R. E. Kelley, and L. A. Bullington, Petrol. Refin.*, 36(2):127, 1957; 36(3):288, 1957.]

The amount of liquid on the plate increases with the weir height and with the flow rate of liquid; but it decreases slightly with increasing vapor flow rate, because this decreases the density of the froth. The liquid holdup also depends on the physical properties of liquid and vapor, and only approximate methods of predicting the holdup are available. A simple method of estimating h_l uses the weir height h_w, the calculated height of clear liquid over the weir h_{ow}, and an empirical correlation factor ε_o:

$$h_l = \varepsilon_o (h_w + h_{ow}) \tag{12.2-3}$$

The height over the weir is calculated from a form of the Francis equation, which for a straight segmental weir is

$$h_{ow} = 43.4 \left(\frac{L_s}{l_w}\right)^{2/3} \tag{12.2-4}$$

where h_{ow} = height, mm
L_s = flow rate of clear liquid, m³/min
l_w = length of weir, m

If L_s/l_w is in gallons per minute per inch, a coefficient of 0.48 in Eq. (12.2-4) gives h_{ow} in inches.

The actual height of froth over the weir is greater than h_{ow}, since the vapor has only partially separated from the liquid, making the volumetric flow rate at the weir greater than that of the liquid alone. However, the actual height over the weir is not needed in estimating h_l, since the effect of froth density is included in the correlation factor ε_o. For typical weir heights of 25 to 50 mm and the normal range of vapor velocities, values of ε_o are 0.4 to 0.7. The change in ε_o with flow rates of vapor and liquid is complex, and there is no generally accepted correlation. For design purposes, a value of $\varepsilon_o = 0.6$ can be used in Eq. (12.2-3), and some error can be tolerated since most of the pressure drop at high vapor flows is due to the holes.

When h_{ow} is small relative to h_w, Eq. (12.2-3) shows that h_l may be less than h_w, which means less liquid on the tray than corresponds to the weir height. This is a fairly common situation.

For F1-type of valve-tray, the pressure drop across one plate can also be calculated from Eq.(12.2-1) and the pressure drop through the holes can be determined by the following empirical equation:

If the valve is not opened fully ($u_o \leqslant u_{oc}$),
$$h_c = 19.9 \frac{u_0^{0.175}}{\rho_L} \tag{12.2-5}$$

If the valve is opened fully ($u_o \geqslant u_{oc}$),
$$h_c = 5.34 \frac{\rho_V u_0^2}{2\rho_L g} \tag{12.2-6}$$

where h_c is in meters of liquid, and units of other parameters are the same as above.

The critical hole velocity u_{oc} is solved from Eqs.(12.2-5) and (12.2-6):

$$u_{oc} = \left(\frac{73.1}{\rho_V}\right)^{1.825} \tag{12.2-7}$$

where u_{oc} is in m/s, and ρ_V in kg/m³.

For the pressure drop due to the holdup of liquid on the valve-tray h_l, Eq.(12.2-3) still applies.

Downcomer level

The level of liquid in the downcomer must be considerably greater than that on the plate because of the pressure drop across the plate. Referring to Fig. 12.14, note that the top of the downcomer for plate n is at the same pressure as plate $n - 1$. Therefore, the equivalent level in the downcomer must exceed that on the plate by an amount h_p plus any friction losses in the liquid h_d. The total height of clear liquid H_d is

$$H_d = \varepsilon_o(h_w + h_{ow}) + h_p + h_d \tag{12.2-8}$$

Using Eqs. (12.2-1) and (12.2-3) for h_p gives

$$H_d = 2\varepsilon_o(h_w + h_{ow}) + h_c + h_d \tag{12.2-9}$$

The contributions to H_d are shown in Fig. 12.14. Note that an increase in h_w or h_{ow} comes in twice, since it increases the level of liquid on the plate and increases the pressure drop for vapor flow. The term for h_d is usually small, corresponding to one to two velocity heads based on the liquid velocity under the bottom of the downcomer.

For the plate without underflow weir, h_d is determined by the following empirical equation:

$$h_d = 0.153 \left(\frac{L_s}{l_W h_o} \right)^2 = 0.153(u'_o)^2 \qquad (12.2\text{-}10)$$

For the plate with underflow weir,

$$h_d = 0.2 \left(\frac{L_s}{l_W h_o} \right)^2 = 0.2(u'_o)^2 \qquad (12.2\text{-}11)$$

where L_s = flow rate of clear liquid, m³/s
l_W = length of weir, m
h_o = spacing between downcomer bottom and tray, m
u'_o = liquid velocity of downcomer bottom, m/s

The actual level of aerated liquid Z in the downcomer is greater than H_d because of the entrained bubbles. If the average volume fraction liquid is ϕ, the level is

$$Z = H_d / \phi \qquad (12.2\text{-}12)$$

When the height of aerated liquid becomes as great as or greater than the plate spacing, the flow over the weir on the next plate is hindered, and the column becomes flooded. For conservative design, a value of $\phi = 0.5$ is assumed, and the plate spacing and operating conditions are chosen so that Z is less than the plate spacing.

Operating limits and capacity chart

Weeping and hydraulic gradient. At low vapor velocities, the pressure drop is not great enough to prevent liquid from flowing down through some of the holes. This condition is called *weeping* and is more likely to occur if there is a slight gradient in liquid head (hydraulic gradient) across the plate. Hydraulic gradient on plate is caused by the liquid flow resistance across the plate, including friction loss through plate surface, local resistance of parts (bubble-cap, floating valve) on plate and resistance from vapor flow. With such a gradient, vapor will tend to flow through the region where there is less liquid and therefore less resistance to flow, and liquid will flow through the section where the depth is greatest. Weeping decreases the plate efficiency, since some liquid passes to the next plate without contacting the vapor. The lower limit of operation could be extended by using smaller holes or a lower fraction open area, but these changes would increase the pressure drop and reduce the maximum flow rate. A sieve tray can usually be operated over a three- to four fold range of flow rates between the weeping and flooding points. If a greater range is desired, other types of plates such as valve trays can be used.

The weeping amount should not be greater than 10% of the liquid flow rate. The weeping amount is also determined by kinetic energy factor F (*F factor*). *F factor* is defined as follows:

$$F = u\sqrt{\rho_V} \qquad (12.2\text{-}13)$$

where u is the vapor velocity in meters per second.
F factor is the index of dynamic pressure of vapor. For valve-tray, *F factor* in the opening of valve-tray must be greater than 5~6 (m/s)(kg/m³)$^{0.5}$ in order to avoid excessive weeping.

Excessive entrainment and flooding. The upper limit of the vapor velocity in a sieve-tray column is usually determined by the velocity at which entrainment becomes excessive, causing a large drop in plate efficiency for a small increase in vapor rate. This limit is called the flooding point, or more properly, the *entrainment flooding point*, since the allowable vapor velocity is sometimes limited by other factors. When the pressure drop on a plate is too high and the liquid in the downcomer backs up to the plate above, the flow from that plate is inhibited. This leads to an increase in liquid level and a further increase in pressure drop. This phenomenon is called *downcomer flooding;* it can occur before entrainment becomes excessive. For liquids with low surface tension, the limit may be the velocity that makes the froth height equal to the plate spacing, leading to a large carryover of liquid to the plate above.

The early correlations for the flooding limit focused on entrainment and the settling velocity of liquid drops. For large drops, the terminal velocity varies with $\sqrt{(\rho_L - \rho_V)/\rho_V}$ (see Eq. (3.2-19), and this term is included in the correlation, even though $(\rho_L - \rho_V)$ is practically the same as ρ_L in most cases. Figure 12.16 is an empirical correlation[5] for sieve plates that is widely quoted and said to apply to valve trays and bubble-cap trays as well. For a given value of $L/V(\rho_V/\rho_L)^{0.5}$ and a selected plate spacing, the correlation gives a value of K_v which is used to calculate the maximum permissible vapor velocity.

$$u_c = K_v \sqrt{\frac{\rho_L - \rho_V}{\rho_V}} \left(\frac{\sigma}{20}\right)^{0.2} \qquad (12.2\text{-}14)$$

where u_c = maximum vapor velocity based on bubbling area, ft/s
 σ = surface tension of liquid, dyn/cm

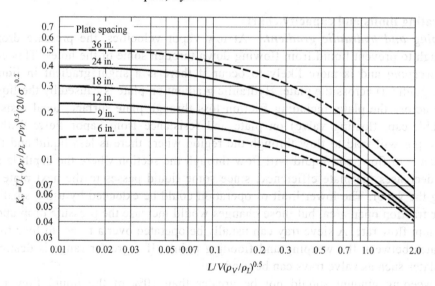

Fig. 12.16 Values of K_v at flooding conditions for sieve plates; L/V=ratio of mass flow rate of liquid to vapor, u is in feet per second, and σ is in dynes per centimeter. [*J. R. Fair, Petrol. Chem. Eng.,* 33(10):45, 1961. *Courtesy Petroleum Engineer.*]

Note that the effects of plate spacing and other variables in the correlation are similar to the effects predicted for other types of flooding. Increasing the plate spacing delays the onset of downcomer flooding and froth height flooding, in addition to decreasing the chance for entrained drops to reach the next plate. An increase in the vapor density decreases u_c in Eq. (12.2-13); it also increases the dry-plate pressure drop, which can lead to downcomer flooding. The surface tension term, which was not in the original correlation, was added because of the lower flooding velocities observed with some organic liquids. A value of $\sigma =$ 20 dyn/cm is typical for organic liquids, and the correlation shows that the flooding velocity for such liquids is about 20 percent lower than for water, for which σ is about 72 dyn/cm. The correlation may not be reliable for liquids with a very low surface tension, in which cases the velocity might be limited by the froth height.

The effect of pressure is reflected in ρ_V, which is included in both the abscissa and ordinate of Fig. 12.16. For most distillations at atmospheric pressure, $L/V(\rho_V /\rho_L)^{0.5}$ is quite small, K_v is nearly constant, and u_c varies with $\rho_V^{-0.5}$. For high-pressure operation, where K_v decreases as ρ_V increases, there is a greater effect of ρ_V on u_c. The flooding velocity is much lower for columns operated at high pressure than for those at atmospheric pressure.

The correlation in Fig. 12.16 does not include the effect of weir height, which can range from 1/2 to 4 in. Increasing h_w increases the depth of liquid on the plate, which probably decreases u_c, but few data are available. A detailed discussion of flooding and an alternate correlation for entrainment flooding is given by Kister.

Flooding is the operation limit of countercurrent flow between liquid and vapor and must be avoided. Operating limit equation of *downcomer flooding* is as follows:

$$H_d \leqslant \Phi(H_T + h_w) \qquad (12.2\text{-}15)$$

where H_d = equivalent height of clear liquid in downcomer, m
Φ = correct coefficient considering aeration and safe operation
H_T = plate spacing (distance between plates), m

Some workers define the excessive entrainment as that the amount of vapor entrainment is greater than 10 percent. It can also be determined by the percentage flooding, that is, the actual velocity is usually about 75% the flooding velocity.

Other detrimental operations include the *vapor bubble entrainment* caused by liquid overflowing into down-comer and "coning" which occurs at low liquid rates. When velocity of liquid flow into down-comer is too high to separate the vapor entrained by froth from liquid, vapor is brought into the lower plate and plate efficiencies decreased.

Operating limit of vapor bubble entrainment is determined by the residence time of liquid in downcomer:

$$\theta \geqslant \frac{A_f H_T}{L_{s,\max}} = 3 \sim 5 \qquad (12.2\text{-}16)$$

where θ = residence time of liquid in downcomer, s
A_f = cross sectional area of downcomer, m^2
$L_{s,\max}$ = maximum volumetric liquid rate, m^3/s

At low liquid rates, the vapor will push the liquid back from the holes and jet upward, leading a poor liquid and vapor contact and little liquid on the tray with no liquid seal. In order to avoid the "coning" phenomenon, the liquid height over the weir h_{ow} must be over 6 mm.

Operating range and capacity chart. For certain vapor-liquid system, there exists a area of satisfactory operation for vapor and liquid loadings within a plate column. This is described by capacity performance chart. A typical capacity chart for a valve-tray is shown in Fig. 12.17.

The upper limit to vapor flow is set by the

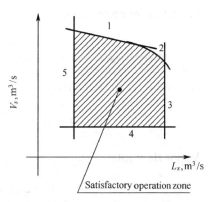

Fig. 12.17 Valve-tray capacity performance chart.

condition of flooding or excessive carry over of liquid to the next plate by entrainment. At flooding there is a sharp drop in plate efficiency and increase in pressure drop. Line 1 is drawn from the percentage flooding limit which stands for excessive froth entrainment and curve 2 from the operating limit equation of *downcomer flooding* (Eq. (12.2-5)).

Line 4 is the lower limit of the vapor flow which is set by the condition of excessive weeping, when $F_o = u_o\sqrt{\rho_V} = 5 \sim 6$. Here u_o is the vapor velocity of the floating-valve hole.

Line 3 is the downcomer back-up limitation drawn from Eq.(12.2-16) and line 5 the lower limit of volumetric liquid flow rate when the liquid height over the weir h_{ow} is taken as 6 mm.

The capacity performance chart is usually used for Checking whether the design of plate column is satisfactory or not, determining the turndown ratio of the column and judging the potential of capacity.

EXAMPLE 12.3 A sieve-plate column operating at atmospheric pressure is to produce nearly pure methanol from an aqueous feed containing 40 mol percent methanol. The distillate product rate is 5800 kg/h. (*a*) For a reflux ratio of 3.5 and a plate spacing of 18 in., calculate the allowable vapor velocity and the column diameter. (*b*) Calculate the pressure drop per plate if each sieve tray is 1/8 in. thick with 1/4-in. holes on a 3/4-in. triangular spacing and a weir height of 2 in. (*c*) What is the froth height in the downcomer?
Solution. Calculate u_c at the top of the column, because flooding is more likely here where the vapor density is higher than it is at the bottom.

Physical properties of methanol. Molecular weight is 32, normal boiling point is 65°C, and the density of vapor is

$$\rho_V = \frac{32 \times 273}{22.4 \times 338} = 1.15 \text{kg/m}^3$$

From Perry, *Chemical Engineers' Handbook*, 6th ed., page 3–188, the density of liquid methanol is 810 kg/m³ at 0°C and 792 kg/m³ at 20°C. At 65°C, the estimated density ρ_L is 750 kg/m³. Lange's *Handbook of Chemistry*, 9th ed., 1956, page 1650, gives the surface tension of methanol at 20 and 100°C. By interpolation, at 65°C, σ = 19 dyn/cm.

(*a*) Vapor velocity and column diameter. In Fig. 12.16 the abscissa is

$$\frac{L}{V}\left(\frac{\rho_V}{\rho_L}\right)^{1/2} = \frac{3.5}{4.5}\left(\frac{1.15}{750}\right)^{1/2} = 3.04 \times 10^{-2}$$

For 18-in. plate spacing,

$$K_v = 0.29 = u_c\sqrt{\frac{\rho_V}{\rho_L - \rho_V}}\left(\frac{20}{\sigma}\right)^{0.2}$$

Allowable vapor velocity:

$$u_c = 0.29\sqrt{\frac{750-1.15}{1.15}}\left(\frac{19}{20}\right)^{0.2} = 7.32 \text{ft/s or 2.23 m/s}$$

Vapor flow rate:

$$V = D(R+1) = 4.5 = \frac{5800 \times 4.5}{3600 \times 1.15} = 6.30 \text{m}^3/\text{s}$$

Cross-sectional area of column:

$$\text{Bubbling area } = \frac{6.30}{2.23} = 2.83 \text{m}^2$$

If the bubbling area is 0.7 of the total column area,

$$\text{Column area } = \frac{2.83}{0.7} = 4.04 \text{m}^2$$

Column diameter:

$$D_c = \left(\frac{4 \times 4.04}{\pi}\right)^{1/2} = 2.27 \text{m}$$

(b) Pressure drop. The plate area of one unit of three holes on a triangular 3/4-in. pitch is $\frac{1}{2} \times \frac{3}{4}\left(\frac{3}{4} \times \sqrt{3}/2\right) = 9\sqrt{3}/64$ in.^{2}. The hole area in this section (half a hole) is $\frac{1}{2} \times \frac{\pi}{4}\left(\frac{1}{4}\right)^2 = \pi/128$ in.^{2}. Thus the hole area is $(\pi/128) \times 64/(9\sqrt{3}) = 0.1008$, or 10.08 percent of the bubbling area.

Vapor velocity through holes:

$$u_o = \frac{2.23}{0.1008} = 22.1 \text{m/s}$$

Use Eq. (12.2-2) for the pressure drop through the holes. From Fig. 12.15, $C_0 = 0.73$. Hence

$$h_c = 51.0\left(\frac{22.1^2}{0.73^2}\right)\left(\frac{1.15}{750}\right) = 71.7 \text{mm methanol}$$

Head of liquid on plate:

Weir height: $h_w = 2 \times 25.4 = 50.8$ mm

Height of liquid above weir: Assume the downcomer area is 15 percent of the column area on each side of the column. From Perry, 6th ed., page 1–26, the chord length for such a segmental downcomer is 1.62 times the radius of the column, so

$$l_w = 1.62 \times 2.27/2 = 1.84 \text{ m}$$

Liquid flow rate:

$$L_s = \frac{5800 \times 3.5}{750 \times 60} = 0.45 \text{m}^3/\text{min}$$

From Eq. (12.2-4),

$$h_{ow} = 43.4 \times \left(\frac{0.45}{1.84}\right)^{2/3} = 17.0 \text{ mm}$$

From Eq. (12.2-3), with $\varepsilon_o = 0.6$,

$$h_l = \varepsilon_o(h_w + h_{ow}) = 0.6 \times (50.8 + 17.0) = 40.7 \text{ mm}$$

Total head of liquid [from Eq. (12.2-1)]:

$$h_p = 71.7 + 40.7 = 112.4 \text{ mm}$$

(c) Froth height in downcomer. Use Eq. (12.2-9). Estimate $h_d = 10$ mm methanol.

Then
$$H_d = 2\varepsilon_o(h_w + h_{ow}) + h_p + h_d = 2 \times 40.7 + 71.7 + 10 = 163.1 \text{ mm}$$

From Eq. (12.2-12),
$$Z = \frac{163.1}{0.5} = 326 \text{ mm}(12.8 \text{ in.})$$

Since Z is less than the plate spacing, downcomer flooding should not occur. For conservative design a column diameter of 2.5 m might be used so that the vapor velocity would be 20 percent less than u_c.

12.2.3 Plate Efficiencies

In all the previous discussions of theoretical trays or stages in distillation, we assumed that the vapor leaving a tray was in equilibrium with the liquid leaving. However, if the time of contact and the degree of mixing on the tray are insufficient, the streams will not be in equilibrium. As a result the efficiency of the stage or tray will not be 100%. This means that we must use more actual trays for a given separation than the theoretical number of trays determined by calculation. The definitions for three kinds of plate efficiency were discussed in chapter 9. The discussions in this section apply to the distillation tray towers.

The relation between Murphree η_M and local efficiency η' depends on the degree of liquid mixing and whether or not the vapor is mixed before going to the next plate. Calculations have shown only a small difference in efficiency for completely mixed vapor or unmixed vapor, but the effect of no liquid mixing can be quite large. Most studies have assumed complete vapor mixing in order to simplify the calculations for various degrees of liquid mixing. A correlation based on plug flow of liquid across the plate with eddy diffusion in the liquid phase was developed by workers at the University of Delaware and is given in Fig. 12.18. The abscissa is $(mV/L)\eta'$, and the parameter on the graphs is a Peclet number for axial dispersion:

$$Pe = \frac{Z_l^2}{D_E t_L} \tag{12.2-17}$$

where Z_l = length of liquid flow path, m
 D_E = eddy diffusivity, m²/s
 t_L = residence time of liquid on plate, s

For distillation at atmospheric pressure in a column 0.3 m in diameter the Peclet number is about 10, based on empirical correlations for dispersion on bubblecap and sieve trays This is in the range where significant enhancement of the efficiency should result because of gradients on the plate. For a column 1 m or larger in diameter, the Peclet number would be expected to be greater than 20, and the efficiency should be almost as high as it would be with no mixing in the direction of flow. Tests on very large columns, however, sometimes show lower plate efficiencies than for medium-size columns, probably because of departure from plug flow. With a large column and segmental downcomers the liquid flowing around the edge of the bubbling area, as shown in Fig. 12.19, has an appreciably longer flow path than liquid crossing the middle; and a wide distribution of residence times or even some backflow of liquid may result. These effects can be minimized by special plate design.

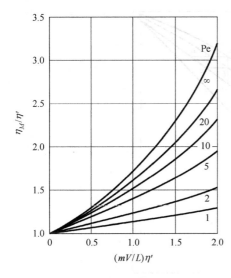

Fig. 12.18 Relationship between Murphree and local efficiencies.

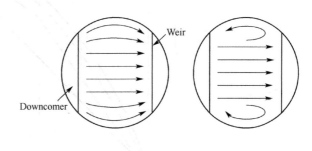

Fig. 12.19 Possible liquid flow patterns in a large column.

Use of Murphree efficiency

When the Murphree efficiency is known, it can readily be used in the McCabe-Thiele diagram. The diagram for an actual plate as compared with that for an ideal plate is shown in Fig. 12.20. Triangle *acd* represents the ideal plate and triangle *abe* the actual plate. The actual plate, instead of enriching the vapor from y_{n+1} to y_n^*, shown by line segment *ac*, accomplished a lesser enrichment $y_n - y_{n+1}$, shown by line segment *ab*. By definition of η_M, the Murphree efficiency is given by the ratio *ab/ac*. To apply a known Murphree efficiency to an entire column, it is necessary only to replace the true equilibrium curve y_e versus x_e by an effective equilibrium curve y_e' versus x_e, whose ordinates are calculated from the equation

$$y_e' = y + \eta_M(y_e - y) \quad (12.2\text{-}18)$$

In Fig. 12.20 an effective equilibrium curve for $\eta_M = 0.60$ is shown. Note that the position of the y_e' versus x_e curve depends on both the operating line and the true equilibrium curve.

Once the effective equilibrium curve has been plotted, the usual step-by-step construction is made and the number of actual plates determined. The reboiler is not subject to a discount for plate efficiency, and the true equilibrium curve is used for the last step in the stripping section.

Factors influencing plate efficiency

Although many studies of plate efficiency have been made and some empirical correlations presented, the best method of predicting efficiency is still in question. Many authors have recommended the O'Connell correlation, a plot which is a rough fit to data from 31 plant columns with bubble-cap trays. The plate efficiency, which ranged from 90 percent to 30 percent, was shown to decrease with increasing values of the feed viscosity μ_L times the relative volatility of the key components. An equation fitting this graphical correlation was given by Lockett.

$$\eta_o = 0.492(\mu_L \alpha)^{-0.245} \quad (12.2\text{-}19)$$

where μ_L = liquid viscosity, cP.

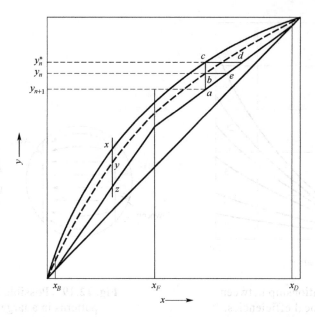

Fig. 12.20 Use of Murphree efficiency on xy diagram. Dashed line is effective equilibrium curve, y'_e, versus x_e for $\eta_M=0.60$; $ba/ca=yz/xz=0.60$.

The efficiency falls as the liquid viscosity increases mainly because of the decrease in diffusivity, which affects the liquid-film resistance. In most distillations the gas film has the controlling resistance, so the effect of a change in liquid viscosity is not large. A high value of α for a binary system means a high value of m at low values of x and a high liquid-film resistance. However, as x approaches 1, $m \cong 1/\alpha$, so the effect of m on the overall efficiency should be small. A correlation with $\mu_L m$ might be an improvement, but it seems unlikely that μ_L and m would have exactly the same effect.

Mass transfer theories have been used with test data to develop complex correlations for local and plate efficiency. These methods allow for the effects of the physical properties of liquid and vapor, flow rates, and plate dimensions. They fit the data better than Eq. (12.2-19), but reliable estimates of efficiency still cannot be obtained. Most columns are designed using efficiencies measured for the same type of plate and similar systems. The theories, however, are useful for predicting the effect of changes in physical properties.

The most important requirement for obtaining satisfactory efficiencies is that the plates operate properly. Adequate and intimate contact between vapor and liquid is essential. Any misoperation of the column, such as excessive foaming or entrainment, poor vapor distribution, or short-circuiting, weeping, or dumping of liquid, lowers the plate efficiency.

Plate efficiency is a function of the rate of mass transfer between liquid and vapor. The prediction of mass-transfer coefficients in sieve trays and their relationship to plate efficiency are discussed later. Some published values of the plate efficiency of a 1.2-m column are shown in Fig. 12.21. This column had sieve trays with 12.7-mm holes and 8.32 percent open area, a 51-mm weir height, and 0.61-m tray spacing. The data are plotted against a flow parameter F(F factor), which tends to cover about the same range for different total pressures, since for constant K_v the flooding velocity varies inversely with $\sqrt{\rho_V}$, as shown by Eq. (12.2-14). Parameter F, generally known as the F factor, is defined as follows:

$$F \equiv u\sqrt{\rho_V} \qquad (12.2\text{-}20)$$

At quite high pressures K_v decreases as ρ_V increases, and the allowable values of F are somewhat reduced, as shown by the data for butane at 11.2 atm in Fig. 12.21. Note that the F factor is similar to K_v but does not include the liquid density or the surface tension.

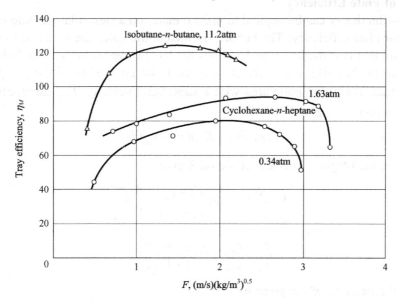

Fig. 12.21 Efficiency of sieve trays in a 1.2-m column. (*From M. Sakata and T. Yanagi, 3rd Int. Symp.Dist.*, p.3.2/21, *ICE*, 1979.)

The normal range of F for sieve trays is 1 to 3 (m/s) $(kg/m^3)^{0.5}$, or 0.82 to 2.46 (ft/s) $(lb/ft^3)^{0.5}$.

EXAMPLE 12.4 What is the F factor at the top of the column in Example 12.3 if the vapor velocity is the maximum allowable?
Solution. From Eq. (12.2-20) the F factor is

$$F \equiv u_c \sqrt{\rho_V} = 2.23\sqrt{1.15} = 2.39 \, (m/s)(kg/m^3)^{0.5}$$

As shown by Fig. 12.21 this is a reasonable value for a sieve-plate column at 1 atm pressure.

The efficiency does not change much with vapor rate in the range between the weeping point and the flooding point. The increase in vapor flow increases the froth height, creating more mass-transfer area, so that the total mass transferred goes up about as fast as the vapor rate. The data for Fig. 12.21 were obtained at total reflux, so the increase in liquid rate also contributed to the increase in interfacial area. The sharp decrease in efficiency near the flooding point is due to entrainment.

The data for cyclohexane-n-heptane show a lower efficiency for operation at a lower pressure, which has been confirmed by tests on other systems. Lowering the pressure decreases the concentration driving force in the vapor phase but increases the vapor diffusivity. It also lowers the temperature, which increases the liquid viscosity and surface tension and decreases the diffusion coefficients in the liquid. The decrease in efficiency is due to a combination of these effects.

Efficiencies greater than 100 percent for the isobutane-butane system show the effect of liquid concentration gradients; the local efficiency for this case is estimated to be 80 percent, based on Fig. 12.18 and $Pe \cong 80$. Another study of the i-butane-n-butane separation in a

2.9-m valve-tray column showed a Murphree plate efficiency of 119 percent and a calculated local efficiency of 82 percent.

Theory of Plate Efficiency

The two-film theory can be applied to mass transfer on a sieve plate to help correlate and extend data for plate efficiency. The bubbles formed at the holes are assumed to rise through a pool of liquid that is vertically mixed and has the local composition x_A. The bubbles change in composition as they rise, and there is assumed to be no mixing of the gas phase in the vertical direction. For a unit plate area with a superficial velocity $\bar{V}_S$, the number of moles transferred in a thin slice dz is

$$\bar{V}_S \rho_M dy_A = K_y a(y_A^* - y_A) dz \tag{12.2-21}$$

Integrating over the height of the aerated liquid Z gives

$$\int_{y_{A1}}^{y_{A2}} \frac{dy_A}{y_A^* - y_A} = \ln \frac{y_A^* - y_{A1}}{y_A^* - y_{A2}} = \frac{K_y a Z}{\bar{V}_S \rho_M} \tag{12.2-22}$$

or

$$\frac{y_A^* - y_{A2}}{y_A^* - y_{A1}} = \exp\left(-\frac{K_y a Z}{\bar{V}_S \rho_M}\right) \tag{12.2-23}$$

The local efficiency η' is given by

$$\eta' = \frac{y_{A2} - y_{A1}}{y_A^* - y_{A1}} \tag{12.2-24}$$

and

$$1 - \eta' = \frac{y_A^* - y_{A1} - y_{A2} + y_{A1}}{y_A^* - y_{A1}} \tag{12.2-25}$$

From Eq. (12.2-23),

$$1 - \eta' = \exp\left(-\frac{K_y a Z}{\bar{V}_S \rho_M}\right) = e^{-N_{Oy}} \tag{12.2-26}$$

where N_{Oy} is the number of overall gas-phase transfer units. For distillation of low-viscosity liquids such as water, alcohol, or benzene at about 100°C, the value of N_{Oy} is about 1 to 2, nearly independent of the gas velocity over the normal operating range of the column. This gives a local efficiency of 63 to 86 percent, and the Murphree efficiency could be higher or lower depending on the degree of lateral mixing on the plate and the amount of entrainment.

The relative importance of the gas and liquid resistances can be estimated by assuming that the penetration theory applies to both phases and with the same contact time. Since the penetration theory [Eq. (6.3-16)] gives k_c, and k_y and k_x equal $k_c \rho_{My}$ and $k_c \rho_{Mx}$, respectively,

$$\frac{k_y}{k_x} = \left(\frac{D_{vy}}{D_{vx}}\right)^{1/2} \frac{\rho_{My}}{\rho_{Mx}} \tag{12.2-27}$$

EXAMPLE 12.5 (*a*) Use the penetration theory to estimate the fraction of the total resistance that is in the gas film in the distillation of a benzene-toluene mixture at 110°C (230°F) and 1 atm pressure. The liquid viscosity μ is 0.26 cP. The diffusivities and densities are, for liquid,

$$D_{vx} = 6.74 \times 10^{-5} \text{ cm}^2/\text{s} \qquad \rho_{Mx} = 8.47 \text{ mol/L}$$

and, for vapor,

$$D_{vy} = 0.0494 \text{ cm}^2/\text{s} \qquad \rho_{My} = 0.0318 \text{ mol/L}$$

(b) How would a fourfold reduction in total pressure change the local efficiency and the relative importance of the gas-film and liquid-film resistances?

Solution

(a) Substitution into Eq. (12.2-27) gives

$$\frac{k_y}{k_x} = \left(\frac{0.0494}{6.74 \times 10^{-5}}\right)^{1/2} \frac{0.0318}{8.47} = 0.102$$

Thus the gas-film coefficient is predicted to be only 10 percent of the liquid-film coefficient, and if $m=1$, about 90 percent of the overall resistance to mass transfer would be in the gas film.

(b) Assume that the column is operated at the same F factor and that this gives the same interfacial area a and froth height Z. The boiling temperature of toluene at 0.25 atm is 68°C, or 341 K, compared to 383 K at 1 atm.

Gas film. Since $D_{vy} \propto T^{1.81}/P$, the new value of D_{vy} is

$$D'_{vy} = \left(\frac{341}{383}\right)^{1.81} \frac{D_{vy}}{0.25} = 3.24 \quad \text{times old value}$$

Assuming that the penetration theory holds with the same t_T, k_c increases by $\sqrt{3.24}$, or 1.8, but at 0.25 atm and 68°C, ρ_{My} is 0.00894 mol/L, so k_y changes by $1.8 \times 0.00894/0.0318 = 0.506$.

Liquid film. Here $D_{vx} \propto T/\mu$, and since $\mu = 0.35$ cP at 68°C, the new value of D_{vx} is

$$D'_{vx} = \frac{341}{383}\left(\frac{0.26 D_{vx}}{0.35}\right) = 0.66 \quad \text{times old value}$$

Thus k_c decreases by $\sqrt{0.66} = 0.81$, and considering the small change in molar density to 8.92 mol/L, k_x changes by $0.81 \times 8.92/8.47 = 0.86$.

If the local efficiency at 1 atm pressure is 0.78, corresponding to 1.5 transfer units, and if the relative values of k_x and k_y are estimated as in part (a), the new value of K'_y is obtained as follows:

$$k'_y = 0.506 k_y \qquad k'_x = 0.86 k_x$$

At 1 atm, $k_y = 0.102 k_x$ and $k_y = 0.907 k_y$. Thus

$$k'_x = \frac{0.86}{0.102} k_y = 8.43 k_y$$

For $m = 1$,

$$\frac{1}{K'_y} = \frac{1}{k'_y} + \frac{1}{k'_x} = \frac{1}{0.506 k_y} + \frac{1}{8.43 k_y} = \frac{2.10}{k_y}$$

$$K'_y = 0.476 k_y$$

The ratio of the number of transfer units is the ratio of the overall coefficients divided by the molar flow rate. If the column is operated at the same F factor, $\sqrt{\rho_y}$ changes by $[(383 \times 0.25)/341]^{0.5} = 0.53$, and $\bar{V}_S$ changes by $1/0.53$. If a, the area per unit volume, is the same, the new value of N'_{oy} is

$$N'_{oy} = 1.5 \times \frac{0.476}{0.53} = 1.35 \qquad \eta' = 1 - e^{-1.35} = 0.74$$

Thus the local efficiency is predicted to drop from 78 to 74 percent, with 94 percent of the total resistance in the gas phase. Close agreement with the actual values of efficiency is not expected because of the assumptions made to simplify the analysis, but the trend is correct, as shown in Fig. 12.21, and it is clear that the gas-film resistance is increasingly important at low pressures. For distillation at high pressures, k_y and k_x are more nearly equal.

名 词 术 语

英文 / 中文

active area / 工作区
annular tubes with multi-holes / 多孔环管式
bubble-cap tray / 泡罩塔板
bulk density / 堆积密度
capacity performance chart / 负荷性能图
channeling / 沟流
coning / 干吹
detrimental operation / 不正常操作
downcomer / 降液管
dumped packings / 乱堆填料
equivalent level of clear liquid in the downcomer / 降液管中清液层高度
excessive froth entrainment / 过量雾沫夹带
flooding / 液泛
flooding point / 泛点
flooding velocity 液泛速度
gas inlet / 气体进口
grid plate / 栅板
height equivalent to a theoretical plate (HETP) / 等板高度
height of liquid over the weir / 堰上液层高度
hydraulic gradient / 液面落差
kinetic energy factor F (F factor) / 动能因数
liquid distributor / 液体分布器
liquid redistributor / 液体再分布器
loading point / 载点
local efficiency / 点效率
murphree efficiency / 默弗里单板效率
overall efficiency / 总板效率
overflow pipes / 溢流管

英文 / 中文

packed tower(column) / 填料塔
packing / 填料
packing depth (packed section) / 填料层
packing factor / 填料因子
packing supports / 填料支撑装置
percentage flooding / 泛点率
plate spacing / 板间距
plate tower(column) / 板式塔
porosity (void fraction) / 空隙率
pressure drop / 压降
riser / 升气管
shower nozzle type / 莲蓬头式
specific surface area / 比表面积
spray density / 喷淋密度
stacked packings / 整砌填料
strong liquor / 浓溶液
structured (ordered) packings / 规整填料
superficial gas velocity based on empty tower / 空塔气速
tower (column) diameter / 塔径
turndown ratio / 操作弹性
underflow weir / 进口堰
valve-tray / 浮阀塔板
vapor bubble entrainment / 气泡夹带
weak liquor / 稀溶液
weeping / 漏液
weir height / 堰高
wettability / 润湿性能
wetting rate / 润湿速率

Appendix

Appendix 1 Unit Conversion

1. Length and Mass

Length			Mass		
m	in	ft	kg	t	lb
1	39.3701	3.2808	1	0.001	2.20462
0.025400	1	0.073333	1000	1	2204.62
0.30480	12	1	0.4536	0.0004536	1

2. Force

N	kgf	lbf	dyn
1	0.102	0.2248	1×10^5
9.80665	1	2.2046	9.80665×10^5
4.448	0.4536	1	4.448×10^5
1×10^{-5}	1.02×10^{-6}	2.243×10^6	1

3. Pressure

Pa	bar	$kgf \cdot cm^{-2}$	atm	mmHg	$lbf \cdot in^{-2}$
1	1×10^{-5}	1.02×10^{-5}	0.99×10^{-5}	0.0075	14.5×10^{-5}
1×10^5	1	1.02	0.9869	750.1	14.5
98.07×10^3	0.9807	1	0.9678	735.56	14.2
1.010325×10^5	1.013	1.0332	1	760	14.697
133.32	1.33×10^{-3}	0.136×10^{-4}	0.00132	1	0.01931
6894.8	0.06895	0.0703	0.068	51.71	1

4. Viscosity

$Pa \cdot s$	P	cP	$kgf \cdot s \cdot m^{-2}$	$lb \cdot ft^{-1} \cdot s^{-1}$
1	10	1×10^3	0.102	0.672
1×10^{-1}	1	1×10^2	0.0102	0.06720
1×10^{-3}	0.01	1	0.102×10^{-3}	6.720×10^{-4}
9.81	98.1	9810	1	6.59
1.4881	14.881	1488.1	0.1519	1

5. Energy

J	$kgf \cdot m$	$kW \cdot h$	kcal	Btu
1	0.102	2.778×10^{-7}	2.39×10^{-4}	9.485×10^{-4}
9.8067	1	2.724×10^{-6}	2.342×10^{-3}	9.296×10^{-3}
3.6×10^6	3.671×10^5	1	860.0	3413
2.685×10^6	273.8×10^3	0.7457	641.33	2544
4.1868×10^3	426.9	1.1622×10^{-3}	1	3.963
1.055×10^3	107.58	2.930×10^{-4}	0.2520	1

6. Power

W	kgf · m · s^{-1}	hp	kcal · s^{-1}	Btu · s^{-1}
1	0.10197	1.341×10^{-3}	0.2389×10^{-3}	0.9486×10^{-3}
9.8067	1	0.01315	0.2342×10^{-2}	0.9293×10^{-2}
745.69	76.0375	1	0.17803	0.70675
4186.8	426.35	5.6135	1	3.9683
1055	107.58	1.4148	0.251996	1

7. Constant Pressure Specific Heat

kJ · kg^{-1} · K^{-1}	kcal · kg^{-1} · ℃$^{-1}$	Btu · ft^{-1} · h^{-1} · ℉$^{-1}$
1	0.2389	0.2389
4.1868	1	1

8. Thermal Conductivity

W · m^{-1} · K^{-1}	kcal · m^{-1} · h^{-1} · ℃$^{-1}$	cal · cm^{-1} · s^{-1} · ℃$^{-1}$	Btu · ft^{-1} · h^{-1} · ℉$^{-1}$
1	0.86	2.398×10^{-3}	0.579
1.163	1	2.778×10^{-3}	0.6720
418.7	360	1	241.9
1.73	1.488	4.134×10^{-3}	1

9. Heat-transfer Coefficient

W · m^{-2} · K^{-1}	kcal · m^{-2} · h^{-1} · ℃$^{-1}$	cal · cm^{-2} · s^{-1} · ℃$^{-1}$	Btu · ft^{-2} · h^{-1} · ℉$^{-1}$
1	0.86	2.389×10^{-5}	0.176
1.163	1	2.778×10^{-5}	0.2048
4.186×10^4	3.6×10^4	1	7374
5.678	4.882	1.356×10^{-4}	1

10. Temperature

$$K = 273.2 + ℃ \quad ℃ = (℉ - 32) \times \frac{5}{9} \quad ℉ = ℃ \times \frac{9}{5} + 32℃$$

11. Mol Gas Constant

R = 8.314 J · mol^{-1} · K^{-1} = 1.987 cal · mol^{-1} · K^{-1} = 0.08206 atm · m^3 · kmol^{-1} · K^{-1}
= 848 kgf · m^{-1} · kmol^{-1} · K^{-1} = 82.06 atm · cm^3 · kmol^{-1} · ℃$^{-1}$

Appendix 2 Dimensionless Groups

Symbol	Name	Definition	Symbol	Name	Definition
Bi	Biot number	$\dfrac{hs}{k}$ for slab	F_r	Froude number	$\dfrac{u^2}{gL}$
		$\dfrac{hr_m}{k}$ for cylinder or sphere	f	Fanning friction factor	$\dfrac{\Delta p_{sc} d}{2L\rho u^2}$
C_D	Drag coefficient	$\dfrac{2F_{Dc}}{\rho u_0^2 A_P}$	Gr	Grashof number	$\dfrac{L^3 \rho^2 \beta g \Delta T}{\mu^2}$
F_o	Fourier number	$\dfrac{\alpha t}{r^2}$	Gz	Graetz number	$\dfrac{\dot{m} c_p}{kL}$

Symbol	Name	Definition	Symbol	Name	Definition
Gz'	Graetz number for mass transfer	$\dfrac{\dot{m}}{\rho D_v L}$	Pe	Peclet number	$\dfrac{du}{\alpha}$ or $\dfrac{du_0}{D_v}$
j_H	Heat-transfer factor	$\dfrac{h}{c_p G}\left(\dfrac{c_p \mu}{k}\right)^{2/3}\left(\dfrac{\mu_w}{\mu}\right)^{0.14}$	Pr	Prandtl number	$\dfrac{c_p \mu}{k}$
j_M	Mass-transfer factor	$\dfrac{kM}{G}\left(\dfrac{\mu}{D_v \rho}\right)^{2/3}$	Re	Reynolds number	$\dfrac{du\rho}{\mu}$
Ma	Mach number	$\dfrac{u}{a}$	N_s	Separation number	$\dfrac{u_t u_0}{g d_p}$
N_{Ae}	Aeration number	$\dfrac{q_g}{n d_a^3}$	Sc	Schmidt number	$\dfrac{\mu}{D_v \rho}$
N_P	Power number	$\dfrac{P_c}{\rho n^3 d^5}$	Sh	Sherwood number	$\dfrac{k_c d}{D_v}$
N_Q	Flow number	$\dfrac{q}{n d_a^3}$	We	Weber number	$\dfrac{d \rho u^2}{\sigma}$
Nu	Nusselt number	$\dfrac{hd}{k}$			

Appendix 3 Standard Steel Pipe

Nominal diameter in	Nominal diameter mm	Outside diameter, mm	Schedual No.	Wall thickness, mm	Pipe weigth, kg/m
$\dfrac{1}{8}$	6	10.00	40	2.00	0.36
			80	2.50	0.46
$\dfrac{1}{4}$	8	13.50	40	2.25	0.63
			80	2.75	0.80
$\dfrac{3}{8}$	10	17.00	40	2.25	0.85
			80	2.75	1.10
$\dfrac{1}{2}$	15	21.25	40	2.75	1.26
			80	3.25	1.62
$\dfrac{3}{4}$	20	26.75	40	2.75	1.68
			80	3.50	2.19
1	25	33.50	40	3.25	2.50
			80	4.00	3.23
$1\dfrac{1}{4}$	32	42.25	40	3.25	3.38
			80	4.00	4.46
$1\dfrac{1}{2}$	40	48.00	40	3.50	4.05
			80	4.25	5.40
2	50	60.00	40	3.50	5.43
			80	4.50	7.47

续表

Nominal diameter in	Nominal diameter mm	Outside diameter, mm	Schedual No.	Wall thickness, mm	Pipe weigth, kg/m
$2\frac{1}{2}$	70	75.50	40	3.75	8.62
			80	4.50	11.40
3	80	88.50	40	4.00	11.28
			80	4.75	15.25
4	100	114.00	40	4.00	16.06
			80	5.00	22.29
5	125	140.00	40	4.50	21.76
			80	5.50	30.92
6	150	165.00	40	4.50	28.23
			80	5.50	42.52

Appendix 4 Condenser and Heat-Exchanger Data

Floating-head type condenser and heat-exchanger (ϕ19mm×2mm heat exchanger tube)

Nominal diameter, mm	Number of tube passes	Number of tube Total	Number of tube Center bank of tubes	Circulation area, m^2 Tube pass	Circulation area, m^2 Shell pass (distance between plates is 200mm)	Heat exchange area, m^2 L=3m	Heat exchange area, m^2 L=4.5m	Heat exchange area, m^2 L=6m	Heat exchange area, m^2 L=9m
325	2	60	7	0.0053	0.0372	10.5	15.8		
	4	52	6	0.0023	0.0409	9.1	13.7		
400	2	120	8	0.0106	0.0481	20.9	31.6	42.3	
	4	108	9	0.0005	0.0444	18.8	28.4	38.1	
500	2	206	11	0.0182	0.0564	35.7	54.1	72.5	
	4	192	10	0.0085	0.0601	33.2	50.4	67.6	
600	2	324	14	0.0286	0.0648	55.8	84.8	113.9	
	4	308	14	0.0136	0.0648	53.1	80.7	108.2	
	6	284	14	0.0083	0.0648	48.9	74.4	99.8	
700	2	468	16	0.0414	0.0768	80.4	122.2	164.4	
	4	448	17	0.0198	0.0731	76.9	117	157.1	
	6	382	15	0.0112	0.0805	65.6	99.8	133.9	
800	2	610	19	0.0539	0.0852		158.9	213.5	
	4	588	18	0.0260	0.0879		153.2	205.8	
	6	518	16	0.0152	0.0962		134.9	181.3	
900	2	800	22	0.0707	0.0925		207.6	279.2	
	4	776	21	0.0343	0.0962		201.4	270.8	
	6	720	21	0.0212	0.0962		186.9	251.3	
1000	2	1006	24	0.0890	0.1044		260.6	350.6	
	4	980	23	0.0433	0.1081		253.9	341.6	
	6	892	21	0.0262	0.1154		231.1	311.0	
1100	2	1240	27	0.1100	0.1127		320.3	431.3	
	4	1212	26	0.0536	0.1163		313.1	421.6	
	6	1120	24	0.0329	0.1236		289.3	389.6	

续表

Nominal diameter, mm	Number of tube passes	Number of tube - Total	Number of tube - Center bank of tubes	Circulation area, m² - Tube pass	Circulation area, m² - Shell pass (distance between plates is 200mm)	Heat exchange area, m² - L=3m	Heat exchange area, m² - L=4.5m	Heat exchange area, m² - L=6m	Heat exchange area, m² - L=9m
1200	2	1452	28	0.1290	0.1283		374.4	504.3	764.2
	4	1424	28	0.0629	0.1283		367.2	494.6	749.5
	6	1348	27	0.0396	0.1319		347.6	468.2	709.5
1300	4	1700	31	0.0751	0.1365			589.3	
	6	1616	29	0.0476	0.1438			560.2	

Floating-head type condenser and heat-exchanger
(ϕ25mm×2.5mm heat exchanger tube)

Nominal diameter, mm	Number of tube passes	Number of tube - Total	Number of tube - Center bank of tubes	Circulation area, m² - Tube pass	Circulation area, m² - Shell pass (distance between plates is 200mm)	Heat exchange area, m² - L=3m	Heat exchange area, m² - L=4.5m	Heat exchange area, m² - L=6m	Heat exchange area, m² - L=9m
325	2	32	5	0.0050	0.0388	7.4	11.1		
	4	28	4	0.0022	0.0437	6.4	9.8		
400	2	74	7	0.0116	0.0437	16.9	25.6	34.4	
	4	68	6	0.0053	0.0485	15.6	23.6	31.6	
500	2	124	8	0.0194	0.0582	28.3	42.8	57.4	
	4	116	9	0.0091	0.0534	26.4	40.1	53.7	
600	2	198	11	0.0311	0.0631	44.9	68.2	91.5	
	4	188	10	0.0148	0.0679	42.6	64.8	86.9	
	6	158	10	0.0083	0.0679	35.8	54.4	73.1	
700	2	268	13	0.0421	0.0728	60.6	92.1	123.7	
	4	256	12	0.0201	0.0776	57.8	87.9	118.1	
	6	224	10	0.0116	0.0873	50.6	76.9	103.4	
800	2	366	15	0.0575	0.0825		125.4	168.5	
	4	352	14	0.0276	0.0873		120.6	162.1	
	6	316	14	0.0165	0.0873		108.3	145.5	
900	2	472	17	0.0741	0.0912		161.2	216.8	
	4	456	16	0.0353	0.0960		155.7	209.4	
	6	426	16	0.0223	0.0960		145.5	195.6	
1000	2	606	19	0.0952	0.1008		206.6	277.9	
	4	588	18	0.0462	0.1056		200.4	269.7	
	6	564	18	0.0295	0.1056		192.2	258.7	
1100	2	736	21	0.1160	0.1104		250.2	336.8	
	4	716	20	0.0562	0.1152		243.4	327.7	
	6	692	20	0.0362	0.1152		235.2	316.7	
1200	2	880	22	0.1380	0.1248		298.6	402.2	609.4
	4	860	22	0.0675	0.1248		291.8	393.1	595.6
	6	828	21	0.0434	0.1296		280.9	378.4	573.4
1300	4	1024	24	0.0804	0.1344			467.1	
	6	972	24	0.0509	0.1344			443.3	

Fixed tube sheet type condenser and heat-exchanger
(ϕ19mm×2mm heat exchanger tube)

Nominal diameter, mm	Number of tube passes	Number of tube		Circulation area, m^2		Heat exchange area, m^2					
		Total	Center bank of tubes	Tube pass	Shell pass (distance between plates is 200mm)	L=1.5m	L=2m	L=3m	L=4.5m	L=6m	L=9m
159	1	15	5	0.0027	0.1242	1.3	1.7	2.6			
219	1	33	7	0.0058	0.0167	2.8	3.7	5.7			
273	1	65	9	0.0115	0.0198	5.4	7.4	11.3	17.1	22.9	
	2	56	8	0.0049	0.0235	4.7	6.4	9.7	14.7	19.7	
325	1	99	11	0.0175	0.0225	8.3	11.2	17.1	26.0	34.9	
	2	88	10	0.0078	0.0262	7.4	10.0	15.2	23.1	31.0	
	4	68	11	0.0030	0.0225	5.7	7.7	11.8	17.9	23.9	
400	1	174	14	0.0307	0.0251	14.5	19.7	30.1	45.7	61.3	
	2	164	15	0.0145	0.0223	13.7	18.6	28.4	43.1	57.8	
	4	146	14	0.0065	0.0251	12.2	16.6	25.3	38.3	51.4	
450	1	237	17	0.0419	0.0246	19.8	26.9	41.0	62.2	83.5	
	2	220	16	0.0194	0.0283	18.4	25.0	38.1	57.8	77.5	
	4	200	16	0.0088	0.0283	16.7	22.7	34.6	52.5	70.4	
500	1	275	19	0.0486	0.0270		31.2	47.6	72.2	96.8	
	2	256	18	0.0226	0.0307		29.0	44.3	67.2	90.2	
	4	222	18	0.0098	0.0307		25.2	38.4	58.3	78.2	
600	1	430	22	0.0760	0.0353		48.8	74.4	112.9	151.4	
	2	416	23	0.0368	0.0316		47.2	72.0	109.3	146.5	
	4	370	22	0.0163	0.0353		42.0	64.0	97.2	131.3	
	6	360	20	0.0106	0.0427		40.8	62.3	94.5	126.8	
700	1	607	27	0.1073	0.0363			105.1	159.4	213.8	
	2	574	27	0.0507	0.0363			99.4	150.8	202.1	
	4	542	27	0.0239	0.0363			93.8	142.3	190.9	
	6	518	24	0.0153	0.0473			89.7	136.0	182.4	
800	1	797	31	0.1408	0.0409			138.0	209.3	280.7	
	2	776	31	0.0686	0.0409			134.3	203.8	273.3	
	4	722	31	0.0319	0.0409			125.0	189.8	254.3	
	6	710	30	0.0209	0.0446			122.9	186.5	250.0	
900	1	1009	35	0.1783	0.0452			174.7	265.0	355.3	536.0
	2	988	35	0.0873	0.0451			171.0	259.5	347.9	524.9
	4	938	35	0.0414	0.0451			162.4	246.4	330.3	498.3
	6	914	34	0.0269	0.0488			158.2	240.0	321.9	485.6
1000	1	1267	39	0.2239	0.0497			219.3	332.8	446.2	673.1
	2	1234	39	0.1090	0.0497			213.6	324.1	434.6	655.6
	4	1186	39	0.0524	0.0497			205.3	311.5	417.7	630.1
	6	1148	38	0.0338	0.0534			198.7	301.5	404.3	609.9

Fixed tube sheet type condenser and heat-exchanger
(ϕ25mm×2.5mm heat exchanger tube)

Nominal diameter, mm	Number of tube passes	Number of tube		Circulation area, m²		Heat exchange area, m²					
		Total	Center bank of tubes	Tube pass	Shell pass (distance between plates is 200mm)	L= 1.5m	L= 2m	L= 3m	L= 4.5m	L= 6m	L= 9m
159	1	11	3	0.0035	0.0163	1.2	1.6	2.5			
219	1	25	5	0.0079	0.0182	2.7	3.7	5.7			
273	1	38	6	0.0119	0.0239	4.2	5.7	8.7	13.1	17.6	
	2	32	7	0.0050	0.0190	3.5	4.8	7.3	11.1	14.8	
325	1	57	9	0.0179	0.0194	6.3	8.5	13.0	19.7	26.4	
	2	56	9	0.0088	0.0194	6.2	8.4	12.7	19.3	25.9	
	4	40	9	0.0031	0.0194	4.4	6.0	9.1	13.8	18.5	
400	1	98	12	0.0308	0.0194	10.8	14.6	22.3	33.8	45.4	
	2	94	11	0.0148	0.0243	10.3	14.0	21.4	32.5	43.5	
	4	76	11	0.0060	0.0243	8.4	11.3	17.3	26.3	35.2	
450	1	135	13	0.0424	0.0243	14.8	20.1	30.7	46.6	62.5	
	2	126	12	0.0198	0.0291	13.9	18.8	28.7	43.5	58.4	
	4	106	13	0.0083	0.0243	11.7	15.8	24.1	36.6	49.1	
500	1	174	14	0.0546	0.0291		26.0	39.6	60.1	80.6	
	2	164	15	0.0257	0.0243		24.5	37.3	56.6	76.0	
	4	144	15	0.0113	0.0243		21.4	32.8	49.7	66.7	
600	1	245	17	0.0769	0.0340		36.5	55.8	84.6	113.5	
	2	232	16	0.0364	0.0388		34.6	52.8	80.1	107.5	
	4	222	17	0.0174	0.0340		33.1	50.5	76.7	102.8	
	6	216	16	0.0113	0.0388		32.2	49.2	74.6	100.0	
700	1	355	21	0.1115	0.0340			80.0	122.6	164.4	
	2	342	21	0.0537	0.0340			77.9	118.1	158.4	
	4	322	21	0.0253	0.0340			73.3	111.2	149.1	
	6	304	20	0.0159	0.0388			69.2	105.0	140.8	
800	1	467	23	0.1466	0.0437			106.3	161.3	216.3	
	2	450	23	0.0707	0.0437			102.4	155.4	208.5	
	4	442	23	0.0347	0.0437			100.6	152.7	204.7	
	6	430	24	0.0225	0.0388			97.9	148.5	199.2	
900	1	605	27	0.1900	0.0432			137.8	209.0	280.2	422.7
	2	588	27	0.0923	0.0432			133.9	203.1	272.3	410.8
	4	554	27	0.0435	0.0432			126.1	191.4	256.6	387.1
	6	538	26	0.0282	0.0480			122.5	185.8	249.2	375.9
1000	1	749	30	0.2352	0.0480			170.5	258.7	346.9	523.3
	2	742	29	0.1165	0.0528			168.9	256.3	343.7	518.4
	4	710	29	0.0557	0.0528			161.6	245.2	328.8	496.0
	6	698	30	0.0365	0.0480			158.9	241.1	323.3	487.7

Appendix 5 Properties of Liquid Water

Temperature T, ℃	Viscosity[1] μ, 10^{-3} Pa·s	Thermal conductivity[2] k, W·m^{-1}·K^{-1}	Density[3] ρ, kg·m^{-3}
0	1.794	0.5536	999.8
4	1.546	0.5640	999.8
10	1.31	0.5761	999.8
16	1.129	0.5882	999.0
21	0.982	0.5986	997.9
27	0.862	0.6090	996.6
32	0.764	0.6193	994.9
38	0.682	0.6263	993.1
49	0.559	0.6418	988.5
60	0.47	0.6539	983.2
71	0.401	0.6643	977.1
82	0.347	0.6712	970.4
93	0.305	0.6782	963.2
104	0.27	0.6816	955.2
116	0.242	0.6851	946.7
127	0.218	0.6851	937.5
138	0.199	0.6851	928.1
149	0.185	0.6851	918.0

[1] From *International Critical Tables*, vol. 5, McGraw-Hill Book Company, New York, 1929, p. 10.
[2] From E. Schmidt and W. Sellschopp, *Forsch. Geb. Ingenieurw.*, **3**:277 (1932).
[3] Calculated from J. H. Keenan and F. G. Keyes, *Thermodynamic Properties of Steam*, John Wiley & Sons., Inc.,New York, 1937.

Appendix 6 Properties of Saturated Steam and Water[1]

Temperature T, ℃	Vapor pressure p_A, kPa	Specific volume, m^3/kg		Enthalpy, kJ/kg		
		Liquid v_x	Saturated vapor v_y	Liquid H_x	Vaporization λ	Saturated vapor H_y
0	0.6108	0.0010	206.23	0.0000	2501.2	2501.2
2	0.6889	0.0010	183.96	6.9774	2497.2	2504.2
4	0.8388	0.0010	152.57	18.653	2490.7	2509.3
7	1.0168	0.0010	127.11	30.328	2484.2	2514.4
10	1.2275	0.0010	106.34	42.004	2477.4	2519.5
13	1.4755	0.0010	89.319	53.656	2470.9	2524.7
16	1.7671	0.0010	75.311	65.308	2464.4	2529.8
18	2.1077	0.0010	63.742	76.961	2457.9	2534.9
21	2.5042	0.0010	54.144	88.590	2451.4	2539.8
24	2.9648	0.0010	46.157	100.22	2444.6	2544.9
27	3.4977	0.0010	39.487	111.85	2438.1	2550.0
29	4.1121	0.0010	33.889	123.45	2431.6	2555.1
32	4.8181	0.0010	29.184	135.06	2425.1	2560.0

续表

Temperature T, ℃	Vapor pressure p_A, kPa	Specific volume, m³/kg		Enthalpy, kJ/kg		
		Liquid v_x	Saturated vapor v_y	Liquid H_x	Vaporization λ	Saturated vapor H_y
35	5.6275	0.0010	25.210	146.66	2418.4	2565.1
38	6.5521	0.0010	21.840	158.27	2411.9	2570.0
43	8.7998	0.0010	16.542	181.46	2398.8	2580.0
49	11.683	0.0010	12.667	204.67	2385.1	2589.8
54	15.341	0.0010	9.8074	227.88	2371.9	2599.8
60	19.940	0.0010	7.6677	251.09	2358.4	2609.3
66	25.662	0.0010	6.0522	274.35	2344.6	2619.1
71	32.716	0.0010	4.8192	297.61	2330.9	2628.4
77	41.341	0.0010	3.8700	320.89	2317.0	2637.9
82	51.814	0.0010	3.1325	344.20	2303.0	2647.2
88	64.418	0.0010	2.5553	367.55	2288.8	2656.3
93	79.490	0.0010	2.0985	390.90	2274.4	2665.1
99	97.389	0.0010	1.7360	414.32	2259.7	2674.0
100	101.34	0.0010	1.6723	419.02	2256.7	2675.8
104	118.51	0.0010	1.4446	437.76	2245.1	2682.8
110	143.27	0.0011	1.2097	461.25	2230.0	2691.2
116	172.16	0.0011	1.0188	484.79	2214.9	2699.6
121	205.60	0.0011	0.8627	508.40	2199.3	2707.7
127	244.21	0.0011	0.7343	532.05	2183.5	2715.6
132	288.55	0.0011	0.6281	555.75	2167.6	2723.3
138	339.09	0.0011	0.5398	579.54	2151.1	2730.7
143	396.66	0.0011	0.4659	603.41	2134.6	2737.9
149	461.81	0.0011	0.4039	627.34	2117.4	2744.9
154	535.31	0.0011	0.3514	651.36	2100.2	2751.4
160	617.77	0.0011	0.3069	675.48	2082.3	2757.9
171	813.10	0.0011	0.2366	724.02	2045.5	2769.6
177	927.56	0.0011	0.2088	748.44	2026.5	2774.9
182	1054.4	0.0011	0.1848	772.98	2006.9	2779.8
188	1194.4	0.0011	0.1640	797.66	1986.7	2784.4
193	1348.6	0.0011	0.1460	822.45	1966.2	2788.6
199	1518.2	0.0012	0.1302	847.38	1944.8	2792.4
204	1703.7	0.0012	0.1164	872.45	1923.0	2795.6
210	1906.4	0.0012	0.1044	897.69	1900.6	2798.2
216	2127.0	0.0012	0.0937	923.09	1877.4	2800.5
221	2367.0	0.0012	0.0844	948.67	1853.4	2802.1
227	2628.3	0.0012	0.0761	974.46	1828.8	2803.3
232	2910.3	0.0012	0.0687	1000.6	1803.4	2804.0

① Abstracted from *Steam Tables*, by Joseph H. Keenan, Frederick G. Keyes, Philip G. Hill, and Joan G. Moore, John Wiley & Sons, New York, 1969, with the permission of the publisher.

Appendix 7   Viscosities of Gases[1]

No.	Gas	X	Y	No.	Gas	X	Y
1	Acetic acid	7.7	14.3	29	Freon-113	11.3	14.0
2	Acetone	8.9	13.0	30	Helium	10.9	20.5
3	Acetylene	9.8	14.9	31	Hexane	8.6	11.8
4	Air	11.0	20.0	32	Hydrogen	11.2	12.4
5	Ammonia	8.4	16.0	33	$3H_2 + N_2$	11.2	17.2
6	Argon	10.5	22.4	34	Hydrogen bromide	8.8	20.9
7	Benzene	8.5	13.2	35	Hydrogen chloride	8.8	18.7
8	Bromine	8.9	19.2	36	Hydrogen cyanide	9.8	14.9
9	Butene	9.2	13.7	37	Hydrogen iodide	9.0	21.3
10	Butylene	8.9	13.0	38	Hydrogen sulfide	8.6	18.0
11	Carbon dioxide	9.5	18.7	39	Iodine	9.0	18.4
12	Carbon disulfide	8.0	16.0	40	Mercury	5.3	22.9
13	Carbon monoxide	11.0	20.0	41	Methane	9.9	15.5
14	Chlorine	9.0	18.4	42	Methyl alcohol	8.5	15.6
15	Chloroform	8.9	15.7	43	Nitric oxide	10.9	20.5
16	Cyanogen	9.2	15.2	44	Nitrogen	10.6	20.0
17	Cyclohexane	9.2	12.0	45	Nitrosyl chloride	8.0	17.6
18	Ethane	9.1	14.5	46	Nitrous oxide	8.8	19.0
19	Ethyl acetate	8.5	13.2	47	Oxygen	11.0	21.3
20	Ethyl alcohol	9.2	14.2	48	Pentane	7.0	12.8
21	Ethyl chloride	8.5	15.6	49	Propane	9.7	12.9
22	Ethyl ether	8.9	13.0	50	Propyl alcohol	8.4	13.4
23	Ethylene	9.5	15.1	51	Propylene	9.0	13.8
24	Fluorine	7.3	23.8	52	Sulfur dioxide	9.6	17.0
25	Freon-11	10.6	15.1	53	Toluene	8.6	12.4
26	Freon-12	11.1	16.0	54	2,3,3-Trimethylbutane	9.5	10.5
27	Freon-21	10.8	15.3	55	Water	8.0	16.0
28	Freon-22	10.1	17.0	56	Xenon	9.3	23.0

Coordinates for use with figure on next page.

[1] By permission, from J. H. Perry (ed.), *Chemical Engineers' Handbook,* 5th ed., pp. **3**-210 and **3**-211. Copyright © 1973, McGraw-Hill Book Company, New York.

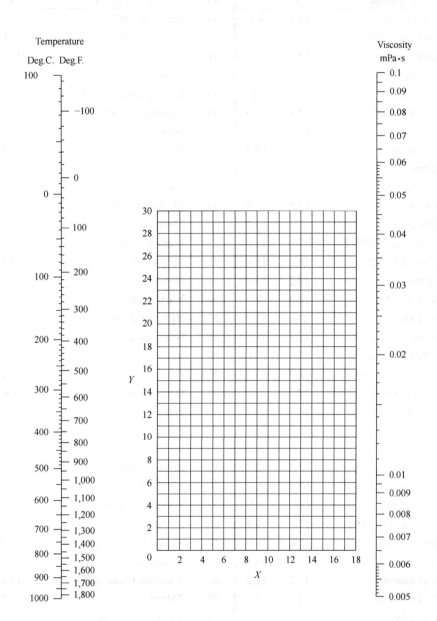

Viscosities of gases and vapors at 1 atm; for coordinates, see table on previous page.

Appendix 8 Viscosities of Liquids[①]

No.	Liquid	X	Y	No.	Liquid	X	Y
1	Acetaldehyde	15.2	4.8	39	Glycerol, 100%	2	30
2	Acetic acid, 100%	12.1	14.2	40	Glycerol, 50%	6.9	19.6
3	Acetic anhydride	12.7	12.8	41	Heptane	14.1	8.4
4	Acetone, 100%	14.5	7.2	42	Hexane	14.7	7
5	Ammonia, 100%	12.6	2	43	Hydrochloric acid, 31.5%	13	16.6
6	Ammonia, 26%	10.1	13.9	44	Isobutyl alcohol	7.1	18
7	Amyl acetate	11.8	12.5	45	Isopropyl alcohol	8.2	16
8	Amyl alcohol	7.5	18.4	46	Kerosene	10.2	16.9
9	Aniline	8.1	18.7	47	Linseed oil, raw	7.5	27.2
10	Anisole	12.3	13.5	48	Mercury	18.4	16.4
11	Benzene	12.5	10.9	49	Methanol, 100%	12.4	10.5
12	Biphenyl	12	18.3	50	Methyl acetate	14.2	8.2
13	Brine, CaCl2, 25%	6.6	15.9	51	Methyl chloride	15	3.8
14	Brine, NaCl, 25%	10.2	16.6	52	Methyl ethyl ketone	13.9	8.6
15	Bromine	14.2	13.2	53	Napthalene	7.9	18.1
16	Butyl acetate	12.3	11	54	Nitric acid, 95%	12.8	13.8
17	Butyl alcohol	8.6	17.2	55	Nitric acid, 60%	10.8	17
18	Carbon dioxide	11.6	0.3	56	Nitrobenzene	10.6	16.2
19	Carbon disulfide	16.1	7.5	57	Nitrotoluene	11	17
20	Carbon tetrachloride	12.7	13.1	58	Octane	13.7	10
21	Chlorobenzene	12.3	12.4	59	Octyl alcohol	6.6	21.1
22	Chloroform	14.4	10.2	60	Pentane	14.9	5.2
23	m-Cresol	2.5	20.8	61	Phenol	6.9	20.8
24	Cyclohexanol	2.9	24.3	62	Sodium	16.4	13.9
25	Dichloroethane	13.2	12.2	63	Sodium hydroxide, 50%	3.2	25.8
26	Dichloromethane	14.6	8.9	64	Sulfur dioxide	15.2	7.1
27	Ethyl acetate	13.7	9.1	65	Sulfuric acid, 98%	7	24.8
28	Ethyl alcohol,100%	10.5	13.8	66	Sulfuric acid, 60%	10.2	21.3
29	Ethyl alcohol, 95%	9.8	14.3	67	Tetrachloroethane	11.9	15.7
30	Ethyl alcohol, 40%	6.5	16.6	68	Tetrachloroethylene	14.2	12.7
31	Ethyl benzene	13.2	11.5	69	Titanium tetrachloride	14.4	12.3
32	Ethyl chloride	14.8	6	70	Toluene	13.7	10.4
33	Ethyl ether	14.5	5.3	71	Trichloroethylene	14.8	10.5
34	Ethyl formate	14.2	8.4	72	Vinyl acetate	14	8.8
35	Ethyl iodide	14.7	10.3	73	Water	10.2	13
36	Ethylene glycol	6	23.6	74	o-Xylene	13.5	12.1
37	Formic acid	10.7	15.8	75	m-Xylene	13.9	10.6
38	Freon-12	16.8	5.6	76	p-Xylene	13.9	10.9

Coordinates for use with figure on next page.

[①] By permission, from J. H. Perry (ed.), *Chemical Engineers' Handbook*, 5th ed., pp. **3**-212 and **3**-213. Copyright © 1973, McGraw-Hill Book Company, New York.

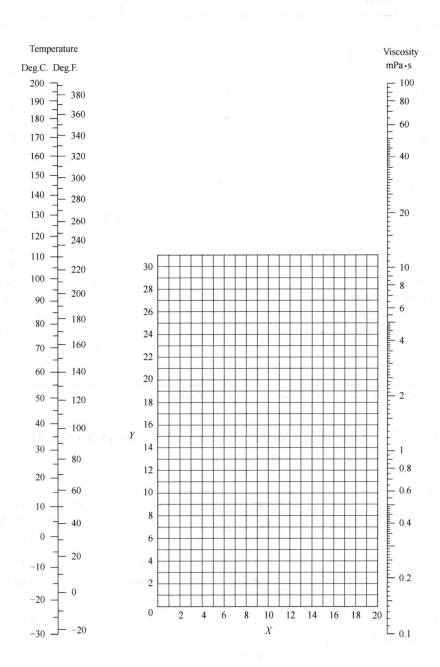

Viscosities of liquids at 1 atm. For coordinates, see table on previous page.

Appendix 9 Thermal Conductivities of Metals[①]

Metal	Thermal conductivity k, W · m^{-1} · K^{-1}		
	0 ℃	17.8 ℃	100 ℃
Aluminum	202.41		205.87
Antimony	18.338		16.781
Brass (70 copper, 30 zinc)	96.88		103.8
Cadmium		92.901	90.306
Copper (pure)	387.52		377.14
Gold		292.37	294.1
Iron (cast)	55.36		51.9
Iron (wrought)		60.377	59.858
Lead	34.6		32.87
Magnesium	159.16	159.16	159.16
Mercury (liquid)	8.304		
Nickel	62.28		58.82
Platinum		69.546	72.487
Silver	418.66		411.74
Sodium (liquid)			84.77
Steel (mild)			44.98
Steel (1% carbon)		45.326	44.807
Steel (stainless, type 304)			16.262
Steel (stainless, type 316)			16.262
Steel (stainless, type 347)			16.089
Tantalum		55.36	
Tin	62.28		58.82
Zinc	112.45		110.72

① Based on W. H. McAdams, *Heat Transmission*, 3rd ed., McGraw-Hill Book Company, New York, 1954, pp. 445–447.

Appendix 10 Thermal Conductivities of Various Solids and Insulating Materials[①]

Material	Apparent density ρ, kg · m^{-3}	Temperature T, ℃	Thermal conductivity k, W · m^{-1} · K^{-1}
Asbestos	465	−200	0.074
	577	0	0.151
	577	400	0.223
Bricks			
Alumina	—	1,315	4.671
Building brickwork	—	20	0.692
Carbon	1549	—	5.190
Fire clay (Missouri)	—	200	1.003
	—	1,000	1.644
	—	1,400	1.765
Kaolin insulating firebrick	304	200	0.087
	304	760	0.195

续表

Material	Apparent density ρ, kg·m^{-3}	Temperature T, °C	Thermal conductivity k, W·m^{-1}·K^{-1}
Silicon carbide, recrystallized	2066	600	18.51
	2066	1,000	13.84
	2066	1,400	10.90
Cardboard, corrugated	—	—	0.640
Concrete			
Clinker	—	—	0.346
Stone	—	—	0.934
1:4 dry	—	—	0.761
Cork, ground	151	30	0.043
Glass			
Borosilicate	2227	30~75	1.090
Window	—	—	0.519~1.055
Granite	—	—	1.73~3.979
Ice	921	0	2.249
Insulating materials			
Fiberglass batts[2]	96	20	0.033
	96	150	0.047
	96	200	0.061
	144	20	0.031
	144	150	0.040
Kapok	14	20	0.035
Polystyrene foam[3]	16	20	0.040
	32~80	20	0.035
Polyurethane foam[3] (made with fluorocarbon gas)	21~48	—	0.024
	64~128	—	0.031
Polyurethane foam[3] (made with CO_2)	21~48	—	0.031
Wall board	237	21	0.048
Magnesia, powdered	796	47	0.606
Paper	—	—	1.298
Porcelain	—	200	1.522
Rubber, soft	—	21	0.130~0.159
Snow	556	0	0.467
Wood (across grain)			
Oak	825	15	0.208
Maple	716	50	0.190
Pine, white	545	15	0.151
Wood (parallel to grain)			
Pine	551	21	0.346

[1] From J. H. Perry (ed.), *Chemical Engineers' Handbook,* 6th ed., McGraw-Hill, NewYork, p. **3**-260, except as noted.
[2] From *Heat Transfer and Fluid Data Book,* vol. 1, Genium Publishing Corp., Schenectady, NY, 1984, sect. 515.24, p. 1.
[3] From *Modern Plastics Encyclopedia,* vol. 65, no. 11, McGraw-Hill Book Co., New York, 1988, p. 657.

Appendix 11 Thermal Conductivities of Gases and Vapors[1]

Substance	Thermal conductivity k, W·m^{-1}·K^{-1}		Substance	Thermal conductivity k, W·m^{-1}·K^{-1}	
	0 ℃	100 ℃		0 ℃	100 ℃
Acetone	0.0099	0.0171	Ethylene	0.0175	0.0279
Acetylene	0.0187	0.0298	Helium	0.1415	0.1709
Air	0.0242	0.0318	Hydrogen	0.1671	0.2145
Ammonia	0.0218	0.0332	Methane	0.0304	0.0441
Benzene		0.0178	Methyl alcohol	0.0144	0.0221
Carbon dioxide	0.0145	0.0221	Nitrogen	0.0240	0.0313
Carbon monoxide	0.0232	0.0304	Nitrous oxide	0.0152	0.0239
Carbon tetrachloride		0.0090	Oxygen	0.0246	0.0325
Chlorine	0.0074		Propane	0.0151	0.0261
Ethane	0.0183	0.0303	Sulfur dioxide	0.0087	0.0119
Ethyl alcohol		0.0215	Water vapor (at 1 atm abs pressure)		0.0235
Ethyl ether	0.0133	0.0227			

[1] Based on W. H. McAdams, *Heat Transmission*, 3rd ed., McGraw-Hill Book Company, New York, 1954, pp. 457–458.

Appendix 12 Thermal Conductivities of Liquids Other Than Water[1]

Liquid	Temperature T, ℃	Thermal conductivity k, W·m^{-1}·K^{-1}	Liquid	Temperature T, ℃	Thermal conductivity k, W·m^{-1}·K^{-1}
Acetic acid	20	0.171	Gasoline	30	0.135
Acetone	30	0.176	Glycerine	20	0.284
Ammonia (anhydrous)	−15~30	0.502	n-Heptane	30	0.140
Aniline	0~20	0.173	Kerosene	20	0.149
Benzene	30	0.159	Methyl alcohol	20	0.215
n-Butyl alcohol	30	0.168	Nitrobenzene	30	0.164
Carbon bisulfide	30	0.161	n-Octane	30	0.144
Carbon tetrachloride	0	0.185	Sulfur dioxide	−15	0.221
Chlorobenzene	10	0.144	Sulfuric acid (90%)	30	0.363
Ethyl acetate	20	0.175	Toluene	30	0.149
Ethyl alcohol (absolute)	20	0.182	Trichloroethylene	50	0.138
Ethyl ether	30	0.138	o-Xylene	20	0.156
Ethylene glycol	0	0.265			

[1] Based on W. H. McAdams, *Heat Transmission*, 3rd ed., McGraw-Hill Book Company, New York, 1954, pp. 455–456.

Appendix 13 Specific Heats of Gases

No.	Gas	Range, ℃	No.	Gas	Range, ℃
10	Acetylene	0~200	1	Hydrogen	0~600
15	Acetylene	200~400	2	Hydrogen	600~1400
16	Acetylene	400~1400	35	Hydrogen bromide	0~1400
27	Air	0~1400	30	Hydrogen chloride	0~1400
12	Ammonia	0~600	20	Hydrogen fluoride	0~1400
14	Ammonia	600~1400	36	Hydrogen iodide	0~1400
18	Carbon dioxide	0~400	19	Hydrogen sulfide	0~700
24	Carbon dioxide	400~1400	21	Hydrogen sulfide	700~1400
26	Carbon monoxide	0~1400	5	Methane	0~300
32	Chlorine	0~200	6	Methane	300~700
34	Chlorine	200~1400	7	Methane	700~1400
3	Ethane	0~200	25	Nitric oxide	0~700
9	Ethane	200~600	28	Nitric oxide	700~1400
8	Ethane	600~1400	26	Nitrogen	0~1400
4	Ethylene	0~200	23	Oxygen	0~500
11	Ethylene	200~600	29	Oxygen	500~1400
13	Ethylene	600~1400	33	Sulfur	300~1400
17B	Freon-11 (CCl_3F)	0~150	22	Sulfur dioxide	0~400
17C	Freon-21 ($CHCl_2F$)	0~150	31	Sulfur dioxide	400~1400
17A	Freon-22 ($CHClF_2$)	0~150	17	Water	0~1400
17D	Freon-113 (CCl_2F-$CClF_2$)	0~150			

528 | Unit Operations of Chemical Engineering

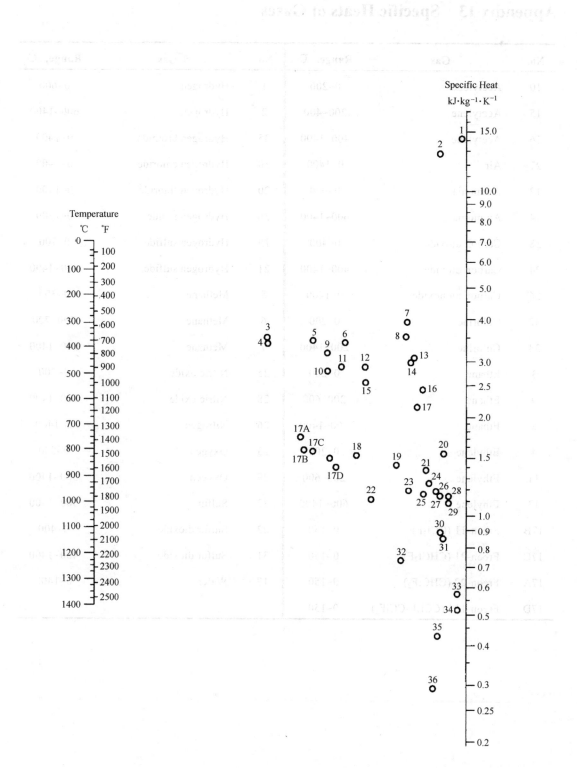

True specific heats c_p of gases and vapors at 1 atm pressure.

Appendix 14 Specific Heats of Liquids

No.	LIQUID	Range, °C	No.	LIQUID	Range, °C
29	Acetic acid, 100%	0~80	7	Ethyl iodide	0~100
32	Acetone	20~50	39	Ethylene glycol	−40~200
52	Ammonia	−70~50	2A	Freon-11 (CCl_3F)	−20~70
37	Amyl alcohol	−50~25	6	Freon-12 (CCl_2F_2)	−40~15
26	Amyl acetate	0~100	4A	Freon-21 ($CHCl_2F$)	−20~70
30	Aniline	0~130	7A	Freon-22 ($CHClF_2$)	−20~60
23	Benzene	10~80	3A	Freon-113 ($CCl_2F\text{-}CClF_2$)	−20~70
27	Benzyl alcohol	−20~30	38	Glycerol	−40~20
10	Benzyl chloride	−30~30	28	Heptane	0~60
49	Brine, $CaCl_2$, 25%	−40~20	35	Hexane	−80~20
51	Brine, NaCl, 25%	−40~20	48	Hydrochloric acid, 30%	20~100
44	Butyl alcohol	0~100	41	Isoamyl alcohol	10~100
2	Carbon disulfide	−100~25	43	Isobutyl alcohol	0~100
3	Carbon tetrachloride	10~60	47	Isopropyl alcohol	−20~50
8	Chlorobenzene	0~100	31	Isopropyl ether	−80~20
4	Chloroform	0~50	40	Methyl alcohol	−40~20
21	Decane	−80~25	13A	Methyl chloride	−80~20
6A	Dichloroethane	−30~60	14	Napthalene	90~200
5	Dichloromethane	−40~50	12	Nitrobenzene	0~100
15	Diphenyl	80~120	34	Nonane	−50~25
22	Diphenylmethane	30~100	33	Octane	−50~25
16	Diphenyl oxide	0~200	3	Perchlorethylene	−30~140
16	Dowtherm A	0~200	45	Propyl alcohol	−20~100
24	Ethyl acetate	−50~25	20	Pyridine	−50~25
42	Ethyl alcohol, 100%	30~80	9	Sulphuric acid 98%	10~45
46	Ethyl alcohol, 95%	20~80	11	Sulphur dioxide	−20~100
50	Ethyl alcohol, 50%	20~80	23	Toluene	0~60
25	Ethyl benzene	0~100	53	Water	10~200
1	Ethyl bromide	5~25	19	o-Xylene	0~100
13	Ethyl chloride	−30~40	18	m-Xylene	0~100
36	Ethyl ether	−100~25	17	p-Xylene	0~100

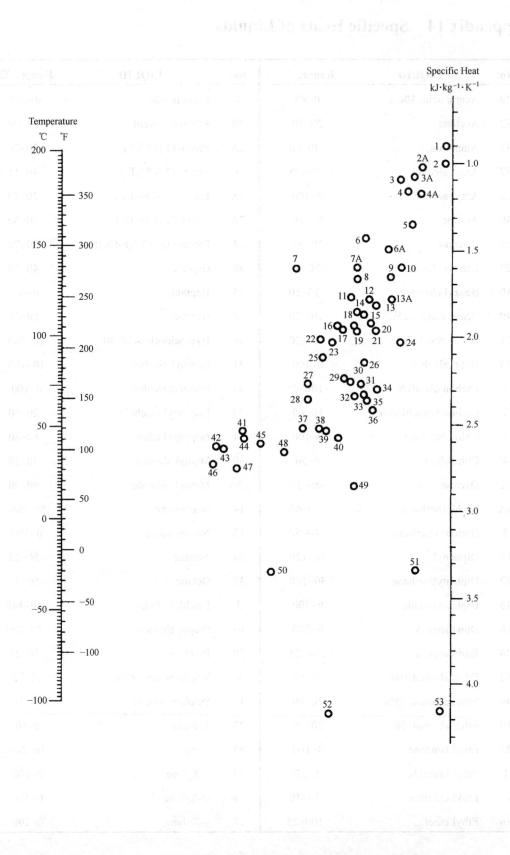

Appendix 15 Prandtl Numbers for Gases at 1 atm and 100 °C[1]

Gas	$Pr = \dfrac{c_p \mu}{k}$	Gas	$Pr = \dfrac{c_p \mu}{k}$
Air	0.69	Hydrogen	0.69
Ammonia	0.86	Methane	0.75
Argon	0.66	Nitric oxide, nitrous oxide	0.72
Carbon dioxide	0.75	Nitrogen	0.7
Carbon monoxide	0.72	Oxygen	0.7
Helium	0.71	Water vapor	1.06

[1] Based on W. H. McAdams, *Heat Transmission*, 3rd ed., McGraw-Hill Book Company, New York, 1954, p. 471.

Appendix 16 Prandtl Numbers for Liquids[1]

Liquid	$Pr = \dfrac{c_p \mu}{k}$ 16 °C	100 °C	Liquid	$Pr = \dfrac{c_p \mu}{k}$ 16 °C	100 °C
Acetic acid	14.5	10.5	Ethyl ether	4	2.3
Acetone	4.5	2.4	Ethylene glycol	350	125
Aniline	69	9.3	n-Heptane	6	4.2
Benzene	7.3	3.8	Methyl alcohol	7.2	3.4
n-Butyl alcohol	43	11.5	Nitrobenzene	19.5	6.5
Carbon tetrachloride	7.5	4.2	n-Octane	5	3.6
Chlorobenzene	9.3	7	Sulfuric acid (98%)	149	15
Ethyl acetate	6.8	5.6	Toluene	6.5	3.8
Ethyl alcohol	15.5	10.1	Water	7.7	1.5

[1] Based on W. H. McAdams, *Heat Transmission*, 3rd ed., McGraw-Hill Book Company, New York, 1954, p. 470.

Appendix 17 Diffusivities and Schmidt Numbers for Gases in Air at 0 °C and 1 atm[1]

Gas	Volumetric diffusivity D_v, cm$^2 \cdot$ s^{-1}	$Sc = \dfrac{\mu}{\rho D_v}$	Gas	Volumetric diffusivity D_v, cm$^2 \cdot$ s^{-1}	$Sc = \dfrac{\mu}{\rho D_v}$
Acetic acid	0.107	1.24	Hydrogen	0.612	0.22
Acetone	0.083	1.6	Methane	0.191	0.69
Ammonia	0.216	0.61	Methyl alcohol	0.133	1
Benzene	0.077	1.71	Naphthalene	0.051	2.57
n-Butyl alcohol	0.070	1.88	Nitrogen	0.181	0.73
Carbon dioxide	0.138	0.96	n-Octane	0.051	2.62
Carbon tetrachloride	0.067	1.97	Oxygen	0.178	0.74
Chlorine	0.111	1.19	Phosgene	0.080	1.65
Chlorobenzene	0.062	2.13	Propane	0.093	1.42
Ethane	0.126	1.04	Sulfur dioxide	0.114	1.16
Ethyl acetate	0.072	1.84	Toluene	0.071	1.86
Ethyl alcohol	0.102	1.3	Water vapor	0.220	0.6
Ethyl ether	0.078	1.7			

[1] By permission, from T. K. Sherwood and R. L. Pigford, *Absorption and Extraction*, 2nd ed., p. 20. Copyright 1952, McGraw-Hill Book Company, New York.

Appendix 18 Collision Integral and Lennard-Jones Force Constants[1]

Collision integral Ω_D

$\dfrac{kT}{\varepsilon_{12}}$	Ω_D	$\dfrac{kT}{\varepsilon_{12}}$	Ω_D	$\dfrac{kT}{\varepsilon_{12}}$	Ω_D
0.30	2.662	1.65	1.1530	4.0	0.8836
0.35	2.476	1.70	1.1400	4.1	0.8788
0.40	2.318	1.75	1.1280	4.2	0.8740
0.45	2.184	1.80	1.1160	4.3	0.8694
0.50	2.066	1.85	1.1050	4.4	0.8652
0.55	1.966	1.90	1.0940	4.5	0.8610
0.60	1.877	1.95	1.0840	4.6	0.8568
0.65	1.798	2.00	1.0750	4.7	0.8530
0.70	1.729	2.10	1.0570	4.8	0.8492
0.75	1.667	2.20	1.0410	4.9	0.8456
0.80	1.612	2.30	1.0260	5.0	0.8422
0.85	1.562	2.40	1.0120	6.0	0.8124
0.90	1.517	2.50	0.9996	7.0	0.7896
0.95	1.476	2.60	0.9878	8.0	0.7712
1.00	1.439	2.70	0.9770	9.0	0.7556
1.05	1.406	2.80	0.9672	10	0.7424
1.10	1.375	2.90	0.9576	20	0.6640
1.15	1.346	3.00	0.9490	30	0.6232
1.20	1.320	3.10	0.9406	40	0.5960
1.25	1.296	3.20	0.9328	50	0.5756
1.30	1.273	3.30	0.9256	60	0.5596
1.35	1.253	3.40	0.9186	70	0.5464
1.40	1.233	3.50	0.9120	80	0.5352
1.45	1.215	3.60	0.9058	90	0.5256
1.50	1.198	3.70	0.8998	100	0.5130
1.55	1.182	3.80	0.8942	200	0.4644
1.60	1.167	3.90	0.8888	400	0.4170

Lennard-Jones force constants

Compound	ε/k, K	σ, Å	Compound	ε/k, K	σ, Å
Acetone	560.2	4.600	i-butane	313.0	5.341
Acetylene	231.8	4.033	Carbon dioxide	195.2	3.941
Air	78.6	3.711	Carbon disulfide	467.0	4.483
Ammonia	558.3	2.900	Carbon monoxide	91.7	3.690
Argon	93.3	3.542	Carbon tetrachloride	322.7	5.947
Benzene	412.3	5.349	Carbonyl sulfide	336.0	4.130
Bromine	507.9	4.296	Chlorine	316.0	4.217
n-butane	310.0	5.339	Chloroform	340.2	5.389

Compound	ε/k, K	σ, Å	Compound	ε/k, K	σ, Å
Cyanogen	348.6	4.361	Methane	148.6	3.758
Cyclohexane	297.1	6.182	Methanol	481.8	3.626
Cyclopropane	248.9	4.807	Methylene chloride	356.3	4.898
Ethane	215.7	4.443	Methyl chloride	350.0	4.182
Ethanol	362.6	4.530	Mercury	750.0	2.969
Ethylene	224.7	4.163	Neon	32.8	2.820
Fluorine	112.6	3.357	Nitric oxide	116.7	3.492
Helium	10.2	2.551	Nitrogen	71.4	3.798
n-Hexane	339.3	5.949	Nitrous oxide	232.4	3.828
Hydrogen	59.7	2.827	Oxygen	106.7	3.467
Hydrogen cyanide	569.1	3.630	n-Pentane	341.1	5.784
Hydrogen chloride	344.7	3.339	Propane	237.1	5.118
Hydrogen iodide	288.7	4.211	n-Propyl alcohol	576.7	4.549
Hydrogen sulfide	301.1	3.623	Propylene	298.9	4.678
Iodine	474.2	5.160	Sulfur dioxide	335.4	4.112
Krypton	178.9	3.655	Water	809.1	2.641

① From J. O. Hirschfelder, C. F. Curtiss, and R. B. Bird, *Molecular Theory of Gases and Liquids,* New York: Wiley, 1954.

Appendix 19 Equilibrium Data for Ethanol-Water System at 101.325 kPa

Liquid		Gas	
mass fraction, %	mole fraction, %	mass fraction, %	mole fraction, %
0.01	0.0040	0.13	0.053
0.03	0.0117	0.39	0.153
0.04	0.0157	0.52	0.204
0.05	0.0196	0.65	0.255
0.06	0.0235	0.78	0.307
0.07	0.0274	0.91	0.358
0.08	0.0313	1.04	0.410
0.09	0.0352	1.17	0.461
0.10	0.0400	1.30	0.51
0.15	0.0550	1.95	0.77
0.20	0.08	2.60	1.03
0.30	0.12	3.80	1.57
0.40	0.16	4.90	1.98
0.50	0.19	6.10	2.48
0.60	0.23	7.10	2.90
0.70	0.27	8.10	3.33
0.80	0.31	9.00	3.73
0.90	0.35	9.90	4.12
1.00	0.39	10.75	4.51
2.00	0.79	19.70	8.76

Liquid		Gas	
mass fraction, %	mole fraction, %	mass fraction, %	mole fraction, %
3.00	1.19	27.20	12.75
4.00	1.61	33.30	16.34
7.00	2.86	44.60	23.96
10.00	4.16	52.20	29.92
13.00	5.51	57.40	34.51
16.00	6.86	61.10	38.06
20.00	8.92	65.00	42.09
24.00	11.00	68.00	45.41
29.00	13.77	70.80	48.68
34.00	16.77	72.90	51.27
39.00	20.00	74.30	53.09
45.00	24.25	75.90	55.22
52.00	29.80	77.50	57.41
57.00	34.16	78.70	59.10
63.00	40.00	80.30	61.44
67.00	44.27	81.30	62.99
71.00	48.92	82.40	64.70
75.00	54.00	83.80	66.92
78.00	58.11	84.90	68.76
81.00	62.52	86.30	71.10
84.00	67.27	87.70	73.61
86.00	70.63	88.90	75.82
88.00	74.15	90.10	78.00
89.00	75.99	90.70	79.26
90.00	77.88	91.30	80.42
91.00	79.82	92.00	81.83
92.00	81.82	92.70	83.25
93.00	83.87	93.40	84.91
94.00	85.97	94.20	86.40
95.00	88.15	95.05	88.25
95.57	89.41	95.57	89.41

Appendix 20 Henry's Law Constants for Gases in Water

Gas	$E \times 10^{-5}$, kPa										
	0℃	10℃	20℃	30℃	40℃	50℃	60℃	70℃	80℃	90℃	100℃
He	130.70	127.60	126.30	125.60	122.60	116.50					
H_2	58.65	64.43	69.19	73.85	76.08	77.49	77.49	77.09	76.48	76.08	75.47
N_2	53.59	67.47	81.45	93.60	105.40	114.50	121.60	126.60	127.60	127.60	127.60
Air	73.76	55.61	67.26	78.10	88.13	95.93	102.30	106.40	108.40	109.40	108.40

Gas	$E\times10^{-5}$, kPa										
	0℃	10℃	20℃	30℃	40℃	50℃	60℃	70℃	80℃	90℃	100℃
CO	35.66	44.77	54.30	62.81	70.50	77.09	83.17	85.60	85.60	85.70	85.70
O_2	25.83	33.13	40.62	48.12	54.20	59.56	63.72	67.16	69.60	70.81	71.00
CH_4	22.69	30.09	39.09	45.48	52.68	58.45	63.41	67.47	69.09	70.10	71.00
NO	17.12	22.08	26.74	31.40	35.66	39.51	42.34	44.37	45.38	45.79	45.99
C_2H_6	12.76	19.15	26.64	34.64	12.85	50.65	57.23	63.11	66.96	69.59	70.10
C_2H_4	5.59	7.78	10.33	12.87							
O_3	0.97	2.51	3.81	3.06	12.16	27.76					
N_2O		1.43	2.01	2.62		2.87					
CO_2	0.74	1.05	1.44	1.88	2.36		3.45				
C_2H_2	0.73	0.97	1.23	1.48							
H_2S	0.27	0.37	0.49	0.62	0.75	0.90	1.04	1.21	1.37	1.46	1.50
Br_2	0.022	0.037	0.06	0.092	0.47	0.19	0.25	0.33	0.41		

Appendix 21 Partial Pressure p of SO_2 in The Gas in Equilibrium with The Mole Fraction x of SO2 in The Liquid at 20℃

Mole fraction of SO_2 in liquid phase x	0.0000	0.0025	0.005	0.0075	0.010	0.0125	0.015	0.0175	0.020
Partial pressure of SO_2 in gas phase p, atm	0.000	0.072	0.141	0.218	0.300	0.386	0.486	0.588	0.700

Appendix 22 Boiling Point for Inorganic Solution at 1 atm

Solution	Mass fraction, %													
	101℃	102℃	103℃	104℃	105℃	107℃	110℃	115℃	120℃	125℃	140℃	160℃	180℃	200℃
$CaCl_2$	5.66	10.31	14.16	17.36	20.00	24.24	29.33	35.68	40.83	54.80	57.89	68.94	75.85	64.91
KOH	4.49	8.51	11.96	14.82	17.01	20.88	25.65	31.97	36.51	40.23	48.05	54.89	60.41	
KCl	8.42	14.31	18.96	23.02	36.57	32.62	36.47	~108.5						
K_2CO_3	10.31	18.37	24.20	28.57	32.24	37.69	43.67	50.86	56.04	60.40	66.94	133.5		
KNO_3	13.19	23.66	32.23	39.20	45.10	54.65	65.34	79.53						
$MgCl_2$	4.67	8.42	11.66	14.31	16.59	20.23	24.41	29.48	33.07	36.02	39.61			
$MgSO_4$	14.31	22.78	28.31	32.23	35.32	42.86	~108							
NaOH	4.12	7.40	10.15	12.51	14.53	18.32	23.08	26.21	33.77	37.58	48.32	60.13	69.97	77.53
NaCl	6.19	11.03	14.67	17.69	20.32	25.09	28.92							
$NaNO_3$	8.26	15.61	21.87	17.53	32.45	40.47	49.87	60.94	68.94					
Na_2SO_4	15.26	24.81	30.73	31.83	~103.2									
Na_2CO_3	9.42	17.22	23.72	29.18	33.66									
$CuSO_4$	26.95	39.98	40.83	44.47	45.12	~104.2								
$ZnSO_4$	20.00	31.22	37.89	42.92	46.15									
NH_4NO_3	9.09	16.66	23.08	29.08	34.21	42.52	51.92	63.24	71.26	77.11	87.09	93.20	69.00	97.61
NH_4Cl	6.10	11.35	15.96	19.80	22.89	28.37	35.98	46.94						

References

[1] Geankoplis C J. Transport Processes and Unit Operations. 3rd ed. Englewood Cliffs, New Jersey : Prentice-Hall, 1993.

[2] Brodkey R S, Hershey H C. Transport Phenomena: A Unified Approach. New York: McGraw-Hill, 1988.

[3] Kern D Q. Process Heat Transfer. New York: McGraw-Hill, 1990.

[4] 柯尔森等. 化学工程. 丁绪淮等译. 第 3 版. 卷 II 单元操作. 北京：化学工业出版社，1997.

[5] 邹华生，钟理，伍钦. 流体力学与传热. 广州：华南理工大学出版社，2004.

[6] Smith JM, Van Ness HC, Abbott MM. Introduction to Chemical Engineering Thermodynamics. 6th ed. Beijing: Chemical Industry Press, 2002.

[7] Munson B R, Young D F, Okiishi T H. Fundamentals of Fluid Mechanics. 5th ed. Beijing: Publishing House of Electronics Industry, 2006.

[8] Cheremisinoff N P. Applied Fluid Flow Measurement: Fundamentals and Technology. New York : M. Dekker, 1979.

[9] Geankoplis C J. Transport Processes and Separation Process Principles (includes unit operations). New Jersey: Prentice Hall PTR, 2003.

[10] Centrifugal Pumps: Basic Concepts of Operation, Maintenance, and Troubleshooting (Part-I). www.cheresources.com/centrifugalpumps1.

[11] Svarovsky L. Solid-Liquid Separation. London: Butterworth &Co Ltd, 1979.

[12] Anthony L H, Robert N M. Mass Transfer: Fundamentals and Applications. New Jersey: Prentice Hall INC, 1985.

[13] Perry R H, Green D W. Perry's Chemical Engineering Handbook. 7th ed. New York: McGraw-Hill, 1997

[14] 伍钦，钟理，邹华生，曾朝霞. 传质与分离工程. 广州：华南理工大学出版社，2005.

[15] Benaroya A. Fundamentals and Applications of Centrifugal Pumps for the Practicing Engineering. Oklahoma: Petroleum Publishing Company, 1978.

[16] 姚玉英. 化工原理. 下册. 天津：天津大学出版社，1999.

[17] Bird R B, StewartW E, Lightfoot E N. Transport Phenomena. New York: John Wiley & Sons, Inc.1960

[18] Coulson J M, Richardson JF. Chemical Engineering: Volume 4: Backhurst J R, Harker J H. Solutions to the Problems in Chemical Engineering. Volume 1. Oxford: Pergamon Press, 1977.

[19] Coulson J M, Richardson J F. Chemical Engineering: Volume 5: Backhurst JR, Harker J H. Solutions to the Problems in Chemical Engineering. Volume 2. Oxford: Pergamon Press, 1979.

[20] 陈敏恒，丛德滋，方图南，齐鸣斋. 化工原理. 第 3 版. 北京：化学工业出版社，2006.

[21] Christij G. Transport Processes and Unit Operations. 3rd ed. Singapore：Prentice-Hall International, 1995

[22] Coulson J M, Richardson J F. Chemical Engineering. 3rd ed. New York:McGraw-Hill, 1977.

国外名校名著（化工专业）

翻译版系列

动量、热量和质量传递原理（原著第四版）
　　Charles E.Wicks 编著　　　　　　　　　　　　　　　马紫峰等译
传递现象（原著第二版）
　　R.Byron Bird 著　　　　　　　　　　　　　　　　　戴干策等译
化工热力学导论（原著第七版）
　　J.M.Smith 著　　　　　　　　　　　　　　　　　　刘洪来等译
流体相平衡的分子热力学（原著第三版）
　　John M.Prausnitz 著　　　　　　　　　　　　　　　陆小华等译
化工过程优化（原著第二版）
　　Thomas F.Edgar 著　　　　　　　　　　　　　　　　张卫东等译
化工过程设计
　　R.Smith 著　　　　　　　　　　　　　　　　　　　王保国等译
绿色工程：环境友好的化工过程设计
　　David T.Allen 著　　　　　　　　　　　　　　　　　李　辉等译
化工过程安全理论及应用（原著第二版）
　　Daniel A.Crowl 著　　　　　　　　　　　　　　　　蒋军成等译

英文版系列

Transport Phenomena (2nd Ed.)	R. Byron Bird
Transport Phenomena (2nd Ed.)	W. J. Beek
Elements of Chemical Reaction Engineering (4th Ed.)	H. Scott Fogler
Chemical and Engineering Thermodynamics (3rd Ed.)	Stanley I. Sandler
Separation Process Principles	J. D. Seader
Unit Operations of Chemical Engineering (7th Ed.) (Adaptation)	**Warren L.McCabe**
Unit Operations of Chemical Engineering (6th Ed.)	Warren L. McCabe
Process Design Principles ——Synthesis, Analysis, and Evaluation	Warren D. Seider

教 师 反 馈 表

　　McGraw-Hill Education, 麦格劳-希尔教育出版公司是美国著名图书出版与教育服务机构，出版了很多著名的计算机、工程类、经管类以及人文社科类图书。作为麦格劳-希尔教育出版公司在中国的重要合作伙伴之一，化学工业出版社与其合作出版了一些工程类专业教材，近年来双方更加注意加强了彼此的交流与合作。

　　我们十分重视对广大教师的服务，开发了教师手册、习题解答等教学课件以及网上资源。如果您确认将本书作为指定教材，请您务必填好以下表格并经系主任签字盖章后寄回我们的联系地址。我公司将免费向您提供英文原版的教师手册或其他的教学课件。

书号/书名：	
所需要的教学资料：	
您的姓名：	
系：	
院/校：	
您所讲授的课程名称：	
每学期学生人数：	_____ 人　_____ 年级　　学时：
您目前采用的教材：	作者：_____　出版社：_____ 书名：_____
您准备何时用此书授课：	
您的联系地址：	
邮政编码：	联系电话（必填）
E-mail：（必填）	
您对本书的建议：	系主任签字 盖章

McGraw-Hill Education

麦格劳-希尔教育出版公司教师服务中心
北京市海淀区-清华科技园 创业大厦 906 室
北京 100084
电话：010-62790298-108
传真：010-62790292
教师服务热线：800-810-1936
教师服务信箱：instructor_china@mcgraw-hill.com
网址：http://www.mcgraw-hill.com.cn